Understanding NMR Spectroscopy
– Second Edition

Understanding NMR Spectroscopy
– Second Edition

James Keeler

Senior Lecturer in the Department of Chemistry and
Fellow of Selwyn College, Cambridge

A John Wiley & Sons, Ltd Publication

This edition first published 2010
© 2010 John Wiley & Sons, Ltd

Registered office
John Wiley & Sons Ltd, The Atrium, Southern Gate, Chichester, West Sussex, PO19 8SQ, United Kingdom

For details of our global editorial offices, for customer services and for information about how to apply for permission to reuse the copyright material in this book please see our website at www.wiley.com.

The right of the author to be identified as the author of this work has been asserted in accordance with the Copyright, Designs and Patents Act 1988.

Reprinted in January 2011
October 2012

Library of Congress Cataloging-in-Publication Data

Keeler, James.
    Understanding NMR spectroscopy / James Keeler. – 2nd ed.
        p. cm.
    Includes bibliographical references and index.
    ISBN 978-0-470-74609-7 (cloth) – ISBN 978-0-470-74608-0 (pbk.)
    1. Nuclear magnetic resonance spectroscopy–Textbooks. I. Title.
    QD96.N8K44 2010
    543'.66–dc22

                                        2009054393

A catalogue record for this book is available from the British Library.

ISBN 978-0-470-74609-7 (cloth)
ISBN 978-0-470-74608-0 (pbk.)

Set in 10pt Sabon by the author using LaTeX

Preface

I am very pleased to have the opportunity to produce a second edition of *Understanding NMR Spectroscopy*, not least as I have been encouraged by the many kind comments that I have received by users of the first edition. For all its undoubted flaws, the book has clearly been found to be useful in helping people to get to grips with the theory of NMR.

I have resisted the temptation to add a great deal of additional material or to make the discussion more technical. However, I have included a new chapter which covers two topics which, in retrospect, seemed to be serious omissions from the first edition. The first topic is how product operators can be extended to describe experiments in AX_2 and AX_3 spin systems, thus making it possible to discuss the important APT, INEPT and DEPT experiments often used in ^{13}C spectroscopy.

The second topic is spin system analysis i.e. how shifts and couplings can be extracted from strongly coupled (second-order) spectra. In the early days of NMR this kind of analysis was all but essential since the low field strengths then available meant that spectra were often strongly coupled. The current use of much higher fields means that strong-coupling effects are less common, but they have not gone away completely. It therefore remains important to be aware of such effects and their consequences for the appearance of spectra. In a related topic, I also discuss how the presence of chemically equivalent spins leads to spectral features which are somewhat unusual and possibly misleading. In contrast to strong-coupling effects, these features are independent of the field strength and so are not mitigated by the move to higher fields.

The chapter on relaxation has been reorganised, and a discussion of chemical exchange effects has been introduced in order to help with the explanation of transverse relaxation. Finally, I have added a short section on double-quantum spectroscopy to Chapter 10.

The use of two-colour printing will, I hope, both improve the clarity of many of the diagrams and improve the appearance of the printed pages.

I am very much indebted to Dr Andrew Pell (now at the École Normale Supérieure de Lyon) who found time between completing his PhD and starting a postdoctoral position to help me in the preparation of this edition. Andy worked on adding colour to the figures, produced some additional simulations for Chapter 9, recorded all of the experimental spectra and commented on the new sections. I am also grateful to Dr Daniel Nietlispach (Department of Biochemistry, University of Cambridge) for once again providing very useful and perceptive comments on the new material.

Cambridge, January 2010

Preface to the first edition

I owe a huge debt of gratitude to Dr Daniel Nietlispach and Dr Katherine Stott (both from the Department of Biochemistry, University of Cambridge) who have read, corrected and commented on drafts of the entire book. Their careful and painstaking work has contributed a great deal to the final form of the text and has, in my view, improved it enormously. I am also grateful to them for their constant enthusiasm, which sustained and encouraged me throughout the project. I could not have wanted for two more constructive and helpful readers.

Special thanks are also due to Professor Nikolaus Loening (Lewis and Clark College, Portland, Oregon) who, at short notice and with great skill, provided all of the experimental spectra in the book. His good humoured response to my pernickety requirements is much appreciated. Andrew Pell (Selwyn College, Cambridge) also deserves special mention and thanks for his skilled assistance in producing the solutions manual for the exercises.

I would like to acknowledge the support and advice from my collaborator and colleague Dr Peter Wothers (Department of Chemistry, University of Cambridge): he remains both my sternest critic and greatest source of encouragement. I am also grateful to Professor Jeremy Sanders (Department of Chemistry, University of Cambridge) for his much valued support and advice.

My appreciation and understanding of NMR, such that it is, has been very much influenced by those I have been fortunate enough to work alongside, both as research students and collaborators; I thank them for their insights. I would also like to thank Professor Malcolm Levitt (Department of Chemistry, University of Southampton), Professor Art Palmer (Columbia University, New York) and Dr David Neuhaus (MRC LMB, Cambridge) for tirelessly answering my many questions.

This book grew out of a series of lecture notes which, over a number of years, I prepared for various summer schools and graduate courses. On the initiative of Dr Rainer Haessner (Technische Universität, Munich), the notes were made available on the web, and since then I have received a great deal of positive feedback about how useful people have found them. It was this, above all, which encouraged me to expand the notes into a book.

The book has been typeset by the author using LaTeX, in the particular implementation distributed as MiKTeX (www.miktex.org). I wish to express my thanks to the many people who develop and maintain the LaTeX system. All of the diagrams have been prepared using *Adobe Illustrator* (Adobe Systems Inc., San Jose, California), sometimes in combination with *Mathematica* (Wolfram Research Inc., Champaign, Illinois).

Finally, I am delighted to be able to dedicate this book to Professor Ray Freeman. I was lucky enough to have started my NMR career in Ray's group, and what I learnt there, both from Ray and my fellow students, has stood me in good stead ever since. My appreciation for Ray has continued as strong as ever since those early days, and I am pleased to have this opportunity to acknowledge the debt I owe him.

Cambridge, July 2005

Dedicated to

Professor Ray Freeman FRS

Contents

1

What this book is about and who should read it

This book is aimed at people who are familiar with the use of routine NMR for structure determination and who wish to deepen their understanding of just exactly how NMR experiments 'work'. It is one of the great virtues of NMR spectroscopy that one can use it, and indeed use it to quite a high level, without having the least idea of how the technique works. For example, we can be taught how to interpret two-dimensional spectra, such as COSY, in a few minutes, and similarly it does not take long to get to grips with the interpretation of NOE (nuclear Overhauser effect) difference spectra. In addition, modern spectrometers can now run quite sophisticated NMR experiments with the minimum of intervention, further obviating the need for any particular understanding on the part of the operator.

You should reach for this book when you feel that the time has come to understand just exactly what is going on. It may be that this is simply out of curiosity, or it may be that for your work you need to employ a less common technique, modify an existing experiment to a new situation or need to understand more fully the limitations of a particular technique. A study of this book should give you the confidence to deal with such problems and also extend your range as an NMR spectroscopist.

One of the difficulties with NMR is that the language and theoretical techniques needed to describe it are rather different from those used for just about all other kinds of spectroscopy. This creates a barrier to understanding, but it is the aim of this book to show you that the barrier is not too difficult to overcome. Indeed, in contrast to other kinds of spectroscopy, we shall see that in NMR it is possible, quite literally on the back of an envelope, to make exact predictions of the outcome of quite sophisticated experiments. Further, once you have got to grips with the theory, you should find it possible not only to analyse existing experiments but also dream up new possibilities.

There is no getting away from the fact that we need quantum mechanics in order to understand NMR spectroscopy. Developing the necessary quantum mechanical ideas from scratch would make this book rather a

Understanding NMR Spectroscopy, Second Edition James Keeler
© 2010 John Wiley & Sons, Ltd

hard read. Luckily, it is not really necessary to introduce such a high level of formality provided we are prepared to accept, on trust, certain quantum mechanical ideas and are prepared to use these techniques more or less as a recipe. A good analogy for this approach is to remember that it is perfectly possible to learn to add up and multiply without appreciating the finer points of number theory.

One of the nice features we will discover is that, despite being rigorous, the quantum mechanical approach still retains many features of the simpler *vector model* often used to describe simple NMR experiments. Once you get used to using the quantum mechanical approach, you will find that it does work in quite an intuitive way and gives you a way of 'thinking' about experiments without always having to make detailed calculations.

Quantum mechanics is, of course, expressed in mathematical language, but the mathematics we will need is not very sophisticated. The only topic which we will need which is perhaps not so familiar is that of complex numbers and the complex exponential. These will be introduced as we go along, and the ideas are also summarized in an appendix.

1.1 How this book is organized

The ideas we need to describe NMR experiments are built up chapter by chapter, and so the text will make most sense if it read from the beginning. Certain sections are not crucial to the development of the argument and so can be safely omitted at a first reading; these sections are clearly marked as such in the margin.

Optional sections are marked like this:

Optional section ⇒

Chapter 6, which explains how quantum mechanics is formulated in a way useful for NMR, is also entirely optional. It provides the background to the product operator formalism, which is described in Chapter 7, but this latter chapter is written in such a way that it does not rely on anything from Chapter 6. At some point, I hope that you will want to find out about what is written in Chapter 6, but if you decide not to tackle it, rest assured that you will still be able to follow what goes on in the rest of the book.

The main sequence of the book really ends with Chapter 8, which is devoted to two-dimensional NMR. You should dip into Chapters 9–13 as and when you feel the need to further your understanding of the topics they cover. This applies particularly to Chapter 10 which discusses a selection of more advanced ideas in two-dimensional NMR, and Chapter 11 which is concerned with the rather 'technical' topic of how to write phase cycles and how field gradient pulses are used.

Quite deliberately, this book starts off at a gentle pace, working through some more-or-less familiar ideas to start with, and then elaborating these as we follow our theme. This means that you might find parts of the discussion rather pedestrian at times, but the aim is always to be clear about what is going on, and not to jump over steps in calculations or arguments. The same philosophy is followed when it comes to the more difficult and/or less familiar topics which are introduced in the later chapters. If you are already familiar with the vector model of pulsed NMR, and are happy with thinking about multiplets in terms of energy levels, then you might wish to jump in at Chapter 6 or Chapter 7.

Each chapter ends with some exercises which are designed to help your understanding of the ideas presented in that chapter. Tackling the exercises will undoubtedly help you to come to grips with the underlying ideas.

1.2   Scope and limitations

In this book we are going to discuss the high-resolution NMR of liquid samples and we will concentrate, almost exclusively, on spin-half nuclei (mainly ^{1}H and ^{13}C). The NMR of solids is an important and fast-developing field, but one which lies outside the scope of this book.

The experiments we will choose to describe are likely to be encountered in the routine NMR of small to medium-sized molecules. Many of the experiments are also applicable to the study of large biomolecules, such as proteins and nucleic acids. The special multi-dimensional experiments which have been devised for the study of proteins will not be described here, but we note that such experiments are built up using the repertoire of pulsed techniques which we are going to look at in detail.

The existence of the chemical shift and scalar coupling is, of course, crucial to the utility of NMR spectroscopy. However, we will simply treat the values of shifts and coupling constants as experimentally derived parameters; we will have nothing to say about their calculation or interpretation – topics which are very well covered elsewhere.

1.3   Context and further reading

This is not a 'how to' book: you will find no advice here on how to select and run a particular experiment, nor on how to interpret the result in terms of a chemical structure. What this book is concerned with is how the experiments work. However, it is not a book of NMR theory for its own sake: rather, the ideas presented, and the theories introduced, have been chosen carefully as those most useful for understanding the kinds of NMR experiments which are actually used.

There are many books which describe how modern NMR spectroscopy is applied in structural studies, and you may wish to consult these alongside this text in order to see how a particular experiment is used in practice. Two useful texts are: J. K. M. Sanders and B. K. Hunter, *Modern NMR Spectroscopy* (2nd edition, OUP, 1993), and T. D. W. Claridge, *High-Resolution NMR Techniques in Organic Chemistry* (Elsevier Science, 1999).

There are also a number of books which are at roughly the same level as this text and which you may wish to consult for further information or an alternative view. Amongst these, R. Freeman, *Spin Choreography* (Spektrum, 1997) and F. J. M. van de Ven, *Multidimensional NMR in Liquids* (VCH, 1995) are particularly useful. If you wish to go further and deeper into the theory of NMR, M. H. Levitt, *Spin Dynamics* (2nd edition, John Wiley & Sons, Ltd, 2008) is an excellent place to start.

The application of NMR to structural studies of biomolecules is a vast area which we will only touch on from time to time. A detailed account of this important area, covering both theoretical and practical matters, can be found in J. Cavanagh, W. J. Fairbrother, A. G. Palmer III, M. Rance and N. J. Skelton, *Protein NMR Spectroscopy* (2nd edition, Academic Press, 2007).

At the end of each chapter you will also find suggestions for further reading. Many of these are directions to particular chapters of the books we have already mentioned.

1.4   On-line resources

A solutions manual for the exercises at the end of each chapter is available on-line via the *spectroscopyNOW* website:

http://www.spectroscopynow.com/nmr
follow the 'Education' link from this page

A list of corrections and amendments will also be available on this site, as well as other additional material. It will also be possible to download all of the figures (in 'jpeg' format) from this book.

1.5   Abbreviations and acronyms

ADC	analogue to digital converter
APT	attached proton test
COSY	correlation spectroscopy
CTP	coherence transfer pathway
DEPT	distortionless enhancement by polarization transfer
DQF COSY	double-quantum filtered COSY
FID	free induction decay
HETCOR	heteronuclear correlation
HMBC	heteronuclear multiple-bond correlation
HMQC	heteronuclear multiple-quantum correlation
HSQC	heteronuclear single-quantum correlation
INEPT	insensitive nuclei enhanced by polarization transfer
NMR	nuclear magnetic resonance
NOE	nuclear Overhauser effect
NOESY	nuclear Overhauser effect spectroscopy
RF	radiofrequency
rx	receiver
ROESY	rotating frame Overhauser effect spectroscopy
SHR	States–Haberkorn–Ruben
SNR	signal-to-noise ratio
TOCSY	total correlation spectroscopy
TPPI	time proportional phase incrementation
TROSY	transverse relaxation optimized spectroscopy
tx	transmitter

2

Setting the scene

You will probably find that much of this chapter covers topics you are familiar with or have at least come across before. The point of the chapter is, as the title says, to set the scene for what follows by reminding you of the basic language of NMR, how we describe NMR spectra and how some important quantities are defined. There is also a section on oscillations and rotations, explaining how these are described and represented mathematically. These are key ideas which we will use extensively in the rest of the book.

2.1 NMR frequencies and chemical shifts

Like all forms of spectroscopy, an NMR spectrum is a plot of the intensity of absorption (or emission) on the vertical axis against frequency on the horizontal axis. NMR spectra are unusual in that they appear at rather low frequencies, typically in the range 10 to 800 MHz, corresponding to wavelengths from 30 m down to 40 cm. This is the radiofrequency (RF) part of the electromagnetic spectrum which is used for radio and TV broadcasts, mobile phones etc.

It is usual in spectroscopy to quote the frequency or wavelength of the observed absorptions; in contrast, in NMR we give the positions of the lines in 'ppm' using the chemical shift scale. The reason for using a shift scale is that it is found that the frequencies of NMR lines are directly proportional to the magnetic field strength. So doubling the field strength doubles the frequency, as shown in Fig. 2.1 on the following page. This field dependence makes it difficult to compare absorption frequencies between spectrometers which operate at different field strengths, and it is to get round this problem that the chemical shift scale is introduced. On this scale, the positions of the peaks are *independent of the field strength*. In this section we will explore the way in which the scale is defined, and also how to convert back and forth between frequencies and ppm – something we will need to do quite often.

Understanding NMR Spectroscopy, Second Edition James Keeler
© 2010 John Wiley & Sons, Ltd

Fig. 2.1 Schematic NMR spectra consisting of two lines. In (a) the magnetic field is such that the two lines appear at 200.0002 and 200.0004 MHz, respectively; their separation is 200 Hz. The spectrum shown in (b) is that expected when the applied magnetic field is doubled. The frequency of each peak is doubled and, as a consequence, the separation between the two peaks has now also doubled to 400 Hz.

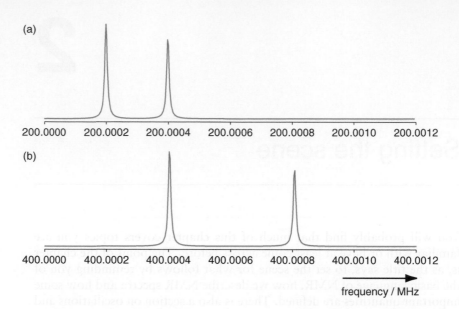

Before we look at the definition of the chemical shift it is worthwhile pointing out that the frequency at which an NMR signal appears also depends on the nuclear isotope (e.g. ^{1}H, ^{13}C, ^{15}N etc.) being studied. Also, for a given field, the NMR absorptions for a particular isotope cover rather a small range of frequencies relative to the absolute frequency of the absorption. In an experiment it is therefore usual only to measure the NMR spectrum from one particular isotope at a time.

2.1.1   Chemical shift scales

The chemical shift scale is set up first by agreeing a simple *reference compound*, a line from which is taken to define zero on the chemical shift scale. For ^{1}H and ^{13}C this reference compound is TMS. The choice of reference compound is arbitrary, but subject to careful international agreement so as to make sure everyone is using the same compound and hence the same origin on their shift scales.

The position of a peak in the spectrum is specified by measuring its frequency separation from the reference peak, and then dividing this difference by the frequency of the reference peak. As we are taking the *ratio* of two frequencies, the field dependence cancels out. The ratio thus specifies the position of a line in a way which is *independent* of the applied field, which is what we require.

Expressed mathematically, the chemical shift δ is given by

$$\delta(\text{ppm}) = 10^6 \times \frac{\upsilon - \upsilon_{\text{ref}}}{\upsilon_{\text{ref}}}, \tag{2.1}$$

where υ is the frequency of the NMR line in question and υ_{ref} is the frequency of the line from the agreed reference compound. Clearly, the line from the reference compound will appear at $\delta = 0$.

It is usual to quote chemical shifts in 'parts per million' (ppm) in order to make the numbers more convenient, and this is why in the definition of

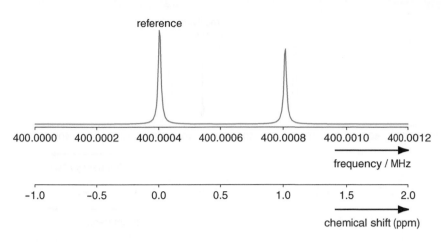

Fig. 2.2 A schematic NMR spectrum consisting of two lines is shown with both a frequency scale and a chemical shift scale, in ppm. The left-hand peak has been chosen as the reference, and so appears at 0 ppm. Note that it is usual for the ppm scale to be plotted increasing to the left, and not to the right as shown here.

δ the ratio is multiplied by 10^6. Figure 2.2 shows the schematic spectrum of Fig. 2.1 (b) on the facing page with both a frequency scale and a chemical shift scale in ppm; the left-hand peak has been chosen as the reference and so appears at 0 ppm. The right-hand peak appears at 1 ppm and it is easy to see that if a ppm scale were to be added to the spectrum of Fig. 2.1 (a), the right-hand peak would still be at 1 ppm.

The frequency scale in Fig. 2.2 increases to the right, which is the natural way to plot it, but as a consequence the ppm scale also increases to the right. This looks unusual since in NMR it is the universal practice to plot spectra with the ppm scale increasing to the left.

2.1.2 Conversion from shifts to frequencies

Sometimes we need to know the frequency separation of two peaks, in Hz. The software used to process and display NMR spectra usually has an option to toggle the scale between Hz and ppm, so measuring the peak separation is quite easy. However, sometimes we will need to make the conversion from ppm to Hz manually.

The definition of the chemical shift, Eq. 2.1 on the preceding page, can be rearranged to

$$10^{-6} \times \delta(\text{ppm}) \times \upsilon_{\text{ref}} = \upsilon - \upsilon_{\text{ref}}.$$

From this it is clear that a peak at δ ppm is separated from the reference peak by $10^{-6} \times \delta \times \upsilon_{\text{ref}}$. It follows that two peaks at shifts δ_1 and δ_2 are separated by a frequency of $10^{-6} \times (\delta_1 - \delta_2) \times \upsilon_{\text{ref}}$.

It is usual to express the frequency of the reference peak in MHz $(= 10^6$ Hz). When this is done the factor of 10^{-6} cancels and the frequency separation in Hz is simply

$$\text{frequency separation in Hz} = (\delta_1 - \delta_2) \times \upsilon_{\text{ref}}(\text{in MHz}). \qquad (2.2)$$

So, for example, if the reference frequency is 500 MHz, then two peaks at 2.3 and 1.8 ppm are separated by $(2.3 - 1.8) \times 500 = 250$ Hz.

Put even more simply, if the frequency of the reference peak is 500 MHz then 1 ppm corresponds to 500 Hz; if the reference peak is at 800 MHz, 1 ppm corresponds to 800 Hz. The conversion from ppm to Hz is therefore rather simple.

Fig. 2.3 Our two-line spectrum is shown with both a frequency scale and an offset frequency scale. The receiver reference frequency has been chosen as 400.0007 MHz, as indicated by the arrow. As a result, the right-hand peak has an offset frequency of +100 Hz and the left-hand peak has an offset of −300 Hz. It is important to realize that the choice of the receiver reference frequency is entirely arbitrary and is not related to the frequency of the resonance from the reference compound.

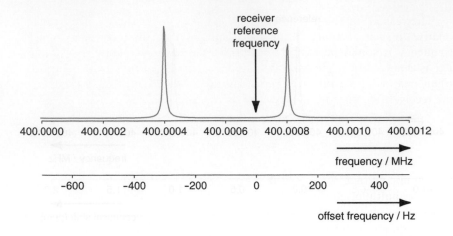

2.1.3   The receiver reference frequency and the offset frequency

The RF circuits in virtually all NMR spectrometers are arranged in such a way that the frequencies of the peaks in the spectrum are not measured absolutely but are determined relative to the *receiver reference frequency*. This reference frequency can be set quite arbitrarily by the operator of the spectrometer; typically, it is placed somewhere in the middle of the peaks of interest.

It is important to realize that this receiver reference frequency has got *nothing* to do with the resonance from the reference compound; the receiver reference can be placed anywhere we like. The usual arrangement is that when the full spectrum is displayed, the receiver reference frequency is in the middle of the displayed region, so the frequencies of the peaks can be positive or negative, depending on which side of the reference frequency they fall.

The frequency of a peak relative to the receiver reference frequency is called the *offset frequency* (or, for short, the offset) of the peak. This offset frequency v_{offset} is given by

$$v_{\text{offset}} = v - v_{\text{rx}},$$

where v is the frequency of the peak of interest and v_{rx} is the receiver reference frequency ('rx' is the traditional abbreviation for receiver). We see from this definition that the offset frequency can be positive or negative, as exemplified in Fig. 2.3.

When calculating the chemical shift using Eq. 2.1 on page 6 it is usually sufficiently accurate to divide, not by the frequency of the line from the reference compound (v_{ref}), but by the receiver reference frequency, v_{rx}:

$$\delta(\text{ppm}) = 10^6 \times \frac{v - v_{\text{ref}}}{v_{\text{rx}}}. \tag{2.3}$$

The reason for this is that NMR resonances cover such a small range relative to their absolute frequencies that, provided the receiver reference frequency is somewhere in the spectrum, the difference between v_{rx} and v_{ref} is completely negligible. Similarly, when converting from shifts to frequencies (Eq. 2.2 on page 7), it is generally sufficiently accurate to use the receiver reference frequency in place of v_{ref}.

2.2 Linewidths, lineshapes and integrals

We cannot extract much useful information from a spectrum unless the peaks or multiplets are clearly separated from one another – an observation which is as true for the most complex multi-dimensional spectrum as it is for the simplest conventional ^{1}H spectrum. Whether or not two peaks are resolved will depend on the separation between them relative to their *linewidth* and, to an extent, their *lineshape*. These two properties are thus of paramount importance in NMR.

It is not uncommon for lines in NMR spectra of small to medium-sized molecules to have widths of a few Hz. Thus, compared with their absolute frequencies, NMR lines are very narrow indeed. However, what we should really be comparing with the linewidth is the *spread* of frequencies over which NMR lines are found for a given nucleus. This spread is generally rather small, so relatively speaking the lines are not as narrow as we might suppose. Indeed, NMR experiments on complex molecules are primarily limited by the linewidths of the resonances involved.

The basic lineshape seen in simple NMR experiments is the *absorption mode* lineshape, illustrated in Fig. 2.4. The lineshape is entirely positive and is symmetrical about the maximum. The breadth of the line is specified by quoting its width at half of the peak height, as is also shown in the figure.

When we first learn about proton NMR we are told that the area under a peak or multiplet, i.e. the integral, is proportional to the number of protons giving rise to that feature. It therefore follows that if two peaks are both associated with single protons, they must have the same integral and hence if one of the lines is broader it will have reduced peak height; this is illustrated in Fig. 2.5. This reduction in peak height also means that the signal-to-noise ratio of the spectrum is reduced.

As two lines get closer and closer together, they begin to overlap and eventually will merge completely so that it is no longer possible to see the two separate lines; the process is illustrated in Fig. 2.6 on the next page. The diagram shows that, by the time the separation falls somewhat below the linewidth, the merging of the two lines is complete so that they are no longer distinct. The exact point at which the lines merge depends on the lineshape.

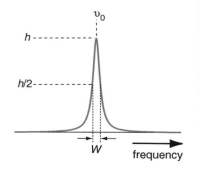

Fig. 2.4 An absorption mode lineshape. The peak is centred at v_0 and is of height h; the width of the peak is specified by giving the width W measured at half the peak height ($h/2$).

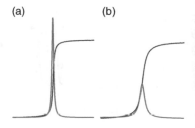

Fig. 2.5 Illustration of how the area or integral of a peak corresponding to a certain number of protons is fixed. The peak shown in (b) is three times broader than that shown in (a); however, they have the same integral (shown by the grey line). As a result, the peak height of the broader line is reduced, also by a factor of three.

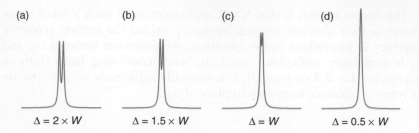

$\Delta = 2 \times W$ $\Delta = 1.5 \times W$ $\Delta = W$ $\Delta = 0.5 \times W$

Fig. 2.6 Illustration of how the ability to resolve two lines depends on their separation relative to the linewidth. In (a) the separation Δ is twice the linewidth, W; the two peaks are clearly resolved. In (b) the separation has decreased so that it is equal to 1.5 times the linewidth and as a result the 'dip' between the two lines is less pronounced. Further reduction in the separation makes the dip even smaller, as in (c) where the separation is equal to the linewidth. Finally, in (d) where the separation is half of the linewidth, the two peaks are no longer distinct and a single line is seen.

2.3 Scalar coupling

Scalar or J coupling between nuclei is mediated by chemical bonds and is therefore very useful in establishing which nuclei are close to one another on the bonding framework. The presence of such coupling gives rise to *multiplets* in the spectrum; for example, as shown in Fig. 2.7, if two spin-half nuclei are coupled, the resonance from each spin splits symmetrically about the chemical shift into two lines, called a *doublet*.

Each doublet is split by the same amount, a quantity referred to as the *coupling constant*, J. It is found that the values of coupling constants are *independent* of the field strength; they are always quoted in Hz.

One way of thinking about the two lines of a doublet is to associate them with different *spin states* of the coupled spin. The idea here is that a spin-half nucleus can be in one of two spin states, described as 'up' and 'down' (in Chapter 3 we will have a lot more to say about what up and down actually mean). So, for the doublet centred at the chemical shift of the *first* spin, one line is associated with the *second* spin being in the up spin state, and the other line is associated with the second spin being in the down spin state; Fig. 2.7 (b) illustrates the idea. Similarly, for the doublet centred at the shift of the *second* spin, one line is associated with the *first* spin being up and the other line with the first spin being down.

In terms of frequencies, the line associated with the coupled spin being in the up state is shifted by $\frac{1}{2}J_{12}$ to the left, and the line associated with the coupled spin being in the down state is shifted by $\frac{1}{2}J_{12}$ to the right. So, the two lines of the doublet are separated by J_{12}, and placed symmetrically about the chemical shift.

2.3.1 Tree diagrams

If there are couplings present to further spins, the form of the multiplets can be predicted using 'tree' diagrams, of the type shown in Fig. 2.8 on the facing page. Multiplet (a) is the doublet arising from the first spin due to

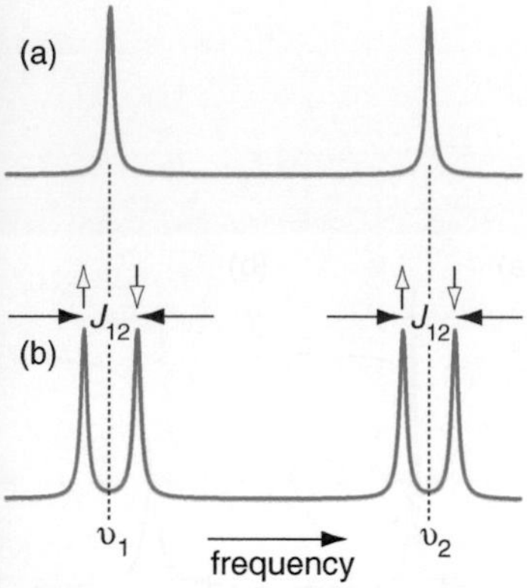

Fig. 2.7 Spectrum (a) shows two lines, at frequencies v_1 and v_2, from two different spins. If there is a *scalar coupling* between the spins, each line splits symmetrically into two, giving two *doublets*, as shown in (b). The splitting of the two lines in each doublet is the coupling constant, J_{12}. One way of thinking about the two lines of a doublet is to associate one line with the coupled spin being in the 'up' spin state, and the other line with the coupled spin being in the 'down' spin state; these spin states are indicated by the open-headed arrows.

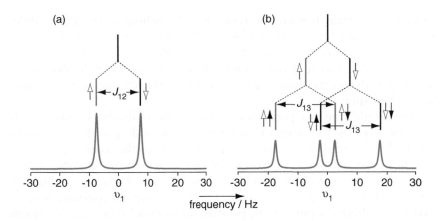

Fig. 2.8 Illustration of how multiplets are built up as a result of scalar coupling. In (a) we see a doublet which arises from the coupling of the first spin to a second spin; the coupling constant is J_{12}. The doublet can be built up using a tree diagram in which the original line at v_1 is split symmetrically into two; the left-hand line is associated with the second spin being up (indicated by an upward pointing arrow), and the right-hand line is associated with the second spin being 'down' (a downward arrow). If the first spin is also coupled to a third spin, with coupling constant J_{13}, each line of the doublet is split once more, as is shown in (b); the resulting multiplet is called a doublet of doublets. The first branching of the tree diagram represents the coupling to the second spin and is the same as in (a). The second layer represents the coupling to the third spin: again, the line which splits to the left is associated with the third spin being up, and the one which splits to the right is associated with the third spin being down. The spin states of the second spin are shown using arrows with open heads, whereas those of the third spin have filled heads. Each line of the doublet of doublets is thus associated with particular spin states of the two coupled spins. The parameters chosen to draw the diagram were: $v_1 = 0$, $J_{12} = 15$ Hz and $J_{13} = 20$ Hz.

its coupling to the second, and over the multiplet is shown the tree diagram which can be used to construct it. At the top, we start with a line at v_1. In the next layer down this line splits symmetrically into two: one shifted by $\frac{1}{2}J_{12}$ to the left, and one by $\frac{1}{2}J_{12}$ to the right, thus producing the doublet. These two branches can be associated with the second spin being in the up and down spin states, respectively.

If a third spin is now introduced which also has a coupling (of size J_{13}) to the first spin, we have to add another layer of branching to the tree diagram; this is shown in Fig. 2.8 (b). As before, we start with a line at v_1. The first layer of the branching is due to the coupling to the second spin, and is exactly the same as in (a). To construct the second layer, each line from the first is split symmetrically into two but this time with the splitting being J_{13}. As before, the branch which splits to the left is associated with the third spin being up and the branch which splits to the right is associated with the third spin being down. Overall, the result is a four line multiplet, called a *doublet of doublets*.

If we assume that a branching to the left reduces the frequency of the line, and that a branching to the right increases it, we can work out the frequencies of each of the four lines of the doublet of doublets simply by noting whether they are the result of a branching to the left or right. So, the left-most line of the doublet of doublets shown in (b) must have frequency $(\upsilon_1 - \frac{1}{2}J_{12} - \frac{1}{2}J_{13})$, whereas the next line along has frequency $(\upsilon_1 + \frac{1}{2}J_{12} - \frac{1}{2}J_{13})$ as it derives from a branching to the right due to the coupling to the second spin and a branching to the left due to the coupling to the third spin.

You should convince yourself that the doublet of doublets looks exactly the same if, in the tree diagram, you first split according to the coupling to the third spin and then according to the coupling to the second spin.

The question arises as to how we know that it is the up spin state which is associated with the line which splits to the left. In fact, whether it is the up or down state depends on the *sign* of the coupling constant; here we have chosen both couplings to be positive. In section 3.6 on page 38 we will return to the influence which the sign of the coupling has on the spectrum. However, for the moment we will simply note that the appearance of the multiplet is unaffected by the sign of the coupling.

The final thing to note from Fig. 2.8 on the previous page is that since the doublet and the doublet of doublets are both from one spin, the integral of both must be the same. So, adding the second splitting to form the doublet of doublets reduces the intensity of the lines by a factor of two.

2.3.2   Weak and strong coupling

All we have said so far about the multiplets which arise from scalar coupling is applicable only in the *weak coupling* limit. This limit is when the frequency separation of the two coupled spins is much larger in magnitude than the magnitude of the scalar coupling between the two spins.

For example, suppose that we record a proton spectrum at 500 MHz and that there are two protons whose resonances are separated by 2 ppm and which have a coupling of 5 Hz between them. As explained in section 2.1.2 on page 7, the frequency separation between the two lines is $2 \times 500 = 1000$ Hz. This is two hundred times greater than the coupling constant, so we can be sure that we are in the weak coupling limit. The coupling between different isotopes (e.g. ^{13}C and ^1H) is always in the weak coupling limit on account of the very large frequency separation between the resonance frequencies of different isotopes (usually of the order of several MHz).

On the other hand, if the frequency separation of the resonances from two coupled spins is comparable with the coupling constant between them, we have what is called *strong coupling*. In this limit, both the frequencies and intensities of the lines are perturbed from the simple weak coupling prediction. We will return to a more detailed discussion of the effects of strong coupling in section 12.1 on page 442 and section 12.7 on page 468.

Unless we say otherwise, everything described in this book applies only to weakly coupled spin systems. This is something of a limitation, but for strongly coupled systems the calculations for all but the simplest experiments become very much more complex and the resulting spectra are rather hard to interpret, so little is to be gained by such an analysis. In practice, therefore, we need not be too worried by this limitation to weak coupling.

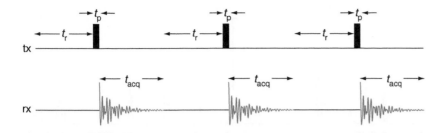

Fig. 2.9 Timing diagram showing how a basic NMR spectrum is recorded. The line marked 'tx' shows the location of high-power RF pulses; tx is the traditional abbreviation for an RF transmitter. The NMR signal is detected by a receiver during the times shown on the line marked 'rx'. During time t_r the spins come to equilibrium. A very short RF pulse is applied for time t_p and then the resulting FID is recorded for time t_{acq}. In order to improve the signal-to-noise ratio, the whole process is repeated several times over and the FIDs are added together; this process is called time averaging. Here, the experiment is repeated three times.

Confusion can arise as the term strong coupling is sometimes used to mean a coupling constant with a large *size*. Strictly, this is an erroneous use of the term.

2.4 The basic NMR experiment

The way we actually record an NMR spectrum using a pulsed experiment is shown in Fig. 2.9. First, a delay is left in order to allow the spins to come to equilibrium; this is called the *relaxation delay*, t_r. Typically this delay is of the order of a few seconds.

Next, a very short burst, typically lasting no more that 20 μs, of high power RF is applied. This excites a transient signal known as a *free induction decay* or FID, which is then recorded for a time called the *acquisition time*, t_{acq} which usually lasts between 50 ms and a few seconds. Finally, Fourier transformation of the FID gives us the familiar spectrum.

The NMR signal tends to be rather weak, so that it is almost never the case that the spectrum from a single FID has sufficient signal-to-noise to be useful. In order to improve the signal-to-noise ratio we use *time averaging*. The idea here is to repeat the experiment many times and then add together the resulting FIDs. The signal part of the FID simply adds up so that after N experiments the signal will be N times stronger. However, the noise, because it is random, adds up more slowly – usually increasing as $\sqrt{N}$. Overall, then, repeating the experiment N times gives an improvement in the signal-to-noise ratio by a factor of $\sqrt{N}$. We usually describe this by saying that N 'transients' or 'scans' were recorded. Calling each experiment a scan is something of a misnomer, but it is an historic usage which has stuck firmly.

Figure 2.10 on the following page shows the proton spectrum of quinine, whose structure is shown in Fig. 2.11 on the next page. Throughout the rest of the text, we will be using spectra of this molecule to illustrate various different experiments.

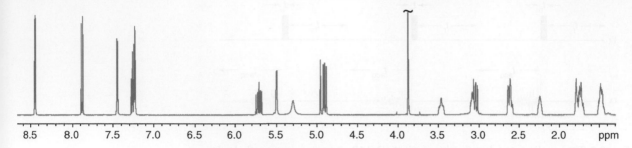

Fig. 2.10 500 MHz proton spectrum of quinine (in CDCl$_3$ solution), whose structure is shown in Fig. 2.11. The group of multiplets between 7 and 9 ppm are clearly from the aromatic ring, while those between 4.5 and 6 ppm include the protons on the double bond. The intense peak at 3.8 ppm (which has been truncated) is from the OCH$_3$ group.

2.4.1   Heteronuclear NMR and broadband decoupling

In an NMR experiment we can usually only observe one kind of nucleus at a time, such as proton, ^{13}C or ^{15}N. Historically, proton NMR was the first to be exploited widely, and it is still the most recorded nucleus. As a result, all nuclei which are not protons are grouped together and called *heteronuclei*.

Scalar couplings can occur between any magnetic nuclei which are reasonably close on the bonding network. It is usual to distinguish between *homonuclear* couplings, which are couplings between nuclei of the same type, and *heteronuclear* couplings, which are couplings between nuclei of different types.

While couplings certainly provide useful information, at times they can be troublesome as the presence of many couplings will result in complex broad multiplets. This is particularly the case when observing ^{13}C spectra of organic molecules in which any one ^{13}C is likely to be coupled to several protons.

The effect of all of these ^{13}C–^{1}H couplings can be removed if, while the ^{13}C spectrum is recorded, the protons are irradiated with a *broadband decoupling* sequence. Such sequences generally involve continuous irradiation of the protons with a carefully designed repeating set of pulses of particular phases and flip angles. The most commonly employed sequence is called WALTZ–16, although there are many more which can be used. Such broadband decoupling essentially sets all of the ^{13}C–^{1}H couplings to zero, so that in the ^{13}C spectrum there is a single peak at each shift. The simplification achieved is very significant, and in addition the signal-to-noise ratio is improved as all of the intensity appears in a single line rather than being spread across a multiplet. This is well illustrated by the comparison of Fig. 2.12 and Fig. 2.13 on the facing page, which are the coupled and decoupled ^{13}C spectra of quinine.

The main issue with broadband decoupling sequences is that, as they are applied continuously during data acquisition, the sample itself may be heated to a significant degree simply by absorbing the RF power. The wider the range of chemical shifts of the nucleus being irradiated, the more power is needed and hence the more serious the heating effect. For protons,

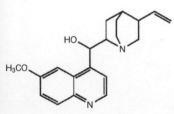

Fig. 2.11 The structure of quinine.

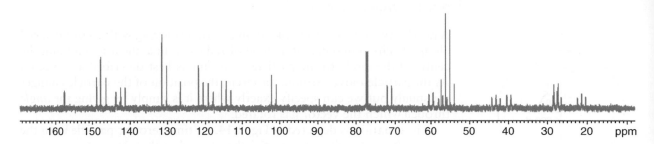

Fig. 2.12 500 MHz ^{13}C spectrum of quinine recorded *without* broadband proton decoupling. The presence of both the large one-bond ^{13}C–^{1}H couplings, and numerous long-range couplings, makes for a rather complex spectrum. The 1:1:1 triplet at 77 ppm is from the CDCl$_3$ solvent.

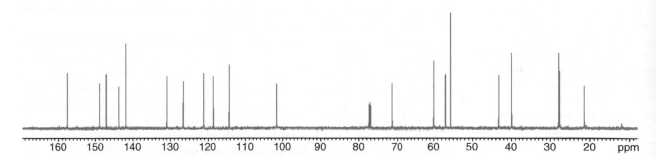

Fig. 2.13 500 MHz ^{13}C spectrum of quinine recorded *with* broadband proton decoupling. The resulting collapse of all the multiplets means that each ^{13}C gives rise to a single line. Compared with the coupled spectrum, Fig. 2.12, both the resolution and the signal-to-noise ratio has improved greatly.

with their modest shift range, this is generally not a problem. However, if we want to observe protons and decouple ^{13}C, the wide range of ^{13}C shifts means that more power is required and so heating can be more of a problem.

2.5 Frequency, oscillations and rotations

Quantities with the dimensions of frequency occur a great deal in NMR: for example the frequencies of the lines themselves, offset frequencies and coupling constants. Often, when we specify a frequency we are thinking of it in relation to some kind of oscillation or, as we shall see is more common in NMR, a *rotation*. In this section we will look at how we specify frequencies, how they are related to rotations and how the resulting motion or oscillation can be expressed mathematically.

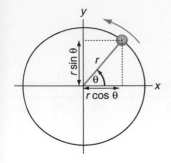

Fig. 2.14 Imagine a particle following a circular path about the origin; here the rotation is anti-clockwise. At time zero, the particle starts on the *x*-axis and after a certain time it has rotated through an angle θ, measured from the *x*-axis. If the circle is of radius *r*, the *x*- and *y*-coordinates are $r\cos\theta$ and $r\sin\theta$, respectively. These coordinates are also the *x*- and *y*-components of the vector from the origin to the particle.

2.5.1   Motion in a circle

A good way to start is to think about a particle moving with constant speed along the circumference of a circle of radius *r*. Imagine a line joining the centre of the circle to the particle – this line is best described as a vector. As the particle moves around the circle, the position of the particle changes constantly but we can specify exactly where it is simply by giving the *angle* through which the vector has rotated.

The situation is depicted in Fig. 2.14. At time zero the particle is on the *x*-axis; after some time, the particle has moved so that the vector joining it to the origin makes an angle θ to the *x*-axis. From the diagram, it is clear that the *x*-coordinate of the particle is $r\cos\theta$ and the *y*-coordinate is $r\sin\theta$.

Another way of looking at this is to say that the *x*-component of the vector is $r\cos\theta$, where by 'component' we mean the projection of the vector onto the axis. This projection is found by drawing a line from the tip of the vector and which is perpendicular to the *x*-axis; where this line cuts the *x*-axis gives the component along that axis.

Figure 2.15 shows these *x*- and *y*-components plotted against the angle θ; above the graph is shown the corresponding position of the particle. Rather than specifying the angle in degrees we have chosen to give it in *radians*. Recall that there are 2π radians in a complete revolution i.e. 360°. So, $\theta = \pi/2$ corresponds to one quarter of a revolution or 90°. Similarly, $\theta = 3\pi/2$ corresponds to three quarters of a revolution, or 270°.

The rotation of the particle, which is simply a steady increase in the angle θ, gives rise to *x*- and *y*-components which are *oscillating* as cosine and sine functions. We see that there is thus a strong connection between rotational and oscillatory motion.

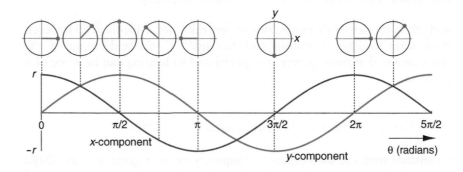

Fig. 2.15 Illustration of how the *x*- and *y*-components of the vector specifying the position of a particle moving round a circular path vary with the angle θ, as defined in Fig. 2.14. The *x*- and *y*-components vary as $\cos\theta$ and $\sin\theta$, and are shown by grey and blue lines, respectively. For some selected angles, the position of the particle is shown above the graph; the horizontal axis gives the angle θ in radians, expressed as multiple of π. Note that the *x*- and *y*-components can be positive and negative, and that after a complete revolution, $\theta = 2\pi$, the pattern repeats.

2.5.2 Frequency

We started out by supposing that the particle was moving around the circle at a constant speed. Suppose that it takes a time T to complete one revolution – this is called the *period*. This period is the same for the sine and cosine waves which describe the position of the particle and it is the time after which they repeat.

The *frequency* of the rotation or oscillation, υ is simply

$$\upsilon = \frac{1}{T}.$$

So a fast oscillation or rotation, which has a short period, corresponds to a high frequency. Another way of thinking about the frequency is to see it as the number of cycles or oscillations per unit time.

The period is specified in seconds (s), so the frequency has units s^{-1} or, equivalently, Hertz (symbol Hz). Thus an oscillation or rotation with a period of 0.0013 s corresponds to a frequency of 769 Hz. In words, this means that in 1 s (that is, a unit of time) the oscillation goes through 769 complete cycles.

2.5.3 Angular frequency

There is another way of expressing the frequency, which is to give it in radians per second; this is called the *angular frequency*, ω. Suppose that the period is T. During this time the angle θ increases by 2π radians so the angular frequency is

$$\omega = \frac{2\pi}{T}.$$

If the time is in seconds, ω will be in radians per second i.e. rad s^{-1}.

Given that the frequency is related to the period by $\upsilon = 1/T$, it follows that frequency and angular frequency are related by

$$\omega = 2\pi \times \upsilon \qquad \text{or} \qquad \upsilon = \frac{\omega}{2\pi}. \qquad (2.4)$$

When we begin to explore the theory of NMR in more detail it will be more convenient to use rad s^{-1} rather than Hz as the unit for frequencies. However, the frequency scales you will find on spectra and the values of coupling constants are invariably quoted in Hz, so we will often need to use Eq. 2.4 to swap back and forth between these two frequency units.

The usual convention is to use the symbols f, F or υ (Greek 'nu') to represent frequencies in Hz, whilst ω and Ω (Greek lower and upper case 'omega') are used to represent frequencies in rad s^{-1}.

2.5.4 Phase

Suppose that our particle depicted in Fig. 2.14 on the facing page starts on the x-axis and then rotates through an angle ϕ, as shown in Fig. 2.16. We describe this situation by saying that the particle has 'acquired a phase ϕ'. In this context, phase is just a way of saying how far the particle has proceeded around the circle. If we think about the sine and cosine functions

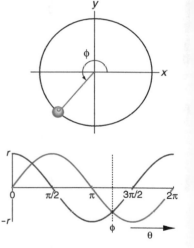

Fig. 2.16 The position of the particle is described by the angle ϕ which is called the *phase*. Here, the phase, measured from the x-axis, is 225° or 1.25π radians. Thought of in terms of the oscillating x- and y-components, the phase tells us how far along the wave we have travelled; on the graph, the vertical dashed line shows the phase, ϕ, corresponding to the diagram above. As before, the grey and blue lines represent the x- and y-components, respectively.

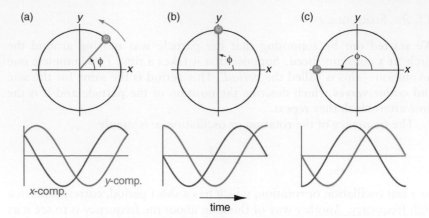

Fig. 2.17 Illustration of how a phase angle can be used to describe the starting position of our particle. The top row shows the starting positions at time zero; in (a) the phase, ϕ, is $\pi/4$ radians or 45°, in (b) the phase is $\pi/2$ radians or 90°, and in (c) the phase is π radians or 180°. The bottom row shows the time dependence of the x- and y-components (grey and blue lines, respectively) as the particle moves from its starting position. Note that in all cases theses components are just those shown in Fig. 2.15 on page 16 shifted to the left by differing amounts.

which represent the x- and y-components, the phase just tells us how far along the oscillation we have proceeded.

If the particle is moving at a constant angular frequency of ω rad s^{-1}, then after time t the angle (in radians) through which the particle has moved is just

$$\text{phase after time } t \;=\; \text{angular frequency} \times \text{time}$$
$$=\; \omega t.$$

For a particle moving at constant speed, therefore, the phase angle simply increases linearly with time.

So, what we have plotted along the horizontal axis of Fig. 2.15 on page 16 is the phase acquired after a certain time. The axis could just as well be labelled with time, as, for a constant speed, phase and time are directly proportional to one another.

We can also use the idea of phase to specify the *starting* position of the particle. So far, we have assumed that at time zero the particle is on the x-axis, but this is not necessarily the case. The more general situation is where at time zero the particle starts at a position described by a phase ϕ (measured, by convention, from the x-axis). Figure 2.17 illustrates this idea for three different phases.

In (a) the starting phase is $\pi/4$ radians. So, at $t = 0$ the x-component is $\cos(\pi/4)r$ which is $0.707\,r$; similarly the y-component at time zero is $\sin(\pi/4)r$ which is also $0.707\,r$. Then, as time proceeds the x- and y-components oscillate in the familiar way. However, neither of these components is a sine or cosine wave – rather, they are sine and cosine waves which have been shifted 'to the left' by our starting phase of $\pi/4$ radians.

Mathematically we can write the two components in the following way:

$$x\text{-component} = r\cos(\omega t + \phi)$$
$$y\text{-component} = r\sin(\omega t + \phi).$$

Note that when $t = 0$ we have $r\cos\phi$ and $r\sin\phi$, as expected.

Although the initial phase ϕ can have any value, there are some special cases which will be of interest. The first is when $\phi = \pi/2$ radians or $90°$, as depicted in Fig. 2.17 (b) on the facing page. For this phase the graph of the y-component is clearly a cosine wave and, after a bit of thought, it is also clear that the graph of the x-component is *minus* a sine wave. If we think about the particle rotating from the initial position shown in (b) then clearly the x-component starts at zero and then at first becomes negative; similarly the y-component starts at its maximum and initially decreases. These are the properties of minus a sine function and a cosine function, respectively.

Mathematically we can write the two components as

$$x\text{-component} = r\cos(\omega t + \pi/2)$$
$$y\text{-component} = r\sin(\omega t + \pi/2).$$

These expressions can be tidied up using the standard identities

$$\cos(A + B) = \cos(A)\cos(B) - \sin(A)\sin(B)$$
$$\sin(A + B) \equiv \sin(A)\cos(B) + \cos(A)\sin(B).$$

Applying the first of these to the expression for the x-component we find:

$$
\begin{aligned}
x\text{-component} &= r\cos(\omega t + \pi/2) \\
&= r\cos(\omega t)\cos(\pi/2) - r\sin(\omega t)\sin(\pi/2) \\
&= -r\sin(\omega t),
\end{aligned}
$$

where on the last line we have used the fact that $\cos(\pi/2) = 0$ and $\sin(\pi/2) = 1$. So, as expected from the diagram, a phase shift of $\pi/2$ does indeed give us an x-component of the form $-r\sin(\omega t)$.

The y-component can be treated in a similar way using the second identity:

$$
\begin{aligned}
y\text{-component} &= r\sin(\omega t + \pi/2) \\
&= r\sin(\omega t)\cos(\pi/2) + r\cos(\omega t)\sin(\pi/2) \\
&= r\cos(\omega t).
\end{aligned}
$$

As expected, we see that the y-component is a cosine wave.

Finally Fig. 2.17 (c) on the preceding page shows the case where the initial phase is π radians or $180°$. From the graphs of the x- and y-components it is clear that all that has happened is that both have changed sign relative to the case where the initial phase is zero. We can demonstrate this mathematically:

$$
\begin{aligned}
x\text{-component} &= r\cos(\omega t + \pi) \\
&= r\cos(\omega t)\cos(\pi) - r\sin(\omega t)\sin(\pi) \\
&= -r\cos(\omega t),
\end{aligned}
$$

where on the last line we have used the fact that $\cos(\pi) = -1$ and $\sin(\pi) = 0$. A similar argument shows that the y-component is $-r\cos(\omega t)$. Thus, an initial phase of π simply causes the x- and y-components to change sign.

It is also common to describe the situations shown in Fig. 2.17 on page 18 as being the result of a *phase shift*. So, (a) is a phase shift of $\pi/4$, (b) a shift of $\pi/2$ and (c) a shift of π. We will encounter this language often.

2.5.5 Representation using complex numbers

The position of a particle on a circle is conveniently represented using the complex exponential $\exp(i\theta)$ or $e^{i\theta}$. This function follows the identity

$$\exp(i\theta) \equiv \underbrace{\cos\theta}_{\text{real}} + i\underbrace{\sin\theta}_{\text{imag.}}.$$

In words, the real and imaginary parts of $\exp(i\theta)$ are $\cos\theta$ and $\sin\theta$, respectively. This immediately makes us think of the x- and y-components of our rotating particle, Fig. 2.14 on page 16. Indeed, if we redraw this diagram and label the axes 'real' and 'imaginary', rather than x and y (Fig. 2.18), we see that the position of the particle is described exactly by $r\exp(i\theta)$, where r is the radius.

Recalling that if the particle is moving at an angular velocity ω then after time t the angle θ is equal to ωt, so the position of the particle is given by

$$r\exp(i\omega t).$$

Similarly, if the phase at time $t = 0$ is ϕ, the position of the particle becomes

$$r\exp(i[\omega t + \phi]) \equiv r\exp(i\omega t)\exp(i\phi).$$

The identity comes about because of the property of exponentials i.e. $e^{A+B} \equiv e^A \times e^B$.

This representation of rotational motion or oscillations using complex exponentials turns out to be very convenient, and we shall use it often.

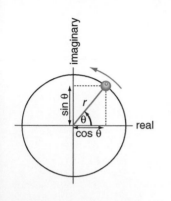

Fig. 2.18 If we imagine the x- and y-axes as corresponding to the real and imaginary parts of a complex number, then the position of the particle is described using the complex exponential as $r\exp(i\theta)$.

2.6 Photons

Electromagnetic radiation can, for some purposes, be thought of as consisting of particles called *photons*. The energy of a photon is related to its frequency, υ, according to

$$E = h\upsilon,$$

where h is a universal constant known as Planck's constant. From this equation we see that the higher the frequency, the more energetic the photon.

If the frequency is expressed in angular units (ω in rad s^{-1}) then, recalling Eq. 2.4 on page 17, we have $\upsilon = \omega/2\pi$ and so the energy is

$$\begin{aligned} E &= h(\omega/2\pi) \\ &= \hbar\omega. \end{aligned}$$

$\hbar$ is $h/2\pi$, a quantity which will appear often in our calculations. It is pronounced 'h bar' or 'h cross'.

2.7 Moving on

The scene is now set, and we are ready to start our description of NMR proper. The first topic we will explore is how energy levels and selection rules are useful in thinking about NMR spectra, and how quantum mechanics can be used to find these energy levels.

2.8 Further reading

Chemical shifts, scalar couplings and their effect on spectra:

Chapters 2 and 3 from P. J. Hore, *Nuclear Magnetic Resonance* (Oxford University Press, 1995).

Broadband decoupling:

Chapter 7 from R. Freeman, *Spin Choreography* (Spektrum, 1997).

Complex numbers and the complex exponential:

Chapter 7 from D. S. Sivia and S. G. Rawlings, *Foundations of Science Mathematics* (Oxford University Press, 1999).

2.9 Exercises

2.1 In a ^{1}H NMR spectrum the peak from TMS is found to occur at 500.134 271 MHz. Two other peaks in the spectrum are found at 500.135 021 and 500.137 921 MHz; compute the chemical shifts of these two peaks in ppm.

Given that the receiver reference frequency is 500.135 271 MHz, recompute the chemical shifts of the two peaks using Eq. 2.3 on page 8; comment on your answers.

What would the frequency separation, in Hz and in rad s^{-1}, be between these two peaks if the spectrum were recorded using a different spectrometer which operates at 400 MHz for protons? The receiver reference frequency for this spectrometer is 400.130 000 MHz.

2.2 Following the approach described in section 2.3.1 on page 10, use a tree diagram to predict the form of the multiplet expected for spin A when it is coupled to two other spins, B and C, with coupling constants $J_{AB} = 10$ Hz and $J_{AC} = 2$ Hz. Work out the frequency of each line and label it with the spin states of the coupled spins; assume that the multiplet is centred at 0 Hz.

Repeat the process for the cases (a) $J_{AB} = 10$ Hz and $J_{AC} = 12$ Hz, and (b) $J_{AB} = 10$ Hz and $J_{AC} = 10$ Hz. What special feature arises in the latter case?

Predict the form of the A spin multiplet expected when a fourth spin D is introduced, using the coupling constants $J_{AB} = 10$ Hz, $J_{AC} = 2$ Hz and $J_{AD} = 5$ Hz.

2.3 A rotation has a period of 2.5×10^{-9} s; compute the corresponding frequency in both Hz and rad s^{-1}.

Compute how long it will take the object to rotate through an angle of: (a) 90°; (b) $3\pi/2$ radians; and (c) 720°.

An oscillation has a angular frequency of 7.85×10^4 rad s^{-1}. Compute the corresponding frequency in Hz and the period in s.

2.4 Following the style of Fig. 2.17 on page 18, make sketch graphs of the x- and y-components of a rotating particle as a function of time for the case where the starting phase ϕ is: (a) 0°; (b) 135°; (c) 2π radians; (d) $3\pi/2$ radians. In each case, comment on the form of your graphs, noting particularly whether they are simple sine or cosine functions.

2.5 In Fig. 2.17 (c) on page 18 the y-component is $r \sin(\omega t + \pi)$. Using the same approach as on page 19, show that this y-component can be written more simply as $-r \sin(\omega t)$.

3

Energy levels and NMR spectra

In this chapter we will look at how energy levels can be used to understand simple NMR spectra. This approach is of somewhat limited utility when it comes to understanding how NMR experiments work, but it is nevertheless worthwhile exploring as it is a good vehicle for introducing quantum mechanics, gives us some useful ways of thinking about NMR spectra, and introduces the idea of multiple-quantum transitions.

The usual explanation given for the appearance of lines in a spectrum is that they arise as a result of transitions between a set of energy levels possessed by the molecule. The existence of these energy levels is a consequence of the quantization of the energy. In favourable cases, we can use quantum mechanics to calculate what these energy levels are, and find the set of labels or quantum numbers which characterize each level.

The molecule can absorb photons whose energies match the *difference* in energy between two of these quantized energy levels, as is illustrated in Fig. 3.1. Here we see two energy levels, with energies E_{upper} and E_{lower}, separated by ΔE:

$$\Delta E = E_{upper} - E_{lower}.$$

A photon can only be absorbed if its energy, $h\upsilon$, matches the energy separation of the two levels i.e.

$$h\upsilon = \Delta E.$$

The spectrum thus consists of a series of lines whose frequencies depend on the energy separation between the levels.

A photon of the correct energy will only be absorbed if the transition between the two energy levels is allowed according to the quantum mechanical selection rules which apply to the system. These rules are usually expressed in terms of the quantum numbers of the levels involved, and typically require that, in going from one level to another, a particular quantum number must change by a specified amount.

This description of how spectra arise is deceptively simple and, for NMR, not really adequate. In the first section in this chapter we will tease out what the problem is and therefore discover the limitations of such an

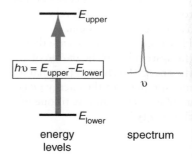

Fig. 3.1 The basic description of spectroscopy in terms of energy levels. A photon may be absorbed provided its energy, given by $h\upsilon$, matches the energy separation between two energy levels, here $E_{upper} - E_{lower}$. The result is an absorption line in the spectrum, at frequency υ.

Understanding NMR Spectroscopy, Second Edition James Keeler
© 2010 John Wiley & Sons, Ltd

approach. Nevertheless, despite these difficulties, we will see that there are many aspects of NMR which can be understood by thinking about energy levels, and the rest of the chapter is therefore devoted to explaining how these levels are found, and how we can use them to predict the form of spectra.

3.1   The problem with the energy level approach

The above description of how a spectrum arises implies that the molecule sits in one energy level and then the absorption of a photon causes it to move to another level. This is certainly not an uncommon or contentious thing to imply. Indeed, in any elementary course of quantum mechanics or spectroscopy we are usually told that 'the energy is quantized' and it is then either stated, or implied, that it follows that the molecule must be 'in one of the energy levels'.

The problem is that quantum mechanics emphatically does *not* say that a molecule must be in one of the energy levels. What quantum mechanics says about the molecule and its energy is rather more complicated and, on the face of it, rather surprising.

One of the fundamental postulates of quantum mechanics is to do with what happens when we make a measurement. This postulate raises all sorts of practical and philosophical problems, but as far as we know it is correct so we will simply use it and not worry too much about the philosophy. Adapted for the present purpose, the postulate implies that, if we measure the energy of a molecule, the value we will obtain will *always* correspond to one of the energy levels. Figure 3.2 attempts to illustrate this idea.

In essence, spectroscopy is a way of measuring the energy of a molecule, since we determine the frequency, and hence the energy, of the photons which are absorbed. Strictly, it is an energy *difference* that we are measuring, but we can think of this as two successive measurements of the energy. As described in the previous paragraph, each measurement of the energy will give a value which corresponds to one of the energy levels, so in spectroscopy it appears that transitions take place between these levels.

3.1.1   Wavefunctions and mixed states

You might be forgiven for thinking that it is splitting hairs to make a distinction between a molecule actually being in one of the energy levels as opposed to it *appearing* to be in one of the levels. However, the distinction becomes significant when we start to think about the *wavefunction* which describes the molecule.

In quantum mechanics, the wavefunction carries within it all of the information needed to compute the properties of the molecule. Later on we will see some examples of such functions, but for now we will just take it that such functions exist. Each energy level has associated with it a *different* wavefunction and, just as it is commonly implied that the molecule must be in one of the energy levels, it is also often implied that the wavefunction of the molecule must be the one associated with that level.

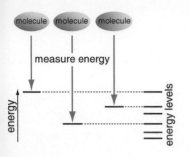

Fig. 3.2 An illustration of the process of measurement in quantum mechanics. A postulate of the theory is that a measurement of the energy yields a value which corresponds to one of the energy levels available to the system. However, it is not the case that the system has to be *in* one of the energy levels, it is just that a measurement of the energy gives a value corresponding to one of the energy levels.

However, just as it is not true that the molecule must be in one of the energy levels, it is also *not* true that the wavefunction must be one associated with an energy level. This is a hard idea to come to terms with, as any book on quantum mechanics always has nice diagrams showing the energy levels alongside pictures of the associated wavefunctions (for example, as in Fig. 3.3). We are enticed into slotting the molecule into one of the levels and giving it the associated wavefunction! However, this is not a correct description.

In fact, the wavefunction for the molecule is generally a mixture of the wavefunctions associated with the energy levels: such a wavefunction is often called a *mixed state* or a *superposition state*. We will see as the book progresses that NMR experiments, even the simplest ones, work by manipulating and exploiting the properties of these mixed states.

We can legitimately ask why it is not necessary to be concerned about these mixed states when thinking about IR or UV spectroscopy, since for these kinds of spectroscopy the energy level approach is sufficient. There are two reasons for this.

First, mixed states arising from molecular vibrational and electronic energy levels have very short lifetimes so that it is hard (if not impossible) to manipulate them. The reason for these short lifetimes is that the states are changed and perturbed by the very frequent collisions which occur between molecules. In contrast, nuclei lie deeply buried inside the molecules and are little affected by collisions, so the mixed states arising from nuclear spins can be rather long lived. We can thus manipulate and observe them at our leisure.

The second reason why mixed states are important in NMR, and not in IR or UV spectroscopy, is that in NMR we have a straightforward way of manipulating these states using RF pulses. The equivalent for other forms of spectroscopy would be extremely short high-powered bursts of laser radiation. Although such sources are available, they are hardly routine.

Thus we really do not need concern ourselves with mixed states when it comes to the description of most kinds of spectroscopy other than NMR. Simply concentrating on the energy levels and supposing, albeit somewhat incorrectly, that the molecules are in one or other of the energy levels will give us a perfectly adequate description of the spectrum. In contrast, in NMR, we do need to take into account the existence and behaviour of these mixed states.

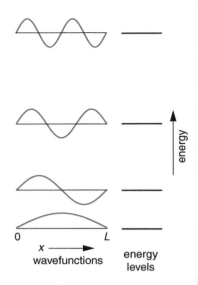

Fig. 3.3 The energy levels and associated wavefunctions for a particle constrained so that it can only move between $x = 0$ and $x = L$ – the so-called 'particle in a box'. We are tempted to slot the particle into one of the energy levels and hence give it the associated wavefunction. However, this is not what quantum mechanics specifies; rather the wavefunction which describes the particle is in general a *mixture* of the wavefunctions corresponding to each energy level.

3.1.2 Energy levels in NMR

Despite the cautions and caveats of the previous section, energy levels and their associated wavefunctions do play a very important part in the theory of NMR. To start with, the *frequencies* of the lines in an NMR spectrum can always be predicted by thinking about the allowed transitions between the energy levels – indeed, the rest of this chapter is devoted to this topic.

Later on we will discover that the energy levels determine how the mixed states evolve over time. In NMR, time evolution is of central importance as we detect the FID as a function of time, and multiple-pulse sequences are all about manipulating the spins through different time periods. So, energy levels remain central to our discussion of NMR, even to the most sophisticated level.

3.1.3   The way ahead

The time has come to flesh out some of the ideas which have been introduced in this section, and to do this we will need to introduce quantum mechanics. In particular, we need to understand what a wavefunction is, and learn how energy levels and the associated wavefunctions can be found in particular cases. In fact, the quantum mechanics of NMR is surprisingly easy: we rarely, if ever, need to evaluate any integrals or differentials, and most problems can be solved by rather simpler algebra.

To start with we will develop sufficient quantum mechanics to be able to find the energy levels and associated wavefunctions of a single spin one-half. After that, we will discuss the resulting spectrum before moving on to more complex arrangements of spins.

3.2   Introducing quantum mechanics

Quantum mechanics is a powerful theoretical framework which provides an essentially complete description of the microscopic world. Quantum mechanical calculations on 'real' systems are often so complex that they can only be tackled numerically using powerful computers. This is why elementary courses of quantum mechanics concentrate on very simple model systems, such as the 'harmonic oscillator' or 'rigid rotor', for which it is possible to make calculations using a pen and paper. Luckily for us, the quantum mechanical description of NMR is particularly straightforward and it turns out to be possible to use it to analyse just about any experiment without the need to resort to computer-based calculations.

To develop the quantum mechanics we need from first principles would be both time consuming and laborious, so we shall not do it. Rather, we will simply state the key ideas and then use them to work forward to the practical results we need.

Wavefunctions and operators are of central importance in quantum mechanics, so we will start out by discussing them in turn.

3.2.1   Wavefunctions

A *wavefunction* is a mathematical function which contains a complete description of the system: if we know the wavefunction we can deduce from it anything we wish to know, such as the position of a particle or its energy.

For example, one of the possible wavefunctions for an electron in a hydrogen atom is

$$\psi(r) = \exp(-ar).$$

Here, r is the distance from the electron to the nucleus and a is a constant, which is known; the function is illustrated in Fig. 3.4. The wavefunction is written $\psi(r)$ to remind us that the function depends on the variable r. We could substitute any value of r into this expression and then evaluate it to give a number: this is what is meant by ψ being a function of r.

We said that the wavefunction tells us everything about the system, but where is this information held and how do we extract it? To do this, we need to introduce quantum mechanical *operators*.

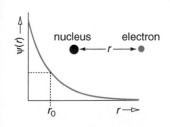

Fig. 3.4 An example of a possible wavefunction for the electron in a hydrogen atom; r is the distance between the nucleus and the electron, and $\psi(r)$ is the wavefunction. For a particular value of r, r_0, the wavefunction evaluates to a number.

3.2.2 Operators

Mathematically, an operator is something which acts on a function to produce a new function. A good example is the operator d/dx, which means 'differentiate with respect to the variable x'; let us apply this operator to the function $\sin x$ and see what happens. Recalling that the differential of $\sin x$ is $\cos x$, the effect of the operator is

$$\underbrace{\frac{d}{dx}}_{\text{operator}} \underbrace{\sin x}_{\text{function}} = \underbrace{\cos x}_{\text{new function}}.$$

The new function, generated by the action of the operator, is therefore $\cos x$.

Operators are important in quantum mechanics as in this theory they 'represent observables'. Observables are things we can measure, for example energy – which is our main concern here. The wavefunction contains all the information, but we need the appropriate operator to 'extract' this information from the wavefunction. At this stage, we do not need to know the mathematical process by which an operator is used to find the value of some observable quantity, but we will return to this point later on.

A key point about operators is that the *order* in which they act is important. This is in contrast to functions and numbers, which can be reordered freely – for example

$$2 \times \cos x \times \sin x \equiv \sin x \times 2 \times \cos x.$$

The first point to make is that in general the order of operators and functions cannot be changed. For example consider the operator d/dx and the function $\sin x$:

$$\frac{d}{dx} \sin x \neq \sin x \frac{d}{dx}.$$

Here we see that changing the order of the operator and the function gives quite a different result.

The second point is that if we have two operators acting one after another on a function the result depends on the order in which the operators act. For example, consider the two operators d/dx, which means 'differentiate with respect to x', and x, which means simply 'multiply by x'. Applying these operators in this order to $\sin x$ gives:

$$\underbrace{x}_{\text{2nd operator}} \underbrace{\frac{d}{dx}}_{\text{1st operator}} \overbrace{\sin x}^{\text{function}} = \underbrace{x}_{\text{second operator}} \cos x$$

$$= x \cos x.$$

Note that the operator immediately to the left of the function is the one which acts first.

Now let us apply the operators in the other order

$$\underbrace{\frac{d}{dx}}_{\text{2nd operator}} \underbrace{x}_{\text{1st operator}} \overbrace{\sin x}^{\text{function}} = \underbrace{\frac{d}{dx}}_{\text{2nd operator}} x \sin x$$

$$= \sin x + x \cos x.$$

We see that applying the operators in the reverse order gives quite a different result.

Sometimes, the order of operators does not matter, in which case the operators are said to *commute*. When the order does matter, the operators are said not to commute.

3.2.3   Eigenfunctions and eigenvalues of operators

We introduced operators by saying that their effect on a function was to change it to a new function. However, there are some functions which, when a particular operator acts on them, remain unchanged apart from multiplication by a constant.

For example, consider the operator d/dx acting on the function $\exp(Ax)$, where A is a constant:

$$\underbrace{\frac{d}{dx}}_{\text{operator}} \underbrace{\exp(Ax)}_{\text{function}} = \underbrace{A}_{\text{constant}} \times \underbrace{\exp(Ax)}_{\text{same function}}. \tag{3.1}$$

In contrast to the case where the operator acted on the function $\sin x$, the result of it operating on $\exp(Ax)$ is for the function to be unchanged with the exception of multiplication by a constant. Functions which have this property are said to be *eigenfunctions* of the operator, and the multiplying constant is called the *eigenvalue*. Generally there are several eigenfunctions for a given operator, each with a corresponding eigenvalue.

The eigenfunctions and eigenvalues of an operator have the following relation between them:

(operator) *acting on* (eigenfunction) = (eigenvalue) × (eigenfunction).

This is called the *eigenvalue equation*. Equation 3.1 is of the form of the eigenvalue equation, so we can say that $\exp(Ax)$ is an eigenfunction of the operator d/dx, with eigenvalue A.

3.2.4   Measurement

We mentioned in section 3.1 on page 24 the quantum mechanical postulate which implies that a measurement of the energy of a molecule will *always* give a value corresponding to one of the energy levels. A more general statement of the postulate is as follows:

> If we measure the value of some observable, the result will *always* be one of the *eigenvalues* of the particular operator which represents this observable.

To take a concrete example, if we measure the energy of a system, then the result must be one of the eigenvalues of the operator for energy. It is these eigenvalues which are the 'energy levels' of the system which we have been talking about. Further, the eigenfunction associated with a particular eigenvalue is the wavefunction which corresponds to the energy level.

The eigenvalues and eigenfunctions of the energy operator are thus very important as they give us the energy levels of the system and the associated wavefunctions. To find the energy eigenfunctions we first need to know the energy operator, which is what the next section is about.

3.2.5 Hamiltonians and angular momentum

The operator which represents the observable quantity energy is so important in quantum mechanics that it has its own name – it is called the *Hamiltonian operator*. Usually the name is shortened to 'the Hamiltonian' and it is commonly represented using the symbols H and $\mathcal{H}$, appropriately festooned with sub- and superscripts as required. A 'hat' is often added to remind us that the symbol represents an operator (rather than a function): $\hat{H}$.

Constructing Hamiltonians is something of an art form which will not be discussed here. Rather, we will simply state the form of the various Hamiltonians we need and leave it at that.

For a nuclear spin in a magnetic field of strength B_0 applied along the z-axis, the Hamiltonian which represents the energy of interaction between the spin and the magnetic field is

$$\hat{H}_{\text{one spin}} = -\gamma B_0 \hat{I}_z \qquad (3.2)$$

where γ is the *gyromagnetic ratio*, which is a fundamental property of the nucleus in question.

$\hat{I}_z$ is an operator which represents the z-component of the *nuclear spin angular momentum*. Angular momentum is a concept from classical physics where it is associated with rotational motion. For example, a mass following a circular path has angular momentum, which turns out to be a vector quantity having both a magnitude and a direction, as is illustrated in Fig. 3.5.

Some nuclei appear to possess an intrinsic source of angular momentum which is usually called *nuclear spin angular momentum*. The name is something of a problem, as it makes it sound as if the angular momentum arises from a literal spinning of the nucleus, which it certainly does not. Rather, the angular momentum is an intrinsic property of the nucleus, just like its mass or its charge.

Like classical angular momentum, the nuclear spin angular momentum is a vector quantity, having both a direction and a magnitude. The operator $\hat{I}_z$ represents the z-component of this angular momentum and it is this component which interacts with the applied magnetic field which is also along the z-direction.

3.2.6 Eigenfunctions and eigenvalues of $\hat{I}_z$

Angular momentum operators, such as $\hat{I}_z$, turn out to be very important in the theory of NMR, and we will come across them again and again. At this stage, it is the eigenfunctions and eigenvalues of $\hat{I}_z$ which are of particular importance. Finding these requires a rather subtle argument which we will not go into here; rather we will simply state the result and go on to explore the consequences.

The number of these eigenvalues depends on the spin of the nucleus in question, and this in turn is specified by a quantum number I, called the

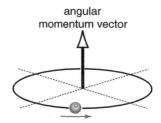

angular
momentum vector

Fig. 3.5 In classical physics, a mass moving around a circular path possesses *angular momentum* which is a vector quantity, having both a magnitude and a direction. For motion in a circular path, the angular momentum vector points in a direction perpendicular to the plane of rotation.

nuclear spin angular momentum quantum number or, more succinctly, the spin quantum number. I can be integer $(0, 1, 2 \ldots)$ or half-integer $(\frac{1}{2}, \frac{3}{2} \ldots)$.

It turns out that the operator $\hat{I}_z$ has $(2I + 1)$ eigenfunctions (with associated eigenvalues), each of which is characterized by another quantum number m, which can only take values between $-I$ and $+I$ in integer steps. For example, if $I = 1$, m can be -1, 0 or $+1$; if $I = \frac{1}{2}$, m can be $-\frac{1}{2}$ or $+\frac{1}{2}$.

From now on we will restrict ourselves to a spin-half nucleus, for which $\hat{I}_z$ has just two eigenfunctions, characterized by $m = +\frac{1}{2}$ and $m = -\frac{1}{2}$. The two corresponding eigenfunctions are $\psi_{+\frac{1}{2}}$ and $\psi_{-\frac{1}{2}}$, and these obey the eigenvalue equations:

$$\hat{I}_z \psi_{+\frac{1}{2}} = +\tfrac{1}{2}\hbar \psi_{+\frac{1}{2}} \qquad \hat{I}_z \psi_{-\frac{1}{2}} = -\tfrac{1}{2}\hbar \psi_{-\frac{1}{2}}. \qquad (3.3)$$

$\psi_{+\frac{1}{2}}$ thus has eigenvalue $+\frac{1}{2}\hbar$, and $\psi_{-\frac{1}{2}}$ has eigenvalue $-\frac{1}{2}\hbar$.

These two eigenvalue equations can be written more compactly as

$$\hat{I}_z \psi_m = m\hbar \psi_m,$$

where, as above, $m = \pm\frac{1}{2}$. We see that the quantum number m is not only a label for the eigenfunctions, but also gives the eigenvalue as $m\hbar$.

You will have noticed that we have not said what the eigenfunctions ψ_m actually are – the reason for this is that this is a piece of information that we never need to know. All we need to know is that these functions *exist* and what the associated eigenvalues are. Armed with these eigenfunctions we can now go on to use them to find the eigenvalues of the Hamiltonian for one spin, and hence the energy levels.

3.2.7   Eigenvalues for the one-spin Hamiltonian

Not surprisingly, the functions $\psi_{+\frac{1}{2}}$ and $\psi_{-\frac{1}{2}}$ are also the eigenfunctions of the Hamiltonian for a single spin, Eq. 3.2 on the previous page:

$$\hat{H}_{\text{one spin}} = -\gamma B_0 \hat{I}_z.$$

The reason why this is not surprising is that $\psi_{\pm\frac{1}{2}}$ are eigenfunctions of $\hat{I}_z$, and the only difference between this operator and $\hat{H}_{\text{one spin}}$ is multiplication by $-\gamma B_0$.

To show that $\psi_{+\frac{1}{2}}$ is indeed an eigenfunction, and to find the corresponding eigenvalue, we need to work out what the effect of the Hamiltonian is on $\psi_{+\frac{1}{2}}$:

$$\begin{aligned}
\hat{H}_{\text{one spin}} \psi_{+\frac{1}{2}} &= -\gamma B_0 \left[\hat{I}_z \psi_{+\frac{1}{2}} \right] \\
&= -\gamma B_0 \left[\tfrac{1}{2}\hbar \psi_{+\frac{1}{2}} \right] \\
&= -\tfrac{1}{2}\hbar \gamma B_0 \psi_{+\frac{1}{2}}.
\end{aligned}$$

On the first line we note that the expression in square brackets is $\hat{I}_z$ acting on one of its eigenfunctions. We can therefore use Eq. 3.3 to replace $\hat{I}_z \psi_{+\frac{1}{2}}$ with $\frac{1}{2}\hbar \psi_{+\frac{1}{2}}$; this brings us to the second line. On the third line we have just tidied things up by moving the constants to the left.

What we have shown is that when $\hat{H}_{\text{one spin}}$ acts on the function $\psi_{+\frac{1}{2}}$ the result is to regenerate the function multiplied by a constant $(-\frac{1}{2}\hbar\gamma B_0)$. This is exactly the property of an eigenfunction, so

$\psi_{+\frac{1}{2}}$ is an eigenfunction of $\hat{H}_{\text{one spin}}$ with eigenvalue $-\frac{1}{2}\hbar\gamma B_0$.

Using the same approach, it is easy to show that $\psi_{-\frac{1}{2}}$ is also an eigenfunction, with eigenvalue $+\frac{1}{2}\hbar\gamma B_0$.

The only difference between $\hat{H}_{\text{one spin}}$ and $\hat{I}_z$ is multiplication by some constants, the presence of which does not stop an eigenfunction of $\hat{I}_z$ also being an eigenfunction of $\hat{H}_{\text{one spin}}$. The eigenvalues of this Hamiltonian are just those for $\hat{I}_z$ multiplied by the constant factor $-\gamma B_0$.

3.2.8 Summary

We have covered quite a lot of ground in this introduction to quantum mechanics, so it is worthwhile pausing to summarize the key results.

- Operators represent observable quantities; particularly important is the Hamiltonian operator which represents the energy.

- The eigenvalues of the Hamiltonian are the energy levels available to the system; the eigenfunctions are the associated wavefunctions.

- For a spin one-half, the operator for the z-component of the angular momentum, $\hat{I}_z$, has two eigenfunctions labelled by $m = \pm\frac{1}{2}$:

$$\hat{I}_z\psi_m = m\hbar\psi_m.$$

- The Hamiltonian for one spin in a magnetic field is

$$\hat{H}_{\text{one spin}} = -\gamma B_0 \hat{I}_z.$$

- The eigenfunctions of $\hat{I}_z$ are also eigenfunctions of this Hamiltonian, with eigenvalues E_m where

$$E_m = -m\hbar\gamma B_0.$$

These are the two energy levels of a single spin-half in a magnetic field.

3.3   The spectrum from one spin

3.3.1   Energy levels

As we have shown in the last section, the two energy levels or states available to a single spin-half are

$$E_m = -m\hbar\gamma B_0 \qquad m = +\frac{1}{2} \text{ or } -\frac{1}{2}. \tag{3.4}$$

B_0 is the magnetic field strength, usually given in Tesla (symbol T), and the gyromagnetic ratio, γ, is usually given in rad s^{-1} T^{-1}. With these units, the energy comes out in Joules, as expected.

For spin-half nuclei it is traditional (and compact) to give the energy level with $m = +\frac{1}{2}$ the label α and the level with $m = -\frac{1}{2}$ the label β:

$$E_\alpha = -\tfrac{1}{2}\hbar\gamma B_0 \qquad E_\beta = +\tfrac{1}{2}\hbar\gamma B_0. \qquad (3.5)$$

The α state with $m = +\frac{1}{2}$ is often described as 'spin up', and the β state with $m = -\frac{1}{2}$ as 'spin down'.

Quantum mechanics tells us that the allowed transitions are ones in which m changes by $+1$ or -1. So, if we go from the α state, with $m = +\frac{1}{2}$, to the β state, with $m = -\frac{1}{2}$, the change in m, Δm, is

$$\begin{aligned}
\Delta m_{\alpha\to\beta} &= -\tfrac{1}{2} - \left(+\tfrac{1}{2}\right) \\
&= -1.
\end{aligned}$$

So, the transition from α to β is allowed, and it is easy to work out that for the transition from β to α, $\Delta m = +1$, so this is allowed too.

Which out of the α or β state is the lower in energy depends on the sign of the gyromagnetic ratio, γ. For ^{1}H and ^{13}C, γ is found to be positive, but for ^{15}N γ is negative. From now on, unless specifically stated otherwise, we will assume that γ is positive; for such a nucleus the energy levels are as shown in Fig. 3.6.

The energy of the allowed transition from α to β is

$$\begin{aligned}
\Delta E_{\alpha\to\beta} &= E_\beta - E_\alpha \\
&= \tfrac{1}{2}\hbar\gamma B_0 - \left(-\tfrac{1}{2}\hbar\gamma B_0\right) \\
&= \hbar\gamma B_0 \\
&= (h/2\pi)\gamma B_0.
\end{aligned}$$

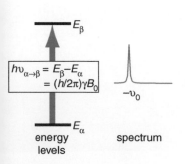

$h\nu_{\alpha\to\beta} = E_\beta - E_\alpha$
$= (h/2\pi)\gamma B_0$

E_β

E_α

$-\nu_0$

energy levels   spectrum

Fig. 3.6 The two energy levels for a single spin-half nucleus having a positive gyromagnetic ratio. The allowed transition between them gives rise to a single line at minus the Larmor frequency.

Note that to compute ΔE we have taken the energy of the upper state minus that of the lower state, thus making ΔE a positive quantity. We will stick to this convention throughout all that follows.

Recalling that the energy of a photon of frequency ν is $h\nu$, it follows that the frequency of the photon corresponding to the above energy gap, $\nu_{\alpha\to\beta}$, is found simply by dividing the energy gap by h:

$$\begin{aligned}
\nu_{\alpha\to\beta} &= \Delta E_{\alpha\to\beta}/h \\
&= \gamma B_0/2\pi. \qquad (3.6)
\end{aligned}$$

Our prediction is for a line at frequency $(\gamma B_0)/2\pi$ Hz.

3.3.2 The Larmor frequency

We now define the *Larmor frequency* of the spin, ω_0 (in rad s^{-1}), in the following way:

definition of Larmor frequency: $\omega_0 = -\gamma B_0$ in rad s^{-1}.   (3.7)

The Larmor frequency in Hz, ν_0, is simply $\omega_0/2\pi$:

definition of Larmor frequency: $\nu_0 = -\gamma B_0/2\pi$ in Hz.   (3.8)

The minus sign in these definitions seems a bit awkward, but it does have a reason, which will become clear when we look at the vector model in the following chapter.

Comparing Eq. 3.6 and Eq. 3.8 we see that the transition from α to β occurs at *minus* the Larmor frequency:

$$\upsilon_{\alpha \to \beta} = -\upsilon_0.$$

This is a nice simple result. For a single spin, there is one allowed transition which results in a line in the spectrum at minus the Larmor frequency. From Eq. 3.8 on the facing page we see that this frequency is proportional to the magnetic field strength, as we expect, with the constant of proportionality being the gyromagnetic ratio.

We know that nuclei of the same isotope (e.g. protons), but in different chemical environments, give lines at slightly different frequencies on account of the chemical shift. We could accommodate this by allowing the gyromagnetic ratio to be different for different protons, but this is not really convenient. A better way is to keep γ the same for all nuclei of the same isotope and redefine the Larmor frequency to include the chemical shift:

$$\omega_0 = -\gamma \left(1 + 10^{-6} \times \delta\right) B_0 \quad \text{or} \quad \upsilon_0 = -\gamma \left(1 + 10^{-6} \times \delta\right) B_0 / 2\pi.$$

In these expressions δ is the chemical shift in ppm.

The tabulated values for γ are usually for a bare nucleus i.e. in the absence of the influence of the surrounding electrons. However, it is usual to measure chemical shifts relative to an agreed reference compound, so $\delta = 0$ does not correspond to a bare nucleus. As a result, these expressions for the Larmor frequency cannot be used directly. In practice this is of little consequence since the Larmor frequency of a nucleus in a particular environment is invariably determined by experiment.

3.3.3 Writing the energies in frequency units

As we are always going to want to convert the energy differences between levels to frequencies, we might just as well express the original energies in frequency units to start with. It might seem strange at first to write an energy in Hz or rad s^{-1}, but remember that energy is in direct proportion to frequency so going from one to the other simply involves multiplication by a constant factor.

As we saw in section 2.6 on page 20, to convert energies from Joules to Hz, all we need to do is divide by Planck's constant, h, and to convert from Joules to rad s^{-1} we divide by $\hbar$. So the energies given by Eq. 3.4 on page 31 can be written in frequency units as:

$$\text{in Hz:} \quad E_m \;=\; -m\gamma B_0 / 2\pi$$
$$\text{in rad s}^{-1}: \quad E_m \;=\; -m\gamma B_0.$$

These can be made even simpler by introducing the Larmor frequencies, as defined in Eq. 3.7 and Eq. 3.8 on the preceding page

$$\text{in Hz:} \quad E_m \;=\; m\upsilon_0$$
$$\text{in rad s}^{-1}: \quad E_m \;=\; m\omega_0.$$

Note that the minus sign has disappeared as the definition of the Larmor frequencies introduces a further negative.

Now the energies, measured in Hz, of the α and β states are simply

$$E_\alpha = \tfrac{1}{2} \upsilon_0 \qquad E_\beta = -\tfrac{1}{2} \upsilon_0,$$

and so the frequency of the $\alpha \to \beta$ transition is

$$
\begin{aligned}
\upsilon_{\alpha \to \beta} &= E_\beta - E_\alpha \\
&= -\tfrac{1}{2} \upsilon_0 - \left(\tfrac{1}{2} \upsilon_0 \right) \\
&= -\upsilon_0.
\end{aligned}
$$

This is exactly the same result we found before, but you can see that writing the energies in frequency units is much simpler and more direct than working in energy units.

3.4   Writing the Hamiltonian in frequency units

Just as it is convenient to write the energy levels in frequency units, it would also simplify things if when we find the eigenvalues of the Hamiltonian they came out directly in frequency units, rather than in Joules. If we take an energy in Joules and divide by $\hbar$, we obtain a frequency in rad s^{-1}, so shifting the Hamiltonian from energy to angular frequency units is simply a matter of 'losing' a factor of $\hbar$.

This is normally done by removing the factor of $\hbar$ from the eigenvalues of the operator $\hat{I}_z$. So, rather than Eq. 3.3 on page 30, we have

$$\hat{I}_z \psi_{+\frac{1}{2}} = +\tfrac{1}{2} \psi_{+\frac{1}{2}} \qquad \hat{I}_z \psi_{-\frac{1}{2}} = -\tfrac{1}{2} \psi_{-\frac{1}{2}}. \tag{3.9}$$

What we are doing is expressing the eigenvalues of $\hat{I}_z$ in units of $\hbar$.

Recall that the Hamiltonian for one spin is

$$\hat{H}_{\text{one spin}} = -\gamma B_0 \hat{I}_z.$$

Using Eq. 3.9, we can show that $\psi_{\frac{1}{2}}$ is an eigenfunction of this Hamiltonian in the same way as we did in section 3.2.7 on page 30:

$$
\begin{aligned}
\hat{H}_{\text{one spin}} \psi_{+\frac{1}{2}} &= -\gamma B_0 \hat{I}_z \psi_{+\frac{1}{2}} \\
&= -\tfrac{1}{2} \gamma B_0 \psi_{+\frac{1}{2}} \\
&= +\tfrac{1}{2} \omega_0 \psi_{+\frac{1}{2}}.
\end{aligned}
$$

The eigenvalue is thus $-\tfrac{1}{2} \gamma B_0$ or, if we introduce the Larmor frequency defined in Eq. 3.7 on page 32, $+\tfrac{1}{2} \omega_0$. Using the same approach we can show that the other eigenfunction is $\psi_{-\frac{1}{2}}$, with eigenvalue $-\tfrac{1}{2} \omega_0$.

We can also use the definition of the Larmor frequency to write the one-spin Hamiltonian as

$$\hat{H}_{\text{one spin}} = \omega_0 \hat{I}_z \qquad \text{in rad s}^{-1}. \tag{3.10}$$

If we want the frequencies in Hz, all we would need to do is divide everything by 2π:

$$\hat{H}_{\text{one spin}} = \upsilon_0 \hat{I}_z \qquad \text{in Hz,}$$

where we have used $v_0 = \omega_0/2\pi$. The eigenfunctions of this Hamiltonian are the same as before, but the eigenvalues are $\pm\frac{1}{2}v_0$. The eigenvalues written in various units are summarized in the following table:

m	wavefunction	in energy units	eigenvalues in frequency units (rad s^{-1})	in frequency units (Hz)
$+\frac{1}{2}$	$\psi_{+\frac{1}{2}}$ or ψ_α	$-\frac{1}{2}\hbar\gamma B_0$	$\frac{1}{2}\omega_0$	$\frac{1}{2}v_0$
$-\frac{1}{2}$	$\psi_{-\frac{1}{2}}$ or ψ_β	$\frac{1}{2}\hbar\gamma B_0$	$-\frac{1}{2}\omega_0$	$-\frac{1}{2}v_0$

3.5 The energy levels for two coupled spins

It is fairly straightforward to extend our treatment of one spin to two. The first thing to do is to write down the Hamiltonian. For one spin it was (in units of Hz)

$$\hat{H}_{\text{one spin}} = v_0 \hat{I}_z.$$

To extend this to two spins we simply add a similar term for the second spin:

$$\hat{H}_{\text{two spins, no coupl.}} = \underbrace{v_{0,1}\hat{I}_{1z}}_{\text{spin 1}} + \underbrace{v_{0,2}\hat{I}_{2z}}_{\text{spin 2}}.$$

This needs some explanation. $\hat{I}_{1z}$ is the operator for the z-component of angular momentum of the *first* spin, and $\hat{I}_{2z}$ is a similar operator referring to the *second* spin. Similarly, $v_{0,1}$ is the Larmor frequency of the first spin, and $v_{0,2}$ that of the second spin (they need not be the same). It is important to realize that separate operators are needed for each spin. For the moment we will leave out the coupling between the two spins, but this will be remedied shortly.

Having found the Hamiltonian, we now need to find the eigenfunctions, and this turns out to be rather easy as they are just *products* of the eigenfunctions of $\hat{I}_z$ for each spin. For the first spin, the two eigenfunctions of $\hat{I}_{1z}$ are given by Eq. 3.9 on the facing page

$$\hat{I}_{1z}\psi_{\alpha,1} = +\frac{1}{2}\psi_{\alpha,1} \qquad \hat{I}_{1z}\psi_{\beta,1} = -\frac{1}{2}\psi_{\beta,1}. \tag{3.11}$$

Note how the subscript 1 has been added to the eigenfunction to indicate that it is an eigenfunction of the spin one operator. Two similar eigenfunctions exist for $\hat{I}_{2z}$:

$$\hat{I}_{2z}\psi_{\alpha,2} = +\frac{1}{2}\psi_{\alpha,2} \qquad \hat{I}_{2z}\psi_{\beta,2} = -\frac{1}{2}\psi_{\beta,2}. \tag{3.12}$$

We will now show that $\psi_{\alpha,1}\psi_{\beta,2}$ is an eigenfunction of $\hat{H}_{\text{two spins, no coupl.}}$ and, as before, we do this by acting on the function with the Hamiltonian operator. There are several steps in the calculation:

$$\begin{aligned}
\hat{H}_{\text{two spins, no coupl.}}\psi_{\alpha,1}\psi_{\beta,2} &= \left(v_{0,1}\hat{I}_{1z} + v_{0,2}\hat{I}_{2z}\right)\psi_{\alpha,1}\psi_{\beta,2} \\
&= v_{0,1}\hat{I}_{1z}\psi_{\alpha,1}\psi_{\beta,2} + v_{0,2}\underbrace{\hat{I}_{2z}\psi_{\alpha,1}}_{\text{swap order}}\psi_{\beta,2} \\
&= v_{0,1}\left[\hat{I}_{1z}\psi_{\alpha,1}\right]\psi_{\beta,2} + v_{0,2}\psi_{\alpha,1}\left[\hat{I}_{2z}\psi_{\beta,2}\right].
\end{aligned}$$

Table 3.1 Eigenfunctions and eigenvalues (in Hz) for two spins, without a coupling between them. Each eigenfunction is labelled with the m value for each spin; in addition the spin-state labels are also given.

m_1	m_2	spin states	eigenfunction	eigenvalue (energy)
$+\frac{1}{2}$	$+\frac{1}{2}$	$\alpha\alpha$	$\psi_{\alpha,1}\psi_{\alpha,2}$	$+\frac{1}{2}\upsilon_{0,1} + \frac{1}{2}\upsilon_{0,2}$
$+\frac{1}{2}$	$-\frac{1}{2}$	$\alpha\beta$	$\psi_{\alpha,1}\psi_{\beta,2}$	$+\frac{1}{2}\upsilon_{0,1} - \frac{1}{2}\upsilon_{0,2}$
$-\frac{1}{2}$	$+\frac{1}{2}$	$\beta\alpha$	$\psi_{\beta,1}\psi_{\alpha,2}$	$-\frac{1}{2}\upsilon_{0,1} + \frac{1}{2}\upsilon_{0,2}$
$-\frac{1}{2}$	$-\frac{1}{2}$	$\beta\beta$	$\psi_{\beta,1}\psi_{\beta,2}$	$-\frac{1}{2}\upsilon_{0,1} - \frac{1}{2}\upsilon_{0,2}$

To get to the second line we have just multiplied out the bracket. On the third line, the first term is the same but, compared with the previous line, in the second term the *order* of $\hat{I}_{2z}$ and $\psi_{\alpha,1}$ has been swapped. As we noted in section 3.2.2 on page 27, we are not normally allowed to reorder operators and functions, but in this case it is permissible as the operator refers to spin two whereas the function refers to spin one. The operator thus has no effect on the function, so reordering is permissible. It would not, however, be permissible to reorder $\hat{I}_{1z}$ and $\psi_{\alpha,1}$ as both refer to the same spin.

Now look at the last line. Using Eq. 3.11 on the preceding page, the first square bracket can be rewritten as $+\frac{1}{2}\psi_{\alpha,1}$, and using Eq. 3.12 on the previous page the second square bracket can be rewritten as $-\frac{1}{2}\psi_{\beta,2}$. Using these substitutions we find

$$
\begin{aligned}
\hat{H}_{\text{two spins, no coupl.}}\psi_{\alpha,1}\psi_{\beta,2} &= \upsilon_{0,1}\left[\hat{I}_{1z}\psi_{\alpha,1}\right]\psi_{\beta,2} + \upsilon_{0,2}\psi_{\alpha,1}\left[\hat{I}_{2z}\psi_{\beta,2}\right] \\
&= \tfrac{1}{2}\upsilon_{0,1}\psi_{\alpha,1}\psi_{\beta,2} - \tfrac{1}{2}\upsilon_{0,2}\psi_{\alpha,1}\psi_{\beta,2} \\
&= \underbrace{\left[\tfrac{1}{2}\upsilon_{0,1} - \tfrac{1}{2}\upsilon_{0,2}\right]}_{\text{eigenvalue}} \underbrace{\psi_{\alpha,1}\psi_{\beta,2}}_{\text{eigenfunction}}.
\end{aligned}
$$

To go to the final line we have simply factored out $\psi_{\alpha,1}\psi_{\beta,2}$.

What we have shown is that when $\hat{H}_{\text{two spins, no coupl.}}$ acts on the function $\psi_{\alpha,1}\psi_{\beta,2}$ that function is regenerated multiplied by a constant $\frac{1}{2}\upsilon_{0,1} - \frac{1}{2}\upsilon_{0,2}$. In other words, $\psi_{\alpha,1}\psi_{\beta,2}$ is an eigenfunction of the Hamiltonian and $\frac{1}{2}\upsilon_{0,1} - \frac{1}{2}\upsilon_{0,2}$ is the eigenvalue i.e. the energy.

It is easy to go on to use the same method to show that there are three more possible eigenfunctions, each consisting of a product of an eigenfunction of $\hat{I}_{1z}$ with an eigenfunction of $\hat{I}_{2z}$. The results are summarized in Table 3.1.

As well as labelling the wavefunctions as α or β for each spin, we have also given the value of m for the first and second spins, m_1 and m_2. The table also lists the 'spin states' which is a shorthand way of describing the wavefunction: the first letter gives the spin state of the first spin and the second the spin state of the second spin.

If you look at this table it is easy to spot that the general expression for the eigenvalues (the energy levels) is

$$
E_{m_1,m_2} = m_1\upsilon_{0,1} + m_2\upsilon_{0,2} \qquad \text{in Hz.}
$$

3.5.1 Introducing scalar coupling

A scalar coupling between the two spins adds a third term to the Hamiltonian:

$$\hat{H}_{\text{two spins}} = \upsilon_{0,1}\hat{I}_{1z} + \upsilon_{0,2}\hat{I}_{2z} + \underbrace{J_{12}\hat{I}_{1z}\hat{I}_{2z}}_{\text{coupling term}}.$$

where J_{12} is the scalar coupling between spins one and two, in Hz (the Larmor frequencies are also in Hz).

It turns out that the four wavefunctions listed in the Table 3.1 on the preceding page are *still* eigenfunctions of this Hamiltonian, although with different eigenvalues. As an example we will take the function corresponding to the $\alpha\beta$ spin states and show that this is an eigenfunction of just the coupling term, $J_{12}\hat{I}_{1z}\hat{I}_{2z}$. The procedure is, as before, to apply the operators to the wavefunction:

$$J_{12}\hat{I}_{1z} \underbrace{\hat{I}_{2z}\,\psi_{\alpha,1}}_{\text{reorder}} \psi_{\beta,2} = J_{12}\left[\hat{I}_{1z}\,\psi_{\alpha,1}\right]\left[\hat{I}_{2z}\,\psi_{\beta,2}\right]$$

$$= J_{12}\left[\tfrac{1}{2}\psi_{\alpha,1}\right]\left[-\tfrac{1}{2}\psi_{\beta,2}\right]$$

$$= -\tfrac{1}{4}J_{12}\,\psi_{\alpha,1}\psi_{\beta,2}.$$

On the first line we have reordered the two terms $\hat{I}_{2z}\psi_{\alpha,1}$ indicated by the underbrace; as before, we can do this as the operator and function refer to separate spins. Having made this reordering, we recognize that the two expressions in square braces can be substituted using Eq. 3.11 and Eq. 3.12 on page 35. Making these substitutions gives us the second line. Finally, some simple tidying up gives us the third line.

We have shown that operating on $\psi_{\alpha,1}\psi_{\beta,2}$ with the coupling term of the Hamiltonian regenerates the original function times the constant $-\tfrac{1}{4}J_{12}$. So, $\psi_{\alpha,1}\psi_{\beta,2}$ is an eigenfunction with eigenvalue $-\tfrac{1}{4}J_{12}$. To complete the calculation we need to show that $\psi_{\alpha,1}\psi_{\beta,2}$ is an eigenfunction of $\hat{H}_{\text{two spins}}$, not just of the coupling term. However, as we have already shown that this product function is an eigenfunction of $J_{12}\hat{I}_{1z}\hat{I}_{2z}$ and also of $(\upsilon_{0,1}\hat{I}_{1z}+\upsilon_{0,2}\hat{I}_{2z})$, it follows that it is also an eigenfunction of the sum of these terms. The eigenvalue is, not surprisingly, the sum of the eigenvalues of the separate terms: $+\tfrac{1}{2}\upsilon_{0,1} - \tfrac{1}{2}\upsilon_{0,2} - \tfrac{1}{4}J_{12}$.

In fact, it is no coincidence that these 'product' functions, such as $\psi_{\alpha,1}\psi_{\beta,2}$ are eigenfunctions of the coupling term in the Hamiltonian. The underlying reason is that the coupling term commutes with the other terms (see section 3.2.2 on page 27), and there is a theorem in quantum mechanics which states that two commuting operators will have common eigenfunctions.

Applying the same procedure to the other three product functions gives us the complete set of energies shown in Table 3.2 on the following page; for future convenience the levels have been numbered.

Once more it is easy to see that in general the eigenvalues (energies) obey:

$$E_{m_1,m_2} = m_1\upsilon_{0,1} + m_2\upsilon_{0,2} + m_1 m_2 J_{12} \qquad \text{in Hz.}$$

If the two spins are of the same type (e.g. both protons) then we have a *homonuclear* spin system and $\upsilon_{0,1} \approx \upsilon_{0,2}$. Under these circumstances the

Table 3.2 Eigenfunctions and eigenvalues (energies, in Hz) for two coupled spins.

number	m_1	m_2	spin states	eigenfunction	eigenvalue (energy)
1	$+\frac{1}{2}$	$+\frac{1}{2}$	$\alpha\alpha$	$\psi_{\alpha,1}\psi_{\alpha,2}$	$+\frac{1}{2}\upsilon_{0,1} + \frac{1}{2}\upsilon_{0,2} + \frac{1}{4}J_{12}$
2	$+\frac{1}{2}$	$-\frac{1}{2}$	$\alpha\beta$	$\psi_{\alpha,1}\psi_{\beta,2}$	$+\frac{1}{2}\upsilon_{0,1} - \frac{1}{2}\upsilon_{0,2} - \frac{1}{4}J_{12}$
3	$-\frac{1}{2}$	$+\frac{1}{2}$	$\beta\alpha$	$\psi_{\beta,1}\psi_{\alpha,2}$	$-\frac{1}{2}\upsilon_{0,1} + \frac{1}{2}\upsilon_{0,2} - \frac{1}{4}J_{12}$
4	$-\frac{1}{2}$	$-\frac{1}{2}$	$\beta\beta$	$\psi_{\beta,1}\psi_{\beta,2}$	$-\frac{1}{2}\upsilon_{0,1} - \frac{1}{2}\upsilon_{0,2} + \frac{1}{4}J_{12}$

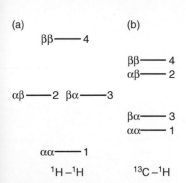

Fig. 3.7 Energy levels, drawn approximately to scale, for two different two-spin systems. (a) The energy levels of a homonuclear system (two protons); on this scale the $\alpha\beta$ and $\beta\alpha$ states have the same energy. (b) The energy levels of a ^{13}C–^1H pair. The Larmor frequency of a proton is about four times that of ^{13}C, and this leads to the $\alpha\beta$ and $\beta\alpha$ states having substantially different energies.

$\alpha\beta$ and $\beta\alpha$ levels (levels 2 and 3) are very similar in energy, and lie more or less mid way in energy between the other two levels, as is depicted in Fig. 3.7 (a). In contrast, if the two spins are of different types (e.g. one proton and one ^{13}C), all four levels have markedly different energies, as shown in Fig. 3.7 (b).

3.6   The spectrum from two coupled spins

The selection rule for allowed transitions in the case of two spins is just the same as it was for one spin i.e. m can only change by ±1. However, when applied to two spins the rule has to be supplemented somewhat to say that m of only *one* of the spins can change by ±1. So, the quantum number of the first spin, m_1, may change by ±1, or that of the second spin, m_2, may change by the same amount.

Applying these rules means that the allowed transitions are between levels 1 and 2, 3 and 4, 1 and 3, and 2 and 4. The resulting frequencies are easily worked out, as is shown here for that of the 1–2 transition:

$$
\begin{aligned}
\upsilon_{12} &= E_2 - E_1 \\
&= +\tfrac{1}{2}\upsilon_{0,1} - \tfrac{1}{2}\upsilon_{0,2} - \tfrac{1}{4}J_{12} - (\tfrac{1}{2}\upsilon_{0,1} + \tfrac{1}{2}\upsilon_{0,2} + \tfrac{1}{4}J_{12}) \\
&= -\upsilon_{0,2} - \tfrac{1}{2}J_{12}.
\end{aligned}
$$

The complete set of transitions are:

transition	spin states	frequency
$1 \to 2$	$\alpha\alpha \to \alpha\beta$	$-\upsilon_{0,2} - \frac{1}{2}J_{12}$
$3 \to 4$	$\beta\alpha \to \beta\beta$	$-\upsilon_{0,2} + \frac{1}{2}J_{12}$
$1 \to 3$	$\alpha\alpha \to \beta\alpha$	$-\upsilon_{0,1} - \frac{1}{2}J_{12}$
$2 \to 4$	$\alpha\beta \to \beta\beta$	$-\upsilon_{0,1} + \frac{1}{2}J_{12}$

The energy levels and corresponding schematic spectrum are shown in Fig. 3.8 on the next page. As expected, the spectrum consists of two doublets, each split by J_{12} and centred at the Larmor frequencies of spins one and two.

As a result of the selection rule, each allowed transition corresponds to one of the spins 'flipping' from one spin state to the other, while the state of the other spin remains fixed. For example, transition 1–2 involves a spin two going from α to β whilst spin one remains in the α state. In this

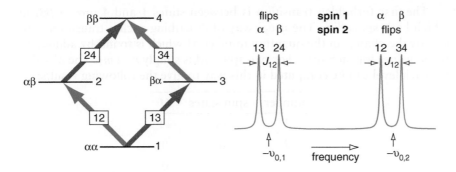

Fig. 3.8 On the left, the energy levels of a two-spin system; the arrows show the allowed transitions: grey arrows indicate transitions in which spin one flips and blue arrows indicate those in which spin two flips. On the right, the corresponding spectrum; each line is marked according to the two energy levels involved, which spin flips (the active spin) and the spin state of the passive spin. It is assumed that the Larmor frequency of spin two is greater in magnitude than that of spin one, and that the coupling J_{12} is positive.

transition we say that spin two is *active* and spin one is *passive* and in the α spin state. As spin two flips in this transition, it is not surprising that the transition forms one part of the doublet for spin two.

Transition 3–4 is similar to 1–2 except that the passive spin (spin one) is in the β state; this transition forms the second line of the doublet for spin two. This discussion illustrates a very important point, which is that the lines of a multiplet can be associated with different spin states of the coupled (passive) spins. We will use this kind of interpretation very often, especially when considering two-dimensional spectra.

The two transitions in which spin one flips are 1–3 and 2–4, and these are associated with spin two being in the α and β spin states, respectively. Which spin flips and the spin states of the passive spins are shown in Fig. 3.8 for each transition.

If the coupling J_{12} is negative, working through the calculation gives the same four lines at identical positions as for a positive coupling of the same magnitude. This result is in accord with the observation we made earlier in section 2.3 on page 10 concerning the effect (or lack of it) of changing the sign of the coupling.

However, what does change when the coupling becomes negative are the *labels* of the lines, as is illustrated in Fig. 3.9. For example, transition 1–2 is now the right-hand line of the doublet, rather than the left line. From the point of view of the spectrum, what swaps over is the spin state of the passive spin associated with each line of the multiplet.

3.6.1   Multiple-quantum transitions

There are two more transitions in our two-spin system which are not allowed by the usual selection rule and so do not appear in the spectrum; the transitions are illustrated in Fig. 3.10 on the next page. We will discover later on that, using two-dimensional NMR, we can detect these transitions indirectly.

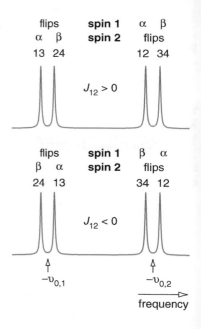

Fig. 3.9 Illustration of the effect of changing the sign of the coupling on the spectrum of two coupled spins. The top spectrum is for a positive coupling, whereas the lower is for a negative coupling of the same magnitude. Note that the appearance of the spectrum is identical, but that the labels on the transitions change.

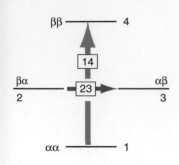

Fig. 3.10 In a two-spin system there is one double quantum transition (1–4) and one zero-quantum transition (2–3). The frequency of neither of these transitions are affected by the size of the coupling between the two spins.

The first forbidden transition is between states 1 and 4 ($\alpha\alpha \rightarrow \beta\beta$) in which both spins flip. The usual way of describing such a transition is to specify the change in the quantum number M which is found by adding up the m values for each spin. For two spins, M is simply $m_1 + m_2$. The M value for each level can be computed in this way to give the following results:

number	spin states	M
1	$\alpha\alpha$	+1
2	$\alpha\beta$	0
3	$\beta\alpha$	0
4	$\beta\beta$	−1

The change in M, ΔM, for the 1–4 transition is thus −2 and so this transition is called a *double-quantum* transition. Using the same approach, all of the allowed transitions described above have $\Delta M = \pm 1$ and so are called *single-quantum* transitions.

From the table of energy levels (Table 3.2 on page 38) it is easy to work out that the frequency of the 1–4 transition is $(-\upsilon_{0,1} - \upsilon_{0,2})$ i.e. the sum of the Larmor frequencies. Note that the coupling has no effect on the frequency of this line.

The second forbidden transition is between states 2 and 3 ($\beta\alpha \rightarrow \alpha\beta$); again, both spins flip. The ΔM value is 0, so this is called a *zero-quantum* transition, and its frequency is $(-\upsilon_{0,1} + \upsilon_{0,2})$ i.e. the difference of the Larmor frequencies. As with the double-quantum transition, the coupling has no effect on the frequency of this line.

Optional section ⇒

3.7 Three spins

3.7.1 The Hamiltonian and energy levels

Now we are getting used to writing Hamiltonians, the following should come as no surprise:

$$
\begin{aligned}
\hat{H}_{\text{three spins}} \;=\; & \upsilon_{0,1}\hat{I}_{1z} + \upsilon_{0,2}\hat{I}_{2z} + \upsilon_{0,3}\hat{I}_{3z} \\
& + \underbrace{J_{12}\hat{I}_{1z}\hat{I}_{2z}}_{\text{1–2 coupling}} + \underbrace{J_{13}\hat{I}_{1z}\hat{I}_{3z}}_{\text{1–3 coupling}} + \underbrace{J_{23}\hat{I}_{2z}\hat{I}_{3z}}_{\text{2–3 coupling}}.
\end{aligned}
$$

The first three terms represent the interactions of spins 1, 2 and 3 with the magnetic field; note that there are three different Larmor frequencies. The next three terms represent all of the possible couplings, with each term being the product of the $\hat{I}_z$ operator for the relevant two spins.

The eigenfunctions of this Hamiltonian turn out to be products of the eigenfunctions of $\hat{I}_z$, ψ_α and ψ_β, for each spin. For example, one such product function is $\psi_{\alpha,1}\psi_{\alpha,2}\psi_{\beta,3}$. Since each spin can be α or β, there are a total of *eight* separate product functions.

We will not go through the process of showing that these are indeed the eigenfunctions and finding the corresponding eigenvalues, but simply state the general form of the eigenvalues, that is the energies:

$$
E_{m_1 m_2 m_3} = m_1 \upsilon_{0,1} + m_2 \upsilon_{0,2} + m_3 \upsilon_{0,3} + m_1 m_2 J_{12} + m_1 m_3 J_{13} + m_2 m_3 J_{23}.
$$

Table 3.3 Eigenfunctions and corresponding eigenvalues (energies, in Hz) for three coupled spins. The first four levels all have the third spin in the α state, whereas for the second four, the third spin is in the β state.

number	spin states	M	energy
1	$\alpha\alpha\alpha$	$\frac{3}{2}$	$+\frac{1}{2}\nu_{0,1} + \frac{1}{2}\nu_{0,2} + \frac{1}{2}\nu_{0,3} + \frac{1}{4}J_{12} + \frac{1}{4}J_{13} + \frac{1}{4}J_{23}$
2	$\alpha\beta\alpha$	$\frac{1}{2}$	$+\frac{1}{2}\nu_{0,1} - \frac{1}{2}\nu_{0,2} + \frac{1}{2}\nu_{0,3} - \frac{1}{4}J_{12} + \frac{1}{4}J_{13} - \frac{1}{4}J_{23}$
3	$\beta\alpha\alpha$	$\frac{1}{2}$	$-\frac{1}{2}\nu_{0,1} + \frac{1}{2}\nu_{0,2} + \frac{1}{2}\nu_{0,3} - \frac{1}{4}J_{12} - \frac{1}{4}J_{13} + \frac{1}{4}J_{23}$
4	$\beta\beta\alpha$	$-\frac{1}{2}$	$-\frac{1}{2}\nu_{0,1} - \frac{1}{2}\nu_{0,2} + \frac{1}{2}\nu_{0,3} + \frac{1}{4}J_{12} - \frac{1}{4}J_{13} - \frac{1}{4}J_{23}$
5	$\alpha\alpha\beta$	$\frac{1}{2}$	$+\frac{1}{2}\nu_{0,1} + \frac{1}{2}\nu_{0,2} - \frac{1}{2}\nu_{0,3} + \frac{1}{4}J_{12} - \frac{1}{4}J_{13} - \frac{1}{4}J_{23}$
6	$\alpha\beta\beta$	$-\frac{1}{2}$	$+\frac{1}{2}\nu_{0,1} - \frac{1}{2}\nu_{0,2} - \frac{1}{2}\nu_{0,3} - \frac{1}{4}J_{12} - \frac{1}{4}J_{13} + \frac{1}{4}J_{23}$
7	$\beta\alpha\beta$	$-\frac{1}{2}$	$-\frac{1}{2}\nu_{0,1} + \frac{1}{2}\nu_{0,2} - \frac{1}{2}\nu_{0,3} - \frac{1}{4}J_{12} + \frac{1}{4}J_{13} - \frac{1}{4}J_{23}$
8	$\beta\beta\beta$	$-\frac{3}{2}$	$-\frac{1}{2}\nu_{0,1} - \frac{1}{2}\nu_{0,2} - \frac{1}{2}\nu_{0,3} + \frac{1}{4}J_{12} + \frac{1}{4}J_{13} + \frac{1}{4}J_{23}$

In this expression, m_1, m_2 and m_3 can each be $+\frac{1}{2}$ (the α state) or $-\frac{1}{2}$ (the β state). The energies and corresponding M values ($= m_1 + m_2 + m_3$) are shown in Table 3.3.

In the table, we have grouped the energy levels into two groups of four: the first group all have spin three in the α state and the second have spin three in the β state. The energy levels (for a homonuclear system) are shown schematically in Fig. 3.11.

3.7.2 Single-quantum spectrum

As before, the selection rule is that m of just one spin can change by $+1$, which means that $\Delta M = \pm 1$. Applying this rule we see that there are four allowed transitions in which spin one flips: 1–3, 2–4, 5–7 and 6–8. The frequencies of these lines can easily be worked out from the energy levels given in Table 3.3, and are shown on the following page along with the spin states of the passive spins (two and three in this case).

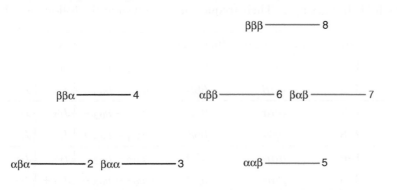

Fig. 3.11 Energy levels for a homonuclear three-spin system. The levels can be grouped into two sets of four: those with spin three in the α state (shown on the left with black lines) and those with spin three in the β state (shown on the right with blue lines).

Fig. 3.12 Energy levels for a three-spin system showing by the arrows the four allowed transitions which result in the doublet of doublets at the shift of spin one. The schematic multiplet is shown on the right, where it has been assumed that $\nu_{0,1} = -100$ Hz, $J_{12} = 10$ Hz and $J_{13} = 2$ Hz. The multiplet is labelled with the spin states of the passive spins.

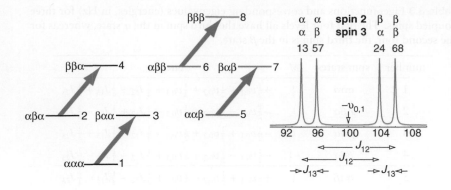

transition	state of spin two	state of spin three	frequency
1–3	α	α	$-\nu_{0,1} - \frac{1}{2}J_{12} - \frac{1}{2}J_{13}$
2–4	β	α	$-\nu_{0,1} + \frac{1}{2}J_{12} - \frac{1}{2}J_{13}$
5–7	α	β	$-\nu_{0,1} - \frac{1}{2}J_{12} + \frac{1}{2}J_{13}$
6–8	β	β	$-\nu_{0,1} + \frac{1}{2}J_{12} + \frac{1}{2}J_{13}$

These four transitions form the four lines of the multiplet (a doublet of doublets) centred at the Larmor frequency of spin one. The schematic spectrum is illustrated in Fig. 3.12. As in the case of a two-spin system, we can label each line of the multiplet with the spin states of the passive spins – in the case of the multiplet from spin one, this means the spin states of spins two and three. In the same way, we can identify the four transitions which contribute to the multiplet from spin two (1–2, 3–4, 5–6 and 7–8) and the four which contribute to that from spin three (1–5, 3–7, 2–6 and 4–8).

3.7.3 Multiple-quantum transitions

There are six double-quantum transitions in which two spins flip and in which M changes by 2. Their frequencies are given in the following table.

transition	initial state	final state	frequency
1–4	$\alpha\alpha\alpha$	$\beta\beta\alpha$	$-\nu_{0,1} - \nu_{0,2} - \frac{1}{2}J_{13} - \frac{1}{2}J_{23}$
5–8	$\alpha\alpha\beta$	$\beta\beta\beta$	$-\nu_{0,1} - \nu_{0,2} + \frac{1}{2}J_{13} + \frac{1}{2}J_{23}$
1–7	$\alpha\alpha\alpha$	$\beta\alpha\beta$	$-\nu_{0,1} - \nu_{0,3} - \frac{1}{2}J_{12} - \frac{1}{2}J_{23}$
2–8	$\alpha\beta\alpha$	$\beta\beta\beta$	$-\nu_{0,1} - \nu_{0,3} + \frac{1}{2}J_{12} + \frac{1}{2}J_{23}$
1–6	$\alpha\alpha\alpha$	$\alpha\beta\beta$	$-\nu_{0,2} - \nu_{0,3} - \frac{1}{2}J_{12} - \frac{1}{2}J_{13}$
3–8	$\beta\alpha\alpha$	$\beta\beta\beta$	$-\nu_{0,2} - \nu_{0,3} + \frac{1}{2}J_{12} + \frac{1}{2}J_{13}$

These transitions come in three pairs. Transitions 1–4 and 5–8 are centred at the sum of the Larmor frequencies of spins one and two; this is not surprising as in these transitions it is the spin states of both spins one and two which flip. The two transitions are separated by the *sum* of the

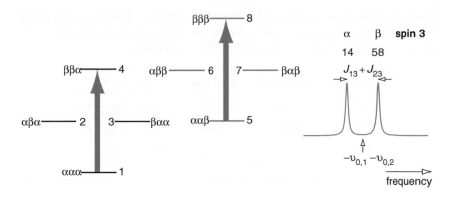

Fig. 3.13 There are two double-quantum transitions in which spins one and two both flip (transitions 1–4 and 5–8). The two resulting lines form a doublet which is centred at the sum of the Larmor frequencies of spins one and two and which is split by the sum of the couplings to spin three. As with the single-quantum spectra, we can associate the two lines of the doublet with different spin states of the third spin.

couplings to spin three ($J_{13} + J_{23}$), but they are unaffected by the coupling J_{12} which is between the two active spins.

We can describe these transitions as a kind of double quantum doublet. spins one and two are both active in these transitions, and spin three is passive. Just as we did before, we can associate one line with spin three being in the α state (transition 1–4) and one with it being in the β state (transition 5–8). A schematic representation of the spectrum is shown in Fig. 3.13.

There are also six zero-quantum transitions in which M does not change. Like the double quantum transitions these group in three pairs, but this time centred around the *difference* in the Larmor frequencies of two of the spins. These zero-quantum doublets are split by the *difference* of the couplings to the spin which does not flip in the transitions. There are thus many similarities between the double- and zero-quantum spectra.

In a three-spin system there is one triple-quantum transition, in which M changes by 3, between levels 1 ($\alpha\alpha\alpha$) and 8 ($\beta\beta\beta$). In this transition all of the spins flip, and from the table of energies we can easily work out that its frequency is $-\upsilon_{0,1} - \upsilon_{0,2} - \upsilon_{0,3}$, i.e. the sum of the Larmor frequencies.

We see that the single-quantum spectrum consists of three doublets of doublets, the double-quantum spectrum of three doublets and the triple-quantum spectrum of a single line. This illustrates the idea that as we move to higher orders of multiple quantum, the corresponding spectra become simpler. This feature has been used in the analysis of some complex spin systems.

3.7.4   Combination lines

There are three more transitions which we have not yet described. For these, M changes by 1 but all three spins flip; they are called *combination lines*. Such lines are not seen in normal spectra but, like multiple quantum transitions, they can be detected indirectly using two-dimensional spectra. These lines may become observable in strongly coupled spectra. The following table gives the frequencies of these three lines:

transition	initial state	final state	frequency
2–7	$\alpha\beta\alpha$	$\beta\alpha\beta$	$-v_{0,1} + v_{0,2} - v_{0,3}$
3–6	$\beta\alpha\alpha$	$\alpha\beta\beta$	$+v_{0,1} - v_{0,2} - v_{0,3}$
4–5	$\beta\beta\alpha$	$\alpha\alpha\beta$	$+v_{0,1} + v_{0,2} - v_{0,3}$

Notice that the frequencies of these lines are not affected by any of the couplings.

3.8   Summary

In this chapter we have seen precisely what we mean by 'energy levels' and why they are important in NMR. We have seen that these energy levels are the eigenvalues of the relevant Hamiltonian, and we have looked at how Hamiltonians are written and how the eigenvalues (and eigenfunctions) can be found.

Armed with these energy levels and the necessary selection rules, we can predict the spectra we expect from two and three coupled spins. By working from the energy levels, we see where the idea that lines can be associated with particular spin states of coupled spins come from.

Finally, we saw that for coupled spins we have the possibility of multiple quantum transitions which are not allowed in simple experiments but which can be detected indirectly. Like normal spectra, such multiple quantum spectra can contain multiplets, which can be interpreted in a similar way to those in conventional spectra.

3.9   Further reading

The origins of nuclear spin:

Chapter 1 from M. H. Levitt, *Spin Dynamics* (2nd edition, John Wiley & Sons, Ltd, 2008).

Operators and wavefunctions:

Chapter 1 from N. J. B. Green, *Quantum Mechanics 1: Foundations* (Oxford University Press, 1997).

3.10 Exercises

3.1 Following the same approach used in section 3.2.7 on page 30, show that $\psi_{-\frac{1}{2}}$ is an eigenfunction of $\hat{H}_{\text{one spin}}$ with eigenvalue $+\frac{1}{2}\hbar\gamma B_0$.

3.2 Calculate the Larmor frequency (in Hz, MHz and rad s^{-1}) of ^{13}C at a magnetic field strength of 9.4 T; the gyromagnetic ratio of ^{13}C is $+6.7283 \times 10^7$ rad s^{-1} T^{-1} and you assume that the chemical shift is zero.

3.3 Show that $\psi_{\pm\frac{1}{2}}$ are eigenfunctions of the one-spin Hamiltonian when it is written in angular frequency units

$$\hat{H}_{\text{one spin}} = \omega_0 \hat{I}_z;$$

hence find the corresponding eigenvalues. You should assume that

$$\hat{I}_z \psi_{+\frac{1}{2}} = +\tfrac{1}{2}\psi_{+\frac{1}{2}} \qquad \hat{I}_z \psi_{-\frac{1}{2}} = -\tfrac{1}{2}\psi_{-\frac{1}{2}}$$

i.e. that $\psi_{\pm\frac{1}{2}}$ are eigenfunctions of $\hat{I}_z$ with eigenvalues $\pm\frac{1}{2}$.

3.4 Following the approach of section 3.5 on page 35, show that $\psi_{\alpha,1}\psi_{\alpha,2}$ is an eigenfunction of the Hamiltonian for two spins with no coupling between them:

$$\hat{H}_{\text{two spins, no coupl.}} = \upsilon_{0,1} \hat{I}_{1z} + \upsilon_{0,2} \hat{I}_{2z}.$$

Hence find the corresponding eigenvalue (the energy); make sure that each step in your argument is clear and justified.

Show that $\psi_{\alpha,1}\psi_{\alpha,2}$ is also an eigenfunction of the coupling term $J_{12}\hat{I}_{1z}\hat{I}_{2z}$; find the corresponding energy.

Without further detailed calculations explain why $\psi_{\alpha,1}\psi_{\alpha,2}$ is an eigenfunction of the Hamiltonian for two spins with coupling

$$\hat{H}_{\text{two spins}} = \upsilon_{0,1} \hat{I}_{1z} + \upsilon_{0,2} \hat{I}_{2z} + J_{12}\hat{I}_{1z}\hat{I}_{2z};$$

state the corresponding eigenvalue (energy).

3.5 Larmor frequencies are usually tens or hundreds of MHz, but to make the numbers easier to handle in this problem we will assume that the Larmor frequencies are very much smaller.

Consider a system of two coupled spins. Let the Larmor frequency of the first spin be -100 Hz and that of the second spin be -200 Hz, and let the coupling between the two spins be 5 Hz. Compute the energies (in Hz) of the four energy levels, according to Table 3.2 on page 38.

Using these energies, compute the frequencies of the four allowed transitions; make a sketch of the spectrum, roughly to scale, and label each line with the energy levels involved (i.e. 1–2 etc.). Also, indicate for each line which spin flips and the spin state of the passive spin.

Repeat your calculation, and redraw the sketch, for the case where the coupling is −5 Hz. Comment on the effect of changing the sign of the coupling.

3.6 For a three-spin system, use Table 3.3 on page 41 to work out the frequencies of the four allowed transitions in which spin two flips.

Then, taking $v_{0,2} = -200$ Hz, $J_{23} = 4$ Hz and the rest of the parameters as in Fig. 3.12 on page 42, compute the frequencies of the lines which comprise the spin two multiplet. Make a sketch of the multiplet (roughly to scale) and label the lines in the same way as is done in Fig. 3.12 on page 42.

How would these labels change if $J_{23} = -4$ Hz?

3.7 For a three-spin system, use Table 3.3 on page 41 to compute the frequencies of the six zero-quantum transitions; mark these transitions on an energy level diagram.

Explain why these six transitions fall into three groups of two.

How would you describe the zero-quantum spectrum?

4

The vector model

In the last chapter we made quite a lot of progress towards understanding the form of NMR spectra by working from the energy levels and selection rules. However, this has brought us no closer to understanding how even the simplest pulse–acquire NMR experiment *actually* works. Ultimately it is only quantum mechanics which will give us the complete understanding we are looking for. However, before we embark on the full rigours of that approach we will spend some time exploring the much simpler *vector model*.

Strictly speaking, the vector model only applies to uncoupled spins, so you might think that it is of little use. However, the model gives us an excellent start towards understanding RF pulses, and also is a convenient way of thinking about some key experiments, such as the spin echo. In due course, we will discover that many features of the vector model have direct counterparts in the full quantum mechanical treatment, and this common ground between the two approaches will help us to come to grips with the more complex quantum mechanical approach. The final reason for spending some time with the vector model is that much of the language used to talk about pulsed NMR is derived from this model.

So, although the vector model has its limitations, it is very worthwhile to know where the model comes from and how to use it.

4.1 The bulk magnetization

In section 3.2.5 on page 29 we described how some nuclei appear to contain a source of spin angular momentum. It turns out that associated with this angular momentum there is always a *nuclear spin magnetic moment*; what this means is that the nucleus generates a small magnetic field, just as if it were a tiny bar magnet.

When the nucleus is placed in a magnetic field (as we always do to record an NMR spectrum), there is an interaction between the nuclear magnetic moment and the applied field. The energy of the interaction depends on the angle between the magnetic moment and the applied field.

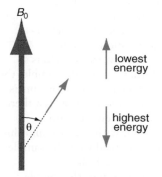

Fig. 4.1 The energy of interaction between a magnetic moment, represented by the small arrow, and an applied magnetic field, B_0, depends on the angle θ between the magnetic moment and the field direction. The lowest energy arrangement is when the magnetic moment is parallel to the field ($\theta = 0$), and the highest energy arrangement is when the moment is opposed to the field ($\theta = \pi$ radians).

Understanding NMR Spectroscopy, Second Edition James Keeler
© 2010 John Wiley & Sons, Ltd

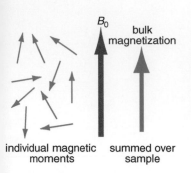

Fig. 4.2 As a result of the disruption due to thermal motion, the individual magnetic moments are not all able to adopt the lowest energy arrangement in which they align with the field. For nuclear magnetic moments the interaction with the field is so small that, across the sample, the arrangement of the moments is almost random. However, there is a small preference for alignment with the field and this, when averaged over the sample, gives rise to a bulk magnetization of the sample, parallel to the field direction. This magnetization can be represented by a vector, called the bulk magnetization vector.

The lowest energy arrangement is when this angle is zero i.e. the magnetic moment is parallel to the field, and the highest energy is when the magnetic moment is opposed to the field (see Fig. 4.1 on the preceding page).

The energy of the spins in our sample is thus minimized if all of the individual magnetic moments align with the field. However, this alignment is opposed by the random thermal motion of the molecules which is trying to drive the system to state where the magnetic moments have random orientations. The energy of this thermal motion is very much greater than the energy of interaction between a nuclear magnetic moment and the applied field, and so the thermal motion easily disrupts the alignment of the magnetic moments.

However, the randomizing effect of the thermal motion is not complete since, as we have described, there is a very small energetic advantage for the magnetic moment to be aligned with the field. As a consequence, the magnetic moments are aligned in such a way that, averaged over our sample, there is a slight net alignment of the moments parallel to the magnetic field. One way of describing this alignment is to say that out of 10^5 spins it is as if just *one* magnetic moment is aligned with the field and the rest are aligned randomly – you can see why the alignment is described as 'slight'.

As a result of this net alignment, the NMR sample becomes *magnetized*, which means that the sample *as a whole* acquires a magnetic moment, just as each spin has a magnetic moment; this is illustrated in Fig. 4.2. This magnetization of the sample is along the direction of the applied magnetic field and is represented by a *bulk magnetization vector*. The description 'bulk' is there to remind us that the magnetization is a property of the *whole* sample. Note that the magnetization is a vector quantity, having a magnitude and a direction.

The vector model is only concerned with what happens to this magnetization vector. The nice thing about the model is that the behaviour of the vector is completely classical – we do not need any quantum mechanics to work out what will happen. The model just involves rotations of the magnetization vector in normal space, and so is intuitive and quite easy to understand. Later on we will discover that for uncoupled spins the predictions of the model are identical to those from a quantum mechanical treatment. For such systems the vector model is exact.

4.1.1   Surely the spins can only be 'up' or 'down'?

Figure 4.2 might cause you some difficulties as it shows the magnetic moments from individual spins pointing in all directions, whereas you will read in many elementary accounts of NMR that 'the spins are either up (state α) or down (state β)'. The problem is that this statement, although oft repeated, is simply not true, for the reasons set out in section 3.1.1 on page 24.

In that section we described how spins are generally in what are called mixed or superposition states which are combinations of the wavefunctions for the α and β states. A consequence of this is that the magnetic moment can point anywhere between up, where it would point for a pure α state, and down, where it would point for a pure β state. We will return to this point when we examine the quantum mechanics of a single spin in more detail in Chapter 6.

4.1.2 Axis systems

The rest of this chapter is going to be concerned with the motion of the magnetization vector in three-dimensional space. This is a convenient moment, therefore, to describe the axis system we are going to use, which is the *right-handed* axis set illustrated in Fig. 4.3.

The axes are described as right-handed as if you grasp the z-axis with your right hand, with the thumb pointing along the $+z$ direction, then your fingers curl from the x- to the y-axis. The curl of your fingers also gives the sense of a positive rotation in such an axis system. So, looking down the z-axis from $+z$ towards the origin, a positive rotation is anti-clockwise and takes us from x to y. Similarly, a positive rotation about the x-axis takes us from z to $-y$ since if you imagine grasping the x-axis with your right hand you will find that the curl of your fingers takes you from z to $-y$.

4.1.3 The equilibrium magnetization

When a sample is first placed in a magnetic field there is no bulk magnetization along the z-axis; rather, it takes a finite time for this magnetization to build up. If we wait long enough, the magnetization reaches a steady value, and at this point we say that the *equilibrium magnetization* has been established.

The process by which the sample comes to equilibrium is illustrated in Fig. 4.4 on the next page. In the absence of a magnetic field the moments are oriented randomly because all orientations have the same energy. Consequently, there is no net magnetization of the sample. When a magnetic field is applied there is an energetic preference for the magnetic moments to be oriented parallel to the field, but to start with the magnetic moments are still oriented randomly so there is no net magnetization.

Over time, the random molecular motion ensures that the lower-energy orientations are preferentially populated and, as was described above, this leads to the growth of the net magnetization vector along the z-axis. As more moments adopt lower energy orientations the magnetization grows until it reaches a steady value. At this point there is no further change and the system is at equilibrium.

The process by which the spins come to equilibrium in a magnetic field is called *relaxation*. For nuclear spins it is a relatively slow process – it can easily take several seconds for the equilibrium net magnetization to build up. In Chapter 9 we will look into the details as to how random molecular motion leads to relaxation.

As we have seen (Fig. 4.1 on page 47), the energy of a particular magnetic moment depends on the angle between it and the applied field. However, it turns out that the energy is independent of the orientation of the moment in the xy-plane. As a result there is no energetic preference for any particular orientation in the xy-plane, so at equilibrium we expect that the x- and y-components of the individual magnetic moments will be distributed randomly. Averaged over the sample these individual

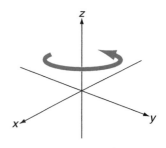

Fig. 4.3 The right-handed axis set which will be used throughout this book; the axes are right-handed in the sense that if the z-axis is grasped with the right hand and with the thumb pointing along $+z$, the fingers curl from x to y. In such an axis system, a positive rotation about a particular axis is defined by the curl of the fingers if that axis is grasped with the right hand and with the thumb pointing in the positive direction. A positive rotation about the $+z$-axis is shown.

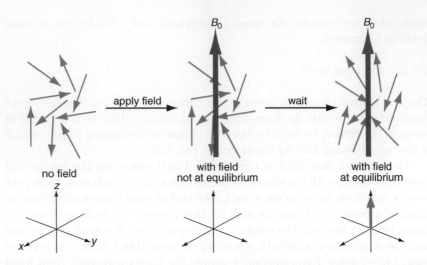

Fig. 4.4 Illustration of how the equilibrium magnetization builds up. On the left is shown the case where no magnetic field is applied: the individual magnetic moments are at random orientations so that, summed over the sample, there is no net magnetization. In the presence of a magnetic field, there is an energetic preference for moments to be aligned with the field, but it takes time for these orientations to be populated. So, when the magnetic field is first applied there is still no magnetization. However, after waiting sufficient time, the orientations with the moment parallel to the field become more populated as shown (greatly exaggerated) on the right. The result is that, summed over the sample, there is a net magnetization along the field direction.

transverse components will cancel one another out such that there is no bulk magnetization in the transverse plane. Therefore, at equilibrium, the bulk magnetization vector is aligned along the z-axis. We will see in the following sections how bulk transverse magnetization can be created by the application of RF pulses.

4.2   Larmor precession

Once it has formed, the equilibrium magnetization vector is fixed in size and direction – it does not vary over time, which is a particular property of the equilibrium state. However, suppose that by some means (which we will go into later) the magnetization vector has been tipped away from the z-axis, such that the vector makes an angle β to that axis. It turns out that what then happens is that the magnetization vector rotates about the direction of the magnetic field sweeping out a cone with a constant angle β, as shown in Fig. 4.5. This kind of motion is called *precession* – the vector is said to *precess* about the field.

If the magnetic field strength is B_0, then it turns out that the frequency of the precession is

$$\omega_0 = -\gamma B_0 \text{ in rad s}^{-1} \qquad \text{or} \qquad \upsilon_0 = -\gamma B_0/2\pi \text{ in Hz}$$

where γ is the gyromagnetic ratio. The precession frequency *is* the Larmor

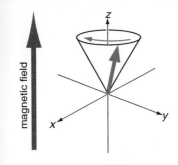

Fig. 4.5 Once tilted away from the z-axis, the magnetization vector rotates about the field direction, sweeping out a cone of constant angle to the z-axis; this motion is called precession. The direction of precession shown is for a nucleus with a positive gyromagnetic ratio and hence a negative Larmor frequency.

frequency which we encountered in section 3.3.2 on page 32. For a single spin, we saw that the allowed transition occurs at the Larmor frequency which is exactly the same as the frequency at which the magnetization vector precesses about the applied field; this is not a coincidence. The precession of the magnetization about the field is sometimes called Larmor precession.

For nuclei which have a positive gyromagnetic ratio the Larmor frequency is negative (see section 3.3.2 on page 32) which means that the precession of the magnetization vector about the field corresponds to a negative rotation. Recalling section 4.1.2 on page 49, a negative rotation about z is clockwise when viewed looking down the z-axis from $+z$ to the origin; this is the sense of rotation shown in Fig. 4.5 on the preceding page.

The magnetization will only execute this precessional motion when it is at an angle to the magnetic field direction. So, at equilibrium, where the magnetization is along the z-axis and hence parallel to the field B_0, the magnetization remains stationary.

4.3 Detection

The precession of the magnetization vector is what we actually detect in a pulsed NMR experiment. All we have to do is to mount a small coil of wire round the sample, with the axis of the coil aligned in the xy-plane, as is illustrated in Fig. 4.6. As the precessing magnetization vector 'cuts' the coil a current is induced which we can amplify and then record – this is the *free induction* signal or, more usually, free induction decay (FID).

The detection process is analogous to the way in which an electric current can be produced by induction. You may recall a demonstration in which a bar magnet thrust into a coil of wire leads to the generation of a current in the wire. If the magnet is moved rhythmically in and out of the coil, or rotated inside the coil, the result is an oscillating current. The magnetization of the sample behaves just like a magnet, so as the vector precesses it leads to the generation of an oscillating current in the coil.

A coil wound round the x-axis detects the x-component of the magnetization, and we can work out what this will be using some simple geometry. Suppose that the equilibrium magnetization vector is of size M_0. If this has been tilted through an angle β towards the x-axis, the x-component is $M_0 \sin\beta$, as shown in Fig. 4.7.

Although the magnetization vector precesses on a cone, we can visualize what happens to the x- and y-components much more simply by just thinking about the projection of the vector onto the xy-plane. At time zero the x-component of this vector is $M_0 \sin\beta$. Then, as the vector precesses at ω_0 about the z-axis, this initial x-component rotates at the same frequency in the xy-plane.

This rotation of a vector is exactly of the type we discussed in section 2.5 on page 15. Here the vector is of length $r = M_0 \sin\beta$ and starts out at time zero along the x-axis. The vector rotates at frequency ω_0 so that at time t the angle through which the vector has rotated is $\omega_0 t$. The x- and y-components are thus $r \cos\omega_0 t$ and $r \sin\omega_0 t$, respectively, as is illustrated in Fig. 4.8 on the next page. The components of the magnetization, denoted

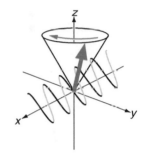

Fig. 4.6 The precessing magnetization will cut a coil wound round the x-axis, thereby inducing a current in the coil. This current is amplified and then detected to give the free induction signal.

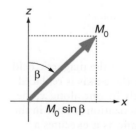

Fig. 4.7 Tilting the magnetization through an angle β towards the x-axis gives an x-component of size $M_0 \sin\beta$.

Fig. 4.8 At time zero the magnetization is positioned such that its x-component is $M_0 \sin \beta$ and its y-component is zero. After time t, the angle through which the vector has rotated is $\omega_0 t$, so the x- and y-components of the magnetization are $M_0 \sin \beta \cos \omega_0 t$ and $-M_0 \sin \beta \sin \omega_0 t$, respectively. Plots of these components of the magnetization are given in the lower part of the figure; it is assumed that the Larmor frequency, ω_0, is negative.

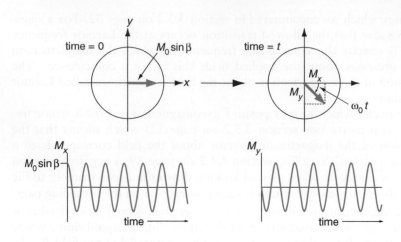

M_x and M_y, are given by

$$M_x = M_0 \sin \beta \cos \omega_0 t \qquad M_y = -M_0 \sin \beta \sin \omega_0 t. \qquad (4.1)$$

These are simply oscillations at the Larmor frequency and are plotted in Fig. 4.8. Fourier transformation of these signals gives us the familiar spectrum, which in this case is a single line at ω_0. The details of how the Fourier transform works will be covered in Chapter 5.

4.4 Pulses

We now turn to the important question as to how we can rotate the magnetization away from its equilibrium position along the z-axis. Conceptually it is easy to see what we have to do. All that is required is to replace the magnetic field along the z-axis with one in the xy-plane (say along the x-axis), as shown in Fig. 4.9. The magnetization is now no longer aligned with the field, and so the magnetization will precess about the field. In this case the precession will be in the yz-plane which will bring the magnetization towards the transverse plane, which is what we require.

Unfortunately it is all but impossible to switch the magnetic field suddenly in this way. Remember that the main magnetic field is supplied by a powerful superconducting magnet, and there is no way that this can be switched off quickly. We will need to find another approach, and it turns out that the key is to use the idea of resonance.

The idea is to apply a very small magnetic field along the x-axis but – crucially – to make this field oscillate at or near the Larmor frequency; in other words, the oscillating field is *resonant* with the Larmor precession frequency. We will show that even though B_0 is many times greater in size than the oscillating field, the latter can make the magnetization move away from the z-axis provided that the resonance condition is met.

The coil used to detect the precessing magnetization (Fig. 4.6 on the preceding page) can also be used to generate the oscillating magnetic field. All we do is feed some RF power to the coil and the resulting oscillating current will create the required oscillating magnetic field along

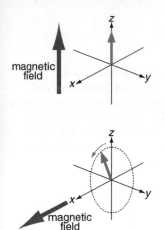

Fig. 4.9 If the magnetic field along the z-axis is replaced quickly by one along x, the magnetization rotates in the yz-plane as it executes a precessional motion about the field. As a result, the magnetization moves towards the transverse plane.

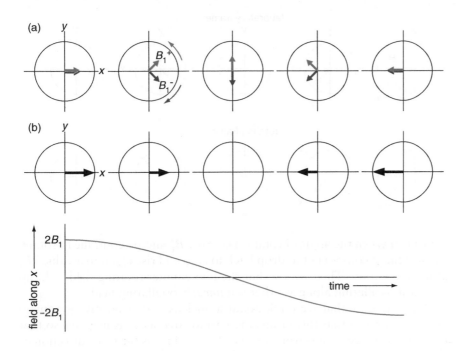

Fig. 4.10 Illustration of how two counter-rotating fields B_1^+ and B_1^-, shown in (a), add together to give a field which is oscillating along the x-axis, shown in (b). The graph shows how the field along x varies with time.

the x-direction. This field is called the *radiofrequency* or *RF field*. To understand how this weak RF field can rotate the magnetization we need to introduce the idea of the *rotating frame*.

4.4.1   Rotating frame

When RF power is applied to the coil wound along the x-axis, the result is a magnetic field which oscillates along the x-axis. What we mean by this is that the magnetic field starts out pointing along $+x$, gradually shrinks to zero and then increases along the $-x$ direction. It then shrinks back to zero and finally increases back to its original value along $+x$. We will take the frequency of this oscillation to be ω_{tx} (in rad s^{-1}) and the size of the magnetic field to be $2B_1$; the reason for the 2 will become apparent later. ω_{tx} is called the *transmitter frequency* for the reason that an RF transmitter is used to produce the power.

'tx' is the traditional abbreviation for transmitter.

It turns out to be a lot easier to work out the effect of this oscillating field if we replace it, in our minds, with two counter-rotating fields, as is illustrated in Fig. 4.10. The two counter-rotating fields have the same magnitude B_1. One, denoted B_1^+, rotates in the positive sense (from x to y) and the other, denoted B_1^-, rotates in the negative sense; both are rotating at the transmitter frequency ω_{tx}.

At time zero, B_1^+ and B_1^- are both aligned along the x-axis, and so add up to give a total field of $2B_1$ along x. As time proceeds, the vectors rotate away from the x-axis and in opposite directions. Since the two vectors have the same magnitude and are rotating at the same frequency, their y-components *always* cancel one another out. However, their x-components shrink towards zero as the angle through which the vectors have rotated approaches 90°. Then, as the angle increases beyond this point, the x-components grow once more, but this time along the $-x$-axis, reaching a

Fig. 4.11 The top row shows the motion of the field B_1^- when viewed in a fixed axis system, or *laboratory frame*. The field is rotating at $-\omega_{tx}$ i.e. in the negative sense, which is clockwise in this view. The bottom row shows the same field, but this time viewed in an axis system which is rotating at $-\omega_{tx}$ about the z-axis; in this *rotating frame*, the field B_1^- appears to be static.

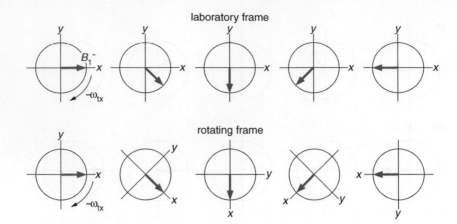

maximum when the angle of rotation is $180°$. B_1^+ and B_1^- continue to rotate, causing the x-component to drop back to zero and rise again to a value $2B_1$ along the x-axis. Thus we see that the two counter-rotating fields add up to a field oscillating along x – that is a *linearly* oscillating field.

Suppose now that we think about a nucleus with a positive gyromagnetic ratio; recall that this means the Larmor frequency is negative so that the sense of precession is from x towards $-y$, which is the same direction as the rotation of B_1^-. It turns out that the other field, B_1^+, which is rotating in the opposite sense to the Larmor precession, has no significant interaction with the magnetization and so we will ignore it.

We now employ a 'trick' and move to a co-ordinate system which, rather than being static (the *laboratory frame*), is rotating about the z-axis in the same direction and at the same rate as B_1^- (i.e. at $-\omega_{tx}$). As is shown in Fig. 4.11, in this rotating set of axes B_1^- appears to be static and directed along the x-axis. This is a very nice result as, when viewed in this *rotating frame*, the RF field is neither oscillating nor rotating, but is simply stationary. Moving to the rotating frame has therefore removed the time dependence of the field, and this makes it much easier to work out the effect that the field has on the magnetization.

An analogy to help you understand how the rotating frame simplifies things by removing time dependence is shown in Fig. 4.12. In (a) the anxious parent is watching their child riding on a merry-go-round (or carousel). The child's motion is rather complicated: not only is the horse going round and round, but it is also going up and down.

In (b) the parent is standing on the carousel, and so is going around with the child – in other words the parent has entered a rotating frame going at the same speed as the merry-go-round. Now, from the point of view of the parent, the child is just executing a simple up–down motion. Thus by moving to an appropriate frame of reference, the description of a complex motion is simplified.

It is important to realize that the purpose of the rotating frame is to remove the time dependence of the RF field, B_1^-. In the laboratory frame this field is rotating about the z-axis at $-\omega_{tx}$, but in a rotating frame also moving at $-\omega_{tx}$, B_1^- appears to be stationary. This is the *only* choice of rotating frame frequency which will make B_1^- stationary. The magnetic field due to the applied RF is often called 'the radiofrequency field' or 'the B_1 field'.

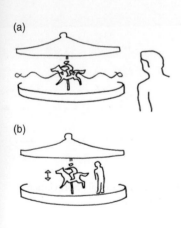

Fig. 4.12 An analogy for the rotating frame. In (a) a child riding on a merry-go-round executes a rather complex motion as seen by a fixed observer. However, if the observer joins the child on the merry-go-round, as in (b), the child appears to be executing a simple up–down motion.

We will use the rotating frame to help us work out what effect the RF field has on the magnetization, but before we do this we need to consider how the Larmor precession is affected by viewing things in the rotating frame.

4.4.2   Larmor precession in the rotating frame

Suppose that we choose the rotating frame such that it rotates at the same frequency and in the same sense as the Larmor precession. Viewed in this frame, the magnetization will appear to be stationary i.e. the *apparent* Larmor frequency is zero. Of course the precession has not really stopped, it is just that we are viewing it differently.

More generally, if the rotating frame frequency is $\omega_{\text{rot. frame}}$, the apparent frequency of the Larmor precession in such a frame will be $(\omega_0 - \omega_{\text{rot. frame}})$. This difference frequency is called the *offset*, and is given the symbol Ω:

$$\Omega = \omega_0 - \omega_{\text{rot. frame}}.$$

The frequency ω at which magnetization precesses around a magnetic field B is given by

$$\omega = -\gamma B.$$

Another way of looking at this relationship is to say that if we know the precessional frequency we can work out the magnetic field B using $B = -\omega/\gamma$. Following this line of argument we can say that if in the rotating frame the precessional frequency appears to be Ω, then the apparent magnetic field ΔB is given by

$$\Delta B = -\frac{\Omega}{\gamma}.$$

ΔB is called the *reduced field* in the rotating frame. Clearly if the offset is zero, then so too is the reduced field.

The important conclusion from this section is that, when viewed in the rotating frame, the apparent magnetic field along the z-axis (the reduced field) can be much smaller than the applied magnetic field, B_0. Indeed, if the rotating frame is at the Larmor frequency, the apparent field is zero. When the reduced field becomes comparable with the B_1 field, the latter can start to influence the motion of the magnetization even though it is very much smaller than the applied field B_0. To understand the details of how this works we need to introduce the concept of the effective field.

4.4.3   The effective field

From the discussion so far we can see that when RF power is applied there are two magnetic fields in the rotating frame. First, there is the RF field which, provided we choose $\omega_{\text{rot. frame}} = -\omega_{\text{tx}}$, gives rise to a static field B_1 along the x-axis.

Secondly, there is the reduced field, ΔB, given by $(-\Omega/\gamma)$. Since $\Omega = (\omega_0 - \omega_{\text{rot. frame}})$ and $\omega_{\text{rot. frame}} = -\omega_{\text{tx}}$ it follows that

$$\begin{aligned}\Omega &= \omega_0 - (-\omega_{\text{tx}}) \\ &= \omega_0 + \omega_{\text{tx}}.\end{aligned} \tag{4.2}$$

In this discussion we will assume that the gyromagnetic ratio is positive, so that the Larmor frequency is negative.

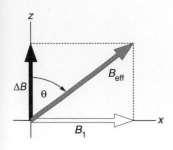

Fig. 4.13 In the rotating frame the effective field B_{eff} is the vector sum of the reduced field ΔB and the B_1 field. The tilt angle, θ, is defined as the angle between B_{eff} and ΔB.

This looks rather strange, but recall that ω_0 is negative, so if the transmitter frequency and the Larmor frequency are comparable in magnitude, the offset will be small.

In the rotating frame, the reduced field (which is along z) and the B_1 field (which is along x) add vectorially to give an effective field B_{eff} as illustrated in Fig. 4.13. The size of this effective field is given by simple geometry as:

$$B_{\text{eff}} = \sqrt{B_1^2 + (\Delta B)^2}. \tag{4.3}$$

The crucial point is that the magnetization precesses around the *effective field*, just as in the laboratory frame the Larmor precession takes place around the B_0 field. As usual, the precessional frequency about the effective field ω_{eff} is proportional to the field:

$$\omega_{\text{eff}} = |\gamma| \, B_{\text{eff}}.$$

The vertical lines ($||$) indicate that we should take the *absolute value* of the quantity they enclose i.e. ignore the sign. Thus, regardless of the sign of γ, ω_{eff} is always positive.

If the offset is small the effective field will lie close to the x-axis, and so the equilibrium magnetization will be rotated away from the z-axis, which is exactly what we want to achieve. The key point is that, although B_0 is very much larger than B_1, we can eliminate the effect of the B_0 field by setting the transmitter frequency close to the Larmor frequency i.e. by making the offset small. With this condition the reduced field ΔB is small and it is then possible for the small B_1 field to begin to exert an influence. In the limit that the offset is zero, ΔB disappears and the B_1 field is the only one left in the rotating frame.

The angle between ΔB and B_{eff} is called the *tilt angle* and is usually given the symbol θ. From Figure 4.13 we can see that:

$$\sin\theta = \frac{B_1}{B_{\text{eff}}} \qquad \cos\theta = \frac{\Delta B}{B_{\text{eff}}} \qquad \tan\theta = \frac{B_1}{\Delta B}.$$

All three definitions are equivalent. When the offset is zero, $\theta = \pi/2$; when the offset is large, θ approaches zero or π, depending on the sign of the offset and hence of ΔB.

4.4.4 The effective field in frequency units

For practical purposes the thing that is important is the precession frequency about the effective field, ω_{eff}. It is therefore convenient to think about the construction of the effective field not in terms of magnetic fields, but in terms of the precession frequencies that they cause.

For each field the precession frequency is proportional to the magnetic field with the constant of proportion being γ, the gyromagnetic ratio. For example, we have already seen that in the rotating frame the apparent Larmor precession frequency, Ω, depends on the reduced field:

$$\Omega = -\gamma \Delta B.$$

We define ω_1 as the precessional frequency about the B_1 field (note that the absolute value of γ is taken, so ω_1 is always positive):

$$\omega_1 = |\gamma| \, B_1,$$

and we already have

$$\omega_{\text{eff}} = |\gamma| \, B_{\text{eff}}.$$

Using these definitions, Eq. 4.3 on the facing page can be rewritten in terms of frequencies to give the following expression for ω_{eff}

$$\omega_{\text{eff}} = \sqrt{\omega_1^2 + \Omega^2}. \tag{4.4}$$

Figure 4.13 on the preceding page can be relabelled with frequencies, as shown in Fig. 4.14, and the tilt angle can also be expressed in terms of frequencies:

$$\sin\theta = \frac{\omega_1}{\omega_{\text{eff}}} \qquad \cos\theta = \frac{\Omega}{\omega_{\text{eff}}} \qquad \tan\theta = \frac{\omega_1}{\Omega}.$$

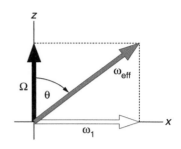

Fig. 4.14 The effective field can be thought of in terms of frequencies instead of the fields used in Fig. 4.13.

4.4.5 Summary

This has been rather a long and involved section which has introduced many new ideas, so it is well to end by summarizing the key points.

- RF power is supplied to a small coil wound round the sample in such a way that the oscillating current in the coil creates an oscillating transverse magnetic field e.g. along the x-axis.

- This linearly oscillating field can be decomposed into two counter-rotating fields. Only the field which is rotating in the same sense as the Larmor precession need be considered.

- The rotating field can be made static by moving to an appropriate rotating frame.

- In the rotating frame, the Larmor precession is modified; we interpret this as the B_0 field being replaced by the reduced field ΔB.

- The magnetization rotates about the effective field which is a vector sum of the RF field B_1 and the reduced field.

- The effective field will lie close to the x-axis if the offset is small i.e. the transmitter frequency is close to the Larmor frequency.

4.5 On-resonance pulses

The simplest case to deal with is when the transmitter frequency is exactly the same as the Larmor frequency – called an *on-resonance* pulse. Under these circumstances the offset Ω is zero and so, referring to Fig. 4.14, we see that the effective field lies along the x-axis and is of size ω_1. The tilt angle θ of the effective field is $\pi/2$ or $90°$.

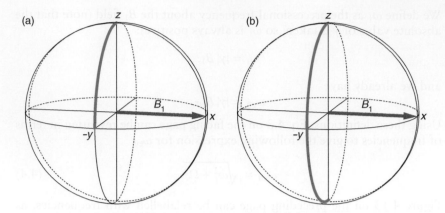

Fig. 4.15 Three-dimensional representations of the motion of the magnetization vector during an on-resonance pulse. The thick blue line is the path followed by the tip of the vector, which is assumed to start on the $+z$-axis. The effective field, shown by the grey arrow, is the same as the B_1 field and lies along the x-axis; the magnetization therefore precesses in the yz-plane. The rotation is in the positive sense about x, so the magnetization moves toward the $-y$-axis. In (a) the flip angle is $90°$ and so the magnetization ends up along $-y$; in (b) the pulse flip angle is $180°$ which places the magnetization along $-z$.

For an on-resonance pulse the motion of the equilibrium magnetization vector is very simple; all that happens is that it is rotated from the z-axis and in the yz-plane, just as shown in Fig. 4.9 on page 52. The precession frequency is ω_1 and so if the RF field is applied for a time t_p, the angle β through which the magnetization has been rotated will be given by

$$\beta = \omega_1 t_p \qquad (4.5)$$

β is called the *flip angle* of the pulse. By altering the time for which the RF is applied we can alter the angle through which the magnetization is rotated.

The most commonly used flip angles are $\pi/2$ ($90°$) and π ($180°$), and the motion of the magnetization vector during such on-resonance pulses is shown in Fig. 4.15. The $90°$ pulse rotates the magnetization from the equilibrium position to the $-y$-axis. The magnetization ends up along $-y$ because the rotation is in a positive sense about the x-axis. Imagine grasping the x-axis with your right hand and with the thumb pointing along the $+x$-direction; your fingers then curl in the sense of a positive rotation, which takes the vector from z towards $-y$.

If the pulse flip angle is set to $180°$, the magnetization is taken all the way from $+z$ to $-z$; this is called an *inversion pulse*. In general, for a flip angle β simple geometry, illustrated in Fig. 4.16, tells us that the z- and y-components are

$$M_z = M_0 \cos\beta \qquad M_y = -M_0 \sin\beta.$$

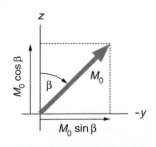

Fig. 4.16 If an on-resonance pulse of flip angle β is applied to equilibrium magnetization we can use simple geometry to work out the resulting y- and z-components.

For the pulses we have been describing so far the RF (B_1) field is aligned along the x-axis, so such a pulse is properly described as 'an x pulse' or 'a pulse about x'; this is often written 90°_x or $90^\circ(x)$. Later on, in section 4.10 on page 66, we will describe the effects of pulses about other axes. Generally the convention in NMR is that, unless otherwise specified, a pulse is assumed to be about the x-axis. However, to start with we will specify the phase of all of the pulses in order to avoid confusion.

4.5.1 Hard pulses

In practical NMR spectroscopy we usually have several resonances in the spectrum, each of which has a different Larmor frequency; we cannot therefore be on-resonance with all of the lines in the spectrum. However, it is possible to make the RF (B_1) field strength sufficiently large that for a range of resonances the effective field lies very close to the x-axis. Then, to all intents and purposes, the magnetization behaves as if the pulse is on resonance.

This is best illustrated using an example. Suppose that we have a proton spectrum covering a range of about 10 ppm, and that we place the transmitter frequency in the middle of the spectrum. The largest possible offset is 5 ppm which, at a Larmor frequency of 500 MHz, translates to an offset of 2500 Hz. Converting to rad s^{-1} this gives $\Omega = 1.57 \times 10^4$ rad s^{-1}.

A typical value for the length of a 90° pulse might be about 12 μs, and from this we can work out the field strength, ω_1, since this is related to the flip angle according to Eq. 4.5 on the facing page

$$\beta = \omega_1 t_p \qquad \text{hence} \qquad \omega_1 = \frac{\beta}{t_p}.$$

In this case we know that for a 90° pulse $\beta = \pi/2$ and the duration, t_p, is 12×10^{-6} s, so ω_1 can be computed as

$$
\begin{aligned}
\omega_1 &= \frac{\pi/2}{12 \times 10^{-6}} \\
&= 1.31 \times 10^5 \text{ rad s}^{-1}.
\end{aligned}
$$

The tangent of tilt angle is therefore computed as

$$
\begin{aligned}
\tan\theta &= \frac{\omega_1}{\Omega} \\
&= \frac{1.31 \times 10^5}{1.57 \times 10^4} \\
&= 8.34,
\end{aligned}
$$

which corresponds to a tilt angle of 83°, as illustrated in Fig. 4.17. We can use Eq. 4.4 on page 57 to calculate ω_{eff} as 1.32×10^5 rad s^{-1}; as expected, this value is very close to ω_1 as $\omega_1 \gg \Omega$.

Recall that for an on-resonance pulse the tilt angle is 90°. What we have shown in this example is that for a peak at the edge of the spectrum the tilt angle is within a few degrees of that for an on-resonance pulse and so, to a good approximation, we can assume that the magnetization vector will behave as it does for an on-resonance pulse. Such a pulse is called a *hard pulse* or a *non-selective* pulse.

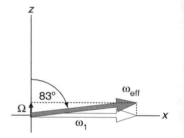

Fig. 4.17 Illustration of the position of the effective field for the example given in the text where the offset is 2500 Hz and the field strength corresponds to a 90° pulse of duration 12 μs. These conditions result in the RF field strength ω_1 being much greater than the offset Ω, and so the tilt angle is close to 90°. To all intents and purposes we can assume that the pulse is in fact on resonance, with the effective field lying along the x-axis and of strength ω_1.

In summary, the condition for a hard pulse is that the RF field strength, ω_1, must be much greater than the offset, Ω:

$$\text{hard pulse}: \quad \omega_1 \gg |\Omega|.$$

We need the modulus signs as the offset might be negative. If this condition holds, the effective field is the same as ω_1 and, like ω_1, lies along the x-axis.

A spectrometer is designed to have sufficient RF power to create hard pulses over the normal range of shifts of a given nucleus. As the main magnetic field B_0 becomes greater, the range of possible offsets increases (recall that these scale with the field), and so for a pulse to be considered 'hard' the B_1 field must be proportionately stronger i.e. more RF power is needed.

4.6   Detection in the rotating frame

It was explained in section 2.1.3 on page 8 that, due to the way most NMR spectrometers are constructed, the frequencies of the lines in the spectrum are measured relative to the receiver reference frequency. The commonest arrangement is to make the receiver reference frequency the *same* as the transmitter frequency. Recall that we also need to make the rotating frame frequency equal to the transmitter frequency in order to make the B_1 field static. If the transmitter, rotating frame and receiver reference frequencies are all the same then the offset frequency described in section 2.1.3 on page 8 and the offset defined in Eq. 4.2 on page 55 will have the *same* value.

Consider a peak which is 100 Hz from the transmitter frequency, and hence 100 Hz from the receiver reference frequency. In a rotating frame at the transmitter frequency, the magnetization vector from this peak will precess at 100 Hz. One way of looking at this is to say that we are detecting the precession of the magnetization *in the rotating frame*. So, instead of the detected signal oscillating at the Larmor frequency, the oscillation is at 100 Hz – that is the offset frequency.

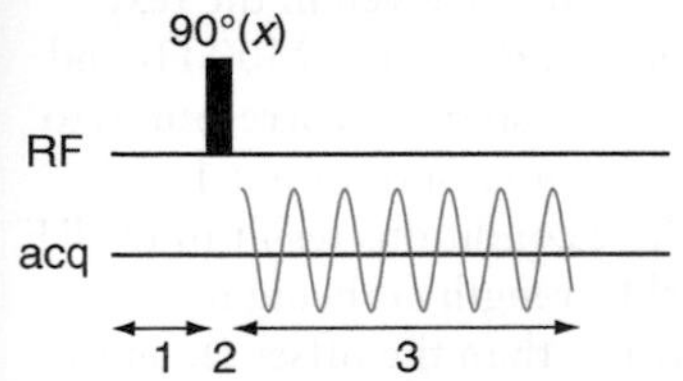

From now on we will make the assumption that the receiver reference and transmitter frequencies are the same, and that we are detecting the signal in the rotating frame. With these assumptions the x- and y-magnetizations given in Eq. 4.1 on page 52 are rewritten

$$M_x = M_0 \sin\beta \cos\Omega t \qquad M_y = M_0 \sin\beta \sin\Omega t,$$

where Ω is the offset.

Fig. 4.18 Timing diagram or pulse sequence for the simple pulse–acquire experiment. The line marked 'RF' shows the location of the radiofrequency pulses, and the line marked 'acq' shows when the signal is recorded or *acquired*. The pulse sequence can be divided up into three periods, as shown on the lower line.

4.7   The basic pulse–acquire experiment

At last we are in a position to understand how the simple pulse–acquire NMR experiment, introduced in section 2.4 on page 13, actually works. The timing diagram – or *pulse sequence* as it is usually known – is shown in Fig. 4.18.

The pulse sequence can be divided up into three periods, as shown in the diagram; we will assume that the RF pulse is 'hard'. During

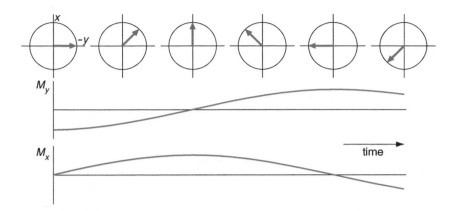

Fig. 4.19 Evolution of the
magnetization during the
acquisition time (period 3) of
the pulse–acquire experiment;
the xy-plane of the rotating
frame is shown. The
magnetization starts out
along $-y$ and evolves at the
offset frequency Ω (here
assumed to be positive). The
resulting x- and
y-magnetizations are shown
in the graphs in the lower
part of the figure.

period 1 equilibrium magnetization builds up along the z-axis. As was described in section 4.5 on page 57, during period 2 the $90°(x)$ pulse rotates this magnetization onto the $-y$-axis. During period 3 the magnetization precesses in the transverse plane at the offset Ω, as is illustrated in Fig. 4.19.

Some simple geometry, shown in Fig. 4.20, enables us to deduce how the x- and y-magnetizations vary with time. The offset is Ω, so after time t the vector has precessed through an angle Ωt; the x- and y-components are thus proportional to the sine and cosine of this angle:

$$M_y = -M_0 \cos \Omega t \qquad M_x = M_0 \sin \Omega t.$$

As we commented on before, Fourier transformation of the detected signals arising from these magnetizations will give a spectrum, with a peak appearing at frequency Ω.

4.7.1   Spectrum with several lines

If the spectrum has more than one line, then we can associate a separate magnetization vector with each. If the pulse is hard, then each vector will be rotated into the transverse plane, and then they will precess at their individual offsets. The detected signal will be the sum of contributions from each vector; for example the y-component will be

$$M_y = -M_{0,1} \cos \Omega_1 t - M_{0,2} \cos \Omega_2 t - M_{0,3} \cos \Omega_3 t \ \ldots$$

where $M_{0,1}$ is the equilibrium magnetization associated with the first line, Ω_1 is its offset, and so on for the other lines. Fourier transformation of the resulting detected signal will produce a spectrum with lines at Ω_1, Ω_2 etc.

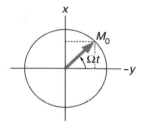

Fig. 4.20 After a $90°(x)$ pulse the
magnetization starts out along
the $-y$-axis. It then rotates
through an angle Ωt during time
t; the x-component is
proportional to $\sin \Omega t$ and the
y-component to $\cos \Omega t$.

4.8   Pulse calibration

It is crucial that the pulses we use in NMR experiments have the correct flip angles. For example, to obtain the maximum intensity in the pulse–acquire experiment we must use a $90°$ pulse, and if we wish to invert magnetization we must use a $180°$ pulse – other flip angles will simply not give the required result. Pulse calibration is therefore an important preliminary to

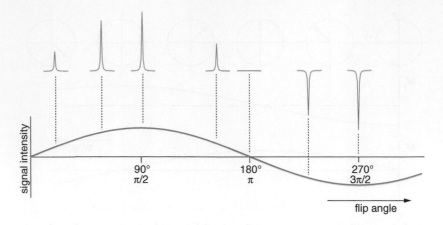

Fig. 4.21 Illustration of how pulse calibration is achieved using a pulse–acquire experiment in which the flip angle of the pulse, β, is varied. In such an experiment, the signal intensity is proportional to $\sin\beta$, as shown by the graph. Along the top are the spectra which would be expected for various different flip angles (indicated by the dashed lines). The signal is a maximum for a flip angle of 90°, goes through a null at 180°, and after that goes negative. Pulse calibration is achieved by increasing the duration of the pulse until the signal goes through a null; this time corresponds to a 180° pulse.

any experiment and is usually carried out using a modified pulse–acquire experiment.

We have already shown that, for a hard or on-resonance pulse applied to equilibrium magnetization, the y-component of magnetization after a pulse of flip angle β is proportional to $\sin\beta$ (Fig. 4.16 on page 58). Recalling that it is this transverse magnetization which is detected, it follows that the intensity of the signal in a pulse–acquire experiment will vary as $\sin\beta$. A typical outcome of such an experiment in which the pulse flip angle is varied is shown in Fig. 4.21.

The normal practice is to increase the flip angle of the pulse (by increasing its length) until a null is found; the flip angle is then 180°. There is a maximum in the signal when the flip angle is 90°, but the maximum is broad and hence rather difficult to locate precisely. In contrast, the null at 180° is sharper and easier to locate.

Once the duration of a 180° pulse is found, simply halving the time gives a 90° pulse. The pulse length for other flip angles can be found by simple proportion.

Suppose that the 180° pulse was found to be of duration t_{180}. Since the flip angle is given by $\beta = \omega_1 t_p$ (Eq. 4.5 on page 58) it follows that

$$\pi = \omega_1 t_{180}$$
$$\text{hence } \omega_1 = \frac{\pi}{t_{180}},$$

where we have remembered to write the flip angle β in radians. Using this expression, we can determine ω_1, the RF field strength, from the duration of a pulse of known flip angle.

Sometimes we want to quote the field strength not in rad s^{-1} but in Hz, in which case all we need to do is divide the above result by 2π:

$$(\omega_1/2\pi) = \frac{1}{2\,t_{180}}\ \text{Hz}.$$

For example, let us suppose that we found a null in the signal at a pulse length of 15.5 μs. The RF field strength is therefore given by

$$
\begin{aligned}
\omega_1 &= \frac{\pi}{t_{180}} \\
&= \frac{\pi}{15.5 \times 10^{-6}} \\
&= 2.03 \times 10^5\ \text{rad s}^{-1}.
\end{aligned}
$$

In frequency units the calculation is

$$
\begin{aligned}
(\omega_1/2\pi) &= \frac{1}{2\,t_{180}} \\
&= \frac{1}{2 \times 15.5 \times 10^{-6}} \\
&= 32.3\ \text{kHz}.
\end{aligned}
$$

This result is often expressed in words by saying 'the B_1 field is 32.3 kHz'. At first sight this is rather a strange thing to say, as surely B_1 is a magnetic field, not a frequency. When we specify the B_1 field in Hz, what we are in fact doing is giving the frequency at which the magnetization will precess about the field. In practice, this is a more useful thing to know than the size of the field in Tesla.

4.9 The spin echo

We are now able to analyse the most famous pulsed NMR experiment, the *spin echo*, which is a component of a very large number of more complex experiments. The pulse sequence is shown in Fig. 4.22.

The special thing about the spin echo sequence is that, at the end of the second τ delay, the magnetization ends up along the *same* axis, *regardless* of the length of τ or the size of the offset Ω. We describe this outcome by saying that 'the offset has been refocused', meaning that at the end of the sequence it is just as if the magnetization has not evolved at all i.e. as if the offset is zero.

4.9.1   180° pulses as refocusing pulses

To understand how the spin echo sequence works we first need to understand the effect of the 180° pulse when it is applied to *transverse magnetization*. Figure 4.23 on the following page shows the effect of a 180°(x) pulse on three magnetization vectors at different positions in the transverse plane. We see that each vector is carried through a different arc but all end up in mirror image positions with respect to the xz-plane.

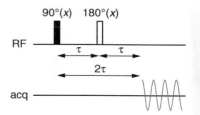

Fig. 4.22 Pulse sequence for the *spin echo* experiment. The 180° pulse (indicated by an open rectangle as opposed to the closed one for a 90° pulse) is in the centre of a delay of duration 2τ, thus separating the sequence into two equal periods, τ. The signal is not acquired until after the second delay τ, or put another way, until 2τ after the beginning of the sequence. The durations of the pulses are in practice very much shorter than the delays τ, but for clarity the length of the pulses has been exaggerated.

Fig. 4.23 Illustration of the effect of a 180°(*x*) pulse on three vectors (coloured in different shades of blue) which start out at different angles from the −*y*-axis. All three are rotated by 180° about the *x*-axis on the trajectories indicated by the thick lines which dip into the southern hemisphere. After the pulse, the vectors end up in mirror image positions with respect to the *xz*-plane.

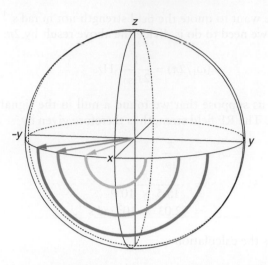

A second view of what is going on is given in Fig. 4.24. The diagram describes the fate of a magnetization vector which has precessed away from the −*y*-axis through an angle ϕ i.e. it has acquired a phase ϕ. To work out the effect of the 180° pulse it is convenient first to resolve the magnetization vector into its components along the *x*- and *y*-axes, as is shown in the diagram.

The *x*-component is unaffected by the 180°(*x*) pulse as this component is aligned along the same axis as the B_1 field. The *y*-component is simply rotated to the opposite axis, in this case from −*y* to +*y*. This is analogous to a 180° pulse rotating the equilibrium magnetization from the +*z*-axis to the −*z*-axis.

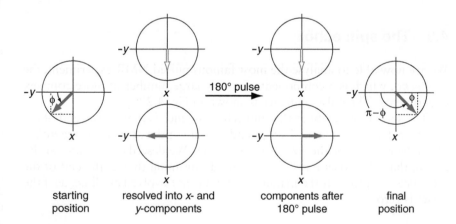

| starting position | resolved into *x*- and *y*-components | 180° pulse | components after 180° pulse | final position |

Fig. 4.24 Illustration of the effect of a 180°(*x*) pulse on a magnetization vector in the transverse plane; the initial position of the vector is described by a phase angle ϕ, measured from the −*y*-axis. The effect of the pulse is best visualized by resolving the vector into its *x*- and *y*-components. As the former is aligned with the B_1 field, it is unaffected by the pulse; the *y*-component is simply reversed in direction. The final position of the vector is found by recombining the *x*- and *y*-components. We see that the vector ends up in a mirror image position with respect to the *xz*-plane, with a phase angle $(\pi - \phi)$ radians, measured from the −*y*-axis.

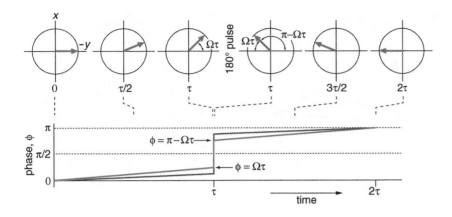

Fig. 4.25 Illustration of how the spin echo refocuses the evolution of the offset. The sequence starts with a $90°(x)$ pulse (not shown) which places the magnetization along $-y$. The upper part of the figure shows the position of the magnetization vector at various times; note how the $180°(x)$ pulse moves the magnetization to a mirror image position with respect to the xz-plane. The bottom of the diagram shows a graph of how the phase accrued by the magnetization vector (measured from its starting position on the $-y$-axis) varies throughout the sequence. The $180°$ pulse causes a discontinuity in the phase, changing its value from $\Omega\tau$ to $(\pi - \Omega\tau)$. The evolution of the phase for two different offsets is shown by the blue and dark grey lines. Regardless of the offset or the delay τ, the magnetization ends along the y-axis with a phase of π.

When the two components are recombined to give the final position of the vector, we find that it has been moved to a mirror image position with respect to the xz-plane. Expressed in terms of the phase angle, the effect of the $180°(x)$ pulse is to move the vector from a position described by a phase ϕ (measured from $-y$), to one with a phase $(\pi - \phi)$.

4.9.2 How the spin echo works

Figure 4.25 illustrates the motion of a typical magnetization vector during the spin echo. The diagram commences after the initial $90°(x)$ pulse has placed the magnetization along $-y$, and the position of the magnetization vector at selected times is shown. In addition, there is a graph giving the phase ϕ which the magnetization vector acquires as the sequence proceeds. This phase is measured anti-clockwise from the starting position of the vector on the $-y$-axis.

During the first delay τ the vector precesses from $-y$ towards the x-axis (we are assuming that the offset is positive). The angle through which the vector rotates is simply $\Omega\tau$, which is therefore its phase.

As has already been explained, the effect of the $180°(x)$ pulse is to move the vector to a mirror image position with respect to the xz-plane. So, after the $180°$ pulse the vector is at an angle $\Omega\tau$ to the y-axis rather than being at $\Omega\tau$ to the $-y$-axis. Measured from the $-y$-axis, the phase is now $(\pi - \Omega\tau)$.

During the second delay τ the vector continues to evolve and will once again rotate through an angle of $\Omega\tau$; the additional phase acquired is

Fig. 4.26 Three-dimensional representations of the effect on equilibrium magnetization of (a) a 90° pulse about the y-axis and (b) a 90° pulse about the $-x$-axis. The magnetization starts from the $+z$-axis and the path followed by the tip of the magnetization vector is shown by the blue line. The pulses are assumed to be on resonance (or hard), with the B_1 fields in the positions shown.

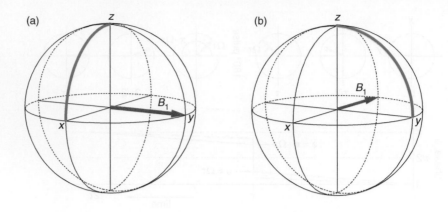

thus $\Omega\tau$. Just after the 180° pulse the phase is $(\pi - \Omega\tau)$, and adding to this the phase $\Omega\tau$ acquired during the second delay τ gives us the phase at the end of the sequence as $(\pi - \Omega\tau) + \Omega\tau = \pi$ i.e. the magnetization vector is aligned along the y-axis.

Clearly, the final position of the vector is independent of the offset Ω and the delay τ. This is the special feature of a spin echo and why it is described as refocusing the offset. The refocusing comes about because first the phase acquired during the two τ delays is the same, and secondly the intervention of the 180° pulse between them causes the phases to cancel one another out.

The 180° pulse is called a *refocusing pulse* because it causes the evolution during the first delay τ to be undone by the second delay. It is interesting to note that the spin echo sequence gives exactly the same result as the sequence $90°(x) - 180°(x)$ with the delays omitted.

Figure 4.25 on the previous page also shows the phase throughout the sequence. Note that there is a discontinuity when the 180° pulse is applied since at this point the phase changes from $\Omega\tau$ to $(\pi - \Omega\tau)$. If the offset is smaller the phase evolution is different, as shown by the dark grey line, but the phase at the very end of the sequence is still π.

4.10   Pulses of different phases

So far we have only allowed the RF (B_1) field to be along the x-axis, but the field can just as well be in any direction in the transverse plane. Commonly, pulses with the field aligned along the four cardinal directions are used i.e. x, y, $-x$ and $-y$. For example, a pulse with the field along the y-axis is referred to as 'a y pulse' or a 'pulse about y'. A y pulse is sometimes called a pulse 'phase shifted by 90°', indicating that it is about an axis shifted by 90° from the x-axis (which is taken as the reference point). Thus a 90° pulse about the $-y$-axis, written $90°(-y)$, can be described as a pulse phase shifted by 270°.

If we apply a 90° pulse about the y-axis to equilibrium magnetization we find that the vector rotates in the xz-plane such that the magnetization ends up along the x-axis, as is illustrated in Fig. 4.26 (a). As before, we

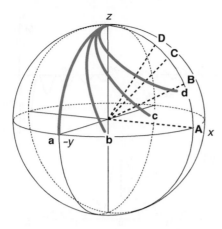

Fig. 4.27 Three-dimensional representation showing the path followed during an x pulse for various different resonance offsets. The duration is chosen so that, on resonance, the flip angle is 90°; the blue lines show the path followed by the tip of the magnetization vector which is assumed to start on $+z$. Path **a** is for the on-resonance case; the effective field lies along x and is indicated by the dashed line **A**. Path **b** is for the case where the offset is half the RF field strength ($\Omega = \omega_1/2$); the effective field is marked **B**. Paths **c** and **d** are for offsets equal to and 1.5 times the RF field strength, respectively; the effective field directions are labelled **C** and **D**.

can determine the effect of such a pulse by thinking of it as a positive rotation about the y-axis. A 90° pulse about $-x$ rotates the equilibrium magnetization to the y-axis, as is shown in Fig. 4.26 (b) on the preceding page.

As we described in section 4.9.1 on page 63, a 180°(x) pulse causes the vectors to move to mirror image positions with respect to the xz-plane. In a similar way, a 180°(y) pulse causes the vectors to move to mirror image positions with respect to the yz-plane. Finally, it is interesting to note that a 180° pulse about *any* axis in the transverse plane will rotate magnetization from $+z$ to $-z$.

4.11 Off-resonance effects and soft pulses

⇐ Optional section

So far we have only dealt with the case where the pulse is either on resonance or where the RF field strength is large compared with the offset (a hard pulse, section 4.5.1 on page 59), which is in effect the same situation. We now turn to the case where the offset Ω is comparable in size with the RF field strength ω_1. The consequences of this are sometimes a problem to us, but they can also be turned to our advantage for *selective excitation*.

Referring to Fig. 4.14 on page 57, we see that as the offset becomes comparable with the RF field strength, the effective field begins to move up from the x-axis towards the z-axis (assuming that the offset is positive). As a consequence, rather than the equilibrium magnetization simply being rotated in the yz-plane from z to $-y$, the magnetization follows a more complex curved path. A series of such paths for increasing offsets are

Fig. 4.28 Plots of the magnetization produced by a $90°(x)$ pulse to equilibrium magnetization (assumed to be of size 1) as a function of the offset; the pulse length has been adjusted so that on resonance the flip angle is $90°$. The horizontal axes of the plots are the offset Ω expressed as a ratio of the RF field strength ω_1. Plots (a), (b) and (c) are, respectively, the x-magnetization, y-magnetization, and the absolute value of the transverse magnetization.

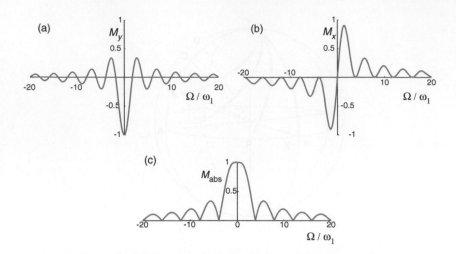

shown in Fig. 4.27 on the previous page. In this figure the duration of the pulse has been set so that the flip angle is $90°$ on resonance.

There are two things to note from this diagram. First, although on resonance the magnetization vector ends up along $-y$, for the off-resonance case the vectors stop short of the transverse plane. This means that the signal we would observe after such a pulse will be weaker than for the on-resonance case, simply because less transverse magnetization has been generated.

The second thing to note is that, whereas the on-resonance pulse produces only y-magnetization, the off-resonance pulses also produce some x-magnetization. We will see later on in section 5.3.4 on page 88 that this leads to phase errors in the spectrum.

In the limit that the offset becomes very much larger than the RF field strength, the effective field lies very close to the z-axis, and so the pulse is incapable of rotating the equilibrium magnetization away from the z-axis. Such a pulse is so far off resonance that it has no effect. For example, pulses applied to ^{13}C have no effect on protons as the resonance offset of any protons would be tens if not hundreds of MHz from the ^{13}C transmitter frequency. For typical RF field strengths of the order of kHz, it therefore follows that proton resonances will not be affected by pulses to ^{13}C.

We can see more clearly what is going on if we plot the x- and y-magnetizations, produced by a nominal $90°(x)$ pulse applied to equilibrium magnetization, as a function of the offset; these graphs are shown in Fig. 4.28. In (a) we see the y-magnetization and, as expected, on resonance the equilibrium magnetization ends up entirely along $-y$. However, as the offset increases the amount of y-magnetization generally decreases but there is an oscillation imposed on this overall decrease. At some offsets the magnetization is zero and at others it is positive. The plot of the x-magnetization, (b), shows a similar story with the magnetization generally falling off as the offset increases, but again with a strong oscillation.

Plot (c) is of the *magnitude* of the magnetization, which is given by

$$M_{abs} = \sqrt{M_x^2 + M_y^2}.$$

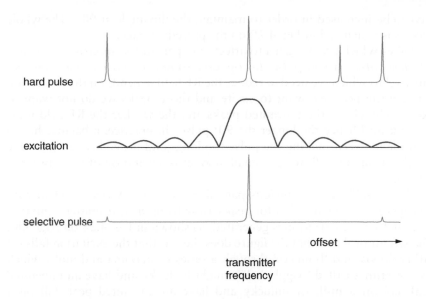

Fig. 4.29 Visualization of the selective excitation of just one line in the spectrum. At the top is shown the spectrum that would be excited using a hard pulse. If the transmitter is placed on resonance with one line and the strength of the RF field reduced, then the pattern of excitation we expect is as shown in the middle (see the plot of M_{abs} in Fig. 4.28 on the preceding page). As a result, the peaks at non-zero offsets are attenuated and the spectrum which is excited will be as shown at the bottom.

This gives the total transverse magnetization in any direction, and it is, of course, always positive. We see from this plot the characteristic nulls and subsidiary maxima in the amount of magnetization as the offset increases.

What plot (c) tells us is that although a pulse can excite magnetization over a wide range of offsets, the region over which it does so efficiently is really rather small. If we want at least 90% of the magnetization to be rotated to the transverse plane (i.e. $M_{abs} \geq 0.9$), the offset must be less than about 1.6 times the RF field strength.

4.11.1 Excitation of a range of shifts

There are some immediate practical consequences of these off-resonance effects for RF pulses. Suppose that we are trying to record the full range of ^{13}C shifts (200 ppm) on a spectrometer whose magnetic field gives a proton Larmor frequency of 800 MHz and hence a ^{13}C Larmor frequency of 200 MHz. If we place the transmitter frequency at 100 ppm, the maximum offset that a peak can have is 100 ppm which, at this Larmor frequency, translates to 20 kHz. According to our criterion above, if we are willing to accept a reduction to 90% of the full intensity at the edges of the spectrum we would need an RF field strength of $20/1.6 \approx 12.5$ kHz. This corresponds to a 90° pulse width of 20 μs. If the spectrometer has insufficient power to produce this pulse width, the excitation at the edges of the spectrum will fall below the 90% mark.

4.11.2 Selective excitation

Sometimes we want to excite just a portion of the spectrum, for example a single line or just the lines of one multiplet. We can achieve this by putting the transmitter on resonance with the line we want to excite (or in the middle of the multiplet), and then reducing the RF field strength until the degree of excitation of the rest of the spectrum is sufficiently small. At the same time as the RF field strength is reduced, the duration of the pulse will have to be increased in order to maintain the flip angle at 90°. The whole process is visualized in Fig. 4.29 on the preceding page.

Pulses which are designed to affect only part of the spectrum are called *selective pulses* or *soft pulses* (as opposed to non-selective or hard pulses). The level to which we need to reduce the RF field depends on the separation between the peak we want to excite and those peaks we do not want to excite. The closer the unwanted peaks are, the weaker the RF field must be made and hence the longer the 90° pulse. In practice, a balance has to be made between making the pulse too long (and hence losing signal due to relaxation) and allowing a small amount of excitation of the unwanted signals.

Figure 4.29 on the previous page does not portray one problem with this approach, which is that for peaks away from the transmitter a mixture of x- and y-magnetization is generated (as shown in Fig. 4.28 on page 68). The second problem that the figure does show is that the excitation falls off rather slowly and 'bounces' through a series of maxima and nulls, which are sometimes called 'wiggles'. We might be lucky and have an unwanted peak fall on a null, or unlucky and have an unwanted peak fall on a maximum.

Much effort has been put into getting round both of these problems. The key feature of all of the successful solutions is to 'shape' the envelope of the RF pulse i.e. not just switch it on and off abruptly, but with a smooth variation. Such pulses are called *shaped pulses*. The simplest of these are basically bell-shaped (like a gaussian function, for example). These suppress the wiggles at large offsets and give just a smooth decay, but they do not improve the phase properties. To attack this part of the problem requires an altogether more sophisticated approach (see *Further reading*).

4.11.3 Selective inversion

Sometimes we want to invert the magnetization associated with just one resonance while leaving all the others in the spectrum unaffected; such a pulse would be called a *selective inversion* pulse. Just as for selective excitation, all we need to do is to place the transmitter on resonance with the line we wish to invert, and reduce the RF field strength until the other resonances in the spectrum are not affected significantly. Of course we will need to lengthen the pulse so that the on-resonance flip angle is maintained.

Figure 4.30 on the next page shows the z-magnetization generated as a function of offset for such an inversion pulse. Compared with the behaviour of a 90° excitation pulse (Fig. 4.28 on page 68), we see that the range over which there is significant inversion is somewhat smaller and that the off-resonance oscillations are smaller in amplitude.

This observation has two consequences: one 'good' and one 'bad'. The good consequence is that a selective 180° pulse is, for a given field strength, more selective than a corresponding 90° pulse. In particular, the weaker off-resonance wiggles are a useful feature. The bad consequence is that, when it comes to hard 180° pulses, the range of offsets over which there is anything like complete inversion is much more limited than the range of offsets over which a 90° pulse gives significant excitation, something which can be seen by comparing Fig. 4.30 with Fig. 4.28 on page 68. Thus, 180° pulses are often the source of problems in spectra with large offset ranges.

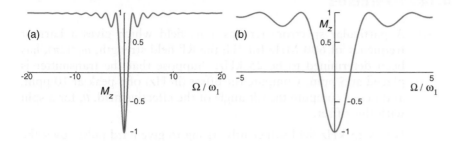

Fig. 4.30 Plots of the z-magnetization produced by a pulse applied to equilibrium magnetization as a function of the offset; the flip angle on resonance has been set to 180°. Plot (b) covers a narrower range of offsets than plot (a). Comparing these plots with those of Fig. 4.28 on page 68, we see that both show characteristic wiggles as we go off resonance; however, the range of offsets over which inversion is 90% complete is much less than that over which 90% excitation is achieved.

4.12 Moving on

Several times now we have referred to the fact that a Fourier transform can be used to turn the measured FID into a spectrum. The next chapter explores this process in more detail and, along the way, introduces some of the useful manipulations which we can subject the FID to before Fourier transforming it. We will also explore in more detail how phase errors manifest themselves in spectra.

4.13 Further reading

The origins of microscopic and macroscopic nuclear magnetism:

Chapter 2 from M. H. Levitt, *Spin Dynamics* (2nd edition, John Wiley & Sons, Ltd, 2008).

The vector model, spin echoes and selective pulses:

Chapters 2, 4 and 5 from R. Freeman, *Spin Choreography* (Spektrum, 1997).

Shaped selective pulses:

R. Freeman, *Progress in Nuclear Magnetic Resonance Spectroscopy*, **32**, 59–106 (1998).

4.14   Exercises

4.1   A particular spectrometer has a B_0 field which gives a Larmor frequency of 600 MHz for ^{1}H; the RF field strength, $\omega_1/(2\pi)$, has been determined to be 25 kHz. Suppose that the transmitter is placed at 5 ppm. Compute the offset (in Hz) of a peak at 10 ppm, and hence compute the tilt angle of the effective field, θ, for a spin with this offset.

Is this 25 kHz field sufficiently strong to give hard pulses over the full range of ^{1}H chemical shifts?

Repeat the calculation for a Larmor frequency of 900 MHz and comment on your result.

4.2   Explain why it is that the maximum signal in a pulse–acquire experiment is seen when the flip angle of the pulse is 90°. What would you expect to see in such an experiment if the flip angle of the pulse were set to: (a) 180°; (b) 270°?

4.3   In an experiment to determine the pulse length, an operator observed a positive signal for pulse widths of 5 and 10 μs; as the pulse was lengthened further the intensity decreased going through a null at 20.5 μs and then becomes negative.

Explain what is happening in this experiment and use the data to determine the RF field strength in Hz and in rad s^{-1}, and the length of a 90° pulse.

A further null in the signal was seen at 41.0 μs; to what do you attribute this?

4.4   Using an approach similar to that of Fig. 4.24 on page 64, show that a 180° pulse about the y-axis rotates vectors to mirror image positions with respect to the yz-plane.

[Hint: as in this figure, resolve the vector into its x- and y-components; however, for the case of a 180°(y) pulse, it is the y-component which is unaffected and the x-component which is inverted by the pulse.]

4.5   Use vector diagrams, similar to those of Fig. 4.25 on page 65, to show what happens during the spin echo sequence

$$90°(x) - \tau - 180°(y) - \tau -$$

Also, draw up a phase evolution diagram appropriate for this sequence. In what way does the result differ from a spin echo in which the 180° pulse is about the x-axis?

Without drawing up further detailed diagrams, state what the effect of applying the refocusing pulse about the $-x$-axis would be.

4.6 The gyromagnetic ratio of ^{31}P is 1.08×10^8 rad s^{-1} T^{-1}. This nucleus shows a wide range of shifts, covering some 700 ppm.

Estimate the minimum 90° pulse length you would need to excite peaks over this complete range to within 90% of their theoretical maximum for a spectrometer with a B_0 field strength of 9.4 T.

[Hint: see section 4.11 on page 67.]

4.7 A spectrometer operates at a Larmor frequency of 400 MHz for ^{1}H and hence 100 MHz for ^{13}C. Suppose that a 90° pulse of length 10 μs is applied to the protons. Does this have a significant effect on the ^{13}C nuclei? Explain your answer carefully.

4.8 From the plots of Fig. 4.28 on page 68 we see that there are some offsets at which the transverse magnetization goes to zero. Recall that during the pulse the magnetization starts on $+z$ and is rotated about the *effective field*; the nulls in the excitation are when the magnetization has been rotated all the way back to $+z$ i.e. when the rotation about the effective field is through 2π radians, or some multiple of this angle. We can work out the offset at which this occurs in the following way.

The effective field is given by

$$\omega_{\text{eff}} = \sqrt{\omega_1^2 + \Omega^2}.$$

To simplify things, we will express the offset as a multiple κ of the RF field strength:

$$\Omega = \kappa \omega_1.$$

Show that, using this expression for Ω, ω_{eff} is given by:

$$\omega_{\text{eff}} = \omega_1 \sqrt{1 + \kappa^2}.$$

The null condition is when the rotation is 2π:

$$\text{null condition} \qquad \omega_{\text{eff}} t_{\text{p}} = 2\pi,$$

where t_{p} is the length of the pulse. The final thing to note is that the on-resonance flip angle is $\pi/2$; this means that

$$\text{on resonance} \qquad \omega_1 t_{\text{p}} = \pi/2.$$

Combine the last three equations to show that the null occurs when

$$4 = \sqrt{1 + \kappa^2} \qquad \text{i.e. } \kappa = \sqrt{15}.$$

The predicted null is at $\kappa = \Omega/\omega_1 = \sqrt{15}$ i.e. $\Omega = \sqrt{15}\,\omega_1$. Does this agree with Fig. 4.28 on page 68?

Predict the value of κ at which the next null will occur.

Further nulls continue to occur at larger offsets; show that at large offsets, which means $\kappa \gg 1$, the nulls occur at $\kappa = 4n$, where n is an integer.

[Hint: the nulls occur at rotation angles of $2n\pi$; for $\kappa \gg 1$, $\sqrt{1 + \kappa^2}$ can be approximated.]

4.9   When calibrating a pulse by looking for the null produced by a 180° rotation, why is it important to choose a line which is close to the transmitter frequency (i.e. one with a small offset)?

4.10   Use vector diagrams to predict the outcome of the sequence:

$$90°(x) - \tau - 90°(x)$$

when applied to equilibrium magnetization. In your answer, explain how the x-, y- and z-magnetizations depend on the delay τ and the offset Ω.

For a fixed delay, sketch a graph of the x- and y-magnetization as a function of the offset. At what values of $\Omega\tau$ do any nulls occur?

4.11   Consider the spin echo sequence to which a 90° pulse has been added at the end:

$$90°(x) - \tau - 180°(x) - \tau - 90°(\phi).$$

The axis about which the pulse is applied is given in brackets after the flip angle. Explain in what way the outcome is different depending on whether the phase ϕ of the pulse is chosen to be x, y, $-x$ or $-y$.

4.12   The so-called 1–$\bar{1}$ sequence is:

$$90°(x) - \tau - 90°(-x)$$

For a peak which is on resonance the sequence does not excite any observable magnetization. However, for a peak with an offset such that $\Omega\tau = \pi/2$ the sequence results in all of the equilibrium magnetization appearing along the x-axis. Further, if the delay is such that $\Omega\tau = \pi$ no transverse magnetization is excited.

Use vector diagrams to explain these observations, and make a sketch graph of the amount of transverse magnetization generated as a function of the offset for a fixed delay τ.

The sequence has been used for suppressing strong solvent signals which might otherwise overwhelm the spectrum. The solvent is placed on resonance, and so is not excited; τ is chosen so that the peaks of interest are excited. How does one go about choosing the value for τ?

4.13   The so-called 1–1 sequence is:

$$90°(x) - \tau - 90°(y).$$

Describe the excitation that this sequence produces as a function of offset. How could it be used for observing spectra in the presence of strong solvent signals?

4.14   If there are two peaks in the spectrum, we can work out the effect of a pulse sequence by treating the two lines separately. *There is a separate magnetization vector for each line.*

Suppose that a spectrum has two lines, A and B. Suppose also that line A is on resonance with the transmitter and that the offset of line B is 100 Hz.

Starting from equilibrium, we apply the following pulse sequence:

$$90°(x) - \tau - 90°(x)$$

Using the vector model, work out what happens to the magnetization from line A.

Assuming that the delay τ is set to 5 ms (1 ms = 10^{-3} s), work out what happens to the magnetization vector from line B.

Suppose now that we move the transmitter so that it is exactly between the two lines. The offset of line A is now +50 Hz and of line B is −50 Hz.

Starting from equilibrium, we apply the following pulse sequence:

$$90°(x) - \tau$$

Assuming that the delay τ is set to 5 ms, work out what happens to the two magnetization vectors.

5

Fourier transformation and data processing

In pulsed NMR spectroscopy we measure what is called a *time-domain* signal – that is, the measured signal is a function of time. This is in contrast to most other kinds of spectroscopy, in which measurements of absorption or emission are made directly as a function of frequency to give the usual spectrum or *frequency-domain* signal.

Except in the simplest cases, the time-domain representation is virtually uninterpretable by eye, so it is vital to have a way of generating the frequency-domain representation (i.e. a normal spectrum) from the time-domain signal. This is where the Fourier transform comes in: it is a relatively simple mathematical procedure, well suited to implementation on a computer, which generates the spectrum from the FID. The process is illustrated in Fig. 5.1.

In this chapter we will first explore how the Fourier transform works, and then look in more detail at the relationship between the time and frequency domains. This will naturally bring us to the important topics of lineshapes and the phases of lines in NMR spectra. The chapter closes with a description of the manipulations which can be performed on the FID, prior to Fourier transformation, in order to improve the signal-to-noise ratio or the resolution of the spectrum.

Before proceeding, there is one point we need to clarify about the way in which the FID is actually measured. For a computer to be able to manipulate the time-domain signal (the FID), it has to be converted into a digital form. This *digitization* process involves measuring the size of the signal, converting this value to a number and then storing it away in computer memory.

In Chapter 13 we will look at this digitization process in more detail, but suffice it to say that digitization involves sampling the signal at regular intervals chosen so as to give a good approximation to the smoothly varying signal. The digital representation of the FID thus consists of several data points, evenly spaced in time, as shown in Fig. 5.2. In this chapter we will, for convenience, represent the FID as if it were a continuous smooth

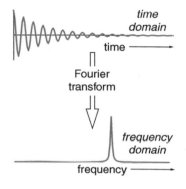

Fig. 5.1 The Fourier transform is a mathematical process which turns a time-domain signal, the FID, into a frequency-domain signal, the spectrum.

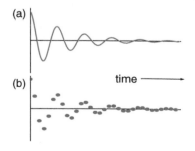

Fig. 5.2 The amplitude of the FID, as shown in (a), varies smoothly as a function of time. In order to be able to manipulate this time-domain signal in a computer, the signal has to be digitized at regular intervals, as shown in (b).

Understanding NMR Spectroscopy, Second Edition James Keeler
© 2010 John Wiley & Sons, Ltd

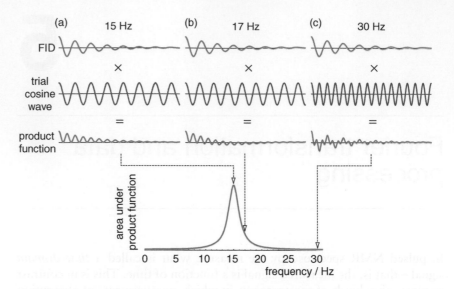

Fig. 5.3 Illustration of how the Fourier transform works. The FID, shown along the top, is multiplied by trial cosine functions of known frequency to give a product function. The area under this product function corresponds to the intensity in the spectrum at the frequency of the cosine wave. Three cases are shown. In (a) the trial cosine is at 15 Hz which matches the oscillation in the FID; as a result the product function is always positive and the area under it is a maximum. In (b) the trial frequency is 17 Hz; now the product function has positive and negative excursions, but on account of the decay of the FID, the area under the trial function is positive, although smaller than the area in case (a). The intensity of the spectrum at 17 Hz is therefore less than at 15 Hz. Finally, in (c) the trial frequency is 30 Hz; the product function oscillates quite rapidly about zero so that the area under it is essentially zero. As a result, the intensity in the spectrum at this frequency is zero. The spectrum is generated by plotting the area under the product function against the frequency of the corresponding trial cosine wave.

function, rather than being sampled at discrete intervals. Most of the time this distinction makes no practical difference, but when it does we will point this out.

5.1   How the Fourier transform works

The best way to see how the Fourier transform works is to start out by considering a particular example. Let us imagine that we have a single resonance at an offset of 15 Hz; the measured FID will thus consist of an oscillation at 15 Hz which decays away over time, as shown across the top of Fig. 5.3. We have assumed that the oscillation is simply a cosine wave – later in this chapter we will discuss other possibilities.

To transform this FID into a conventional spectrum we need to find out the frequency of the oscillation, as this gives the position of the peak in the spectrum. Of course, in this simple case we could determine the frequency from the period of the waveform, but such an approach would

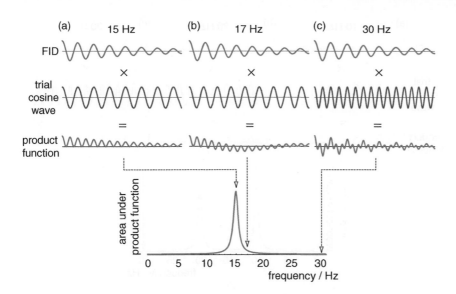

Fig. 5.4 This diagram is essentially the same as Fig. 5.3 on the facing page with the exception that the FID has a slower decay. As a result, for the case where the trial cosine wave is at 17 Hz, the area under the product function is smaller, and so is the corresponding intensity in the spectrum. The overall result is that the line is narrower.

clearly be inapplicable to more complex waveforms. The Fourier transform is a general solution to this problem; it involves comparing the FID with a series of cosine waves in the following way.

Suppose we take our FID and multiply it by a trial cosine wave with a frequency of 15 Hz. As both the FID and the cosine wave have the same frequency, at any time both functions have the *same* sign and so the product of the two functions is therefore *always positive*, as shown in Fig. 5.3 (a) on the facing page. The area under this product function represents the 'amount' of oscillation at 15 Hz there is in the FID, in other words the area represents the intensity of the spectrum at 15 Hz.

Now suppose we make the frequency of the trial cosine wave 30 Hz. Again this is multiplied by the FID to give a product function, as shown in Fig. 5.3 (c). This time, the FID and the trial cosine do not always have the same sign, and so the product function oscillates about zero. If we recall that an area enclosed by the curve above the horizontal axis is counted as positive and that below is counted as negative, it is clear that the area under the product function is very close to zero. So we say that there is no oscillation at 30 Hz present in the FID i.e. the intensity in the spectrum at 30 Hz is zero.

Finally consider the case, shown in Fig. 5.3 (b), where the frequency of the trial cosine function is 17 Hz. The positive and negative excursions of this function and the FID almost match, so to start with the product function is positive. However, eventually they do get out of step and the product function goes negative. The important point is that because of the decay on the FID the area under the product function is not zero – the positive part at the beginning outweighs the negative part at the end. However, the area under the product function is not as great as when the trial cosine wave is at 15 Hz, so the intensity of the spectrum at 17 Hz is significant, but not as great as that at 15 Hz.

The whole process is repeated with more trial cosine waves covering a range of frequencies. The spectrum is then constructed by plotting the area under the product function against the frequency of the corresponding trial

Fig. 5.5 This diagram is essentially the same as Fig. 5.3 on page 78 with the exception that the FID is appropriate for a situation where there are two resonances, with offsets of 10 Hz and 20 Hz. Multiplying by trial cosine waves at 10 and 20 Hz, (a) and (b), gives product functions which clearly have non-zero areas; these areas correspond to intensity in the spectrum at 10 and 20 Hz. However, multiplying by a cosine wave at 30 Hz, (c), gives a product function with zero area; this corresponds to their being no intensity in the spectrum at 30 Hz.

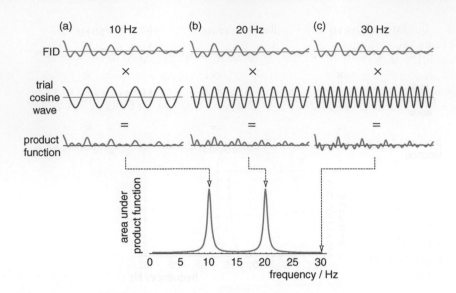

cosine wave, as shown at the bottom of Fig. 5.3. As expected, the spectrum shows a peak centred at 15 Hz, but the peak has a certain width, which is a result of the decay of the FID. We saw above that, when the frequency of the trial cosine function is close to 15 Hz, the area under the product function is still significant, and so there is intensity in the spectrum away from the centre of the peak.

If the decay of the FID is slower, the corresponding peak in the spectrum is narrower; this is illustrated in Fig. 5.4 on the previous page which should be compared with Fig. 5.3 on page 78. In the case of the trial cosine wave at 17 Hz, the negative part of the product function is more significant on account of the slower decay of the FID. As a result, the area under the product function, and hence the intensity in the spectrum at 17 Hz, is reduced. This corresponds to a narrower line.

What happens if the FID is from more than one resonance? In such cases, each gives a separate contribution to the FID so that the time-domain signal is, for example, a sum of decaying cosine waves. A typical situation is illustrated in Fig. 5.5 where two resonances, at 10 Hz and 20 Hz, give rise to a FID with a more complex form. Nevertheless, multiplication by trial cosine waves at either 10 Hz or 20 Hz gives product functions which clearly have finite areas under them, and so there is intensity in the spectrum at these two frequencies. However, multiplication by a trial cosine at 30 Hz gives a product function which oscillates back and forth about zero and so has zero area. There is thus no intensity in the spectrum at 30 Hz.

No matter how complex the original waveform of the FID, this process of multiplying by trial cosine waves will always pick out the intensity at the corresponding frequency. This is the essence of the Fourier transform.

5.1.1   Mathematical formulation of the Fourier transform

Written in words, the procedure we have been describing is

$$\text{intensity at } f \text{ Hz} = \text{area under } [\text{FID} \times \cos{(2\pi f t)}].$$

The quantity in the square bracket is the product function described above. Let us call the intensity at frequency f, $S_{spectrum}(f)$, and the amplitude of the FID at time t is $S_{FID}(t)$: the (f) and (t) remind us that these are functions of frequency and time respectively. Next, recall that in calculus, the area under a function is the same as its *integral*. So, our expression in words can be written more formally as

$$S_{spectrum}(f) = \int_0^{+\infty} S_{FID}(t) \cos(2\pi f t)\, dt. \qquad (5.1)$$

The integral is over time (as shown by dt), which is the horizontal axes of the plots of the FID.

The upper limit of the integral is infinite time. However in practice, as the FID eventually decays to zero, we only need to compute the integral from zero up to some time at which the FID is negligible. Equation 5.1 is one definition of the Fourier transform which allows us to compute the spectrum, $S_{spectrum}(f)$, from the FID, $S_{FID}(t)$. All we need to do is compute the integral for a range of frequencies, thus building up the spectrum.

On the spectrometer the FID is represented by a set of points which have been sampled at regular intervals. To Fourier transform such data, rather than computing the integral we multiply each point in the FID by the value of the trial cosine wave computed for the time corresponding to that point. This gives the product function as a series of data points, and these are then summed to give the intensity in the spectrum at the frequency of the trial cosine wave.

Expressed mathematically the integral of Eq. 5.1 becomes the sum

$$S_{spectrum}(f) = \sum_{i=1}^{i=N} S_{FID}(t_i) \cos(2\pi f t_i),$$

where t_i is the time corresponding to the ith data point and $S_{FID}(t_i)$ is the value of the FID at this time; N is the total number of data points.

5.1.2   Why the Fourier transform works

We have, with the aid of diagrams and words, attempted to explain how the Fourier transform picks out from a complex waveform the component oscillating at a certain frequency. The question is, why does this process work?

The Fourier transform uses two key ideas. The first is that any time-domain function can be represented by the sum of cosine waves of different frequencies and amplitudes. How many cosine waves we need depends on the complexity of the time-domain function. The spectrum, or frequency-domain function, is a plot of the amplitude of these cosine waves as a function of their frequency.

The second point is that these cosine waves are *orthogonal* to one another. What this means is that the integral, taken between $t = 0$ and $t = \infty$, of the product of *any* two cosine waves of different frequencies is zero. It is for this reason that the integral of Eq. 5.1 is able to pick out the contribution at just one frequency. A good analogy here is to think about the x-, y- and z-coordinates of a point in space: because the three axes are orthogonal, the values of the three coordinates are independent of one another. By analogy, the contribution to the spectrum at one frequency is independent of that at another because the cosine functions at two different frequencies are orthogonal.

5.2   Representing the FID

In section 4.7 on page 60 we showed that, in the basic pulse–acquire experiment, the evolution of the x- and y-components of the magnetization generated by a $90°(x)$ pulse can be written

$$M_y = -M_0 \cos \Omega t \qquad M_x = M_0 \sin \Omega t,$$

where M_0 is the equilibrium magnetization and Ω is the offset (in rad s^{-1}). As described in section 4.6 on page 60, the assumption is made that we are detecting the signal in a rotating frame at the transmitter frequency.

If our pulse–acquire experiment had used a $90°$ pulse about y, rather than about x, the equilibrium magnetization would have been rotated onto the x-axis. As shown in Fig. 5.6, the evolution of the x- and y-components are given by

$$M_x = M_0 \cos \Omega t \qquad M_y = M_0 \sin \Omega t.$$

It will suit us for our present purposes to use these forms of M_x and M_y; later on, we will see that the choice we make is in any case arbitrary.

The precession of the magnetization gives rise to a current in the RF coil which, after various manipulations by the RF electronics in the spectrometer, results in a signal voltage which can be digitized and the result stored away in computer memory. The spectrometer is capable of detecting simultaneously both the x- and y-components of the magnetization, each giving rise to separate signals which we will denote S_x and S_y.

These signals are proportional to M_x and M_y, but their absolute size is not generally of any interest so we will simply write the constant of proportion as S_0, the maximum value:

$$S_x = S_0 \cos \Omega t \qquad S_y = S_0 \sin \Omega t.$$

Finally, we need to recognize that the magnetization, and hence the signal, will decay over time. We model this by assuming that the signal decays exponentially:

$$S_x = S_0 \cos \Omega t \exp\left(\frac{-t}{T_2}\right) \qquad S_y = S_0 \sin \Omega t \exp\left(\frac{-t}{T_2}\right). \qquad (5.2)$$

T_2 is a time constant which characterizes the decay; the shorter T_2, the more rapid the decay. As we will see in Chapter 9, this decay is usually due to relaxation processes.

Rather than dealing with the x- and y-components separately, it is convenient to bring them together as a complex signal, with the x-component becoming the real part and the y-component the imaginary part. The complex signal is written $S(t)$ to remind us that it is a function of time.

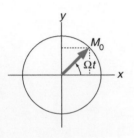

Fig. 5.6 A $90°(y)$ pulse rotates the equilibrium magnetization onto the x-axis; from there it precesses in the transverse plane, creating x- and y-components $M_0 \cos \Omega t$ and $M_0 \sin \Omega t$. The diagram assumes that the offset Ω is positive.

$$
\begin{aligned}
S(t) &= S_x + \mathrm{i}\, S_y \\[4pt]
&= S_0 \cos \Omega t\, \exp\!\left(\frac{-t}{T_2}\right) + \mathrm{i}\left[S_0 \sin \Omega t\, \exp\!\left(\frac{-t}{T_2}\right)\right] \\[4pt]
&= S_0 \left(\cos \Omega t + \mathrm{i}\sin \Omega t\right) \exp\!\left(\frac{-t}{T_2}\right) \\[4pt]
&= S_0 \exp\left(\mathrm{i}\Omega t\right) \exp\!\left(\frac{-t}{T_2}\right).
\end{aligned}
\qquad (5.3)
$$

To go from the third to the fourth line we have used the identity $\cos \theta + \mathrm{i}\sin \theta \equiv \exp(\mathrm{i}\theta)$.

It is convenient to think of this complex signal as being represented by a vector rotating at frequency Ω, as shown in Fig. 5.7. The vector is of length $S_0 \exp(-t/T_2)$ (i.e. it shrinks over time), and the x- and y-components (the real and imaginary parts) are given by Eq. 5.2 on the preceding page.

The division by T_2 in Eq. 5.3 is sometimes rather inconvenient, so we define

$$
R = \frac{1}{T_2}, \qquad R \text{ in units of } \mathrm{s}^{-1} \text{ or Hz}
$$

enabling us to write the time-domain signal as

$$
S(t) = S_0 \exp\left(\mathrm{i}\Omega t\right) \exp\left(-Rt\right).
\qquad (5.4)
$$

R is the (first-order) rate constant for the decay: as R increases the decay becomes more rapid and so the corresponding line becomes broader.

If there are several resonances present, then the complex time-domain signal is a sum of terms such of the type given in Eq. 5.3

$$
S(t) = S_0^{(1)} \exp\left(\mathrm{i}\Omega_1 t\right) \exp\!\left(\frac{-t}{T_2^{(1)}}\right) + S_0^{(2)} \exp\left(\mathrm{i}\Omega_2 t\right) \exp\!\left(\frac{-t}{T_2^{(2)}}\right) + \cdots
$$

where each resonance i has its own intensity $S_0^{(i)}$, frequency Ω_i, and decay constant $T_2^{(i)}$.

5.3 Lineshapes and phase

As has already been outlined, Fourier transformation of the time-domain signal, the FID, gives the required spectrum. In this section we will examine the relationship between the time-domain function and the spectrum in more detail. This will introduce the important topics of the lineshape and phase of the peaks in a spectrum.

5.3.1 Absorption and dispersion lineshapes

In section 5.1 on page 78 the Fourier transform was introduced by thinking about a purely real time-domain function and the result of multiplying it by cosine waves. There is an analogous process for complex time-domain functions where, instead of multiplying by $\cos(2\pi f t)$ and then integrating,

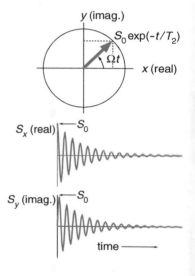

Fig. 5.7 The complex time-domain signal defined in Eq. 5.3 can be thought of as arising from a rotating vector of length $S_0 \exp(-t/T_2)$; the x- and y-components of the vector are given by Eq. 5.2 on the preceding page.

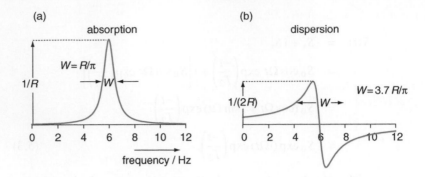

Fig. 5.8 Fourier transformation of an exponentially decaying time-domain signal (Eq. 5.4 on the preceding page) gives a complex frequency-domain signal or spectrum whose real and imaginary parts have the absorption and dispersion mode Lorentzian lineshapes, shown in (a) and (b) respectively. Here, the peak is centred at 6 Hz and the decay constant R has been chosen such that the width (at half height) of the absorption mode line is 1 Hz; note that the heights of both peaks, as well as their widths, are also functions of the decay constant. The dispersion lineshape is considerably broader than the absorption lineshape; it also has both positive and negative parts, which is an undesirable feature. The width W, at half height, of the absorption mode is R/π Hz; the dispersion mode is almost four times wider.

we multiply by the complex exponential $\exp(i2\pi ft)$. As was explained in section 2.5.5 on page 20, $\exp(i2\pi ft)$ is the way of representing an oscillation using complex numbers.

Not surprisingly, if we start with a complex time-domain signal, Fourier transformation gives a complex frequency-domain signal or spectrum. Normally, the software on the spectrometer only displays the real part of this complex spectrum, but it is important to realize that the imaginary part exists, even if it is not displayed.

Fourier transformation of the complex time-domain signal of Eq. 5.4 on the preceding page gives a complex frequency-domain signal, or spectrum, $S(\omega)$:

$$S(t) \quad \overset{\text{FT}}{\to} \quad S(\omega)$$

$$S_0 \exp(i\Omega t)\exp(-Rt) \quad \overset{\text{FT}}{\to} \quad \underbrace{\frac{S_0 R}{R^2 + (\omega - \Omega)^2}}_{\text{real}} + i\underbrace{\frac{-S_0(\omega - \Omega)}{R^2 + (\omega - \Omega)^2}}_{\text{imaginary}}. \quad (5.5)$$

Note that, just as in the FID the running variable is time, in the spectrum the variable is the frequency, ω, here chosen to be in rad s^{-1}.

The real part of the spectrum is a peak with the *absorption mode Lorentzian lineshape*, whereas the imaginary part has the *dispersion mode Lorentzian lineshape*; the lineshapes are illustrated in Fig. 5.8 for the case $S_0 = 1$. You will recognize the real part as similar to the lineshape introduced in section 2.2 on page 9. Note that these particular lineshapes are a result of the assumed exponential decay of the time-domain function.

The factor of S_0 in Eq. 5.5 is just an overall scaling, so it is usual to define the Lorentzian absorption and dispersion mode lineshape functions,

$A(\omega)$ and $D(\omega)$, without this factor:

$$A(\omega) = \frac{R}{R^2 + (\omega - \Omega)^2} \qquad D(\omega) = \frac{-(\omega - \Omega)}{R^2 + (\omega - \Omega)^2}. \qquad (5.6)$$

The absorption lineshape is always positive and is centred at Ω. If we let $\omega = \Omega$ it is easy to see that the height of the peak is $1/R$:

$$\begin{aligned} S(\Omega) &= \frac{R}{R^2 + (\Omega - \Omega)^2} \\ &= \frac{R}{R^2} \\ &= 1/R. \end{aligned}$$

Using these expressions, we can work out the frequency, $\omega_{1/2}$, at which the height of the line has fallen to half this maximum. For example, recalling that the peak height of the absorption lineshape is $1/R$, $\omega_{1/2}$ is found by solving the following equation:

$$\frac{1}{2} \times \frac{1}{R} = \frac{R}{R^2 + (\Omega - \omega_{1/2})^2}.$$

The values for $\omega_{1/2}$ turn out to be $(\Omega + R)$ and $(\Omega - R)$; there are of course two values as the lineshape is symmetrical about frequency Ω. Thus, the width at half height of the absorption mode lineshape is simply $2R$ rad s^{-1} or R/π Hz. These features of the absorption Lorentzian are shown in Fig. 5.8 (a) on the facing page.

These results are exactly what we expected. The larger the decay constant R, the faster the FID decays and hence the broader the line in the spectrum. Also, as R increases the peak height decreases, so as to keep the integral of the line constant (section 2.2 on page 9).

The dispersion lineshape, shown in Fig. 5.8 (b), is not one we would choose for high-resolution spectroscopy. It is broader and half the height of the absorption mode, but the worst feature is that it has both positive and negative parts. A crowded spectrum in which the peaks have the dispersion mode lineshape would very rapidly become quite uninterpretable.

So far we have written the lineshapes in angular frequency units, rad s^{-1}; it is sometimes convenient to rewrite the functions given in Eq. 5.6 in terms of Hz. The results, along with their peak heights and widths are summarized in Table 5.1 on the next page.

Using the definitions of the lineshape functions $A(\omega)$ and $D(\omega)$ from Eq. 5.6, we can write the result of the Fourier transformation of Eq. 5.5 on the preceding page more compactly as

$$S(t) \overset{\text{FT}}{\to} S(\omega)$$

$$S_0 \exp(i\Omega t) \exp(-Rt) \overset{\text{FT}}{\to} S_0 [A(\omega) + iD(\omega)]. \qquad (5.7)$$

5.3.2 Phase

We have already seen that we can influence the form of the x- and y-components of the magnetization by altering the phase of the RF pulse.

Table 5.1 Definitions of the absorption and dispersion mode Lorentzian lineshape functions, A and D, along with their peak heights and widths, given in both rad s^{-1} and Hz. The peaks are centred at Ω rad s^{-1} or F_0 Hz.

	absorption	dispersion
lineshape (rad s^{-1})	$A(\omega) = \dfrac{R}{R^2 + (\omega - \Omega)^2}$	$D(\omega) = \dfrac{-(\omega - \Omega)}{R^2 + (\omega - \Omega)^2}$
lineshape (Hz)	$A(f) = \dfrac{R}{R^2 + 4\pi^2(f - F_0)^2}$	$D(f) = \dfrac{-2\pi(f - F_0)}{R^2 + 4\pi^2(f - F_0)^2}$
peak height	$1/R$	$1/(2R)$
width (rad s^{-1})	$2R$	$2(2 + \sqrt{3})R \approx 7.5R$
width (Hz)	$R/\pi \approx 0.32R$	$(2 + \sqrt{3})R/\pi \approx 1.2R$

What is more, although the receiver in the spectrometer produces two signals which are labelled as deriving from the x- and y-components of the magnetization, due to the way the RF electronics works there is no guarantee that these outputs correspond to the magnetizations measured in the same axis system used to define the phase of the RF pulses.

As a result, the time-domain signal measured by the spectrometer has an essentially arbitrary (and usually unknown) phase ϕ associated with it. We saw in section 2.5.5 on page 20 that mathematically such a phase corresponds to multiplication by $\exp(i\phi)$, so the time-domain signal is

$$S(t) = S_0 \exp(i\Omega t) \exp(-Rt) \times \exp(i\phi).$$

One of the properties of the Fourier transform is that if we multiply the time domain by a constant, the frequency domain is multiplied by the same constant – in other words, the constant propagates through harmlessly. So, following Eq. 5.7 on the previous page, we can just write down the Fourier transform of the time-domain signal with the phase factor as

$$S_0 \exp(i\Omega t) \exp(-Rt) \times \exp(i\phi) \xrightarrow{\text{FT}} S_0 \left[A(\omega) + iD(\omega) \right] \times \exp(i\phi). \quad (5.8)$$

Remember that we usually display just the real part of the frequency domain. To find out what this is, we need to replace $\exp(i\phi)$ with $(\cos\phi + i\sin\phi)$ and then multiply out the brackets:

$$
\begin{aligned}
S_0 \left[A(\omega) + iD(\omega) \right] \exp(i\phi) &= S_0 \left[A(\omega) + iD(\omega) \right] \left[\cos\phi + i\sin\phi \right] \\
&= \underbrace{S_0 \left[\cos\phi\, A(\omega) - \sin\phi\, D(\omega) \right]}_{\text{real}} \\
&+ \underbrace{i S_0 \left[\cos\phi\, D(\omega) + \sin\phi\, A(\omega) \right]}_{\text{imaginary}}. \quad (5.9)
\end{aligned}
$$

What this says is that in general the real part of the spectrum contains a *mixture* of the absorption and dispersion lineshapes; likewise, the imaginary part is also a mixture. The relative contributions of absorption and dispersion depend on the phase ϕ.

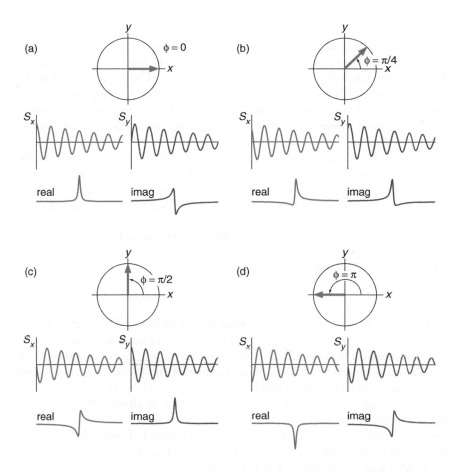

Fig. 5.9 Depiction of the effect of a phase shift on the spectrum. In each case a vector diagram, like that of Fig. 5.7 on page 83, shows how the position of the signal at time zero is specified by a phase angle ϕ measured anti-clockwise from the x-axis. The corresponding x- and y-components of the time-domain signal are shown, along with the real and imaginary parts of the spectrum. In (a) $\phi = 0$ and, as expected, the real part contains the absorption mode lineshape. In (b) a phase shift of $\phi = \pi/4$ results in both the real and imaginary parts of the spectrum being a mixture of absorption and dispersion mode lineshapes. Two special cases are also shown: in (c) a phase shift of $\pi/2$ moves the absorption mode to the imaginary part of the spectrum; in (d) a phase shift of π simply inverts the spectrum when compared with (a).

Figure 5.9 illustrates what the real and imaginary parts of the spectrum will look like for various values of the phase. In (a) the phase is zero, so the real part is the absorption mode, as expected. In (b) the phase is $\pi/4$ radians or 45°; the vector diagram shows where the signal starts at time zero. Note that the x- and y-components of the time domain are no longer simple damped cosine and sine waves, but are phase shifted, which is the same thing as being shifted along the time axis. The real part of the spectrum is clearly neither pure absorption nor pure dispersion, but a mixture of the two.

The figure also illustrates some special cases. In (c) the phase is $\pi/2$ radians or 90°, so $\cos\phi = 0$ and $\sin\phi = 1$. From Eq. 5.9 on the preceding page it is clear that the real part is minus the dispersion lineshape and the imaginary part is the absorption lineshape, which is what is shown in the diagram.

Another special case, shown in (d), is when $\phi = \pi$ radians or 180°. Now $\cos\phi = -1$ and $\sin\phi = 0$, so the real part contains minus the absorption mode lineshape and the imaginary part minus the dispersion mode lineshape. In other words, a phase shift of π or 180° simply inverts the spectrum.

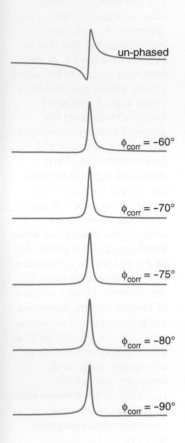

Fig. 5.10 Illustration of how a spectrum is phased by trial and error. The original spectrum is shown at the top, and beneath it are spectra resulting from various phase corrections. In fact, the appropriate phase correction is $-75°$, but at this scale a value within a few degrees of this also looks correctly phased. If we increased the vertical scale, we could inspect the foot of the line in more detail and hence adjust the phase more closely.

5.3.3 Phase correction

Usually, we want to make sure that the real part of the spectrum (the part which is displayed) has the absorption mode lineshape as this is the lineshape which will give us the narrowest peaks. The problem is that the spectrometer produces time-domain data with an arbitrary phase, and so the real part of the resulting spectrum will not have the pure absorption mode – such a spectrum is described as being 'not phased correctly' or 'un-phased'.

Luckily, the solution to the problem is rather simple. Remember that the spectrum is stored in computer memory, so it is easy to have the computer multiply the spectrum by $\exp(i\phi_{corr})$, where ϕ_{corr} is a phase correction. From Eq. 5.8 on page 86, the spectrum will now be described by the function

$$S_0\left[A(\omega) + iD(\omega)\right] \times \exp(i\phi) \times \exp(i\phi_{corr}).$$

Collapsing the two exponentials together gives

$$S_0\left[A(\omega) + iD(\omega)\right] \times \exp(i[\phi + \phi_{corr}]).$$

If we choose $\phi_{corr} = -\phi$, the exponential term will be $\exp(0)$ which is 1, and so the spectrum will become $A(\omega) + iD(\omega)$, which is what we want.

The question is, given that we do not know ϕ, how do we choose ϕ_{corr} to obtain the desired result? The answer is, by trial and error. The software on the spectrometer provides a mode in which ϕ_{corr} can be altered by a 'click and drag' operation, and as the phase correction is altered, the display of the real part of the spectrum is updated continuously. All we do is adjust the phase until the peaks appear to have the required absorption lineshape; at this point, it should be that $\phi_{corr} = -\phi$. The process is called *phasing the spectrum*, and it is illustrated in Fig. 5.10.

Admittedly, the process is somewhat subjective, and no two spectroscopists will come up with the same phase correction for a given spectrum – but the result is close enough for most purposes. Note that a single phase correction is applied to the whole spectrum: the correction is therefore described as a *frequency-independent* or *zero-order* phase correction.

The key message of this section is that, after recording the data, we can adjust the phase in the spectrum to obtain the lineshape we require.

5.3.4 Frequency-dependent phase errors

Sometimes, the phase is *not the same* for all the lines in the spectrum, but varies from one edge of the spectrum to the other. There are many reasons why this might be so, but perhaps the commonest is due to the RF pulse used to excite the spectrum.

Look back at Fig. 4.27 on page 67 and Fig. 4.28 on page 68. What these diagrams show is that, whereas an on-resonance $90°(x)$ pulse rotates the equilibrium magnetization onto the $-y$-axis, as we go off resonance the pulse produces more and more x-magnetization. The position of the magnetization just after the pulse thus varies with the offset; in other words, there is a phase error which increases with the offset. Such a phase error is said to be due to an 'off resonance' effect of the pulse.

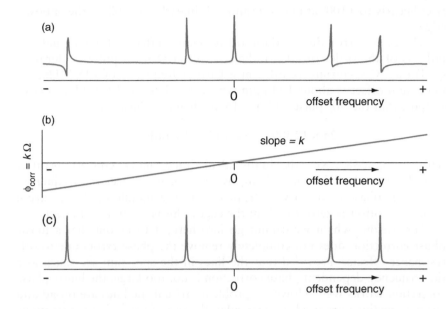

Fig. 5.11 Illustration of how a spectrum which has a frequency-dependent phase error is phased. In (a) the line which is on resonance (at zero frequency) is in pure absorption, but there is a phase error which increases with offset. Often, it is found that the phase error is proportional to the offset, in which case the whole spectrum can be phased by applying a phase correction which varies with the offset in a linear manner, as shown in (b); all we have to do is to choose, by trial and error, constant of proportion k which determines how quickly the phase correction increases with offset i.e. the slope of the line. With the correct choice, a spectrum with all of the lines in phase, shown in (c), is obtained.

When the resulting FID is Fourier transformed we will find that the phases of the lines are not all the same. If an on-resonance peak is phased to the absorption mode, we will find that other lines have dispersion contributions which increase as the offset increases. This is illustrated in Fig. 5.11 (a).

To a good approximation, the phase error due to this off-resonance effect is directly proportional to the offset from the transmitter, Ω. So, if we multiply the spectrum by a phase correction which is proportional to the offset:

$$\phi_{corr} = k \times \Omega,$$

we should be able to phase *all* of the lines in the spectrum by an appropriate choice of the constant k. As before, the choice of k is made by trial and error. Figure 5.11 (b) shows the phase correction and (c) shows the result of applying this, with the correct choice of k, to the spectrum; all of the lines are now in phase. This type of phase correction is often described as being *linear* or *first order*.

The usual convention is to express the frequency-dependent phase correction as the value that the phase takes at the extreme edges of the spectrum. So, for example, a frequency-dependent correction by $100°$ means that the phase correction is zero in the middle (at zero offset) and rises linearly to $+100°$ at one edge and falls linearly to $-100°$ at the opposite edge.

The phase error due to these off-resonance effects of a nominal $90°$ pulse is of the order of (Ωt_p) radians at offset Ω for a pulse of duration t_p. For a ^{13}C spectrum recorded at a Larmor frequency of 125 MHz, the maximum offset is about 100 ppm which translates to 12 500 Hz. Let us suppose that the $90°$ pulse width is 15 μs, then the phase error is

$$2\pi \times 12\,500 \times 15 \times 10^{-6} \approx 1.2 \text{ radians,}$$

which is about $68°$. Note that in the calculation we had to convert the offset from Hz to rad s^{-1} by multiplying by 2π. So, we expect the frequency-dependent phase error to vary from zero in the middle of the spectrum (where the offset is zero) to $68°$ at the edges; this is a significant effect.

For reasons which we cannot go into here, it turns out that a linear phase correction does not completely remove the phase errors due to off-resonance effects. Provided that the lines in the spectrum are sharp and the frequency-dependent phase correction is not too large, the linear phase correction works well. However, problems arise if the lines are broad and large corrections are used; in particular, the use of such phase corrections can give rise to significant distortions in the baseline of the spectrum.

Optional section ⇒

5.4   Manipulating the FID and the spectrum

On the spectrometer, the FID is stored on disc or in memory so it is no trouble at all for the software to perform mathematical manipulations on the time-domain signal prior to Fourier transformation. This section is concerned with some common manipulations of the FID which can be used to enhance the signal-to-noise ratio (SNR) of the spectrum, or to narrow the lines in the spectrum. Correctly applied, these manipulations can be of great utility.

5.4.1   Noise

Inevitably when we record a FID we record noise at the same time. Some of the noise is contributed by the amplifiers and other electronics in the RF receiver, but the major contributor is the thermal noise from the coil used to detect the signal. Reducing the noise contributed by these sources is largely a technical matter which will not concern us here.

It is worthwhile spending a moment describing the nature of the noise recorded by an NMR spectrometer. The first thing is that the noise signal wanders back and forth about zero in a random way (it is noise, after all) but that, averaged over time, the positive and negative excursions cancel one another out so that the mean (average) of the noise is zero.

The second characteristic of the noise is that it has an approximately Gaussian distribution. What this means is that the probability of the noise signal reaching a certain value S_{noise} is proportional to the Gaussian function

$$\exp\left(-\sigma S_{noise}^2\right),$$

where σ is a parameter which characterizes the distribution of the noise. Due to the fact that S_{noise} appears as the square and as the argument to an exponential, it follows that the probability drops off rather quickly as S_{noise} increases. So, although the noise will be 'spiky', large amplitude spikes are much less likely to be found than lower level spikes, whilst small excursions about zero are the most likely. Figure 5.12 shows a typical noise record.

Like the FID, the noise is recorded as a function of time, and then converted to the frequency domain by Fourier transformation. This process does not alter the character of the noise: it retains both the Gaussian distribution and the zero mean. The amplitude of the noise in the spectrum depends on both the amplitude of the noise in the time domain *and* the total time for which the noise was recorded. The longer the acquisition time, the more noise is recorded and this appears as a greater noise amplitude in the spectrum. This property of the noise in the two domains has some practical consequences which are explored in the following two sections.

5.4.2 Effect of the acquisition time

The free induction signal decays over time, but in contrast the noise just goes on and on. Therefore, if we carry on recording data for long after the FID has decayed, we will just measure noise and no signal. Not surprisingly, the resulting spectrum will therefore have a poor SNR.

Figure 5.13 illustrates how the SNR of the spectrum is affected by altering the acquisition time. In (a) data acquisition is carried on long after the NMR signal has decayed into the noise. Simply halving the acquisition time, as shown in (b), still allows us to record the NMR signal but greatly reduces the amount of noise which is recorded; the result is an improvement in the SNR. Shortening the acquisition time still further means that even less noise is recorded. The result, shown in (c), is a significant improvement in the SNR.

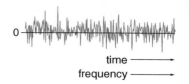

Fig. 5.12 A typical piece of noise, such as would be recorded by an NMR spectrometer. The noise is described as 'Gaussian with zero mean', which means that (1) the time average of the noise is zero, and (2) the probability of a certain amplitude occurring is proportional to a Gaussian function of the amplitude. This means that high amplitude noise spikes occur with lower probability (i.e. less often) than lower amplitude ones.

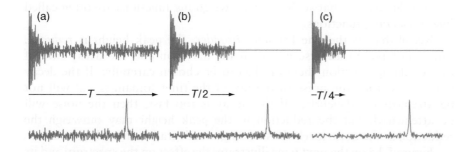

Fig. 5.13 Illustration of the effect of altering the acquisition time on the signal-to-noise ratio (SNR) in the spectrum; at the top are shown a series of FIDs while the corresponding spectra are shown beneath. In (a) the NMR signal has decayed into the noise within the first quarter of the acquisition time T, but the noise carries on unabated throughout the whole time. Shown in (b) is the effect of halving the acquisition time; the SNR of the spectrum improves as a consequence of the fact that we are acquiring less noise but the same amount of signal. In (c) we see that taking the first quarter of the data gives a more substantial improvement in the SNR of the spectrum.

Of course, we must not shorten the acquisition time to the point where we start to miss significant parts of the the NMR signal. Clearly the acquisition time needs to be chosen with care.

5.4.3   Sensitivity enhancement

A typical time-domain signal is strongest at the beginning but, due to the decay, falls away over time; however, the noise remains constant throughout. This gives us the idea that the early parts of the time domain are 'more important' as it is here where the signal is the strongest, whereas the latter parts are 'less important' as the signal has decayed leaving the noise dominant.

We can favour the early parts of the time domain by multiplying it by a function which starts with the value one and then steadily tails away to zero. The idea is that this function will attenuate the later parts of the time domain where the signal is weakest, but leave the early parts unaffected. If the decay rate of the function is chosen carefully, it is possible to improve the SNR of the corresponding spectrum. In the discussion which follows we will use the term 'time-domain signal' to mean the NMR signal plus the noise, whereas the NMR signal alone will be referred to as the FID.

A function used to multiply the time-domain signal is called a *weighting function*. The simplest example is an exponential:

$$W_{\text{LB}}(t) = \exp(-R_{\text{LB}}t), \tag{5.10}$$

where R_{LB} is a rate constant which determines how fast the weighting function decays; we are free to choose the value of R_{LB}. Figure 5.14 shows a sketch of the function.

Multiplication by this particular weighting function will make the envelope of the FID decay more rapidly, and hence the line in the corresponding spectrum will be broader than in the spectrum from the unweighted FID. It is for this reason that such decaying weighting functions are often called *line broadening* functions.

Recall that as the line becomes *broader*, the peak height is *reduced*. This will have an adverse effect on the SNR, and so the decay rate of the weighting function therefore has to be chosen carefully. If the decay is too slow, then the noise in the tail of the time-domain signal will not be attenuated sufficiently. If the decay is too fast, then the noise will be attenuated, but the reduction in the peak height may outweigh the reduction in the noise level, thus leading to a degradation in the SNR.

Figure 5.15 on the next page illustrates the effect on the spectrum and its SNR of different choices of this decay constant. The original time-domain signal, shown in (a), has a rather noisy tail which carries on long after the FID has disappeared into the noise and, as a result, the corresponding spectrum (f) is rather noisy. Multiplication of the time domain by the exponential weighting function (b) gives what is called the weighted time domain, (d). Clearly the noise in the tail has been reduced substantially, whereas the FID is little affected. The corresponding spectrum (g) shows a considerable improvement in SNR when compared with (f); note also the reduction in peak height.

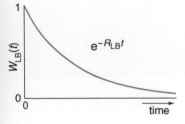

Fig. 5.14 A graph of the decaying exponential function $W_{\text{LB}}(t) = \exp(-R_{\text{LB}}t)$; the decay rate is set by the value of the parameter R_{LB}.

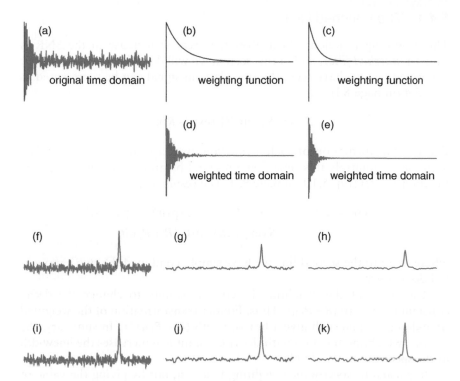

Fig. 5.15 Illustration of how multiplying a time-domain signal by a decaying exponential function (a weighting function) can improve the SNR of the corresponding spectrum. The original time-domain signal is shown in (a); Fourier transformation gives the spectrum (f). Multiplying the time domain by a weighting function (b) gives the weighted time-domain signal (d); (g) is the corresponding spectrum. Note the improvement in SNR of (g) when compared with (f). Multiplying (a) by the more rapidly decaying weighting function (c) gives (e); (h) is the corresponding spectrum, which, when compared with (g), shows the expected reduction in peak height caused by the more rapidly decaying weighting function. Spectra (f)–(h) are all plotted on the same vertical scale so that the decrease in peak height can be seen. The same spectra are plotted in (i)–(k), but this time they have been normalized so that the peak heights are all the same; this shows most clearly the improvement in the SNR and the increase in the linewidth.

If the weighting function decays faster, as shown in (c), the resulting weighted time domain (e) shows not only the reduction of the noise in the tail but also a more rapid decay of the NMR signal. The corresponding spectrum (h) shows a small improvement in SNR over (g), but the change is not as dramatic as that from (f) to (g). There is a clear reduction in the peak height of (h) when compared with (g), once more as a consequence of the line broadening.

The way in which the SNR and linewidths are affected by the choice of decay rate for the weighting function is best seen by comparing spectra (i)–(k); these are the same as (f)–(h), but have been plotted so that the peaks are all the same height. The increase in linewidth as we go from (i) to (j) to (k) is evident; it is also clear that despite the increase in linewidth, spectrum (k) has a slightly better SNR than (j).

5.4.4   The matched filter

The weighting function which gives the greatest increase in the SNR is called a *matched filter*, and in this section we will describe how such a filter is selected. As we have seen, the time-domain signal can be represented as (Eq. 5.4 on page 83)

$$S(t) = S_0 \exp(i\Omega t) \exp(-Rt).$$

Fourier transformation of such a function gives a peak of width R/π Hz. Suppose we multiply this time-domain signal by an exponential weighting function $W_{LB}(t)$ (Eq. 5.10 on page 92). The result is

$$
\begin{aligned}
W_{LB}(t) \times S(t) &= \exp(-R_{LB}t)\left[S_0 \exp(i\Omega t) \exp(-Rt)\right] \\
&= S_0 \exp(i\Omega t) \exp(-[R + R_{LB}]t),
\end{aligned}
$$

where to go to the second line we have simply combined the two exponential decay terms.

The effect of the weighting function is simply to change the decay constant from R to $(R + R_{LB})$. Thus, Fourier transformation of the weighted time-domain signal will give a line of width $(R + R_{LB})/\pi$. In summary, the use of a weighting function with decay constant R_{LB} increases the linewidth in the spectrum by R_{LB}/π Hz.

It is usual to specify the weighting function, not by giving the value of R_{LB}, but by specifying the extra line broadening that it will cause. So, a 'line broadening of 5 Hz' is a function which will *increase* the linewidth in the spectrum by 5 Hz. Such a function would require $R_{LB}/\pi = 5$, which means that $R_{LB} = 15.7 \text{ s}^{-1}$.

It can be shown that the best SNR is obtained by applying a weighting function which gives an extra line broadening equal to the linewidth in the original spectrum – such a weighting function is called a *matched filter*. So, if the linewidth is 2 Hz in the original spectrum, applying an additional line broadening of 2 Hz will give the optimum SNR. If there is a range of linewidths present in the spectrum, then no single value of the line broadening will give the optimum SNR for all the peaks.

The extra line broadening caused by the matched filter may not be acceptable on the grounds of the decrease in resolution it causes. Under these circumstances we have to make a compromise between resolution and SNR, applying as much line broadening as we can tolerate, but perhaps not up to that required for the matched filter. In any case, it is important to use a weighting function which decays sufficiently fast to cut off the noise recorded after the NMR signal has decayed.

5.4.5   Resolution enhancement

In the previous two sections we saw that multiplying the time-domain signal by a decaying function results in an increase in the linewidth. This immediately puts in mind the idea of multiplying the time-domain signal by a weighting function which *increases* over time; this would partly cancel out the decay of the FID and hence *decrease* the linewidth in the spectrum. Weighting functions of this type are said to lead to *resolution enhancement*.

A typical example of such a function is an exponential with a *positive* argument:

$$W_{RE}(t) = \exp(R_{RE}t) \quad R_{RE} > 0.$$

This function starts at one when $t = 0$ and then rises indefinitely; a plot is shown in Fig. 5.16.

The problem with using such a function is that it amplifies the noise in the tail of the time domain, thus degrading the SNR in the corresponding spectrum. To get round this problem it is usual to apply, as well as the rising weighting function, a second *decaying* weighting function whose purpose is to 'clip' the noise at the tail of the time domain. Clearly the decaying function must not decay so fast that it undoes the resolution enhancing effect of the rising function.

A common choice for the decaying function is a Gaussian:

$$W_G(t) = \exp(-\alpha t^2),$$

where α is a parameter which sets the decay rate; the larger α, the faster the decay rate. A Gaussian is a bell-shaped curve, symmetrical about its centre. However, the weighting function uses just the right-hand half of the curve i.e. the part for $t > 0$. Like a simple exponential, the Gaussian starts with the value 1 at $t = 0$ and then tails away smoothly to zero. To start with, a Gaussian decays more slowly than an exponential, but then beyond a certain point the Gaussian drops off more quickly, as is illustrated in Fig. 5.17. With a suitable choice of α, the Gaussian will not at first undo the effect of the rising exponential but will, at longer times, attenuate strongly the noise at the end of the time domain.

The whole process is illustrated in Fig. 5.18 on the following page. The original time-domain signal (a) has been recorded well beyond the point where the signal decays into the noise; the corresponding spectrum is shown in (b). If (a) is multiplied by the rising exponential function plotted in (c), the result is the time-domain signal (d); note how the decay of the FID has been slowed, but the noise in the tail of the time-domain signal has been greatly magnified. Fourier transformation of (d) gives the spectrum (e); the resolution has clearly been improved, but at the expense of a large reduction in the SNR.

Referring now to the bottom part of Fig. 5.18 on the next page we can see the effect of including a Gaussian weighting function. The original time-domain signal (a) is multiplied by the rising exponential (f) and the Gaussian (g); this gives the time-domain signal (h). Note that once again the signal decay has been slowed, but the noise in the tail of the time domain is not as large as it was in (d). Fourier transformation of (h) gives the spectrum (i); the resolution has clearly been improved when compared with (b), but without too great a loss of SNR.

Finally, plot (j) shows the product of the two weighting functions (f) and (g). We can see clearly from this plot how the two functions combine together to first increase the time-domain function and then to attenuate it at longer times. Careful choice of the parameters R_{RE} and α are needed to obtain the optimum result. Usually, a process of trial and error is adopted, with the parameters being adjusted until the best result is obtained.

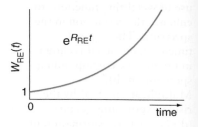

Fig. 5.16 A graph of the rising exponential function $W_{RE}(t) = \exp(R_{RE}t)$; the function starts at one and then rises indefinitely, at a rate set by the parameter R_{RE}.

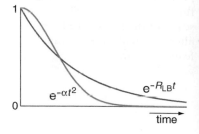

Fig. 5.17 Comparison of the decaying exponential function $\exp(-R_{LB}t)$ and the Gaussian function $\exp(-\alpha t^2)$. To start with, the exponential decays more quickly than the Gaussian, but at long times the situation is reversed. R_{LB} and α have been chosen so that the functions have decayed to half their initial values at the same time.

Fig. 5.18 Illustration of the use of weighting functions to enhance the resolution in the spectrum. The original time-domain signal is shown in (a) and the corresponding spectrum in (b). Multiplication of (a) by the rising exponential (c) gives (d); Fourier transformation of (d) gives the spectrum (e). Comparing (b) and (e) we see that the line has been narrowed, but only at the expense of a serious degradation in the SNR. The situation can be improved by using two weighting functions: a rising exponential (f) and a Gaussian (g); the result of applying both of these is the time-domain signal shown in (h). The corresponding spectrum (i) shows an improvement in resolution without too great a reduction in SNR. Plot (j) is of the product of the two weighting functions (f) and (g).The scales of the plots have been altered to make the relevant features clear.

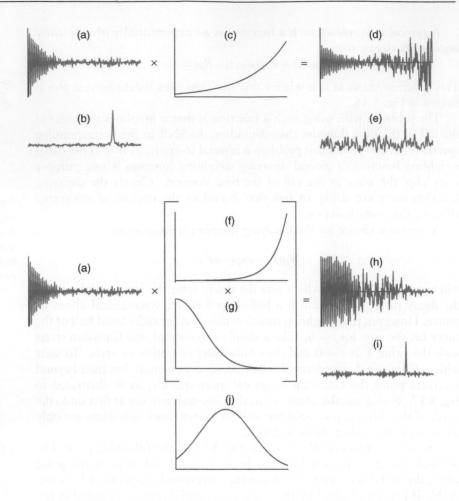

5.4.6   Defining the parameters for sensitivity and resolution enhancement functions

On most spectrometers one does not select the values of R_{LB}, R_{RE} and α directly. Rather these values are computed from some other parameters whose values are perhaps rather more intuitive.

For line broadening using a decaying exponential function, we usually enter the value of the *extra* line broadening (in Hz) which the weighting function will cause. As explained in section 5.4.4 on page 94, the weighting function $\exp(-R_{LB}/t)$ causes an extra line broadening of R_{LB}/π. On the spectrometer we specify this additional linewidth L, from which R_{LB} is computed using $R_{LB} = L\pi$.

For resolution enhancement, the composite weighting function is

$$W_{LG}(t) = \exp(R_{RE}t)\exp(-\alpha t^2). \tag{5.11}$$

If we multiply the time-domain function by this we obtain

$$\begin{aligned} W_{LG}(t) \times S(t) &= \exp(R_{RE}t)\exp(-\alpha t^2)\left[S_0\exp(i\Omega t)\exp(-Rt)\right] \\ &= S_0\exp(i\Omega t)\exp(-[R - R_{RE}]t)\exp(-\alpha t^2). \end{aligned}$$

If we ignore the Gaussian function for the moment (i.e. put $\alpha = 0$), then the overall decay rate is $(R - R_{RE})$, which translates to a linewidth of $(R/\pi - R_{RE}/\pi)$ Hz. Now R/π is the linewidth when no weighting functions are used, so R_{RE}/π is the *reduction* in the linewidth caused by the weighting function.

The way in which R_{RE} is usually specified is to enter the required reduction in the linewidth, L, as a *negative* number; R_{RE} is then computed as $-L/\pi$. This might seem a bit strange, but it is really for compatibility with the line-broadening function: to increase the linewidth by 1 Hz we enter $L = 1$, to decrease the linewidth by the same amount we enter $L = -1$. The composite resolution enhancement weighting function of Eq. 5.11 on the preceding page can thus be written:

$$W_{LG}(t) = \exp(-\pi L t) \exp(-\alpha t^2),$$

where L is entered as a negative number, and is the amount by which the linewidth will be reduced, if the effect of the Gaussian function is ignored.

As shown in Fig. 5.19, the product of the rising exponential and Gaussian gives rise to a maximum in the overall weighting function. It is easy to show that this maximum occurs at a time

$$t_{max} = -\frac{L\pi}{2\alpha};$$

recall that L is negative, so t_{max} is positive. On some spectrometers the value of t_{max} (in seconds) at which you want this maximum to occur is specified, and from this α is computed as

$$\alpha = -\frac{L\pi}{2t_{max}}.$$

On other spectrometers, t_{max} is expressed as a fraction f of the acquisition time: $t_{max} = f t_{acq}$. In this case, once f has been specified, α is computed using

$$\alpha = -\frac{L\pi}{2f t_{acq}}.$$

5.4.7 'Lorentz-to-Gauss' transformation

A weighting function which combines a rising exponential with a Gaussian can be used to change the *lineshape* from a Lorentzian to a Gaussian. These two lineshapes are compared in Fig. 5.20. For the same width at half-height, the Gaussian is more compact, a feature which might be preferred to the Lorentzian line which spreads rather more at the base.

The transformation is achieved by choosing the value of R_{RE} so as to cancel exactly the original decay of the FID. Multiplication by the Gaussian then gives the FID a pure Gaussian envelope, the Fourier transform of which gives a Gaussian lineshape. The width of the Gaussian line is set to the required value by varying the parameter α. The overall process is know colloquially as a *Lorentz-to-Gauss transformation*.

In practice, we would again use trial and error to find the appropriate values of the two parameters; whether or not a complete Lorentz-to-Gauss transformation has actually been achieved is somewhat subjective.

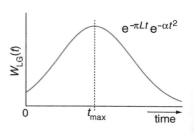

Fig. 5.19 Plot of the product of a rising exponential and a Gaussian function. The maximum occurs at $t_{max} = (-L\pi)/(2\alpha)$ where, as explained in the text, $L < 0$.

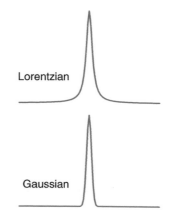

Fig. 5.20 Comparison of the Lorentzian and Gaussian lineshapes; the two peaks have been adjusted so that their peak heights and widths at half-height are equal. The Gaussian is a more compact lineshape.

Fig. 5.21 The top row shows sine bell and the bottom row shows sine bell squared weighting functions for different choices of the phase parameter ϕ, given in radians.

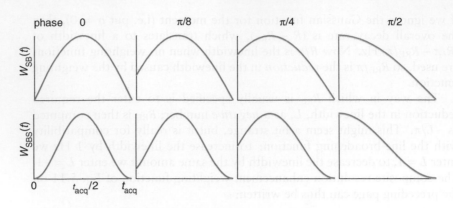

5.4.8   Other weighting functions

Many other weighting functions have been developed for sensitivity enhancement or resolution enhancement. Perhaps the most popular are the *sine bell* functions, which are illustrated in Fig. 5.21.

The basic sine bell is just the function $\sin\theta$ for $\theta = 0$ to $\theta = \pi$. The function is adjusted so that it fits exactly over the acquisition time, as is illustrated in the top left-hand plot of Fig. 5.21. In this form, the function will give resolution enhancement rather like the combination of a rising exponential and a Gaussian function shown in Fig. 5.18 (j) on page 96. Mathematically the weighting function is:

$$W_{SB}(t) = \sin\left(\frac{\pi t}{t_{acq}}\right).$$

Clearly this function goes to zero at $t = t_{acq}$.

The sine bell can be modified by shifting it to the left, as is shown in Fig. 5.21. The further the shift to the left, the smaller the resolution enhancement effect will be, and in the limit that the shift is by $\pi/2$ or $90°$ the function is simply a decaying one and so will broaden the lines. The shift is usually expressed in terms of a phase ϕ. With such a shift the weighting function is:

$$W_{SB}(t) = \sin\left(\frac{(\pi - \phi)t}{t_{acq}} + \phi\right);$$

note that this definition of the function ensures that, regardless of ϕ, it goes to zero at t_{acq}.

The shape of all of these weighting functions are altered subtly by squaring them to give the *sine bell squared* functions; these are shown in the lower part of Fig. 5.21. The weighting function is then

$$W_{SBS}(t) = \sin^2\left(\frac{(\pi - \phi)t}{t_{acq}} + \phi\right).$$

Much of the popularity of these functions probably rests on the fact that there is only one parameter to adjust, rather than two in the case of the combination of an exponential and a Gaussian function.

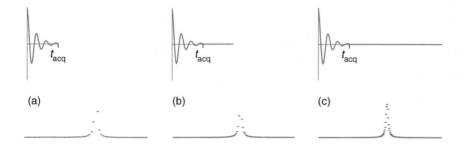

Fig. 5.22 Illustration of the results of zero filling. The time-domain signals along the top row are all the same except that (b) and (c) have been supplemented with increasing numbers of zeroes and so contain more and more data points. Fourier transformation preserves the number of data points, so the line in the corresponding spectrum is represented by more points as more zeroes are added to the end of the FID.

5.5 Zero filling   ⟸ Optional section

We mentioned right at the start of the chapter that the time-domain signal is sampled at regular intervals and stored away in computer memory in a digital form. When these data are subject to a Fourier transform the resulting spectrum is also represented by a set of data points, this time evenly spaced in frequency. Usually, the number of data points in the spectrum is equal to the number in the original time-domain signal. So, although the spectrometer plots a spectrum that appears to be a smooth line, in fact it is joining together a series of closely spaced points.

This is illustrated in Fig. 5.22 (a) which shows the FID and the corresponding spectrum; rather than joining up the points which make up the spectrum we have just plotted the points. Clearly there are only a few data points which define the line.

If we take the original FID and add an equal number of zeroes to it, the corresponding spectrum will have double the number of points, and so the line is represented by more data points. This is illustrated in Fig. 5.22 (b). Adding a set of zeroes equal to the number of data points is called 'zero filling once'.

We can carry on with this zero filling process. For example, having added one set of zeroes, we can add another to double the total number of data points ('zero filling twice'). This results in an even larger number of data points defining the line, as is shown in Fig. 5.22 (c).

Zero filling costs nothing: it is just a manipulation in the computer. This manipulation does not improve the resolution as the measured signal remains the same, but the lines will be better defined in the spectrum. This is desirable, at least for aesthetic reasons if nothing else.

It turns out that the Fourier transform algorithm used by computer programs is most suited to a number of data points which is a power of 2. So, for example, $2^{14} = 16\,384$ is a suitable number of data points to transform, but $15\,000$ is not. In practice, therefore, it is usual to zero fill the time-domain data so that the total number of points is a power of 2. It is always an option, of course, to zero fill beyond this point.

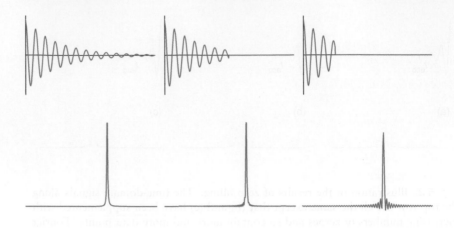

Fig. 5.23 Illustration of how truncation leads to artefacts (called *sinc wiggles*) in the spectrum. The FID on the left has been recorded for sufficient time that it has decayed almost to zero; the corresponding spectrum shows the expected lineshape. However, if data acquisition is stopped before the signal has decayed fully, the corresponding spectra show oscillations around the base of the peak; these oscillations are called sinc wiggles.

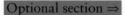
Optional section ⇒

5.6 Truncation

In conventional NMR it is almost always possible to record the FID until it has decayed almost to zero (or into the noise). However, in two-dimensional NMR this may not be the case, simply because of the restrictions on the amount of data which can be recorded, particularly in the indirect dimension (see section 8.1.1 on page 185 for further details). If we stop recording the signal before it has fully decayed the FID is said to be *truncated*; this is illustrated in Fig. 5.23, along with the consequences for the corresponding spectra.

As is shown in the figure, a truncated FID leads to oscillations around the base of the peak in the corresponding spectrum; these are usually called *sinc wiggles* or *truncation artefacts* – the name arises as the peak shape is related to a sinc function $\sin(x)/x$. The more severe the truncation, the larger the sinc wiggles. The separation of successive maxima in these wiggles is $1/t_{acq}$ Hz.

Clearly these oscillations are undesirable as they perturb the baseline and may make it more difficult to discern nearby weaker peaks. Assuming that it is not an option to increase the acquisition time, the only solution is to apply a decaying weighting function to the FID so as to force the signal to go to zero at the end. Unfortunately, this will have the side effect of broadening the lines and reducing the SNR.

Highly truncated time-domain signals are often a feature of multi-dimensional NMR experiments. Much effort has therefore been put into finding alternatives to the Fourier transform which will generate spectra without these truncation artefacts. The popular methods are *linear prediction* and *maximum entropy*. Both have particular merits and drawbacks, and need to be applied with great care.

5.7   Further reading

Fourier transformation and phase correction:

Chapter 5 from M. H. Levitt, *Spin Dynamics* (2nd edition, John Wiley & Sons, Ltd, 2008).

Fourier transformation, weighting functions and sensitivity:

Chapter 4 from R. R. Ernst, G. Bodenhausen and A. Wokaun, *Principles of Nuclear Magnetic Resonance in One and Two Dimensions* (Oxford University Press, 1987).

A comprehensive review on all types of NMR data processing:

J. C. Lindon and A. G. Ferrige, *Progress in Nuclear Magnetic Resonance Spectroscopy*, **14**, 27–66 (1980).

Fourier transformation (a comprehensive text):

R. N. Bracewell, *The Fourier Transform and its Applications* (McGraw–Hill, 1978).

5.8   Exercises

5.1   In a spectrum with just one line, the dispersion mode lineshape might be acceptable – in fact we can think of reasons why it might even be desirable (what might these be?). However, in a spectrum with many lines the dispersion mode lineshape is very undesirable; why?

5.2   An absorption mode Lorentzian line, centred at frequency Ω is given by

$$A(\omega) = \frac{S_0 R}{R^2 + (\omega - \Omega)^2}.$$

We are going to work out the width of the line at half-height. Clearly, this is independent of the position of the line, so we can make the calculation simpler by assuming $\Omega = 0$:

$$A(\omega) = \frac{S_0 R}{R^2 + \omega^2}.$$

Show that the peak height is S_0 and hence, by finding the values of ω (there are two) for which $A(\omega) = S_0/(2R)$, show that the width of the line at half-height is $2R$ rad s^{-1}. Give the corresponding expression for the width in Hz.

5.3   A dispersion Lorentzian line, centred at $\Omega = 0$, is given by

$$D(\omega) = \frac{-\omega}{R^2 + \omega^2}.$$

To find the minima and maxima, we first need to differentiate $D(\omega)$ with respect to ω; show that this differential is

$$\frac{\mathrm{d}D(\omega)}{\mathrm{d}\omega} = \frac{\omega^2 - R^2}{(R^2 + \omega^2)^2}.$$

The minima and maxima are found by solving

$$\frac{\mathrm{d}D(\omega)}{\mathrm{d}\omega} = 0;$$

show that $\omega = \pm R$ are two solutions to this equation. Further show that the peak height at these two frequencies are $\mp 1/(2R)$; draw a sketch of the lineshape and mark on it all of the features you have identified.

There are two frequencies at which the lineshape has half the maximum peak height; these frequencies are found by solving

$$\frac{-\omega}{R^2 + \omega^2} = \frac{1}{4R}.$$

Show that the frequencies are given by

$$\omega = R\left(-2 \pm \sqrt{3}\right).$$

Mark these positions on your sketch.

Similarly show that there are two frequencies at which the peak height is $-1/(4R)$, and hence show that the overall width of the line at half-height is $2(2 + \sqrt{3})R$. Compare this value with that for the absorption mode lineshape.

5.4   Make sketches similar to those of Fig. 5.9 on page 87 for the cases: (a) $\phi = 3\pi/4$; (b) $\phi = 3\pi/2$; (c) $\phi = 2\pi$; (d) $\phi = 5\pi/2$.

5.5   Suppose that we record a spectrum with the simple pulse–acquire sequence using a 90° pulse applied along the x-axis. The resulting FID is Fourier transformed and the spectrum is phased to give an absorption mode lineshape.

We then change the phase of the pulse from x to y, acquire an FID in the same way and phase the spectrum using the *same* phase correction as above. What lineshape would you expect to see in the spectrum? Give the reasons for your answer.

How would the spectrum be affected by: (a) applying the pulse about $-x$; (b) changing the pulse flip angle to 270° about x?

5.6   The gyromagnetic ratio of ^{31}P is 1.08×10^8 rad s^{-1} T^{-1}. This nucleus shows a wide range of shifts, covering some 700 ppm.

Suppose that the transmitter is placed in the middle of the shift range and that a 90° pulse of width 20 μs is used to excite the spectrum; the spectrometer has a B_0 field strength of 9.4 T. Estimate the size of the phase correction (in radians and in degrees) which will be needed at the edges of the spectrum.

5.7   Why is it undesirable to continue to acquire the time-domain signal after the NMR signal has decayed away?

How can weighting functions be used to improve the SNR of a spectrum? In your answer describe how the parameters of a suitable weighting function can be chosen to optimize the SNR. Are there any disadvantages to the use of such weighting functions?

5.8   For a time-domain signal in which the NMR signal has decayed long before the end of the acquisition time, explain why the SNR of the corresponding spectrum can be improved either by shortening the acquisition time, or by applying a suitable weighting function.

5.9   Describe how weighting functions can be used to improve the resolution in a spectrum. In practice, what sets the limit on the improvement that can be obtained? Will zero filling improve the resolution?

5.10 Explain why use of a sine bell weighting function shifted by 45° may enhance the resolution of a spectrum, whereas use of a sine bell shifted by 90° will not.

5.11   In a proton NMR spectrum the peak from TMS was found to show 'wiggles' characteristic of truncation of the FID. However, the other peaks in the spectrum showed no such artefacts. Explain.

How can truncation artefacts be suppressed? Mention any difficulties with your solution to the problem.

6

The quantum mechanics of one spin

To make any further progress in our understanding of multiple-pulse NMR, especially as applied to coupled spin systems, we need to develop some more quantum mechanical tools. There is no avoiding this, as the energy level approach and the vector model are simply not up to the task.

Over the next two chapters we will introduce the *product operator* approach, which is an exact quantum mechanical treatment well-suited to multiple-pulse NMR on coupled spin systems. This approach is relatively simple to apply, and has the advantage of giving results which can be readily interpreted in terms of the appearance of spectra.

This chapter describes the theoretical background which leads up to the product operator formalism, while the next chapter describes how product operators can be used in practice to predict the outcome of experiments. The good news is that you do not need to read this chapter at all, but can jump straight to the next in which you will find a practical description of how to use product operators. There is nothing in this present chapter which you *need* to know in order to understand how to use the product operator method.

However, at some point – perhaps after you have had some experience with making your own calculations – you will probably want to know where this operator approach comes from and why it works. At this point you should read this chapter.

You do not necessarily need to read this chapter, but can jump straight to the next if you wish.

6.1 Introduction

This chapter is concerned with the quantum mechanical description of first a single spin-half nucleus, and then a non-interacting collection of such nuclei. Of course, this is the system to which the vector model, described in Chapter 4, applies exactly, so in a way we are going over the same ground. However, this repetition will be useful as our knowledge of the vector model will help us in understanding and interpreting the quantum mechanics.

Understanding NMR Spectroscopy, Second Edition James Keeler
© 2010 John Wiley & Sons, Ltd

To start with we will work with the wavefunction which describes the spin, and show how the observable magnetization can be computed from a knowledge of this wavefunction. We will then go on to show how the time-dependent Schrödinger equation can be used to predict how the wavefunction, and hence the magnetization, evolves over time.

What we will discover is that, even for this very simple system of non-interacting spins, the quantum mechanical approach seems to be very cumbersome and unintuitive. However, it will be shown that by casting the theory in a different way in which operators are used to represent the motion of the spins, a great deal of simplification is achieved. The practical details of how this simpler operator approach is used, and how it can be extended to coupled spins, is described in the following chapter.

This chapter requires you to have some knowledge of the following mathematical topics: complex numbers, matrices and simple first-order differential equations.

6.2   Superposition states

The central idea of this chapter is that the wavefunction of a single spin, ψ, can always be described as a linear combination of the eigenfunctions of the Hamiltonian for a single spin:

$$\psi = c_{\frac{1}{2}}\psi_{\frac{1}{2}} + c_{-\frac{1}{2}}\psi_{-\frac{1}{2}}. \tag{6.1}$$

Recall from Eq. 3.10 on page 34 that, when written in frequency units, the Hamiltonian for a single spin is $\hat{H}_{\text{one spin}} = \omega_0 \hat{I}_z$. It was shown in section 3.4 on page 34 that the eigenfunctions of the spin operator $\hat{I}_z$, $\psi_{+\frac{1}{2}}$ and $\psi_{-\frac{1}{2}}$:

$$\hat{I}_z \psi_{+\frac{1}{2}} = +\tfrac{1}{2}\psi_{+\frac{1}{2}} \qquad \hat{I}_z \psi_{-\frac{1}{2}} = -\tfrac{1}{2}\psi_{-\frac{1}{2}}, \tag{6.2}$$

are also eigenfunctions of this Hamiltonian. The corresponding eigenvalues are $\pm\tfrac{1}{2}\omega_0$:

$$\hat{H}_{\text{one spin}}\psi_{+\frac{1}{2}} = +\tfrac{1}{2}\omega_0\psi_{+\frac{1}{2}} \qquad \hat{H}_{\text{one spin}}\psi_{-\frac{1}{2}} = -\tfrac{1}{2}\omega_0\psi_{-\frac{1}{2}}.$$

In Eq. 6.1 $c_{\frac{1}{2}}$ and $c_{-\frac{1}{2}}$ are coefficients (just numbers) whose values may change over time. We will see in due course that these coefficients determine the observable magnetization from the spin, and we will also see how to predict the way in which the coefficients vary with time.

Equation 6.1 represents what is called a *superposition of states* or a *mixed state* of the spin. As was explained in section 3.1 on page 24, quantum mechanics does *not* require that the wavefunction of the spin be one corresponding to an energy level i.e. $\psi_{\frac{1}{2}}$ or $\psi_{-\frac{1}{2}}$. It really is very important to grasp the idea that the spin is in a superposition state: everything else in this chapter flows from this central point.

As was explained in section 3.3.1 on page 31, it is common to denote the eigenfunction of $\hat{I}_z$ with eigenvalue $+\tfrac{1}{2}$ as α (spin up), and the state with eigenvalue $-\tfrac{1}{2}$ as β (spin down). We will adopt this notation for the remainder of this chapter, as it is rather more compact; using this, the superposition state is written

$$\psi = c_\alpha \psi_\alpha + c_\beta \psi_\beta. \tag{6.3}$$

6.3   Some quantum mechanical tools

In the rest of this chapter, we are going to need a number of ideas and techniques from quantum mechanics, and so these have been collected together in this section. They may seem a little abstract at this stage, and you might justifiably think 'so what?', but rest assured that all of the topics we cover in this section are going to be useful later on.

6.3.1   Dirac notation

So far we have written the wavefunctions and eigenfunctions using the symbol ψ, adding subscripts when we need to distinguish different functions e.g. $\psi_{1/2}$ or ψ_α. In a way, the ψ is a bit redundant; all it is there for is to give us something to 'hang' the subscripts on.

A more compact and useful notation for wavefunctions was developed by Paul Dirac. In his notation, the wavefunction is written as a 'ket' $|\,\rangle$, and any labels, such as quantum numbers, are written inside the ket. So, for example

$$\psi_{-1/2} \text{ becomes } \left|-\tfrac{1}{2}\right\rangle \quad \text{and} \quad \psi_\beta \text{ becomes } |\beta\rangle.$$

Often, we will need the complex conjugate of a wavefunction, which is denoted by a superscript $\star$ e.g. $\psi_{1/2}^\star$. In the Dirac notation, the complex conjugate is indicated by a 'bra' $\langle\,|$, for example

$$\psi_{1/2}^\star \text{ becomes } \left\langle\tfrac{1}{2}\right| \quad \text{and} \quad \psi_\alpha^\star \text{ becomes } \langle\alpha|.$$

All you have to remember is that the bras and kets are just functions, and so can be manipulated as such.[1]

If a bra appears on the *left* and a ket on the *right*, integration is implied:

$$\underbrace{\langle\alpha|}_{\text{bra}} \ldots \underbrace{|\beta\rangle}_{\text{ket}} \quad \text{implies} \quad \int \psi_\alpha^\star \ldots \psi_\beta \, d\tau.$$

The ... could be other functions, constants or, as is often the case, operators. The integration is with respect to the variable τ, which in quantum mechanics is a shorthand meaning 'the full range of all relevant variables'.

It is important to note that integration is only implied if the bra is on the left and the ket on the right. So, the expression

$$|\beta\rangle \ldots \langle\alpha|$$

does *not* imply an integration.

As we have seen, ψ_α is an eigenfunction of $\hat{I}_z$. This property can be expressed in two different ways:

$$\hat{I}_z \psi_\alpha = \tfrac{1}{2}\psi_\alpha \quad \text{Dirac notation: } \hat{I}_z |\alpha\rangle = \tfrac{1}{2}|\alpha\rangle.$$

To start with, the Dirac notation might appear to add complication for little benefit, but in fact it is a more compact and elegant way of doing many manipulations. For a while we will use both notations, and move over to using only the Dirac form.

---

[1]If you are wondering how to pronounce 'bra' and 'ket', bra rhymes with car, and ket is pronounced just as it is written.

6.3.2   Normalization and orthogonality

A wavefunction is said to be *normalized* if

$$\int \psi^\star \psi \, d\tau = 1 \qquad \text{Dirac notation: } \langle \psi | \psi \rangle = 1, \qquad (6.4)$$

where we have written $|\psi\rangle$ as the Dirac notation for ψ. Admittedly this is a bit redundant, but it does look odd to have an empty ket: $| \; \rangle$.

The eigenfunctions of $\hat{I}_z$, ψ_α (Dirac notation: $|\alpha\rangle$) and ψ_β (Dirac notation: $|\beta\rangle$), are in fact normalized:

$$\int \psi_\alpha^\star \psi_\alpha \, d\tau = 1 \qquad \text{Dirac notation: } \langle \alpha | \alpha \rangle = 1$$

$$\int \psi_\beta^\star \psi_\beta \, d\tau = 1 \qquad \text{Dirac notation: } \langle \beta | \beta \rangle = 1.$$

When a bra and a ket appear next to one another, it is common to leave out one of the vertical lines, so the above become

$$\langle \alpha | \alpha \rangle = 1 \qquad \langle \beta | \beta \rangle = 1. \qquad (6.5)$$

Two wavefunctions, ψ_1 and ψ_2, are said to be *orthogonal* if

$$\int \psi_1^\star \psi_2 \, d\tau = 0 \qquad \text{Dirac notation: } \langle 1 | 2 \rangle = 0;$$

note how the subscripts from the wavefunctions have become the labels in the bra and ket. It turns out that ψ_α and ψ_β are orthogonal

$$\int \psi_\alpha^\star \psi_\beta \, d\tau = 0 \qquad \text{Dirac notation: } \langle \alpha | \beta \rangle = 0; \qquad (6.6)$$

similarly $\langle \beta | \alpha \rangle = 0$.

In the Dirac notation the superposition wavefunction is written

$$|\psi\rangle = c_\alpha |\alpha\rangle + c_\beta |\beta\rangle.$$

The question we want to address is whether or not this wavefunction is normalized. To answer this, we need to compute the integral of Eq. 6.4, for which we need the complex conjugate of ψ:

$$\langle \psi | = c_\alpha^\star \langle \alpha | + c_\beta^\star \langle \beta |.$$

In order to take the complex conjugate of ψ we have had to take the complex conjugates of the coefficients c_α and c_β, as in general these are complex.

In Dirac notation, the integral we need is computed as follows:

$$
\begin{aligned}
\langle \psi | \psi \rangle &= \left[c_\alpha^\star \langle \alpha | + c_\beta^\star \langle \beta | \right] \left[c_\alpha |\alpha\rangle + c_\beta |\beta\rangle \right] \\
&= c_\alpha^\star c_\alpha \underbrace{\langle \alpha | \alpha \rangle}_{=1} + c_\alpha^\star c_\beta \underbrace{\langle \alpha | \beta \rangle}_{=0} + c_\beta^\star c_\alpha \underbrace{\langle \beta | \alpha \rangle}_{=0} + c_\beta^\star c_\beta \underbrace{\langle \beta | \beta \rangle}_{=1} \\
&= c_\alpha^\star c_\alpha + c_\beta^\star c_\beta.
\end{aligned}
$$

To go to the second line we have just multiplied out the square braces, remembering that the bras and kets are just functions. Then, using the fact that $\langle\alpha|$ and $\langle\beta|$ are normalized, Eq. 6.5 on the facing page, we recognize that $\langle\alpha|\alpha\rangle = 1$ and $\langle\beta|\beta\rangle = 1$. Finally, using the fact that $\langle\alpha|$ and $\langle\beta|$ are orthogonal to one another, Eq. 6.6 on the preceding page, we recognize that $\langle\alpha|\beta\rangle = \langle\beta|\alpha\rangle = 0$. Since the integral $\langle\psi|\psi\rangle$ is *not* = 1, the wavefunction $|\psi\rangle$ is not normalized.

We can make the wavefunction normalized by dividing the wavefunction $|\psi\rangle$ by $\sqrt{c_\alpha^\star c_\alpha + c_\beta^\star c_\beta}$, the *normalizing factor*:

$$|\psi\rangle_{\text{normalized}} = \frac{c_\alpha|\alpha\rangle + c_\beta|\beta\rangle}{\sqrt{c_\alpha^\star c_\alpha + c_\beta^\star c_\beta}}.$$

If you repeat the above calculation with this wavefunction, the integral will be = 1.

We will show later on in this chapter that the coefficients c_α and c_β change over time. However, it turns out that, no matter what happens, $(c_\alpha^\star c_\alpha + c_\beta^\star c_\beta)$ remains constant – in other words, the normalizing factor remains constant. All this factor does is simply scale the result of any calculation we make by a constant amount. This is not really a significant effect, so to simplify things we will *assume* that the wavefunction of the superposition state is normalized i.e. c_α and c_β are chosen such that $(c_\alpha^\star c_\alpha + c_\beta^\star c_\beta) = 1$.

6.3.3 Expectation values

In section 3.2.4 on page 28 it was explained that it is a postulate of quantum mechanics that a measurement of some observable quantity will always give one of the eigenvalues of the operator which represents that observable.

For example, if we measure the z-component of the angular momentum of a spin-half nucleus, we will find either the value $+\frac{1}{2}$ or $-\frac{1}{2}$, as these are the two eigenvalues of $\hat{I}_z$, the operator which represents the z-component.

The question is which eigenvalue will we find in a particular measurement? Quantum mechanics does not have a direct answer to this, but does give us a way of finding out what the average result of many measurements will be; this average is called the *expectation value*.

Imagine the following thought experiment. Suppose that the spin is described by a wavefunction ψ, and that we then duplicate this spin many, many times in such a way that each has the *same* wavefunction. Now we make a measurement on each spin, and then compute the average value of these measurements; the process is visualized in Fig. 6.1. It is a postulate of quantum mechanics that this average value, called the *expectation value*, is given by

$$\langle Q\rangle = \frac{\int \psi^\star \hat{Q}\psi \, d\tau}{\int \psi^\star \psi \, d\tau} \qquad \text{Dirac notation: } \langle Q\rangle = \frac{\langle\psi|\hat{Q}|\psi\rangle}{\langle\psi|\psi\rangle}, \qquad (6.7)$$

where $\hat{Q}$ is the operator which represents the observable we are measuring and $\langle Q\rangle$ is the corresponding expectation value.

Recall that for a complex number c, the product $(cc^\star)$ is always real.

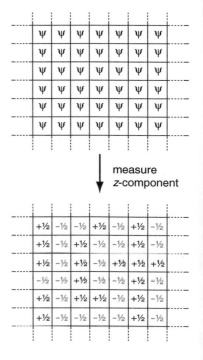

Fig. 6.1 Visualization of the process of taking an expectation value. Suppose that our spin is described by a wavefunction ψ; we imagine duplicating the spin, along with its wavefunction, a very large number of times, as shown at the top of the diagram. Then, we measure the z-component of the angular momentum; quantum mechanics tells us that the outcome of a single measurement is either $+\frac{1}{2}$ or $-\frac{1}{2}$, as these are the eigenvalues of $\hat{I}_z$. The average over a large number of measurements is the expectation value. In the case shown above the expectation value is $(1/42)[19 \times (+\frac{1}{2}) + 23 \times (-\frac{1}{2})] = -0.048$; admittedly, 42 is not really a 'very large number of spins'.

A good way of seeing how this works is to do a particular example. Suppose that the wavefunction is the superposition state

$$|\psi\rangle = c_\alpha|\alpha\rangle + c_\beta|\beta\rangle,$$

and that we are interested in the expectation value of $\hat{I}_z$ i.e. the z-component of angular momentum. As was explained in the previous section, we are going to assume that the wavefunction for the superposition state is normalized i.e. $\langle\psi|\psi\rangle = 1$. As a result, the bottom of the fraction in Eq. 6.7 on the preceding page is equal to one, so the expectation value can be computed as follows:

$$
\begin{aligned}
\langle I_z \rangle &= \langle\psi|\hat{I}_z|\psi\rangle \\
&= \left[c_\alpha^\star\langle\alpha| + c_\beta^\star\langle\beta|\right]\hat{I}_z\left[c_\alpha|\alpha\rangle + c_\beta|\beta\rangle\right] \\
&= c_\alpha^\star c_\alpha\langle\alpha|\underbrace{\hat{I}_z|\alpha\rangle}_{=\frac{1}{2}|\alpha\rangle} + c_\alpha^\star c_\beta\langle\alpha|\underbrace{\hat{I}_z|\beta\rangle}_{=-\frac{1}{2}|\beta\rangle} + c_\beta^\star c_\alpha\langle\beta|\underbrace{\hat{I}_z|\alpha\rangle}_{=\frac{1}{2}|\alpha\rangle} + c_\beta^\star c_\beta\langle\beta|\underbrace{\hat{I}_z|\beta\rangle}_{-\frac{1}{2}|\beta\rangle} \\
&= \tfrac{1}{2}c_\alpha^\star c_\alpha\underbrace{\langle\alpha|\alpha\rangle}_{=1} - \tfrac{1}{2}c_\alpha^\star c_\beta\underbrace{\langle\alpha|\beta\rangle}_{=0} + \tfrac{1}{2}c_\beta^\star c_\alpha\underbrace{\langle\beta|\alpha\rangle}_{=0} - \tfrac{1}{2}c_\beta^\star c_\beta\underbrace{\langle\beta|\beta\rangle}_{=1} \\
&= \tfrac{1}{2}c_\alpha^\star c_\alpha - \tfrac{1}{2}c_\beta^\star c_\beta.
\end{aligned}
$$

$$(6.8)$$

There is quite a lot going on here, so let us take things one line at a time. To go to the second line we have simply substituted $\langle\psi|$ and $|\psi\rangle$ by the superposition states, and then to go to the next line we have multiplied out the square braces, taking care *not* to change the order of the functions and the operator. Recall that in section 3.2.2 on page 27 it was pointed out that, in general, the order of functions and operators must not be changed.

To go to the fourth line, we have simply used the fact that $|\alpha\rangle$ and $|\beta\rangle$ are eigenfunctions of $\hat{I}_z$ (Eq. 6.2 on page 106):

$$\hat{I}_z|\alpha\rangle = \tfrac{1}{2}|\alpha\rangle \qquad \hat{I}_z|\beta\rangle = -\tfrac{1}{2}|\beta\rangle.$$

To go to the last line, we have used the fact that $|\alpha\rangle$ and $|\beta\rangle$ are normalized and orthogonal to one another (Eqs 6.5 and 6.6 on p. 108):

$$\langle\alpha|\alpha\rangle = 1 \qquad \langle\beta|\beta\rangle = 1 \qquad \langle\alpha|\beta\rangle = 0 \qquad \langle\beta|\alpha\rangle = 0.$$

The final result,

$$\langle I_z \rangle = \tfrac{1}{2}c_\alpha^\star c_\alpha - \tfrac{1}{2}c_\beta^\star c_\beta,$$

$$(6.9)$$

can be interpreted in the following way. Each individual measurement of the z-component gives the result $+\tfrac{1}{2}$ or $-\tfrac{1}{2}$; however, when a large number of measurements are taken, the probability of obtaining $+\tfrac{1}{2}$ is $c_\alpha^\star c_\alpha$, and the probability of obtaining the result $-\tfrac{1}{2}$ is $c_\beta^\star c_\beta$. The average value of the z-component is thus

$$
\begin{aligned}
\text{average } z\text{-component} &= \left(\text{probability of obtaining the result } +\tfrac{1}{2}\right) \times \left(+\tfrac{1}{2}\right) \\
&\quad + \left(\text{probability of obtaining the result } -\tfrac{1}{2}\right) \times \left(-\tfrac{1}{2}\right) \\
&= c_\alpha^\star c_\alpha\left(+\tfrac{1}{2}\right) + c_\beta^\star c_\beta\left(-\tfrac{1}{2}\right).
\end{aligned}
$$

This average z-component is thus the same thing as the expectation value.

6.3.4   The x- and y-components of angular momentum

So far we have only referred to the z-component of the angular momentum, represented by the operator $\hat{I}_z$. However, angular momentum is a vector property, and so has components in all three directions. The x-component is represented by the operator $\hat{I}_x$, and the y-component by the operator $\hat{I}_y$.

The important thing to note is that $|\alpha\rangle$ and $|\beta\rangle$ are *not* eigenfunctions of $\hat{I}_x$, nor are they eigenfunctions of $\hat{I}_y$. In fact, it can be shown that these operators have the following effects on $|\alpha\rangle$ and $|\beta\rangle$:

$$\hat{I}_x|\alpha\rangle = \tfrac{1}{2}|\beta\rangle \qquad \hat{I}_x|\beta\rangle = \tfrac{1}{2}|\alpha\rangle \tag{6.10}$$

$$\hat{I}_y|\alpha\rangle = \tfrac{1}{2}\mathrm{i}|\beta\rangle \qquad \hat{I}_y|\beta\rangle = -\tfrac{1}{2}\mathrm{i}|\alpha\rangle, \tag{6.11}$$

where i is the 'complex i'.

Using these, we can work out the expectation value of $\hat{I}_x$ for the case where the wavefunction is our superposition state: $|\psi\rangle = c_\alpha|\alpha\rangle + c_\beta|\beta\rangle$.

$$
\begin{aligned}
\langle I_x \rangle &= \left[c_\alpha^\star \langle\alpha| + c_\beta^\star\langle\beta| \right] \hat{I}_x \left[c_\alpha|\alpha\rangle + c_\beta|\beta\rangle \right] \\
&= c_\alpha^\star c_\alpha \langle\alpha| \underbrace{\hat{I}_x|\alpha\rangle}_{=\frac{1}{2}|\beta\rangle} + c_\alpha^\star c_\beta \langle\alpha| \underbrace{\hat{I}_x|\beta\rangle}_{=\frac{1}{2}|\alpha\rangle} + c_\beta^\star c_\alpha \langle\beta| \underbrace{\hat{I}_x|\alpha\rangle}_{=\frac{1}{2}|\beta\rangle} + c_\beta^\star c_\beta \langle\beta| \underbrace{\hat{I}_x|\beta\rangle}_{\frac{1}{2}|\alpha\rangle} \\
&= \tfrac{1}{2}c_\alpha^\star c_\alpha \underbrace{\langle\alpha|\beta\rangle}_{=0} + \tfrac{1}{2}c_\alpha^\star c_\beta \underbrace{\langle\alpha|\alpha\rangle}_{=1} + \tfrac{1}{2}c_\beta^\star c_\alpha \underbrace{\langle\beta|\beta\rangle}_{=1} + \tfrac{1}{2}c_\beta^\star c_\beta \underbrace{\langle\beta|\alpha\rangle}_{=0} \\
&= \tfrac{1}{2}c_\alpha^\star c_\beta + \tfrac{1}{2}c_\beta^\star c_\alpha .
\end{aligned}
\tag{6.12}
$$

As before, going to the second line is simply a case of multiplying out the square braces, being careful to preserve the ordering of the operators and functions. To go to the third line we use Eq. 6.10 to work out the effect of $\hat{I}_x$ on $|\alpha\rangle$ and $|\beta\rangle$. The final step is to use, as we did before, the fact that $|\alpha\rangle$ and $|\beta\rangle$ are normalized and orthogonal (Eq. 6.5 and Eq. 6.6 on page 108).

A similar approach for $\langle I_y \rangle$, using Eq. 6.11, gives

$$\langle I_y \rangle = \tfrac{1}{2}\mathrm{i}\, c_\beta^\star c_\alpha - \tfrac{1}{2}\mathrm{i}\, c_\alpha^\star c_\beta . \tag{6.13}$$

The feature to note here is that the expectation values of $\hat{I}_x$, $\hat{I}_y$ and $\hat{I}_z$ all depend on the coefficients c_α and c_β which define the superposition state.

6.3.5   Matrix representations

Using the two functions $|\alpha\rangle$ and $|\beta\rangle$ we can construct what is called a *matrix representation* of an operator $\hat{Q}$. In this context, the two functions are called the *basis functions*.

As there are two basis functions, the matrix will have two rows and two columns, i.e. it will be a two-by-two matrix. The element in the ith row and jth column is given by the integral

$$Q_{ij} = \int \psi_i^\star \hat{Q} \psi_j \, \mathrm{d}\tau \qquad \text{Dirac notation: } Q_{ij} = \langle i|\hat{Q}|j\rangle .$$

Let us suppose that the first basis function is $|\alpha\rangle$ and the second is $|\beta\rangle$, then the matrix representation of $\hat{Q}$ is

$$Q = \begin{pmatrix} \langle\alpha|\hat{Q}|\alpha\rangle & \langle\alpha|\hat{Q}|\beta\rangle \\ \langle\beta|\hat{Q}|\alpha\rangle & \langle\beta|\hat{Q}|\beta\rangle \end{pmatrix}. \tag{6.14}$$

For example, the matrix representation of $\hat{I}_z$ is

$$
\begin{aligned}
I_z &= \begin{pmatrix} \langle\alpha|\hat{I}_z|\alpha\rangle & \langle\alpha|\hat{I}_z|\beta\rangle \\ \langle\beta|\hat{I}_z|\alpha\rangle & \langle\beta|\hat{I}_z|\beta\rangle \end{pmatrix} \\
&= \begin{pmatrix} \frac{1}{2}\langle\alpha|\alpha\rangle & -\frac{1}{2}\langle\alpha|\beta\rangle \\ \frac{1}{2}\langle\beta|\alpha\rangle & -\frac{1}{2}\langle\beta|\beta\rangle \end{pmatrix} \\
&= \begin{pmatrix} \frac{1}{2} & 0 \\ 0 & -\frac{1}{2} \end{pmatrix}.
\end{aligned}
\tag{6.15}
$$

To go to the second line we have used the fact that $|\alpha\rangle$ and $|\beta\rangle$ are eigenfunctions of $\hat{I}_z$, and to go to the last line we have used the fact that these two basis functions are normalized and orthogonal.

Using a similar approach we can find the matrix representations of $\hat{I}_x$ and $\hat{I}_y$ as:

$$
I_x = \begin{pmatrix} 0 & \frac{1}{2} \\ \frac{1}{2} & 0 \end{pmatrix} \qquad I_y = \begin{pmatrix} 0 & -\frac{1}{2}i \\ \frac{1}{2}i & 0 \end{pmatrix}.
\tag{6.16}
$$

In due course, we will make much use of these matrix representations.

6.4 Computing the bulk magnetization

To be useful, our quantum mechanical theory must be able to compute what we actually observe in an NMR experiment, which is the transverse magnetization. In this section, we are going to look into how this is achieved, but to start with we will discuss the z-component of the magnetization as the calculation is somewhat simpler.

When we record an NMR signal it is not from one spin but from the extremely large number of spins in our sample. For example 1 cm^3 of a 1 mM solution contains around 6×10^{17} solute molecules – clearly even for this rather dilute solution we are dealing with a large number of spins. Our aim is to be able to compute the bulk x-, y- and z-magnetizations from such a sample.

On the face of it, this appears to be a daunting task as each spin can in principle have a different wavefunction i.e. be in a different superposition state. It would seem that we would need to know all of these wavefunctions – clearly an impossible task. However, as we will see in the following section, the fact that there are so many spins actually makes the calculation rather simple.

6.4.1   The ensemble average

As was explained in section 4.1 on page 47, each spin appears to contain a source of angular momentum; associated with this is a magnetic moment, which can point in any direction. The bulk magnetization of the sample in the z-direction is found by adding up the z-component of the magnetic moment of each spin:

bulk z-magnetization   = z-comp. of the magnetic moment from spin 1

               + z-comp. of the magnetic moment from spin 2

               + z-comp. of the magnetic moment from spin 3

               +   . . .

The z-component of the magnetic moment from a single spin is simply proportional to the z-component of the angular momentum, with the constant of proportionality being the gyromagnetic ratio, γ:

z-component of magnetic moment = $\gamma \times z$-component of angular momentum.

The same applies to the x- and y-components.

As has already been described in section 6.3.3 on page 109, the z-component of the angular momentum is represented by the operator $\hat{I}_z$, and the *average* value of the z-component is given by the expectation value:

$$\langle I_z \rangle = \langle \psi | \hat{I}_z | \psi \rangle.$$

Recall that this expectation value comes from a thought experiment in which we duplicate a single spin, along with its wavefunction, a large number of times and then measure the z-component from each spin; the expectation value is the average of these measurements. However, what we are trying to calculate here is the z-magnetization from a real sample, rather than an imagined set of spins which all have the *same* wavefunction. In general, in a real sample each spin has a *different* wavefunction i.e. the coefficients c_α and c_β in the superposition state are different for each spin.

We extricate ourselves from this difficulty in the following way. Although in principle each spin can have a different wavefunction, for a sample containing 10^{17} spins, there must be a large number of spins which have, to some level of approximation, the same wavefunction. A useful analogy here is to think about the individual weights of all the people in London. If we weighed each person to a precision of 0.25 kg, we would find that there are a lot of people with the 'same' weight; if we made the measurement to a precision of 0.5 kg, the numbers with any particular weight would be even higher. The point is that, for a sufficiently large sample, there will always be many people with the 'same' weight, regardless of how closely we define what 'the same' means.

Consider the first spin, whose superposition state is

$$c_\alpha^{(1)} | \alpha \rangle + c_\beta^{(1)} | \beta \rangle.$$

Following the line above, we argue that in our sample there are many spins which have a wavefunction very similar to this. On average, the contribution of each of these spins to the z-component is given by the expectation value:

$$\langle I_z \rangle^{(1)} = \tfrac{1}{2} c_\alpha^{(1)\star} c_\alpha^{(1)} - \tfrac{1}{2} c_\beta^{(1)\star} c_\beta^{(1)}.$$

The same argument applies to the second spin, whose superposition state is

$$c_\alpha^{(2)} | \alpha \rangle + c_\beta^{(2)} | \beta \rangle$$

and which contributes

$$\langle I_z \rangle^{(2)} = \tfrac{1}{2} c_\alpha^{(2)\star} c_\alpha^{(2)} - \tfrac{1}{2} c_\beta^{(2)\star} c_\beta^{(2)}.$$

The total z-magnetization is the sum of the z-components of each spin, multiplied by γ

$$
\begin{aligned}
M_z &= \gamma\langle I_z\rangle^{(1)} + \gamma\langle I_z\rangle^{(2)} + \gamma\langle I_z\rangle^{(3)} + \ldots && (6.17) \\
&= \gamma\left[\tfrac{1}{2} c_\alpha^{(1)\star} c_\alpha^{(1)} - \tfrac{1}{2} c_\beta^{(1)\star} c_\beta^{(1)}\right] + \gamma\left[\tfrac{1}{2} c_\alpha^{(2)\star} c_\alpha^{(2)} - \tfrac{1}{2} c_\beta^{(2)\star} c_\beta^{(2)}\right] \\
&\quad + \gamma\left[\tfrac{1}{2} c_\alpha^{(3)\star} c_\alpha^{(3)} - \tfrac{1}{2} c_\beta^{(3)\star} c_\beta^{(3)}\right]\ldots && (6.18)
\end{aligned}
$$

Each spin in the sample contributes to this sum.

The usual way of writing this sum is

$$M_z = \tfrac{1}{2}\gamma N \overline{\left(c_\alpha^\star c_\alpha - c_\beta^\star c_\beta\right)}, \qquad (6.19)$$

where the line indicates an *ensemble average* and N is the number of spins in the sample (called the ensemble).

Taking the ensemble average means adding up the contributions from each spin in the ensemble, as shown in Eq. 6.18, and then dividing by the number of spins to obtain an average. To compute the bulk magnetization from the N spins in the sample we need to multiply by N as the ensemble average is the average contribution *per spin*.

Another way of expressing the z-magnetization is to write Eq. 6.17 as

$$M_z = \gamma N \,\overline{\langle I_z\rangle},$$

where the bar indicates the ensemble average of the expectation values of $\hat{I}_z$ from each spin. $\overline{\langle I_z\rangle}$ is found, as before, by adding up the contribution from each spin and dividing by the number of spins in order to obtain an average:

$$\overline{\langle I_z\rangle} = \frac{1}{N}\left(\langle I_z\rangle^{(1)} + \langle I_z\rangle^{(2)} + \langle I_z\rangle^{(3)} + \ldots\right).$$

Using a similar argument, the x- and y-magnetizations are given by the following ensemble averages (for completeness we include the expression for M_z):

$$
\begin{aligned}
M_x &= \tfrac{1}{2}\gamma N \overline{\left(c_\alpha^\star c_\beta + c_\beta^\star c_\alpha\right)} && \text{or} && M_x = \gamma N\,\overline{\langle I_x\rangle} && (6.20) \\
M_y &= \tfrac{1}{2}\mathrm{i}\gamma N \overline{\left(c_\beta^\star c_\alpha - c_\alpha^\star c_\beta\right)} && \text{or} && M_y = \gamma N\,\overline{\langle I_y\rangle} && (6.21) \\
M_z &= \tfrac{1}{2}\gamma N \overline{\left(c_\alpha^\star c_\alpha - c_\beta^\star c_\beta\right)} && \text{or} && M_z = \gamma N\,\overline{\langle I_z\rangle}. && (6.22)
\end{aligned}
$$

The difficulty we still have not got round is the need to know the wavefunction for each spin, as from these three equations it still looks as if we need to know this in order to compute the magnetization. However, in the next section we will find that by introducing the idea of populations we will finally be in a position to compute M_z.

6.4.2 Populations

In section 6.3.3 on page 109 we introduced the interpretation that, when measuring the z-component of angular momentum, the probability of obtaining $+\frac{1}{2}$ is $c_\alpha^\star c_\alpha$, and the probability of obtaining the result $-\frac{1}{2}$ is $c_\beta^\star c_\beta$. These two values of the z-component are associated with the two energy levels of a single spin, and so we can extend this interpretation by saying that $c_\alpha^\star c_\alpha$ is 'the probability of finding the spin in the energy level $|\alpha\rangle$', and $c_\beta^\star c_\beta$ is 'the probability of finding the spin in the energy level $|\beta\rangle$'.

The probability of finding a particular spin in the level $|\alpha\rangle$ is $c_\alpha^\star c_\alpha$. If we add up these probabilities for all the spins in the sample we will find the number of spins which, on measurement, are in the level α. We can interpret this number as the *population* of the level α, n_α:

$$n_\alpha = c_\alpha^{(1)\star} c_\alpha^{(1)} + c_\alpha^{(2)\star} c_\alpha^{(2)} + c_\alpha^{(3)\star} c_\alpha^{(3)} + \ \ldots \ .$$

Similarly the population of level β is

$$n_\alpha = c_\beta^{(1)\star} c_\beta^{(1)} + c_\beta^{(2)\star} c_\beta^{(2)} + c_\beta^{(3)\star} c_\beta^{(3)} + \ \ldots \ .$$

Recalling that the ensemble average of $c_\alpha^\star c_\alpha$ is found by adding up the contribution from each spin and then dividing by N, it follows that

$$n_\alpha = N \overline{c_\alpha c_\alpha^\star} \quad \text{and} \quad n_\beta = N \overline{c_\beta c_\beta^\star}; \tag{6.23}$$

so

$$\overline{c_\alpha c_\alpha^\star - c_\beta c_\beta^\star} = \frac{1}{N}\left(n_\alpha - n_\beta\right).$$

Using this expression in Eq. 6.22 on the preceding page, we can rewrite the bulk z-magnetization as

$$M_z = \tfrac{1}{2}\gamma(n_\alpha - n_\beta). \tag{6.24}$$

In words, the z-magnetization is proportional to the population difference between the two energy levels.

At equilibrium, these populations are predicted by the *Boltzmann distribution*:

$$n_{\alpha,\mathrm{eq}} = \tfrac{1}{2}N \exp\left(-E_\alpha/k_\mathrm{B}T\right) \qquad n_{\beta,\mathrm{eq}} = \tfrac{1}{2}N \exp\left(-E_\beta/k_\mathrm{B}T\right) \tag{6.25}$$

where E_α and E_β are the energies of the two levels (in J), k_B is Boltzmann's constant, T is the temperature and N is the total number of spins. As (for positive γ) the α state has the lower energy, the Boltzmann distribution predicts that $n_{\alpha,\mathrm{eq}} > n_{\beta,\mathrm{eq}}$, and so at equilibrium the sample will have a bulk z-magnetization.

Combining Eqs 6.24 and 6.25 we can compute this equilibrium z-magnetization as

$$\begin{aligned} M_0 &= \tfrac{1}{2}\gamma(n_{\alpha,\mathrm{eq}} - n_{\beta,\mathrm{eq}}) \\ &= \tfrac{1}{4}\gamma N \left[\exp\left(-E_\alpha/k_\mathrm{B}T\right) - \exp\left(-E_\beta/k_\mathrm{B}T\right)\right]. \end{aligned}$$

The key point here is that, as we have a large number of spins, we can use the Boltzmann distribution to find the equilibrium populations and hence the equilibrium z-magnetization. We do not need to know the wavefunction of each spin in the sample.

6.4.3   Transverse magnetization

The next point to address is whether we can say anything useful about the size of the x- and y-magnetizations. Following the same line of argument as above, and using Eq. 6.20 on page 114, the x-magnetization is given by

$$
\begin{aligned}
M_x &= \gamma \langle I_x \rangle^{(1)} + \gamma \langle I_x \rangle^{(2)} + \ldots \\
&= \gamma \left[\tfrac{1}{2} c_\alpha^{(1)\star} c_\beta^{(1)} + \tfrac{1}{2} c_\beta^{(1)\star} c_\alpha^{(1)} \right] + \gamma \left[\tfrac{1}{2} c_\alpha^{(2)\star} c_\beta^{(2)} + \tfrac{1}{2} c_\beta^{(2)\star} c_\alpha^{(2)} \right] + \ldots \\
&= \tfrac{1}{2} \gamma N \overline{\left(c_\alpha^\star c_\beta + c_\beta^\star c_\alpha \right)}.
\end{aligned}
$$

To make further progress, it is convenient to write the complex coefficients c_α and c_β in terms of a (real) magnitude r and a phase ϕ:

$$
c_\alpha = r_\alpha \exp(i\phi_\alpha) \qquad\qquad c_\beta = r_\beta \exp(i\phi_\beta) \tag{6.26}
$$
$$
c_\alpha^\star = r_\alpha \exp(-i\phi_\alpha) \qquad\qquad c_\beta^\star = r_\beta \exp(-i\phi_\beta). \tag{6.27}
$$

Note that the complex conjugates are formed by changing the sign of the argument to the exponential. By definition the r are real and so are unaffected by taking the complex conjugate.

Using this way of writing the coefficients, the expectation value of $\hat{I}_x$ can be written:

$$
\begin{aligned}
\langle I_x \rangle &= \tfrac{1}{2} c_\alpha^\star c_\beta + \tfrac{1}{2} c_\beta^\star c_\alpha \\
&= \tfrac{1}{2} \left[r_\alpha r_\beta \exp(-i\phi_\alpha) \exp(i\phi_\beta) + r_\alpha r_\beta \exp(-i\phi_\beta) \exp(i\phi_\alpha) \right] \\
&= \tfrac{1}{2} r_\alpha r_\beta \left\{ \exp(i[\phi_\beta - \phi_\alpha]) + \exp(-i[\phi_\beta - \phi_\alpha]) \right\}.
\end{aligned}
$$

This can be tidied up further by using the identity

$$
\exp(i\theta) + \exp(-i\theta) \equiv 2\cos\theta
$$

to give

$$
\langle I_x \rangle = r_\alpha r_\beta \cos(\phi_\beta - \phi_\alpha)
$$

Using this, the bulk x-magnetization can be written

$$
\begin{aligned}
M_x &= \gamma \langle I_x \rangle^{(1)} + \gamma \langle I_x \rangle^{(2)} + \ldots \\
&= \gamma r_\alpha^{(1)} r_\beta^{(1)} \cos(\phi_\beta^{(1)} - \phi_\alpha^{(1)}) + \gamma r_\alpha^{(2)} r_\beta^{(2)} \cos(\phi_\beta^{(2)} - \phi_\alpha^{(2)}) + \ldots \\
&= \gamma N \overline{r_\alpha r_\beta \cos(\phi_\beta - \phi_\alpha)}.
\end{aligned}
$$

As before, we have used the overline to indicate an ensemble average.

We now introduce the hypothesis that, *at equilibrium*, the phases ϕ are randomly distributed. As a result $(\phi_\beta - \phi_\alpha)$ is also randomly distributed, and therefore the average of $\cos(\phi_\beta - \phi_\alpha)$ over the sample is zero. This comes about because positive and negative values of the cosine function cancel one another out. Therefore, at equilibrium the x-magnetization is zero.

A similar calculation for the y-magnetization gives

$$
M_y = \gamma N \overline{r_\alpha r_\beta \sin(\phi_\beta - \phi_\alpha)}.
$$

It is clear that this is also zero if the phases are randomly distributed.

Another way of looking at this is to say that, as we know from experiment that at equilibrium there is no x- or y-magnetization, the implication is that the phases are randomly distributed.

Just for completeness we will compute the z-magnetization using this form of the coefficients

$$
\begin{aligned}
\langle I_z \rangle &= \tfrac{1}{2}c_\alpha^\star c_\alpha - \tfrac{1}{2}c_\beta^\star c_\beta \\
&= \tfrac{1}{2}\left[r_\alpha r_\alpha \exp(-i\phi_\alpha)\exp(i\phi_\alpha) - r_\beta r_\beta \exp(-i\phi_\beta)\exp(i\phi_\beta) \right] \\
&= \tfrac{1}{2}\left\{ r_\alpha^2 \exp(i[\phi_\alpha - \phi_\alpha]) - r_\beta^2 \exp(-i[\phi_\beta - \phi_\beta]) \right\} \\
&= \tfrac{1}{2}\left[r_\alpha^2 - r_\beta^2 \right],
\end{aligned}
$$

where to go to the last line we have used $\exp(0) = 1$. In contrast to the case of the x- and y-components, the phases do not affect the value of $\langle I_z \rangle$.

The z-magnetization is thus

$$
M_z = \tfrac{1}{2}\gamma N \overline{\left[r_\alpha^2 - r_\beta^2 \right]}.
$$

We can therefore identify $N\overline{r_\alpha^2}$ as being the population of level $|\alpha\rangle$.

6.4.4 A comment on the units

We need to be a little careful here about the units. Recall from section 3.4 on page 34 that we decided it would be convenient to omit a factor of $\hbar$ and write the eigenvalue equation for $\hat{I}_z$ as

$$
\hat{I}_z |\alpha\rangle = \tfrac{1}{2}|\alpha\rangle
$$

rather than

$$
\hat{I}_z |\alpha\rangle = \tfrac{1}{2}\hbar|\alpha\rangle.
$$

As a result of this choice, our expression for M_z is also missing a factor of $\hbar$, and so is dimensionally incorrect. If we put this factor back into our calculations we find

$$
\langle I_z \rangle = \tfrac{1}{2}\hbar\left(c_\alpha^\star c_\alpha - c_\beta^\star c_\beta \right)
$$

and so

$$
M_z = \tfrac{1}{2}\hbar\gamma N \overline{\left(c_\alpha^\star c_\alpha - c_\beta^\star c_\beta \right)} \quad \text{or} \quad M_z = \hbar\gamma N \overline{\langle I_z \rangle}.
$$

For most purposes, this will make no difference, but it is as well to be aware what is going on. For the remainder of the discussion, we will continue to omit the factor of $\hbar$.

6.5 Summary

We have introduced a lot of new ideas so far, so a quick summary of the really crucial points will not go amiss.

- Each spin is in a superposition state, which can be written as a linear combination of the eigenfunctions of $\hat{I}_z$:

$$
\psi = c_\alpha \psi_\alpha + c_\beta \psi_\beta \qquad \text{Dirac notation: } |\psi\rangle = c_\alpha |\alpha\rangle + c_\beta |\beta\rangle.
$$

The coefficients c_α and c_β are complex numbers which vary with time, and are different for each spin in the sample. The coefficients can be chosen (or scaled) so that the wavefunction ψ is normalized.

- $|\alpha\rangle$ and $|\beta\rangle$ are eigenfunctions of $\hat{I}_z$:

$$\hat{I}_z \, |\alpha\rangle = \tfrac{1}{2}|\alpha\rangle \qquad \hat{I}_z \, |\beta\rangle = -\tfrac{1}{2}|\beta\rangle.$$

The functions are also normalized and orthogonal to one another:

$$\langle\alpha|\alpha\rangle = 1 \qquad \langle\beta|\beta\rangle = 1 \qquad \langle\alpha|\beta\rangle = 0 \qquad \langle\beta|\alpha\rangle = 0.$$

- Repeated observation of the variable corresponding to an operator $\hat{Q}$ gives an average equal to the expectation value $\langle Q\rangle$, where

$$\langle Q\rangle = \frac{\int \psi^\star \hat{Q}\psi \, \mathrm{d}\tau}{\int \psi^\star\psi \, \mathrm{d}\tau} \qquad \text{Dirac notation: } \langle Q\rangle = \frac{\langle\psi|\hat{Q}|\psi\rangle}{\langle\psi|\psi\rangle}.$$

- For the superposition state, the expectation values of the three components of angular momentum are

$$\langle I_z\rangle = \tfrac{1}{2}c_\alpha^\star c_\alpha - \tfrac{1}{2}c_\beta^\star c_\beta \quad \langle I_x\rangle = \tfrac{1}{2}c_\alpha^\star c_\beta + \tfrac{1}{2}c_\beta^\star c_\alpha \quad \langle I_y\rangle = \tfrac{1}{2}\mathrm{i}\,c_\beta^\star c_\alpha - \tfrac{1}{2}\mathrm{i}\,c_\alpha^\star c_\beta.$$

- The components of the bulk magnetization are given by

$$
\begin{aligned}
M_x &= \tfrac{1}{2}\gamma N\overline{\left(c_\alpha^\star c_\beta + c_\beta^\star c_\alpha\right)} &&\text{or}\quad M_x = \gamma N \,\overline{\langle I_x\rangle} \\
M_y &= \tfrac{1}{2}\mathrm{i}\gamma N\overline{\left(c_\beta^\star c_\alpha - c_\alpha^\star c_\beta\right)} &&\text{or}\quad M_y = \gamma N \,\overline{\langle I_y\rangle} \\
M_z &= \tfrac{1}{2}\gamma N\overline{\left(c_\alpha^\star c_\alpha - c_\beta^\star c_\beta\right)} &&\text{or}\quad M_z = \gamma N \,\overline{\langle I_z\rangle},
\end{aligned}
$$

where the overbar indicates an ensemble average, which means adding up the contributions from each spin in the sample and then dividing by the number of spins.

- The z-magnetization can be written in terms of the populations of the two levels, n_α and n_β:

$$M_z = \tfrac{1}{2}\gamma(n_\alpha - n_\beta).$$

- At equilibrium, the z-magnetization depends on the equilibrium populations, which are predicted by the Boltzmann distribution. The equilibrium x- and y-magnetizations are zero.

6.6   Time evolution

So far we have been studiously avoiding the issue of time in quantum mechanics, but we know that time evolution is of central importance in NMR, so it is clearly essential that we extend our understanding of quantum mechanics to include such evolution. The way in which the wavefunction changes over time is predicted by the *time-dependent Schrödinger* equation:

$$\frac{d\psi(t)}{dt} = -i\hat{H}\psi(t) \qquad \text{Dirac notation: } \frac{d|\psi(t)\rangle}{dt} = -i\hat{H}|\psi(t)\rangle. \qquad (6.28)$$

In this equation we have written $\psi(t)$ to remind ourselves that the wavefunction is a function of time.

The derivative on the left tells us how the wavefunction varies with time, and the equation says that this variation depends on the Hamiltonian. We have met the Hamiltonian before (section 3.2.5 on page 29) as the operator for energy; now we see it playing an even more crucial role in determining the time-evolution of the wavefunction.

As the wavefunction varies with time, so do the expectation values we have computed in the previous section. Thus, working out the time evolution of the wavefunction will enable us to find the time evolution of the magnetization.

We will now try to solve the time-dependent Schrödinger equation (TDSE) for two cases which are of great interest to us. The first is for free precession and the second is for an RF pulse.

6.6.1 Free evolution

For a single spin, the Hamiltonian for free evolution (i.e. in the absence of an RF field) is

$$\hat{H} = \Omega \hat{I}_z. \qquad (6.29)$$

This Hamiltonian is written in the rotating frame, so the frequency is the offset Ω, rather than the Larmor frequency ω_0. As was explained in section 4.4.2 on page 55, the offset is the difference between the Larmor frequency and the rotating frame frequency. We will do all of our calculations in the rotating frame as, just as was the case for the vector model, such a choice makes it possible to analyse the effect of an RF pulse. The Hamiltonian is also written in angular frequency units.

The superposition state of our single spin is

$$|\psi(t)\rangle = c_\alpha(t)|\alpha\rangle + c_\beta(t)|\beta\rangle. \qquad (6.30)$$

where we have written $c_\alpha(t)$ to remind ourselves that it is the coefficients $c_\alpha(t)$ and $c_\beta(t)$ which will vary in time. The basis functions $|\alpha\rangle$ and $|\beta\rangle$, are time independent.

Solving the TDSE in this case is not too difficult, but the algebra is perhaps a little intricate at times. To start with simply substitute Eq. 6.30 and the form of the Hamiltonian (Eq. 6.29) into the TDSE:

$$\frac{\mathrm{d}|\psi(t)\rangle}{\mathrm{d}t} = -\mathrm{i}\hat{H}|\psi(t)\rangle$$

$$\frac{\mathrm{d}\left[c_\alpha(t)|\alpha\rangle + c_\beta(t)|\beta\rangle\right]}{\mathrm{d}t} = -\mathrm{i}\Omega\hat{I}_z\left[c_\alpha(t)|\alpha\rangle + c_\beta(t)|\beta\rangle\right]$$

$$\frac{\mathrm{d}c_\alpha(t)}{\mathrm{d}t}|\alpha\rangle + \frac{\mathrm{d}c_\beta(t)}{\mathrm{d}t}|\beta\rangle = -\mathrm{i}\Omega c_\alpha(t)\underbrace{\hat{I}_z|\alpha\rangle}_{=\frac{1}{2}|\alpha\rangle} - \mathrm{i}\Omega c_\beta(t)\underbrace{\hat{I}_z|\beta\rangle}_{=-\frac{1}{2}|\beta\rangle}$$

$$\frac{\mathrm{d}c_\alpha(t)}{\mathrm{d}t}|\alpha\rangle + \frac{\mathrm{d}c_\beta(t)}{\mathrm{d}t}|\beta\rangle = -\frac{1}{2}\mathrm{i}\Omega c_\alpha(t)|\alpha\rangle + \frac{1}{2}\mathrm{i}\Omega c_\beta(t)|\beta\rangle.$$

Let us look at what we have done line by line. To go to the second line we have just substituted $|\psi(t)\rangle$ using Eq. 6.30 on the preceding page, and the Hamiltonian for the free precession, Eq. 6.29. Multiplying out the brackets takes us to the third line. Then we recognize that $|\alpha\rangle$ is an eigenfunction of $\hat{I}_z$, so $\hat{I}_z|\alpha\rangle = \frac{1}{2}|\alpha\rangle$; similarly $\hat{I}_z|\beta\rangle$ can be replaced by $-\frac{1}{2}|\beta\rangle$. This brings us, after a little tidying up, to the fourth line.

We now employ a 'trick' which is commonly used in quantum mechanics: we take this last equation and multiply *from the left* by $\langle\alpha|$.

$$\frac{\mathrm{d}c_\alpha(t)}{\mathrm{d}t}|\alpha\rangle + \frac{\mathrm{d}c_\beta(t)}{\mathrm{d}t}|\beta\rangle = -\frac{1}{2}\mathrm{i}\Omega c_\alpha(t)|\alpha\rangle + \frac{1}{2}\mathrm{i}\Omega c_\beta(t)|\beta\rangle \tag{6.31}$$

$$\langle\alpha|\frac{\mathrm{d}c_\alpha(t)}{\mathrm{d}t}|\alpha\rangle + \langle\alpha|\frac{\mathrm{d}c_\beta(t)}{\mathrm{d}t}|\beta\rangle = \langle\alpha|\left[-\frac{1}{2}\mathrm{i}\Omega c_\alpha(t)\right]|\alpha\rangle + \langle\alpha|\left[\frac{1}{2}\mathrm{i}\Omega c_\beta(t)\right]|\beta\rangle.$$

The derivatives of c_α and c_β, and the quantities in square braces on the right, are all just numbers so we can move them around as we like to give:

$$\frac{\mathrm{d}c_\alpha(t)}{\mathrm{d}t}\underbrace{\langle\alpha|\alpha\rangle}_{=1} + \frac{\mathrm{d}c_\beta(t)}{\mathrm{d}t}\underbrace{\langle\alpha|\beta\rangle}_{=0} = -\frac{1}{2}\mathrm{i}\Omega c_\alpha(t)\underbrace{\langle\alpha|\alpha\rangle}_{=1} + \frac{1}{2}\mathrm{i}\Omega c_\beta(t)\underbrace{\langle\alpha|\beta\rangle}_{=0}$$

$$\frac{\mathrm{d}c_\alpha(t)}{\mathrm{d}t} = -\frac{1}{2}\mathrm{i}\Omega c_\alpha(t). \tag{6.32}$$

To go to the second line we have utilized the by now familiar property that $|\alpha\rangle$ is normalized, hence $\langle\alpha|\alpha\rangle = 1$, and that $|\alpha\rangle$ and $|\beta\rangle$ are orthogonal, hence $\langle\alpha|\beta\rangle = 0$. The result of all of these rather lengthy manipulations is the relatively simple differential equation, Eq. 6.32.

This equation tells us how $c_\alpha(t)$ varies with time, which is what we are trying to work out. The solution to this equation is well known as it occurs in all kinds of physical and mathematical problems; it is

$$c_\alpha(t) = c_\alpha(0)\exp\left(-\frac{1}{2}\mathrm{i}\Omega t\right) \tag{6.33}$$

where $c_\alpha(0)$ is the value of the coefficient at time zero.

To show that this is a solution all we need to do is to substitute the expression for $c_\alpha(t)$ into the left of Eq. 6.32, and then compute the derivative:

$$\frac{dc_\alpha(t)}{dt} = \frac{d}{dt}\left[c_\alpha(0)\exp\left(-\tfrac{1}{2}i\Omega t\right)\right]$$

$$= c_\alpha(0)\left(-\tfrac{1}{2}i\Omega\right)\exp\left(-\tfrac{1}{2}i\Omega t\right)$$

$$= -\tfrac{1}{2}i\Omega\,\underbrace{c_\alpha(0)\exp\left(-\tfrac{1}{2}i\Omega t\right)}_{=c_\alpha(t)}$$

$$= -\tfrac{1}{2}i\Omega\,c_\alpha(t).$$

We have therefore shown that $dc_\alpha(t)/dt$ is equal to $-\tfrac{1}{2}i\Omega c_\alpha(t)$, which is exactly what Eq. 6.32 on the preceding page says, so Eq. 6.33 is the solution to Eq. 6.32.

If we go back to Eq. 6.31 on the facing page and left multiply by $\langle\beta|$ instead of $\langle\alpha|$, then a similar line of argument leads to

$$\frac{dc_\beta(t)}{dt} = \tfrac{1}{2}i\Omega c_\beta(t).$$

The solution to which is easily shown to be

$$c_\beta(t) = c_\beta(0)\exp\left(\tfrac{1}{2}i\Omega t\right).$$

Note that all that is different between the expressions for $c_\alpha(t)$ and $c_\beta(t)$ is the sign in the exponential term.

6.6.2 What this all means

The problem we set out to solve is how the superposition state

$$|\psi(t)\rangle = c_\alpha(t)|\alpha\rangle + c_\beta(t)|\beta\rangle$$

evolves in time under the influence of the free precession Hamiltonian

$$\hat{H} = \Omega\hat{I}_z.$$

The solution is that the coefficients vary according to

$$c_\alpha(t) = c_\alpha(0)\exp\left(-\tfrac{1}{2}i\Omega t\right) \qquad c_\beta(t) = c_\beta(0)\exp\left(\tfrac{1}{2}i\Omega t\right), \tag{6.34}$$

where $c_\alpha(0)$ and $c_\beta(0)$ are the coefficients at time zero. It is useful to remind ourselves that $|\alpha\rangle$ and $|\beta\rangle$ are eigenfunctions of the free precession Hamiltonian, $\Omega\hat{I}_z$:

$$\Omega\hat{I}_z|\alpha\rangle = \underbrace{\tfrac{1}{2}\Omega}_{\text{eigenvalue}}|\alpha\rangle \qquad \Omega\hat{I}_z|\beta\rangle = \underbrace{-\tfrac{1}{2}\Omega}_{\text{eigenvalue}}|\beta\rangle.$$

Recalling these, we can see that Eq. 6.34 says that each coefficient oscillates in phase at a frequency which depends on the corresponding eigenvalue of the Hamiltonian (in fact, the oscillation is as *minus* the corresponding eigenvalue). This phase oscillation is illustrated graphically in Fig. 6.2.

In the next section we will explore the effect that these oscillations have on the expectation values of the different components of angular momentum, and hence on the magnetization.

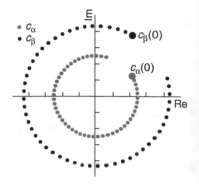

Fig. 6.2 Illustration of the time dependence of the coefficients c_α and c_β during a period of free precession. The real and imaginary parts of the coefficients are plotted along the x- and y-axes, respectively. The blue dots represent the motion of c_α, with the large blue dot representing the value at time zero; the black dots similarly represent c_β. The starting values are arbitrary, and successive dots indicate the coefficients at equal time intervals. Both coefficients follow a circular path which is a result of the phase modulation predicted by Eq. 6.34; each coefficient has constant magnitude, i.e. the distance to the origin, given by $\sqrt{c_\alpha^\star(t)c_\alpha(t)}$ or $\sqrt{c_\beta^\star(t)c_\beta(t)}$, is fixed. Note that the two coefficients proceed in opposite directions on account of the corresponding eigenvalues having different signs.

6.6.3   Effect of free evolution

In section 6.3.3 on page 109 we found that the expectation value of the z-component of angular momentum was (Eq. 6.9 on page 110)

$$\langle I_z \rangle = \tfrac{1}{2} c_\alpha^\star c_\alpha - \tfrac{1}{2} c_\beta^\star c_\beta.$$

We can now substitute in the values of $c_\alpha(t)$ and $c_\beta(t)$ from Eq. 6.34 on the previous page so that we can find out how the expectation value changes with time:

$$
\begin{aligned}
\langle I_z \rangle(t) &= \tfrac{1}{2}\left[c_\alpha(0)\exp\left(-\tfrac{1}{2}\mathrm{i}\Omega t\right)\right]^\star \left[c_\alpha(0)\exp\left(-\tfrac{1}{2}\mathrm{i}\Omega t\right)\right] \\
&\quad -\tfrac{1}{2}\left[c_\beta(0)\exp\left(\tfrac{1}{2}\mathrm{i}\Omega t\right)\right]^\star \left[c_\beta(0)\exp\left(\tfrac{1}{2}\mathrm{i}\Omega t\right)\right] \\
&= \tfrac{1}{2}c_\alpha^\star(0)\exp\left(\tfrac{1}{2}\mathrm{i}\Omega t\right)c_\alpha(0)\exp\left(-\tfrac{1}{2}\mathrm{i}\Omega t\right) \\
&\quad -\tfrac{1}{2}c_\beta^\star(0)\exp\left(-\tfrac{1}{2}\mathrm{i}\Omega t\right)c_\beta(0)\exp\left(+\tfrac{1}{2}\mathrm{i}\Omega t\right) \\
&= \tfrac{1}{2}c_\alpha^\star(0)c_\alpha(0) - \tfrac{1}{2}c_\beta^\star(0)c_\beta(0).
\end{aligned}
$$

To go from the second line to the last we have recognized that $\exp(\theta)\exp(-\theta) = \exp(0) = 1$. The result is that the expectation value of the z-component of angular momentum is *not* affected by free evolution. This is hardly a surprise: as we have seen in the vector model, free evolution is a rotation about the z-axis, and so we do not expect the z-component to be affected by such a rotation.

The x-component behaves in a more interesting way. The expectation value is given by Eq. 6.12 on page 111:

$$
\begin{aligned}
\langle I_x \rangle(t) &= \tfrac{1}{2}c_\alpha^\star c_\beta + \tfrac{1}{2}c_\beta^\star c_\alpha \\
&= \tfrac{1}{2}\left[c_\alpha(0)\exp\left(-\tfrac{1}{2}\mathrm{i}\Omega t\right)\right]^\star \left[c_\beta(0)\exp\left(\tfrac{1}{2}\mathrm{i}\Omega t\right)\right] \\
&\quad +\tfrac{1}{2}\left[c_\beta(0)\exp\left(\tfrac{1}{2}\mathrm{i}\Omega t\right)\right]^\star \left[c_\alpha(0)\exp\left(-\tfrac{1}{2}\mathrm{i}\Omega t\right)\right] \\
&= \tfrac{1}{2}c_\alpha^\star(0)\exp\left(\tfrac{1}{2}\mathrm{i}\Omega t\right)c_\beta(0)\exp\left(\tfrac{1}{2}\mathrm{i}\Omega t\right) \\
&\quad +\tfrac{1}{2}c_\beta^\star(0)\exp\left(-\tfrac{1}{2}\mathrm{i}\Omega t\right)c_\alpha(0)\exp\left(-\tfrac{1}{2}\mathrm{i}\Omega t\right) \\
&= \tfrac{1}{2}c_\alpha^\star(0)c_\beta(0)\exp\left(\mathrm{i}\Omega t\right) + \tfrac{1}{2}c_\beta^\star(0)c_\alpha(0)\exp\left(-\mathrm{i}\Omega t\right). \quad (6.35)
\end{aligned}
$$

In contrast to the z-component, the expectation value of the x-component oscillates at frequency Ω. This is, of course, exactly the frequency of precession in the rotating frame, so it should not come as a surprise.

The final expression for $\langle I_x \rangle(t)$ can be simplified by writing the complex exponentials in terms of sines and cosines using the identities:

$$\exp(\mathrm{i}\theta) \equiv \cos\theta + \mathrm{i}\sin\theta \qquad \exp(-\mathrm{i}\theta) \equiv \cos\theta - \mathrm{i}\sin\theta.$$

Then Eq. 6.35 on the preceding page can be re-expressed as

$$
\begin{aligned}
\langle I_x \rangle (t) &= \tfrac{1}{2} c_\alpha^\star(0) c_\beta(0) \exp(\mathrm{i}\Omega t) + \tfrac{1}{2} c_\beta^\star(0) c_\alpha(0) \exp(-\mathrm{i}\Omega t) \\
&= \tfrac{1}{2} c_\alpha^\star(0) c_\beta(0) \left[\cos(\Omega t) + \mathrm{i}\sin(\Omega t) \right] \\
&\quad + \tfrac{1}{2} c_\beta^\star(0) c_\alpha(0) \left[\cos(\Omega t) - \mathrm{i}\sin(\Omega t) \right] \\
&= \cos(\Omega t) \underbrace{\left[\tfrac{1}{2} c_\alpha^\star(0) c_\beta(0) + \tfrac{1}{2} c_\beta^\star(0) c_\alpha(0) \right]}_{=\langle I_x \rangle(0)} \\
&\quad - \sin(\Omega t) \underbrace{\left[\tfrac{1}{2}\mathrm{i} c_\beta^\star(0) c_\alpha(0) - \tfrac{1}{2}\mathrm{i} c_\alpha^\star(0) c_\beta(0) \right]}_{=\langle I_y \rangle(0)} \\
&= \cos(\Omega t)\langle I_x \rangle(0) - \sin(\Omega t)\langle I_y \rangle(0).
\end{aligned}
$$

On the third line we recognize that the quantity in the first square brace is simply $\langle I_x \rangle$ at time zero (Eq. 6.12 on page 111), and similarly the quantity in the second square brace is $\langle I_y \rangle$ at time zero (Eq. 6.13 on page 111).

This says something entirely familiar, which is that if we start out at time zero with a vector along $-y$, after time t the vector has precessed through an angle Ωt generating an x-component proportional to the sine of this angle.

A similar calculation for the y-component of the angular momentum gives

$$
\langle I_y \rangle(t) = \cos(\Omega t)\langle I_y \rangle(0) + \sin(\Omega t)\langle I_x \rangle(0). \tag{6.36}
$$

Again, this is a familiar result as it shows that a vector aligned initially along x rotates through an angle Ωt towards y; this is visualized in Fig. 6.3.

It is important to realize that these expressions refer to the x- and y-components of a *single spin*. To work out the components of the bulk magnetization we need to multiply by γN and take the ensemble average, as in Eqs 6.20 and 6.21 on p. 114:

$$
\begin{aligned}
\gamma N \overline{\langle I_x \rangle}(t) &= \cos(\Omega t)\gamma N \overline{\langle I_x \rangle}(0) - \sin(\Omega t)\gamma N \overline{\langle I_y \rangle}(0) \\
\text{hence } M_x(t) &= \cos(\Omega t) M_x(0) - \sin(\Omega t) M_y(0) \\
\gamma N \overline{\langle I_y \rangle}(t) &= \cos(\Omega t)\gamma N \overline{\langle I_y \rangle}(0) + \sin(\Omega t)\gamma N \overline{\langle I_x \rangle}(0) \\
\text{hence } M_y(t) &= \cos(\Omega t) M_y(0) + \sin(\Omega t) M_x(0).
\end{aligned}
$$

These relationships show how the components of the magnetization evolve over time, and they predict an oscillatory interchange of the x- and y-components; these predictions are identical to those of the vector model. We now turn our attention to the somewhat more complex case of the effect of an RF pulse.

6.7 RF pulses

As we saw in Chapter 4, an RF pulse results in a transverse magnetic field B_1 appearing in the rotating frame, and if we are on resonance (or the pulse is strong) this is the only field present in the rotating frame. We also saw that the magnetization rotates about this field at a frequency $\omega_1 = |\gamma B_1|$. It should come as no surprise therefore, that the Hamiltonian (in the rotating

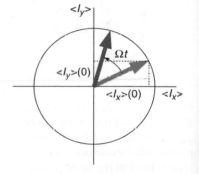

Fig. 6.3 Visualization of the predictions of Eq. 6.36; the horizontal and vertical axes represent the expectation values of the x- and y-components of the angular momentum. The blue vector depicts the initial condition ($t = 0$) in which the expectation values are $\langle I_x \rangle(0)$ and $\langle I_y \rangle(0)$. After time t, the vector has rotated through an angle Ωt to the position given by the dark grey vector. The y-component of this vector clearly depends on both the x- and y-components at time zero, as given by Eq. 6.36.

frame) for an on resonance or strong pulse about the x-axis is

$$\hat{H}_{\text{pulse}} = \omega_1 \hat{I}_x.$$

Comparing this with the free precession Hamiltonian $\hat{H}_{\text{free}} = \Omega \hat{I}_z$ (Eq. 6.29 on page 119), we see that the operator is now $\hat{I}_x$ rather than $\hat{I}_z$; this is because the magnetic field is now along x rather than z. The frequency has changed from the offset Ω to ω_1, the rate at which the RF field rotates the magnetization.

To work out what effect this Hamiltonian has on our superposition state all we have to do is solve the TDSE, just as we did before. However, finding the solution turns out to be rather more complex than before as $|\alpha\rangle$ and $|\beta\rangle$ are not eigenfunctions of $\hat{I}_x$ (and hence of $\hat{H}_{\text{pulse}}$). We will not therefore go into all the details, but simply quote the results.

The differential equations for the coefficients turn out to be

$$\frac{dc_\alpha(t)}{dt} = -\tfrac{1}{2}i\omega_1 c_\beta(t) \qquad \frac{dc_\beta(t)}{dt} = -\tfrac{1}{2}i\omega_1 c_\alpha(t).$$

Comparing these with the equivalent differential equations for free evolution, Eq. 6.32 on page 120, we see that the additional complication for the effect of the pulse is that the rate of change of c_α depends on c_β, and vice versa. The solutions to these equations can be obtained by standard methods and are:

$$c_\alpha(t) = \cos\left(\tfrac{1}{2}\omega_1 t\right) c_\alpha(0) - i\sin\left(\tfrac{1}{2}\omega_1 t\right) c_\beta(0) \tag{6.37}$$

$$c_\beta(t) = \cos\left(\tfrac{1}{2}\omega_1 t\right) c_\beta(0) - i\sin\left(\tfrac{1}{2}\omega_1 t\right) c_\alpha(0). \tag{6.38}$$

Compared with free evolution (Eq. 6.34 on page 121), these results are more complex as the value of c_α at time t depends on *both* c_α and c_β at time zero. What is happening here is that the Hamiltonian is causing a *mixing* between the states $|\alpha\rangle$ and $|\beta\rangle$, resulting in an oscillatory interchange of the two coefficients, which is represented graphically in Fig. 6.4.

It is interesting to compare this figure with the corresponding one for free precession, Fig. 6.2 on page 121. In the case of free precession, the coefficients are simply phase modulated and so proceed on circular paths; their magnitudes, represented by the distance from the origin, remain constant. In contrast, during a pulse the magnitudes of the coefficients change as a result of the oscillatory interchange between the two coefficients.

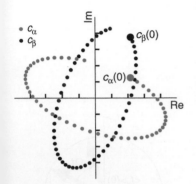

Fig. 6.4 Illustration of the time dependence of the coefficients c_α and c_β during an RF pulse; the motion of the coefficients is represented in the same form as in Fig. 6.2 on page 121. The motion of the coefficients for the case of a pulse is very different to that for free precession shown in Fig. 6.2. First, the paths are not circular; secondly the magnitude of the coefficients change (i.e. their distance from the origin changes), and finally the paths proceed in the same direction. What this diagram represents is an oscillatory interchange between c_α and c_β.

6.7.1   Effect on the components of angular momentum

Now that we have the values of c_α and c_β at any time, we can compute the expectation values of the x-, y- and z-components of the angular momentum, as these are related to the coefficients in the following ways (section 6.3.3 on page 109)

$$\langle I_x\rangle(t) = \tfrac{1}{2}\left[c_\alpha^\star(t)c_\beta(t) + c_\beta^\star(t)c_\alpha(t)\right]$$

$$\langle I_y\rangle(t) = \tfrac{1}{2}i\left[c_\beta^\star(t)c_\alpha(t) - c_\alpha^\star(t)c_\beta(t)\right]$$

$$\langle I_z\rangle(t) = \tfrac{1}{2}\left[c_\alpha^\star(t)c_\alpha(t) - c_\beta^\star(t)c_\beta(t)\right].$$

We have written $\langle I_x \rangle(t)$ etc. to remind ourselves that, as the coefficients depend on time, so will the expectation values.

If we substitute the expression from Eqs 6.37 and 6.38 into the above relationships for $\langle I_x \rangle(t)$ etc., we obtain, after some tedious but essentially straightforward manipulation:

$$
\begin{aligned}
\langle I_x \rangle(t) &= \tfrac{1}{2}\left[c_\alpha^\star(0)c_\beta(0) + c_\beta^\star(0)c_\alpha(0)\right], \\
\langle I_y \rangle(t) &= -\tfrac{1}{2}\mathrm{i}\left[c_\alpha^\star(0)c_\beta(0) - c_\alpha(0)c_\beta^\star(0)\right]\cos\left(\omega_1 t\right) \\
&\quad -\tfrac{1}{2}\left[c_\alpha^\star(0)c_\alpha(0) - c_\beta^\star(0)c_\beta(0)\right]\sin\left(\omega_1 t\right), \qquad (6.39) \\
\langle I_z \rangle(t) &= \tfrac{1}{2}\left[c_\alpha^\star(0)c_\alpha(0) - c_\beta^\star(0)c_\beta(0)\right]\cos\left(\omega_1 t\right) \\
&\quad +\tfrac{1}{2}\mathrm{i}\left[c_\alpha(0)c_\beta^\star(0) - c_\beta(0)c_\alpha^\star(0)\right]\sin\left(\omega_1 t\right).
\end{aligned}
$$

These expressions can be simplified greatly if we recognize that the quantities in square braces are the expectation values of the various components of the angular momentum at time zero, i.e.

$$
\begin{aligned}
\langle I_x \rangle(0) &= \tfrac{1}{2}\left[c_\alpha^\star(0)c_\beta(0) + c_\beta^\star(0)c_\alpha(0)\right], \\
\langle I_y \rangle(0) &= \tfrac{1}{2}\mathrm{i}\left[c_\beta^\star(0)c_\alpha(0) - c_\alpha^\star(0)c_\beta(0)\right], \\
\langle I_z \rangle(0) &= \tfrac{1}{2}\left[c_\alpha^\star(0)c_\alpha(0) - c_\beta^\star(0)c_\beta(0)\right].
\end{aligned}
$$

With these substitutions, things look rather simpler:

$$
\begin{aligned}
\langle I_x \rangle(t) &= \langle I_x \rangle(0), && (6.40) \\
\langle I_y \rangle(t) &= \langle I_y \rangle(0)\cos\left(\omega_1 t\right) - \langle I_z \rangle(0)\sin\left(\omega_1 t\right), && (6.41) \\
\langle I_z \rangle(t) &= \langle I_z \rangle(0)\cos\left(\omega_1 t\right) + \langle I_y \rangle(0)\sin\left(\omega_1 t\right). && (6.42)
\end{aligned}
$$

These relationships are relatively easy to interpret. First, we see that the x-component simply does not change, but retains the same value it had at time zero. Given that we have chosen the pulse to be about the x-axis, it is not surprising that the x-component is unaffected.

The expression for the y-component, Eq. 6.41, predicts an oscillatory interchange of the y- and z-components. For example, if we choose the time such that $\omega_1 t = \pi/2$, then $\langle I_y \rangle(t) = -\langle I_z \rangle(0)$. This looks familiar: the z-component is rotated onto y by a pulse with a flip angle of $\pi/2$ radians or $90°$.

Similarly, the z-component, Eq. 6.42, shows an oscillatory interchange of the y- and z-components. In other words, it represents a vector rotating in the yz-plane at frequency ω_1.

6.7.2 Effect on the components of the bulk magnetization

The results in the previous section tell us how the individual components of the angular momentum from a *single spin* evolve over time. However, what we want to know is how the *bulk magnetization* evolves over time. As was explained in section 6.4 on page 112, we find the bulk magnetization by taking the ensemble averages of the expectation values of the individual components of the angular momentum.

From Eqs 6.20, 6.21 and 6.22 on page 114, we have

$$M_x = \gamma N \langle \overline{I_x} \rangle \qquad M_y = \gamma N \langle \overline{I_y} \rangle \qquad M_z = \gamma N \langle \overline{I_z} \rangle. \qquad (6.43)$$

If we take the expression for $\langle I_y \rangle(t)$, Eq. 6.41 on the preceding page, compute the ensemble average and multiply by γN, we obtain

$$\gamma N \langle \overline{I_y} \rangle(t) \quad = \quad \gamma N \langle \overline{I_y} \rangle(0) \cos(\omega_1 t) - \gamma N \langle \overline{I_z} \rangle(0) \sin(\omega_1 t),$$

which, using Eq. 6.43, we recognize is the same thing as

$$M_y(t) = M_y(0) \cos(\omega_1 t) - M_z(0) \sin(\omega_1 t).$$

Using the same approach for each of the components, we can use Eqs 6.40 to 6.42 to determine the three components of the bulk magnetization as

$$M_x(t) \quad = \quad M_x(0), \qquad (6.44)$$
$$M_y(t) \quad = \quad M_y(0) \cos(\omega_1 t) - M_z(0) \sin(\omega_1 t), \qquad (6.45)$$
$$M_z(t) \quad = \quad M_z(0) \cos(\omega_1 t) + M_y(0) \sin(\omega_1 t). \qquad (6.46)$$

These relationships predict exactly the same outcome of an x-pulse as the vector model: x-magnetization is unaffected, whereas y- and z-magnetization are rotated into one another i.e. the magnetization vector rotates in the yz-plane.

If we assume that the magnetization is at equilibrium at time zero, we can write

$$M_x(0) = 0 \quad M_y(0) = 0 \quad M_z(0) = M_0,$$

where M_0 is the equilibrium magnetization. Using these values in Eqs 6.44 to 6.46, gives

$$M_x(t) = 0 \qquad M_y(t) = -M_0 \sin(\omega_1 t) \qquad M_z(t) = M_0 \cos(\omega_1 t).$$

This is a result we are very familiar with: the equilibrium magnetization is rotated from z through an angle $\omega_1 t$ towards $-y$.

6.8   Making faster progress: the density operator

We have now developed the theory to the point where we can predict the time evolution of the components of the bulk magnetization during free precession and during RF pulses. However, you would be forgiven for thinking that an enormous amount of labour has been needed to derive some essentially trivial results!

The reason that everything is such hard work is that the calculations involve a three-stage process: first, we solve the TDSE using the relevant Hamiltonian; then we compute the components of the angular momentum of a single spin; finally, we compute the ensemble average to find the bulk magnetization.

What we need to find is a way of saving some of this labour. The first step is to notice that the components of the angular momentum are always expressed in terms of particular *products* of the two coefficients c_α and c_β:

$$c_\alpha^\star c_\alpha \quad c_\beta^\star c_\beta \quad c_\alpha^\star c_\beta \quad c_\beta^\star c_\alpha.$$

What is more, when we take the ensemble averages, it is these products which are subject to the averaging process.

These observations lead us to wonder if there is some way of reformulating the theory so that the products of the coefficients, and their ensemble averages, appear in a more convenient way. It turns out that by introducing the *density operator* (also called the *density matrix*) such a simplification can be achieved.

6.8.1 Introducing the density operator

The density operator $\hat{\rho}$ is defined as

$$\hat{\rho} = \overline{|\psi\rangle\langle\psi|}. \tag{6.47}$$

As before, the overbar indicates taking an ensemble average, which means adding up the contributions from each spin in the sample and then dividing by the number of spins.

In the case of our superposition state, $|\psi\rangle$ and $\langle\psi|$ are given by

$$|\psi\rangle = c_\alpha|\alpha\rangle + c_\beta|\beta\rangle \qquad \langle\psi| = c_\alpha^\star\langle\alpha| + c_\beta^\star\langle\beta|.$$

You might think that the definition of Eq. 6.47 makes it look as if $\hat{\rho}$ is a function, not an operator. However, we can see that it is indeed an operator by forming its matrix representation, as described in section 6.3.5 on page 111. Following the general form given in Eq. 6.14 on page 111, the matrix representation of $\hat{\rho}$ is

$$\rho = \begin{pmatrix} \overline{\langle\alpha|\hat{\rho}|\alpha\rangle} & \overline{\langle\alpha|\hat{\rho}|\beta\rangle} \\ \overline{\langle\beta|\hat{\rho}|\alpha\rangle} & \overline{\langle\beta|\hat{\rho}|\beta\rangle} \end{pmatrix} = \begin{pmatrix} \rho_{11} & \rho_{12} \\ \rho_{21} & \rho_{22} \end{pmatrix}.$$

Each of the matrix elements can be evaluated by using Eq. 6.47 and the known properties of $|\alpha\rangle$ and $|\beta\rangle$. As an example, let us work out the top right-hand element i.e. that in row 1 and column 2, the element ρ_{12}. To avoid clutter, we will leave out the overbar which indicates ensemble averaging until the last line.

$$\begin{aligned}
\rho_{12} &= \langle\alpha|\hat{\rho}|\beta\rangle \\
&= \langle\alpha| \left[|\psi\rangle\langle\psi|\right] |\beta\rangle \\
&= \langle\alpha| \left[c_\alpha|\alpha\rangle + c_\beta|\beta\rangle\right] \left[c_\alpha^\star\langle\alpha| + c_\beta^\star\langle\beta|\right] |\beta\rangle \\
&= \langle\alpha| \left[c_\alpha|\alpha\rangle + c_\beta|\beta\rangle\right] \left[c_\alpha^\star \underbrace{\langle\alpha|\beta\rangle}_{=0} + c_\beta^\star \underbrace{\langle\beta|\beta\rangle}_{=1}\right] \\
&= \langle\alpha| \left[c_\alpha|\alpha\rangle + c_\beta|\beta\rangle\right] c_\beta^\star \\
&= \left[c_\alpha \underbrace{\langle\alpha|\alpha\rangle}_{=1} + c_\beta \underbrace{\langle\alpha|\beta\rangle}_{=0}\right] c_\beta^\star \\
&= \overline{c_\alpha c_\beta^\star}.
\end{aligned}$$

To go to the second line, the definition of $\hat{\rho}$ has been inserted, and then on the third line $|\psi\rangle$ and $\langle\psi|$ have been expressed as the superposition of $|\alpha\rangle$ and $|\beta\rangle$. The right-hand square bracket is then multiplied out to give line four, and then we use the properties that $|\alpha\rangle$ and $|\beta\rangle$ are orthogonal and normalized to go to the next line. Repeating the same procedure with the left-hand bracket carries us to the end of the calculation. Finally, we recall that the density operator, and its matrix elements, are defined as an ensemble average, which is why the overbar has been inserted on the final line.

Applying the same procedure gives the other elements as

$$\rho_{21} = \langle \beta | \hat{\rho} | \alpha \rangle = \overline{c_\alpha^\star c_\beta} \quad \rho_{11} = \langle \alpha | \hat{\rho} | \alpha \rangle = \overline{c_\alpha c_\alpha^\star} \quad \rho_{22} = \langle \beta | \hat{\rho} | \beta \rangle = \overline{c_\beta^\star c_\beta}.$$

Thus the complete matrix representation of the density operator is

$$\rho = \begin{pmatrix} \overline{c_\alpha c_\alpha^\star} & \overline{c_\alpha c_\beta^\star} \\ \overline{c_\beta c_\alpha^\star} & \overline{c_\beta c_\beta^\star} \end{pmatrix}. \tag{6.48}$$

Remember that the coefficients c_α and c_β vary with time, so $\hat{\rho}$ is therefore a function of time. Things are already looking hopeful as the elements of this matrix are the ensemble averages of the products of the coefficients – the quantities which always appear when we compute the components of the bulk magnetization.

6.8.2   Calculating the components of the bulk magnetization

The really convenient feature of the density operator is that we can compute the bulk magnetization directly from the matrix form of the operator. For example, the x-magnetization, given by

$$M_x = \tfrac{1}{2}\gamma N \overline{\left(c_\alpha^\star c_\beta + c_\alpha c_\beta^\star \right)}.$$

can be written in terms of the elements of the density operator (Eq. 6.48) as

$$M_x = \tfrac{1}{2}\gamma N (\rho_{21} + \rho_{12}).$$

Similarly, M_y and M_z are

$$M_y = \tfrac{1}{2}i\gamma N (\rho_{12} - \rho_{21}) \qquad M_z = \tfrac{1}{2}\gamma N (\rho_{11} - \rho_{22}).$$

The important point here is that the ensemble averaging is contained *within* the density operator, so we can compute the bulk magnetization *directly* from the density operator. This is an important advantage of this approach.

6.8.3   Equilibrium density operator

Recall from section 6.4.2 on page 115 that $\overline{c_\alpha c_\alpha^\star}$ and $\overline{c_\beta c_\beta^\star}$ are related to the populations, n_α and n_β, of the two states (Eq. 6.23 on page 115):

$$\overline{c_\alpha c_\alpha^\star} = n_\alpha/N \qquad \overline{c_\beta c_\beta^\star} = n_\beta/N.$$

The two diagonal elements of the matrix representation of the density operator are thus n_α/N and n_β/N.

In section 6.4.3 on page 116 we also argued that, at equilibrium, the ensemble averages $\overline{c_\alpha^\star c_\beta}$ and $\overline{c_\beta^\star c_\alpha}$ are zero. So, at equilibrium, the two off-diagonal elements of the matrix form of the density operator are zero.

We can thus write the equilibrium density operator as

$$\hat{\rho}_{eq} = \begin{pmatrix} n_{\alpha,eq}/N & 0 \\ 0 & n_{\beta,eq}/N \end{pmatrix} \tag{6.49}$$

where $n_{\alpha,eq}$ and $n_{\beta,eq}$ are the equilibrium populations.

This is rather a nice result as it gives us a starting point for the calculation since all of our experiments will start with a sample which has come to equilibrium. The final thing we need to know is how the density operator varies with time.

6.8.4 Time evolution of the density operator

Starting from the TDSE, Eq. 6.28 on page 119:

$$\frac{d|\psi(t)\rangle}{dt} = -i\hat{H}|\psi(t)\rangle,$$

it can be shown that the equivalent equation of motion for the density operator is

$$\frac{d\hat{\rho}(t)}{dt} = -i\left(\hat{H}\,\hat{\rho}(t) - \hat{\rho}(t)\,\hat{H}\right); \tag{6.50}$$

this is known as the *Liouville–von Neumann* equation. The order in which operators act is important, so $\hat{H}\,\hat{\rho}$ is *not* the same thing as $\hat{\rho}\,\hat{H}$.

It can be shown that the solution to Eq. 6.50 is

$$\hat{\rho}(t) = \exp(-i\hat{H}t)\,\hat{\rho}(0)\,\exp(i\hat{H}t), \tag{6.51}$$

where $\hat{\rho}(t)$ is the density operator at time t and $\hat{\rho}(0)$ is the density operator at time zero.

You would be forgiven for thinking that this does not look like much of a 'solution' to the problem. However, it turns out that $\exp(\pm i\hat{H}t)$ can be expressed in matrix form, just in the same way as is possible for $\hat{\rho}$. So, the right-hand side of Eq. 6.51 is in fact just the multiplication of three matrices together, which is a trivial task.

However, we are not going to describe how these matrix forms of $\exp(\pm i\hat{H}t)$ are found, as there is another way to approach the use of the density operator which avoids the need for matrices, and results in a much more intuitive method of calculation. This approach casts the whole problem in operators, as described in the next section.

6.8.5 Representing the density operator using a basis of operators

We are familiar with the idea that the position of any point in space can be described by specifying its x-, y- and z-coordinates. Part of the reason why this approach is useful is that the x-, y- and z-directions are all orthogonal to one another i.e. the angles between them are all 90°. This means that, for example, changing the x-component does not change the other components.

Expressed somewhat more formally, we have three orthogonal unit vectors pointing along the x-, y- and z-directions. The vectors have unit length (hence their name), and are denoted $\mathbf{e}_x$, $\mathbf{e}_y$ and $\mathbf{e}_z$. Any vector can be expressed as a linear combination of these three unit vectors:

$$a_x\,\mathbf{e}_x + a_y\,\mathbf{e}_y + a_z\,\mathbf{e}_z,$$

where a_x gives the component in the x-direction and similarly a_y and a_z are the components along y and z, respectively. The vectors $(\mathbf{e}_x, \mathbf{e}_y, \mathbf{e}_z)$ are said to be *basis vectors*.

In a similar way, a matrix can be expressed as a linear combination of a basis set of other matrices. For example, it turns out that the matrix representation of the density operator, for an ensemble of spin-half nuclei, can be expressed as a linear combination of the matrices representing $\hat{I}_x$, $\hat{I}_y$, $\hat{I}_z$ and a fourth matrix which we shall call E. These matrix representations were introduced in section 6.3.5 on page 111 (Eq. 6.15 and Eq. 6.16 on page 112), and are repeated here for convenience, along with the definition of E.

$$I_x = \begin{pmatrix} 0 & \frac{1}{2} \\ \frac{1}{2} & 0 \end{pmatrix} \qquad I_y = \begin{pmatrix} 0 & -\frac{1}{2}\mathrm{i} \\ \frac{1}{2}\mathrm{i} & 0 \end{pmatrix}$$

$$I_z = \begin{pmatrix} \frac{1}{2} & 0 \\ 0 & -\frac{1}{2} \end{pmatrix} \qquad E = \begin{pmatrix} 1 & 0 \\ 0 & 1 \end{pmatrix}.$$

These matrices are orthogonal to one another in an analogous way to the unit vectors along x, y and z. Two matrices, $\mathbf{A}$ and $\mathbf{B}$, are said to be orthogonal if the *trace* of their product is zero:

$$\text{orthogonality:} \quad \text{Tr}\{\mathbf{AB}\} = 0.$$

The trace of a matrix, denoted $\text{Tr}\{\mathbf{M}\}$, is the sum of its diagonal elements.

This is best illustrated by way of an example: let us consider the product of the matrix representations of $\hat{I}_x$ and $\hat{I}_z$:

$$\hat{I}_x\hat{I}_z \;=\; \begin{pmatrix} 0 & \frac{1}{2} \\ \frac{1}{2} & 0 \end{pmatrix} \times \begin{pmatrix} \frac{1}{2} & 0 \\ 0 & -\frac{1}{2} \end{pmatrix}$$

$$=\; \begin{pmatrix} 0 & -\frac{1}{4} \\ \frac{1}{4} & 0 \end{pmatrix}.$$

The trace of this final matrix is the sum of the elements along the diagonal i.e. elements 1,1 and 2,2; clearly in this case the trace is zero, so the matrix representations of $\hat{I}_x$ and $\hat{I}_z$ are orthogonal. Similar calculations will show that any two of the four basis operators are also orthogonal in this sense.

We now write the density operator in terms of a linear combination of the four basis operators:

$$\hat{\rho} = a_E\hat{E} + a_x\hat{I}_x + a_y\hat{I}_y + a_z\hat{I}_z. \tag{6.52}$$

If the operators are written in their matrix forms, this combination is

$$\rho = a_E \begin{pmatrix} 1 & 0 \\ 0 & 1 \end{pmatrix} + a_x \begin{pmatrix} 0 & \frac{1}{2} \\ \frac{1}{2} & 0 \end{pmatrix} + a_y \begin{pmatrix} 0 & -\frac{1}{2}i \\ \frac{1}{2}i & 0 \end{pmatrix} + a_z \begin{pmatrix} \frac{1}{2} & 0 \\ 0 & -\frac{1}{2} \end{pmatrix}.$$

Adding up all of these terms, the matrix form of the density operator is

$$\rho = \frac{1}{2} \begin{pmatrix} 2a_E + a_z & a_x - ia_y \\ a_x + ia_y & 2a_E - a_z \end{pmatrix}. \tag{6.53}$$

Note that a_x and a_y only appear on the off-diagonal elements, whereas a_E and a_z only appear on the diagonal elements.

In section 6.8.2 on page 128 we noted that the really useful feature of the density matrix was that we could compute the bulk magnetizations *directly* from its elements. Repeating the results from that section, it was shown that

$$M_x = \tfrac{1}{2}\gamma N(\rho_{21} + \rho_{12}) \quad M_y = \tfrac{1}{2}i\gamma N(\rho_{12} - \rho_{21}) \quad M_z = \tfrac{1}{2}\gamma N(\rho_{11} - \rho_{22}).$$

Using the elements of ρ from Eq. 6.53, we find

$$M_x = \tfrac{1}{2}\gamma N a_x \quad M_y = \tfrac{1}{2}\gamma N a_y \quad M_z = \tfrac{1}{2}\gamma N a_z. \tag{6.54}$$

Now this really is a very nice result, as it says that if we write the density operator as the linear combination of operators, Eq. 6.52 on the preceding page, we can extract the value of the bulk magnetization just by inspecting the values of the coefficients a_x etc. Note that the coefficient a_E does not contribute to *any* of the magnetizations.

We will see in the next chapter that, by using this operator expansion we can work out the time evolution of the system without solving the TDSE directly, taking any ensemble averages or working with matrices. The resulting approach is thus very convenient to use, and indeed it is the one we will use exclusively in the rest of the book. The practical details of how calculations are actually made using this approach are described in the next chapter.

6.8.6 The equilibrium density operator – again

In section 6.8.3 on page 128 we showed that at equilibrium the matrix representation of the density operator, $\hat{\rho}_{eq}$, was

$$\hat{\rho}_{eq} = \begin{pmatrix} n_{\alpha,eq}/N & 0 \\ 0 & n_{\beta,eq}/N \end{pmatrix},$$

where $n_{\alpha,eq}$ and $n_{\beta,eq}$ are the equilibrium populations of the two levels.

These equilibrium populations can be computed using the Boltzmann distribution (Eq. 6.25 on page 115):[2]

$$n_{\alpha,eq} = \tfrac{1}{2}N\exp(-E_\alpha/k_B T) \quad n_{\beta,eq} = \tfrac{1}{2}N\exp(-E_\beta/k_B T)$$

---

[2]The factor of $\frac{1}{2}$ in these expressions is $1/q$, where q is the partition function. In this case $q = 2$, as there are two accessible states.

The energy of interaction of the spins with the magnetic field is very much less than the thermal energy, which means that $E_m/k_B T$ is very small. As a result, the exponential terms can be well approximated using $\exp(x) = 1 + x$ to give

$$n_{\alpha,eq} = \tfrac{1}{2}N(1 - E_\alpha/k_B T) \qquad n_{\beta,eq} = \tfrac{1}{2}N(1 - E_\beta/k_B T)$$

This can be further simplified by recalling that the energies are (in J)

$$E_\alpha = -\tfrac{1}{2}\hbar\gamma B_0 \qquad E_\beta = +\tfrac{1}{2}\hbar\gamma B_0,$$

so

$$n_{\alpha,eq} = \tfrac{1}{2}N(1 + \frac{\hbar\gamma B_0}{2k_B T}) \qquad n_{\beta,eq} = \tfrac{1}{2}N(1 - \frac{\hbar\gamma B_0}{2k_B T})$$

From these it follows that the average population, $n_{av} = \tfrac{1}{2}(n_{\alpha,eq} + n_{\beta,eq})$, is simply $\tfrac{1}{2}N$, and the difference in the populations, $\Delta n = (n_{\alpha,eq} - n_{\beta,eq})$, is

$$\Delta n = \frac{\hbar\gamma B_0}{2k_B T}.$$

The populations of the two levels can therefore be written

$$n_{\alpha,eq} = n_{av} + \tfrac{1}{2}\Delta n \qquad n_{\beta,eq} = n_{av} - \tfrac{1}{2}\Delta n,$$

and so the equilibrium density operator becomes

$$\rho_{eq} = \frac{1}{N}\begin{pmatrix} n_{av} + \tfrac{1}{2}\Delta n & 0 \\ 0 & n_{av} - \tfrac{1}{2}\Delta n \end{pmatrix}.$$

This equilibrium density matrix can be written in terms of the matrix representations of $\hat{E}$ and $\hat{I}_z$ in the following way:

$$
\begin{aligned}
\hat{\rho}_{eq} &= \frac{n_{av}}{N}\begin{pmatrix} 1 & 0 \\ 0 & 1 \end{pmatrix} + \frac{\Delta n}{N}\begin{pmatrix} \tfrac{1}{2} & 0 \\ 0 & -\tfrac{1}{2} \end{pmatrix} \\
&= \frac{n_{av}}{N}\hat{E} + \frac{\Delta n}{N}\hat{I}_z.
\end{aligned}
$$

It turns out that the matrix $\hat{E}$ never leads to any observable magnetization, so this term can simply be omitted without causing any problems to our calculations. So, the equilibrium density operator (sometimes called the *reduced* density operator on account of the missing $\hat{E}$ term) is simply

$$\hat{\rho}_{eq} = k_I \hat{I}_z$$

where $k_I = \Delta n/N$. The value of the constant k_I depends on the number of spins in the sample, the temperature and the exact spacing between the two energy levels.

Referring back to the operator expansion of $\hat{\rho}$, Eq. 6.52 on page 130, we see that at equilibrium only the coefficient a_z is non-zero:

$$a_{z,eq} = k_I.$$

Using Eq. 6.54 on page 131, we can therefore write the z-magnetization as

$$
\begin{aligned}
M_{z,eq} &= \tfrac{1}{2}\gamma N a_{z,eq} \\
&= \tfrac{1}{2}\gamma N k_I.
\end{aligned}
$$

All experiments start with equilibrium magnetization, so in calculating the result of any experiment every term will be prefaced by the constant factor $(\gamma N k_{\mathrm{I}})$. All this factor does is set the overall size of the equilibrium magnetization and hence the subsequent size of the signal we observe. In NMR, we have no useful way of measuring the absolute size of the signal – rather what we are interested in is how the signal evolves over time. So, for simplicity and convenience we usually ignore the factor k_{I} which sets the size of the equilibrium density matrix, and write:

$$\hat{\rho}_{\mathrm{eq}} = \hat{I}_z.$$

Similarly, when computing the components of the magnetization we ignore the factor γN which simply scales the value. As a result, we can write very simply, but rather informally:

$$M_x = a_x \qquad M_y = a_y \qquad M_z = a_z.$$

6.8.7 Summary

In this section we have shown how the density operator is an alternative to working directly with the wavefunction. The density operator includes the effects of ensemble averaging in such a way that the components of the bulk magnetization can be computed directly. We went on to show that an operator expansion of the density operator makes it particularly straightforward to extract the value of the components of the bulk magnetization. In the next chapter we will go on to show how practical calculations can be performed using this operator approach, and how it can be extended straightforwardly to coupled spin systems.

Here is a summary of the key points.

- The density operator is defined as

$$\hat{\rho} = \overline{|\psi\rangle\langle\psi|},$$

  note that ensemble averaging is included in this definition.

- The density operator evolves in time according to

$$\hat{\rho}(t) = \exp\left(-\mathrm{i}\hat{H}t\right)\hat{\rho}(0)\exp\left(\mathrm{i}\hat{H}t\right).$$

- The density operator can be expanded as a linear combination of basis operators:

$$\hat{\rho} = a_E\hat{E} + a_x\hat{I}_x + a_y\hat{I}_y + a_z\hat{I}_z.$$

- To within a constant scaling factor, the components of the bulk magnetization are given simply by the coefficients in this expansion

$$M_x = a_x \quad M_y = a_y \quad M_z = a_z.$$

- At equilibrium the density operator is simply $\hat{I}_z$.

6.9 Coherence

In section 6.4.3 on page 116 we noted that, at equilibrium, the x- and y-components of the bulk magnetization are zero on account of the randomly distributed phases of the contributions from each spin in the ensemble. For example,

$$M_{y,eq} = \tfrac{1}{2}i\gamma N\overline{\left(c_\beta^\star c_\alpha - c_\alpha^\star c_\beta\right)}_{eq} = 0 \quad \text{or} \quad M_{y,eq} = \gamma N \overline{\langle I_y \rangle}_{eq} = 0.$$

However, we saw in section 6.7 on page 123 that transverse magnetization is generated when an RF pulse is applied to *equilibrium* magnetization. In quantum mechanics this transverse magnetization is described as being the result of the presence of a *coherence* in the sample. In this section we will explore what this term coherence means.

In section 6.7.1 on page 124 we saw that, after an RF pulse, the expectation value of the y-component of angular momentum was given by (Eq. 6.39 on page 125):

$$
\begin{aligned}
\langle I_y \rangle(t) = {} & -\tfrac{1}{2}i\left[c_\alpha^\star(0)c_\beta(0) - c_\alpha(0)c_\beta^\star(0)\right]\cos(\omega_1 t) \\
& -\tfrac{1}{2}\left[c_\alpha^\star(0)c_\alpha(0) - c_\beta^\star(0)c_\beta(0)\right]\sin(\omega_1 t).
\end{aligned}
\tag{6.55}
$$

Remember that this refers to a single spin. The equation predicts that from this single spin there is a y-component, whose size depends on the coefficients c_α and c_β at time zero, i.e. before the pulse.

However, what we are able to detect is the *bulk magnetization*, which is the sum of the contributions from each spin; to compute this sum, we take the ensemble average

$$
\begin{aligned}
M_y(t) = {} & -\tfrac{1}{2}i\,\gamma N\overline{\left[c_\alpha^\star(0)c_\beta(0) - c_\alpha(0)c_\beta^\star(0)\right]}\cos(\omega_1 t) \\
& -\tfrac{1}{2}\gamma N\overline{\left[c_\alpha^\star(0)c_\alpha(0) - c_\beta^\star(0)c_\beta(0)\right]}\sin(\omega_1 t).
\end{aligned}
$$

Now if the spins are at equilibrium at time zero, the first term in square braces is zero on account of the random distribution of the phases at equilibrium. However, the second term in square braces is not zero, but in fact equal to $(n_\alpha - n_\beta)/N$, as was explained in section 6.4.2 on page 115. So, the y-magnetization is

$$M_y(t) = -\tfrac{1}{2}\gamma\left(n_\alpha - n_\beta\right)\sin(\omega_1 t). \tag{6.56}$$

In words, what all of this says is as follows. At equilibrium, the phases of the superposition states are randomly distributed, so there is no bulk transverse magnetization. However, as the two energy levels (α and β) are not equally populated, there is z-magnetization; we can describe this as a *polarization* of the sample along the z-direction. When a pulse is applied, y-magnetization is generated and, according to Eq. 6.56, this magnetization is proportional to the original polarization along the z-axis, $(n_\alpha - n_\beta)$. If there was no polarization along z at time zero (i.e. $n_\alpha = n_\beta$), then the pulse would *not* generate any y-magnetization. All the pulse really does is to rotate the axis along which the polarization is aligned from z to y.

Transverse magnetization is described as being the result of a coherence amongst the spins. Sometimes it is implied that a coherence is an 'alignment of the spins' which is brought about by the pulse – phrases such as 'the pulse brings the spins into alignment' or 'the wavefunctions are aligned by the pulse' are commonly encountered. However, all of these phrases are misleading since, as we have seen, the pulse does not create an alignment or alter the phases of the spins in some magic way: all the pulse does is to rotate a polarization from z to y.

It is important that each spin in the ensemble experiences the same RF field. If this is the case, then each spin behaves in a way described by Eq. 6.55 on the facing page, so that when the ensemble average is taken the second term is indeed the z-magnetization. If, for example, the value of ω_1 were to vary across the sample, then the sine and cosine terms would vary from place to place, and it would be quite easy for the ensemble average of *both* terms to be zero. When there is a coherence present in the sample, this has been generated by the individual spins in the sample all experiencing the *same* interaction with the applied RF field in such a way that a polarization along z has been rotated to another direction. Without the initial polarization along z, a pulse *cannot* generate a coherence.

6.10   Further reading

A detailed discussion of the quantum mechanics of a one-spin system:

Chapters 6–11 from M. H. Levitt, *Spin Dynamics* (2nd edition, John Wiley & Sons, Ltd, 2008).

The postulates of quantum mechanics; angular momentum:

Chapters 1 and 4 from P. W. Atkins and R. Friedman, *Molecular Quantum Mechanics* (4th edition, Oxford University Press, 2005).

6.11 Exercises

To really get to grips with the material in this chapter, it is very instructive to work through *every* line in *all* the equations. By doing this, you can be sure that you have understood the logic of the argument, and each step in a derivation or proof. Just reading the equations and saying 'I see what is going on' is not the same thing as actually working through things for yourself.

6.1   Write the following in Dirac notation:

$$\hat{I}_z \psi_\beta = -\tfrac{1}{2}\psi_\beta \qquad \int \psi_\beta^\star \psi_\alpha \, d\tau \qquad \int \psi_\beta^\star \psi_\beta \, d\tau \qquad \int \psi^\star \hat{Q}\psi \, d\tau.$$

Express the following statements in Dirac notation:
(a) ψ_α is normalized;
(b) ψ_α and ψ_β are orthogonal;
(c) ψ_α is an eigenfunction of $\hat{I}_z$ with eigenvalue $\tfrac{1}{2}$;
(d) ψ can be expressed as a linear combination of ψ_α and ψ_β.

6.2   If the wavefunction for a single spin is given by

$$|\psi\rangle = c_\alpha |\alpha\rangle + c_\beta |\beta\rangle,$$

and assuming that $(c_\alpha^\star c_\alpha + c_\beta^\star c_\beta) = 1$, show that

$$\langle I_y \rangle = \tfrac{1}{2}\mathrm{i}\,c_\beta^\star c_\alpha - \tfrac{1}{2}\mathrm{i}\,c_\alpha^\star c_\beta.$$

What is the interpretation of $\langle I_y \rangle$?

6.3   Using the approach of section 6.3.5 on page 111, show that

$$I_x = \begin{pmatrix} 0 & \tfrac{1}{2} \\ \tfrac{1}{2} & 0 \end{pmatrix} \qquad I_y = \begin{pmatrix} 0 & -\tfrac{1}{2}\mathrm{i} \\ \tfrac{1}{2}\mathrm{i} & 0 \end{pmatrix}.$$

6.4   Given that

$$\langle I_y \rangle = \tfrac{1}{2}\mathrm{i}\,c_\beta^\star c_\alpha - \tfrac{1}{2}\mathrm{i}\,c_\alpha^\star c_\beta,$$

and expressing the coefficients in r/ϕ format:

$$c_\alpha = r_\alpha \exp(\mathrm{i}\phi_\alpha) \quad c_\beta = r_\beta \exp(\mathrm{i}\phi_\beta)$$
$$c_\alpha^\star = r_\alpha \exp(-\mathrm{i}\phi_\alpha) \quad c_\beta^\star = r_\beta \exp(-\mathrm{i}\phi_\beta),$$

show that

$$\langle I_y \rangle = \tfrac{1}{2\mathrm{i}} r_\alpha r_\beta \left\{ \exp(\mathrm{i}[\phi_\beta - \phi_\alpha]) - \exp(-\mathrm{i}[\phi_\beta - \phi_\alpha]) \right\}.$$

Use the identity $\exp(\mathrm{i}\theta) - \exp(-\mathrm{i}\theta) \equiv 2\mathrm{i}\sin\theta$ to tidy this up to

$$\langle I_y \rangle = r_\alpha r_\beta \sin(\phi_\beta - \phi_\alpha).$$

Hence show that the bulk y-magnetization is given by

$$M_y = \gamma N \, \overline{r_\alpha r_\beta \sin(\phi_\beta - \phi_\alpha)}.$$

What value do you expect this to take at equilibrium?

6.5 Using the approach of section 6.6.1 on page 119, show that

$$\frac{dc_\beta(t)}{dt} = \tfrac{1}{2}i\Omega c_\beta(t). \qquad (6.57)$$

[Hint: start from Eq. 6.31 on page 120 and left multiply by $\langle\beta|$.]

Show, by substituting Eq. 6.58 into Eq. 6.57, that the solution to this differential equation is

$$c_\beta(t) = c_\beta(0)\exp\left(\tfrac{1}{2}i\Omega t\right). \qquad (6.58)$$

6.6 Using the approach of section 6.6.3 on page 122, show that, during a period of free evolution, the expectation value of $\hat{I}_y$ evolves according to

$$\langle I_y\rangle(t) = \cos(\Omega t)\langle I_y\rangle(0) + \sin(\Omega t)\langle I_x\rangle(0).$$

Give a graphical interpretation of this result.

6.7 Using the approach described in section 6.8.1 on page 127, show that for a spin described by the wavefunction

$$|\psi\rangle = c_\alpha|\alpha\rangle + c_\beta|\beta\rangle,$$

the matrix representation of the density operator is given by

$$\rho = \begin{pmatrix} \overline{c_\alpha c_\alpha^\star} & \overline{c_\alpha c_\beta^\star} \\ \overline{c_\beta c_\alpha^\star} & \overline{c_\beta c_\beta^\star} \end{pmatrix}.$$

7

Product operators

In this chapter we are going to introduce a quantum-mechanical method for calculating the outcome of a multiple-pulse NMR experiment by representing the state of the spin system using a combination of operators. The background as to how and why this approach works was given in the previous chapter; however, in this chapter we will simply present the method as a recipe which you can apply systematically. The method is exact, and is capable of dealing with the majority of modern pulse sequences.

To start with we will consider a system of uncoupled spins. This is the same system to which the vector model applies, so we will be able to verify our calculations by comparing them with what we found in Chapter 4. However, the real strength of the operator approach is its ability to deal with coupled spin systems, so we will spend some time showing how this is done and illustrating the method with various applications. For coupled spin systems it turns out that we need to use products of operators, hence this approach is usually called the *product operator* method.

Using product operators we will be able to understand most of the important building blocks from which modern multi-pulse NMR experiments are constructed, such as the *J*-modulated spin echo and coherence transfer by pulses. Once we have grasped how these key elements work, the analysis of even rather complex pulse sequences becomes quite straightforward.

7.1 Operators for one spin

We are going to start out by thinking about a *spin system* which consists of just a single spin one-half, without any couplings to other spins. Our sample will contain a very large number of these spin systems, which we will assume are not interacting with one another. This collection of identical spin systems is called an *ensemble*, so our NMR sample can be described as an ensemble of non-interacting single spins.

Understanding NMR Spectroscopy, Second Edition James Keeler
© 2010 John Wiley & Sons, Ltd

Based on the discussion in Chapter 6, we assert that everything about this ensemble can be calculated from a knowledge of the *density operator* $\hat{\rho}$. Furthermore, for a one-spin system, this operator can be expressed as a linear combination of the operators $\hat{I}_x$, $\hat{I}_y$ and $\hat{I}_z$:

$$\hat{\rho}(t) = a_x(t)\hat{I}_x + a_y(t)\hat{I}_y + a_z(t)\hat{I}_z. \tag{7.1}$$

The operator $\hat{I}_z$ represents the z-component of the spin angular momentum; we met this operator before in section 3.2.5 on page 29 where we used it to write the Hamiltonian. Similarly, $\hat{I}_x$ and $\hat{I}_y$ represent the x- and y-components of the spin angular momentum.

The coefficients $a_x(t)$, $a_y(t)$ and $a_z(t)$ are just numbers which vary with time. The really useful thing about writing the density operator in this way is that the amount of x-, y- and z-magnetizations are very simply related to these coefficients:

$$M_x(t) = a_x(t) \qquad M_y(t) = a_y(t) \qquad M_z(t) = a_z(t). \tag{7.2}$$

So, once we know the coefficients we can work out the components of the magnetization from the sample, and how these vary with time: this represents a complete knowledge of the state of the spin system at any time.

To be entirely correct, we should note that M_x is proportional to a_x, not equal to it. However, the constant of proportion is the same for all the components and has no real effect on our calculation other than to scale the answer. In NMR we have no practical way of measuring the absolute size of the magnetization, so this scaling is of no consequence; for simplicity it is therefore usual just to ignore the constant of proportion and write $M_x = a_x$.

We need to know how the density operator evolves in time, and it turns out that this is given by

$$\hat{\rho}(t) = \exp(-i\hat{H}t)\hat{\rho}(0)\exp(i\hat{H}t), \tag{7.3}$$

where $\hat{\rho}(0)$ is the density operator at time $t = 0$, and $\hat{\rho}(t)$ is the operator after time t. $\hat{H}$ is the Hamiltonian which is relevant for the period of time 0 to t; we will specify what these relevant Hamiltonians are in the following section. Although it looks daunting, it turns out that the right-hand side of Eq. 7.3 is quite straightforward to evaluate using a simple recipe which we will introduce shortly.

7.1.1   Hamiltonians for free precession and pulses

The idea of the Hamiltonian as the operator for energy was introduced in section 3.2.5 on page 29. However, the Hamiltonian plays a far more important role than simply determining the energy: it is the operator which determines how the spins evolve in *time*. For pulsed NMR, time evolution is of central importance, so a knowledge of the Hamiltonian is crucial.

A key point to grasp is that the Hamiltonian is different during pulses and periods of free precession. This should not come as a surprise, since during free precession there is simply a magnetic field along the z-direction, whereas during a pulse there is an additional transverse magnetic field. It is these fields which interact with the spin and modify the energy.

In writing our Hamiltonians we will employ the quantum mechanical equivalent of the rotating frame introduced in section 4.4.1 on page 53. Recall that this is an axis system which is rotating about the z-axis at the frequency of the applied RF power, and in the same sense as the Larmor precession. In such a frame, the transverse magnetic field due to a pulse appears to be static and the applied field along the z-axis is reduced in size.

During a period of free precession the Hamiltonian is

$$\hat{H}_{\text{free}} = \Omega \hat{I}_z, \tag{7.4}$$

where, as before (section 4.4.2 on page 55), Ω is the *offset* of the spin, which is the difference between its Larmor frequency and the rotating frame frequency.

During a pulse in which the RF field is applied along the x-axis, there is an additional term in the Hamiltonian which represents this field:

$$\hat{H}_{x,\text{pulse}} = \Omega \hat{I}_z + \omega_1 \hat{I}_x. \tag{7.5}$$

As before, ω_1 is the RF field strength which determines the rate at which the magnetization rotates about the field along the x-axis.

It was shown in section 4.5.1 on page 59 that if the RF field strength ω_1 is much greater in size than the offset Ω, the evolution of the magnetization is dominated by the transverse field. If these conditions hold, we have a hard (or non-selective) pulse, for which the Hamiltonian is simply

$$\hat{H}_{x,\text{hard pulse}} = \omega_1 \hat{I}_x. \tag{7.6}$$

If the pulse is about the y-axis, then the operator simply becomes $\hat{I}_y$:

$$\hat{H}_{y,\text{hard pulse}} = \omega_1 \hat{I}_y. \tag{7.7}$$

Throughout this chapter we will assume that all pulses are hard.

7.1.2 Rotations

To understand how the density operator changes with time we need to consider how we are going to work out the right-hand side of Eq. 7.3 on the facing page. We will illustrate the procedure by considering a specific case, and then go on to generalize the approach.

Imagine at time zero the density operator is simply $\hat{I}_x$ i.e. $a_x = 1$, $a_y = 0$ and $a_z = 0$:

$$\hat{\rho}(0) = \hat{I}_x.$$

Suppose that we now want to work out the effect of a period of free precession, for which the Hamiltonian is given by Eq. 7.4. So, using Eq. 7.3 on the facing page the density operator after evolution for time t is

$$\begin{aligned} \hat{\rho}(t) &= \exp(-i\hat{H}t)\,\hat{\rho}(0)\,\exp(i\hat{H}t) \\ &= \exp(-i\Omega t\hat{I}_z)\,\hat{I}_x\,\exp(i\Omega t\hat{I}_z). \end{aligned} \tag{7.8}$$

The last line can be evaluated using an identity which is well-known in the theory of angular momentum operators:

$$\exp(-i\theta\hat{I}_z)\,\hat{I}_x\,\exp(i\theta\hat{I}_z) \equiv \cos\theta\,\hat{I}_x + \sin\theta\,\hat{I}_y. \tag{7.9}$$

In the present case the angle θ is equal to Ωt. This identity is interpreted as starting with the operator $\hat{I}_x$ and then rotating it through an angle θ about the z-axis. Not surprisingly, this rotation generates a y-component proportional to $\sin\theta$, leaving an x-component proportional to $\cos\theta$. The analogy with x-magnetization rotating towards the y-axis is complete and entirely appropriate.

If we set $\theta = \Omega t$ and apply the identity Eq. 7.9 on the previous page to Eq. 7.8 we find

$$\exp(-i\Omega t\hat{I}_z)\,\hat{I}_x\,\exp(i\Omega t\hat{I}_z) = \cos(\Omega t)\hat{I}_x + \sin(\Omega t)\hat{I}_y.$$

Overall the effect of a period of free precession can be written as

$$\hat{I}_x \xrightarrow{\;\Omega t\hat{I}_z\;} \cos(\Omega t)\hat{I}_x + \sin(\Omega t)\hat{I}_y. \tag{7.10}$$

This is a notation we will use often: the time evolution is represented by a right arrow connecting $\hat{\rho}(0)$ to $\hat{\rho}(t)$, and over the arrow we write $(\hat{H}t)$, where $\hat{H}$ is the relevant Hamiltonian which acts for time t:

$$\hat{\rho}(0) \xrightarrow{\;\hat{H}t\;} \hat{\rho}(t).$$

As there is a one-to-one correspondence between the coefficients in front of the operators and the components of the magnetization (Eq. 7.2 on page 140), our interpretation of Eq. 7.10 is that the magnetization at time zero is along x, and after time t the magnetization has a component $\cos(\Omega t)$ along x and $\sin(\Omega t)$ along y.

So far we have looked at the particular example of $\hat{I}_x$ being rotated about z, but we will now go on to show that the rotation of any operator about any axis essentially behaves in the same way.

Since we are writing the density operator as a linear combination of the operators $\hat{I}_x$, $\hat{I}_y$ and $\hat{I}_z$, and as the Hamiltonians are also expressed in terms of these operators, every time we want to work out what $\exp(-i\hat{H}t)\hat{\rho}(0)\exp(i\hat{H}t)$ is it will boil down to relationships of the type given in Eq. 7.9 on the preceding page. There are only six possible versions of these rotations, and they are all given in Table 7.1 on the next page.

You might notice that there are some combinations missing from this table such as the rotation of $\hat{I}_z$ about the z-axis. However, such a rotation has no effect:

$$\exp(-i\theta\hat{I}_z)\,\hat{I}_z\,\exp(i\theta\hat{I}_z) \equiv \hat{I}_z.$$

Based on what we know from the vector model, this is hardly a surprise: rotating a vector about its own axis does not affect the vector in any way.

Armed with this table of identities and a knowledge of the Hamiltonians, we are in a position to calculate the outcome of any pulse sequence. However, before we do that we should just remind ourselves of the limitations of this approach.

7.1.3 Limitations

The substantial effect which is missing from this operator approach is relaxation. We know that over time the transverse components of the

Table 7.1 Table showing how the rotation through angle θ about a particular axis affects the operators $\hat{I}_x$, $\hat{I}_y$ and $\hat{I}_z$. For convenience, the lines in the table are numbered.

	rotation about	operator	identity
1	x	$\hat{I}_y$	$\exp(-i\theta\hat{I}_x)\,\hat{I}_y\,\exp(i\theta\hat{I}_x) \equiv \cos\theta\,\hat{I}_y + \sin\theta\,\hat{I}_z$
2	x	$\hat{I}_z$	$\exp(-i\theta\hat{I}_x)\,\hat{I}_z\,\exp(i\theta\hat{I}_x) \equiv \cos\theta\,\hat{I}_z - \sin\theta\,\hat{I}_y$
3	y	$\hat{I}_x$	$\exp(-i\theta\hat{I}_y)\,\hat{I}_x\,\exp(i\theta\hat{I}_y) \equiv \cos\theta\,\hat{I}_x - \sin\theta\,\hat{I}_z$
4	y	$\hat{I}_z$	$\exp(-i\theta\hat{I}_y)\,\hat{I}_z\,\exp(i\theta\hat{I}_y) \equiv \cos\theta\,\hat{I}_z + \sin\theta\,\hat{I}_x$
5	z	$\hat{I}_x$	$\exp(-i\theta\hat{I}_z)\,\hat{I}_x\,\exp(i\theta\hat{I}_z) \equiv \cos\theta\,\hat{I}_x + \sin\theta\,\hat{I}_y$
6	z	$\hat{I}_y$	$\exp(-i\theta\hat{I}_z)\,\hat{I}_y\,\exp(i\theta\hat{I}_z) \equiv \cos\theta\,\hat{I}_y - \sin\theta\,\hat{I}_x$

magnetization will decay to zero, and the longitudinal components will return to their equilibrium values; neither of these effects are included. Later on, in Chapter 9, we will look at how the effects of relaxation can be taken into account, but for the moment we will just have to accept this limitation.

The other main limitation of the product operator approach is that it only applies to weakly coupled spin systems. Remember from section 2.3.2 on page 12 that a spin system is weakly coupled if the difference between the Larmor frequencies of two coupled spins is much greater than the size of the coupling constant between them.

7.2 Analysis of pulse sequences for a one-spin system

All of our NMR experiments start with equilibrium magnetization, which is along the z-axis. We will therefore write the equilibrium density operator as $\hat{I}_z$.

7.2.1 Pulse–acquire

The pulse sequence for the basic pulse–acquire experiment is shown in Fig. 7.1. Let us assume that we start at equilibrium, so that the initial density operator is simply $\hat{I}_z$. We then apply an x-pulse of duration t_p using a field strength ω_1, for which the relevant Hamiltonian is given in Eq. 7.6 on page 141:

$$\hat{H}_{x,\text{pulse}} = \omega_1\hat{I}_x.$$

To work out the evolution we need to solve Eq. 7.3 on page 140

$$\hat{\rho}(t_p) = \exp(-i\hat{H}_{x,\text{pulse}}t_p)\,\hat{\rho}(0)\,\exp(i\hat{H}_{x,\text{pulse}}t_p)$$

with $\hat{H}_{x,\text{pulse}} = \omega_1\hat{I}_x$ and $\hat{\rho}(0) = \hat{I}_z$, i. e.

$$\hat{\rho}(t_p) = \exp(-i\omega_1 t_p\hat{I}_x)\,\hat{I}_z\,\exp(i\omega_1 t_p\hat{I}_x).$$

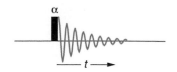

Fig. 7.1 Pulse sequence for the basic pulse–acquire experiment. A pulse of flip angle α and of phase x is applied to equilibrium magnetization. Data acquisition, indicated by the damped cosine-wave FID, starts immediately after the pulse.

The right-hand side can be evaluated using the identity on line 2 of Table 7.1 on the preceding page:

$$\exp(-i\theta\hat{I}_x)\,\hat{I}_z\,\exp(i\theta\hat{I}_x) \equiv \cos\theta\,\hat{I}_z - \sin\theta\,\hat{I}_y,$$

with θ replaced by $\omega_1 t_p$. Doing this gives

$$\hat{\rho}(t_p) = \cos(\omega_1 t_p)\,\hat{I}_z - \sin(\omega_1 t_p)\,\hat{I}_y.$$

Of course, $\omega_1 t_p$ is simply the flip angle α, so the result can be written:

$$\hat{\rho}(t_p) = \cos\alpha\,\hat{I}_z - \sin\alpha\,\hat{I}_y.$$

The result is entirely what we expected: the pulse rotates the magnetization from z towards $-y$, and if the flip angle is $\pi/2$ ($90°$), then the magnetization is rotated entirely onto the $-y$-axis. For other flip angles the y-component is simply $(-\sin\alpha)$.

The effect of this pulse on equilibrium magnetization can be represented using the arrow notation in the following way:

$$\hat{I}_z \xrightarrow{\ \omega_1 t_p \hat{I}_x\ } \cos(\omega_1 t_p)\,\hat{I}_z - \sin(\omega_1 t_p)\,\hat{I}_y.$$

Over the arrow we have written the relevant Hamiltonian multiplied by the time t_p, $\hat{H}_{x,\text{pulse}}t_p$, which in this case is $\omega_1 t_p\hat{I}_x$.

Since $\omega_1 t_p = \alpha$, the term over the arrow can also be written $\alpha\hat{I}_x$, and the sine and cosine can be written in terms of α

$$\hat{I}_z \xrightarrow{\ \alpha\hat{I}_x\ } \cos\alpha\,\hat{I}_z - \sin\alpha\,\hat{I}_y.$$

This arrow notation is so much more compact than writing out all the exponential operators that we will use it from now on.

After the pulse, there is a period of free precession, for which the Hamiltonian is $\hat{H}_{\text{free}} = \Omega\hat{I}_z$. This is a rotation about the z-axis, and so the term $\cos\alpha\,\hat{I}_z$ present after the pulse is not affected by free precession. In contrast, the term in $\hat{I}_y$ is affected, and using the arrow notation the effect of free precession on this term can be written

$$-\sin\alpha\,\hat{I}_y \xrightarrow{\ \Omega t\hat{I}_z\ } -\sin\alpha\left[\cos(\Omega t)\hat{I}_y - \sin(\Omega t)\hat{I}_x\right]$$
$$= -\sin\alpha\cos(\Omega t)\hat{I}_y + \sin\alpha\sin(\Omega t)\hat{I}_x.$$

The term over the arrow is $\hat{H}_{\text{free}}t$, which is $\Omega t\hat{I}_z$.

The way in which this calculation works needs a little more explanation. The factor $(-\sin\alpha)$, which in the initial state is multiplying $\hat{I}_y$, is just a number, and as such is unaffected by the rotation about z. This factor is simply carried forward and multiplies the final result. All we need to do is consider the rotation of $\hat{I}_y$ about z, for which the appropriate identity is (line 6 of Table 7.1 on the previous page)

$$\exp(-i\theta\hat{I}_z)\,\hat{I}_y\,\exp(i\theta\hat{I}_z) \equiv \cos\theta\,\hat{I}_y - \sin\theta\,\hat{I}_x.$$

This identity, with $\theta = \Omega t$, is used on the first line. To go to the second, we have simply multiplied out the bracket.

The final result is that at time t after the pulse the state of our system is

$$\cos \alpha \, \hat{I}_z - \sin \alpha \cos (\Omega t) \hat{I}_y + \sin \alpha \sin (\Omega t) \hat{I}_x. \qquad (7.11)$$

The observable x- and y-components of the magnetization are thus:

$$M_x(t) = \sin \alpha \sin (\Omega t) \qquad M_y(t) = - \sin \alpha \cos (\Omega t).$$

This is in agreement with the predictions we made using the vector model.

7.2.2   The spin echo

Using the vector model we saw that the spin-echo pulse sequence, shown in Fig. 7.2, resulted in the magnetization appearing on the y-axis, regardless of the offset and the time τ. We will now reproduce this result using the operator approach.

The first part of the sequence is simply 90°– delay, which is what we have already computed in the previous section. So we can use the result of Eq. 7.11 with $\alpha \rightarrow \pi/2$ and $t \rightarrow \tau$ to give us the state of the spin system just prior to the π pulse as

$$- \cos (\Omega \tau) \hat{I}_y + \sin (\Omega \tau) \hat{I}_x.$$

The π pulse is about the x-axis, and so the term $\sin (\Omega \tau) \hat{I}_x$ is unaffected. The effect of the pulse, of duration t_π, on the other term is:

$$- \cos (\Omega \tau) \hat{I}_y \xrightarrow{\omega_1 t_\pi \hat{I}_x} - \cos (\Omega \tau) \left[\cos (\omega_1 t_\pi) \hat{I}_y + \sin (\omega_1 t_\pi) \hat{I}_z \right]$$

$$\xrightarrow{\pi \hat{I}_x} - \cos (\Omega \tau) \left[\cos (\pi) \, \hat{I}_y + \sin (\pi) \, \hat{I}_z \right]$$

$$= + \cos (\Omega \tau) \hat{I}_y.$$

On the first line, the term over the arrow is $\hat{H}_{\text{pulse}} t_\pi$, which is equal to $\omega_1 t_\pi \hat{I}_x$. We have also used the identity from line 1 of Table 7.1 on page 143:

$$\exp (-\mathrm{i}\theta \hat{I}_x) \, \hat{I}_y \, \exp (\mathrm{i}\theta \hat{I}_x) \equiv \cos \theta \, \hat{I}_y + \sin \theta \, \hat{I}_z$$

with $\theta = \omega_1 t_\pi$. As before, the factor $- \cos (\Omega \tau)$ is unaffected by the rotations and simply multiplies the result.

To go to the second line, we have used $\omega_1 t_\pi = \pi$ which is the case for a π pulse, and to go to the last line we have used $\cos \pi = -1$ and $\sin \pi = 0$. Overall, the result is that the term $- \cos (\Omega \tau) \hat{I}_y$ is inverted in sign – entirely as expected on the basis of the vector model.

In summary, after the π pulse the state of the spin system is

$$\cos (\Omega \tau) \hat{I}_y + \sin (\Omega \tau) \hat{I}_x.$$

Now we need to consider the evolution during the second delay. Each term has to be considered separately, so we will start with $\cos (\Omega \tau) \hat{I}_y$. Recall that the factor $\cos (\Omega \tau)$ will just multiply our answer, so for simplicity we can set it aside during the calculation and then reintroduce it at the end.

Using this approach, the term $\hat{I}_y$ evolves during the delay τ according to

$$\hat{I}_y \xrightarrow{\Omega \tau \hat{I}_z} \cos (\Omega \tau) \hat{I}_y - \sin (\Omega \tau) \hat{I}_x,$$

Fig. 7.2 Pulse sequence for the spin echo. The filled-in rectangle indicates a 90° pulse, whereas the open rectangle indicates a 180° pulse; unless otherwise indicated, the phase of the pulses is x. In practice, the duration of the pulses is negligible when compared with the delay τ. The sequence ends after the second delay τ, as indicated by the dashed line; the overall delay between the first pulse and the end of the sequence is thus 2τ.

where we have used $\hat{H}_{\text{free}}\tau = \Omega\tau\hat{I}_z$ and the identity from line 6 of Table 7.1 on page 143 with $\theta = \Omega\tau$. Reintroducing the factor $\cos(\Omega\tau)$ the overall result of the evolution of the term $\cos(\Omega\tau)\hat{I}_y$ is

$$\cos(\Omega\tau)\cos(\Omega\tau)\hat{I}_y - \sin(\Omega\tau)\cos(\Omega\tau)\hat{I}_x. \tag{7.12}$$

We now have to consider the evolution of the term $\sin(\Omega\tau)\hat{I}_x$:

$$\hat{I}_x \xrightarrow{\Omega\tau\hat{I}_z} \cos(\Omega\tau)\hat{I}_x + \sin(\Omega\tau)\hat{I}_y.$$

This time we have used the identity from line 5 of Table 7.1 with $\theta = \Omega\tau$.

Reintroducing the factor $\sin(\Omega\tau)$, the overall result of the evolution of the term $\sin(\Omega\tau)\hat{I}_x$ is

$$\cos(\Omega\tau)\sin(\Omega\tau)\hat{I}_x + \sin(\Omega\tau)\sin(\Omega\tau)\hat{I}_y. \tag{7.13}$$

At the end of the spin echo the state of the system is found by adding together Eq. 7.12 and Eq. 7.13. The first thing to note is that the terms in $\hat{I}_x$ cancel one another. The terms in $\hat{I}_y$ are:

$$\cos(\Omega\tau)\cos(\Omega\tau)\hat{I}_y + \sin(\Omega\tau)\sin(\Omega\tau)\hat{I}_y \equiv \left\{[\cos(\Omega\tau)]^2 + [\sin(\Omega\tau)]^2\right\}\hat{I}_y.$$

Using the identity:

$$\cos^2\theta + \sin^2\theta \equiv 1$$

we see that the final state of the spin system is $\hat{I}_y$. As we predicted using the vector model, the magnetization ends up along the y-axis, regardless of the offset and the delay τ.

7.3 Speeding things up

Calculations using this operator approach can become rather laborious, so it is important to simplify things where we can, and to develop strategies for using the rotations from Table 7.1 on page 143 in an efficient way. We introduce two such strategies here.

7.3.1 90° and 180° pulses

The effect of pulses with flip angles of 90° or $\pi/2$ radians is rather simple as when $\omega_1 t_p = \pi/2$, $\cos(\omega_1 t_p) = 0$ and $\sin(\omega_1 t_p) = 1$. Referring to the identities in Table 7.1, we see that 90° pulses about the x- and y-axes simply cause the transformations:

$$\hat{I}_y \xrightarrow{(\pi/2)\hat{I}_x} \hat{I}_z$$

$$\hat{I}_z \xrightarrow{(\pi/2)\hat{I}_x} -\hat{I}_y$$

$$\hat{I}_x \xrightarrow{(\pi/2)\hat{I}_y} -\hat{I}_z$$

$$\hat{I}_z \xrightarrow{(\pi/2)\hat{I}_y} \hat{I}_x.$$

The rotations caused by 180° pulses are even simpler. For such pulses $\omega_1 t_p = \pi$ and so $\cos(\omega_1 t_p) = -1$ and $\sin(\omega_1 t_p) = 0$. So all that happens is that the original term changes sign:

$$\hat{I}_y \xrightarrow{\pi \hat{I}_x} -\hat{I}_y$$

$$\hat{I}_z \xrightarrow{\pi \hat{I}_x} -\hat{I}_z$$

$$\hat{I}_x \xrightarrow{\pi \hat{I}_y} -\hat{I}_x$$

$$\hat{I}_z \xrightarrow{\pi \hat{I}_y} -\hat{I}_z.$$

Note that both 90° and 180° pulses about a particular axis have *no* effect on operators along that axis e.g. a pulse about x has no effect on $\hat{I}_x$.

7.3.2   Diagrammatic representation

The identities in Table 7.1 on page 143 simply represent rotations in a three-dimensional space where the x-, y- and z-axes represent the three operators $\hat{I}_x$, $\hat{I}_y$ and $\hat{I}_z$. For example, the identity

$$\exp(-i\theta\hat{I}_x)\,\hat{I}_z\,\exp(i\theta\hat{I}_x) \equiv \cos\theta\,\hat{I}_z - \sin\theta\,\hat{I}_y$$

can be interpreted as a vector which starts on the z-axis and then is rotated about the x-axis. As a result, the vector moves in the yz-plane, initially setting off towards the $-y$-axis i.e. a positive rotation about x, as is illustrated in Fig. 7.3. All of the identities in Table 7.1 can be interpreted as rotations in this way.

A further feature of these identities is that they all have the same form. Rotation of an operator $\hat{A}$ results in a state which has two terms: the first is $\hat{A}$ multiplied by the cosine of an angle, and second is a 'new' operator, $\hat{B}$, multiplied by the sine of the same angle. In general, the right-hand side takes the form:

$$\cos\theta \times \text{original operator} + \sin\theta \times \text{new operator}.$$

We can work out what the 'new operator' will be by looking at Fig. 7.3. For example, if we start with $\hat{I}_y$ and imagine rotating it in a positive sense about x, the vector initially moves towards z, so the 'new operator' is $\hat{I}_z$. The corresponding identity is therefore

$$\exp(-i\theta\hat{I}_x)\,\hat{I}_y\,\exp(i\theta\hat{I}_x) \equiv \cos\theta\,\hat{I}_y + \sin\theta\,\hat{I}_z,$$

Similarly, if we start with $-\hat{I}_z$ and rotate in a positive sense about x, from the figure we can see that the initial movement is towards y, so the 'new operator' is $\hat{I}_y$, and the corresponding identity is

$$\exp(-i\theta\hat{I}_x)\,(-\hat{I}_z)\,\exp(i\theta\hat{I}_x) \equiv -\cos\theta\,\hat{I}_z + \sin\theta\,\hat{I}_y$$

This identity is not in Table 7.1, but can be found by multiplying the identity from line 2 by -1:

$$-1 \times \left[\exp(-i\theta\hat{I}_x)\,\hat{I}_z\,\exp(i\theta\hat{I}_x)\right] \equiv -\cos\theta\,\hat{I}_z + \sin\theta\,\hat{I}_y.$$

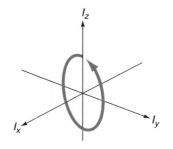

Fig. 7.3 A rotation of the operator $\hat{I}_z$ about x takes it towards $-\hat{I}_y$, which is the same motion as a positive rotation of a vector which starts along $+z$. Remember that the sense of a positive rotation about x is found by grasping the x-axis with your right hand and with your thumb pointing along the $+x$-direction. The curl of your fingers then gives the sense of a positive rotation.

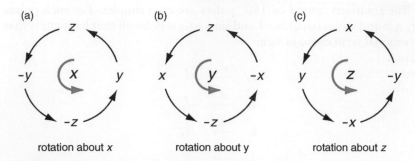

(a)   (b)   (c)

rotation about x   rotation about y   rotation about z

Fig. 7.4 Diagrams for determining the result of rotating any operator about x, y or z. To use the diagrams you simply locate the one for the axis about which the rotation is taking place. The initial operator is then located on the diagram, and the 'new operator' is then found by following the arrow. The result of the rotation is $\cos\theta$ times the original operator plus $\sin\theta$ times the 'new' operator. For a negative rotation about x, we simply use (a) but with the rotation clockwise e.g. z is rotated to y; similarly, (b) can be used for a negative rotation about y.

The effect of any rotation about x can be worked out using using the diagram shown in Fig. 7.4 (a). To use this, you simply locate the initial operator and then move in the sense of the arrow to find the 'new operator'.

Rotations about y and z can be handled in a similar way using diagrams (b) and (c). It is also worth noting that for 90° pulses, the result is a complete transformation to the 'new' operator.

7.3.3   The $1 - \bar{1}$ sequence

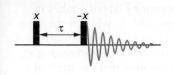

Fig. 7.5 Pulse sequence for the $1 - \bar{1}$ sequence used to suppress a single strong resonance (e.g. a solvent), thus allowing weaker signals from solutes to be observed. It is shown in the text that a line which is on resonance ($\Omega = 0$) is not excited, whereas lines with offsets close to $\pi/(2\tau)$ are excited significantly. By choosing τ, this region of excitation can be adjusted to cover the solute peaks of interest.

To illustrate how we can use the simplifications set out in the previous two sections, we will analyse the $1 - \bar{1}$ sequence, which is used for observing signals in the presence of a very strong solvent resonance.

The pulse sequence is shown in Fig. 7.5. It starts with a 90° pulse about x, followed by a delay τ, and then a 90° pulse about $-x$. Data acquisition follows immediately after the second pulse.

The first pulse simply rotates the equilibrium magnetization $\hat{I}_z$ to $-\hat{I}_y$. Free evolution is a rotation about z, and we see from Fig. 7.4 (c) that $-\hat{I}_y$ is rotated towards $\hat{I}_x$, so that after the delay the state of the system is

$$-\cos(\Omega\tau)\,\hat{I}_y + \sin(\Omega\tau)\,\hat{I}_x.$$

The effect of the final 90° pulse about $-x$ can be worked out by using Fig. 7.4 (a) but, since the pulse is about $-x$, the rotation goes clockwise i.e. in the opposite sense to that shown. Thus, for this 90° rotation, $-\hat{I}_y$ goes to $+\hat{I}_z$; the term in $\hat{I}_x$ is unaffected.

The final result is

$$\cos(\Omega\tau)\,\hat{I}_z + \sin(\Omega\tau)\,\hat{I}_x.$$

What we have here is a pulse sequence which produces transverse magnetization along the x-axis whose size is proportional to $\sin(\Omega\tau)$. A line at zero offset is therefore not excited, whereas a line with an offset such that $(\Omega\tau) = \pi/2$ will be excited to the maximum extent, since $\sin(\pi/2) = 1$.

The sequence can thus be used to suppress an unwanted strong line, such as one from a solvent, by placing it on resonance i.e. with $\Omega = 0$. Lines further off resonance are excited according to the function $\sin(\Omega\tau)$, and by choosing τ such that $(\Omega'\tau) = \pi/2$, lines at (or near) offset Ω' will be excited efficiently. The optimum value for the time τ is therefore $\pi/(2\Omega')$. If the offset is expressed in frequency units, $\Omega' = 2\pi F$, then $\tau = 1/(4F)$.

7.4 Operators for two spins

The product operator method comes into its own when we want to work with coupled spins. Whereas for one spin we only need the three operators $\hat{I}_x$, $\hat{I}_y$ and $\hat{I}_z$, we will need a total of sixteen operators to describe the two-spin system. The density operator can be expressed as a linear combination of these sixteen operators, just as the density operator for a one-spin system could be expressed as a linear combination of $\hat{I}_x$, $\hat{I}_y$ and $\hat{I}_z$.

The sixteen operators are constructed from the following four operators for spin one:

$$\hat{E}_1 \qquad \hat{I}_{1x} \qquad \hat{I}_{1y} \qquad \hat{I}_{1z},$$

and the corresponding four operators for spin two:

$$\hat{E}_2 \qquad \hat{I}_{2x} \qquad \hat{I}_{2y} \qquad \hat{I}_{2z}.$$

Note the subscript 1 to indicate that the operator is for spin one, and the subscript 2 to indicate that the operator is for spin two. $\hat{E}_1$ and $\hat{E}_2$ are 'unit operators' for spins one and two, respectively.

The sixteen product operators needed to describe a two-spin system, are constructed from all possible products consisting of an operator for spin one and an operator for spin two. For example, if the operator for spin one is $\hat{E}_1$, then the four possible products are

$$\hat{E}_1\hat{E}_2 \qquad \hat{E}_1\hat{I}_{2x} \qquad \hat{E}_1\hat{I}_{2y} \qquad \hat{E}_1\hat{I}_{2z}.$$

It turns out that the product $\hat{E}_1\hat{E}_2$ does not give rise to any observable magnetization. $\hat{E}_1\hat{I}_{2x}$ and $\hat{E}_1\hat{I}_{2y}$ correspond to observable x- and y-magnetization on spin two; such terms are also described as *single-quantum coherence*. As we shall see later, this type of magnetization is referred to as *in-phase*. $\hat{E}_1\hat{I}_{2z}$ corresponds to z-magnetization on spin two.

If the operator for spin one is $\hat{I}_{1x}$, then the four products are

$$\hat{I}_{1x}\hat{E}_2 \qquad 2\hat{I}_{1x}\hat{I}_{2x} \qquad 2\hat{I}_{1x}\hat{I}_{2y} \qquad 2\hat{I}_{1x}\hat{I}_{2z}.$$

We have inserted a factor of 2 into any product involving two x, y or z operators. This is for normalization purposes, which we will simply have to accept. The product $\hat{I}_{1x}\hat{E}_2$ is simply in-phase x-magnetization on spin one, just in the same way that $\hat{E}_1\hat{I}_{2x}$ is in-phase x-magnetization on spin two.

Later on, we will show that products such as $2\hat{I}_{1x}\hat{I}_{2x}$ and $2\hat{I}_{1x}\hat{I}_{2y}$, in which *both* operators are x or y (transverse), represent multiple-quantum coherences.

The product $2\hat{I}_{1x}\hat{I}_{2z}$ will turn out to be very important; we will show that it represents what is called *anti-phase magnetization* on spin one. This is observable, and leads to a spin-one doublet in which the two lines are of opposite sign – hence the description 'anti-phase'.

Carrying on, the next four products have $\hat{I}_{1y}$ as the spin-one operator:

$$\hat{I}_{1y}\hat{E}_2 \qquad 2\hat{I}_{1y}\hat{I}_{2x} \qquad 2\hat{I}_{1y}\hat{I}_{2y} \qquad 2\hat{I}_{1y}\hat{I}_{2z}.$$

As before, the first of these terms is in-phase y-magnetization on spin one, the second and third are multiple quantum, and the final term is anti-phase magnetization on spin one, aligned along y.

The final set of four products have $\hat{I}_{1z}$ as the spin-one operator:

$$\hat{I}_{1z}\hat{E}_2 \qquad 2\hat{I}_{1z}\hat{I}_{2x} \qquad 2\hat{I}_{1z}\hat{I}_{2y} \qquad 2\hat{I}_{1z}\hat{I}_{2z}.$$

The first term is simply z-magnetization on spin one; the second and third terms are anti-phase magnetization on spin two, aligned along x and y, respectively. The final term, $2\hat{I}_{1z}\hat{I}_{2z}$, represents a non-equilibrium population distribution which does not lead to observable magnetization.

For brevity, it is usual to omit the operators $\hat{E}_1$ and $\hat{E}_2$. Also, the product $\hat{E}_1\hat{E}_2$ never appears in any practical sequence, so we can set it to one side. This leaves fifteen operators, which can be grouped as follows:

description	operator(s)
z-magnetization on spin one	$\hat{I}_{1z}$
in-phase x- and y-magnetization on spin one	$\hat{I}_{1x}, \hat{I}_{1y}$
z-magnetization on spin two	$\hat{I}_{2z}$
in-phase x- and y-magnetization on spin two	$\hat{I}_{2x}, \hat{I}_{2y}$
anti-phase x- and y-magnetization on spin one	$2\hat{I}_{1x}\hat{I}_{2z}, 2\hat{I}_{1y}\hat{I}_{2z}$
anti-phase x- and y-magnetization on spin two	$2\hat{I}_{1z}\hat{I}_{2x}, 2\hat{I}_{1z}\hat{I}_{2y}$
multiple-quantum coherence	$2\hat{I}_{1x}\hat{I}_{2x}, 2\hat{I}_{1x}\hat{I}_{2y}, 2\hat{I}_{1y}\hat{I}_{2x}, 2\hat{I}_{1y}\hat{I}_{2y}$
non-equilibrium population	$2\hat{I}_{1z}\hat{I}_{2z}$

Our task is now to understand precisely what the difference is between in-phase and anti-phase magnetization: this turns out to be due to the evolution of scalar coupling.

7.4.1 Effect of coupling

Scalar coupling acts, along with offsets, during periods of free evolution. To describe such evolution we need to know the corresponding Hamiltonian, which was introduced in section 3.5.1 on page 37; here it is written in angular frequency units:

$$\hat{H}_{\text{two spins}} = \Omega_1\hat{I}_{1z} + \Omega_2\hat{I}_{2z} + 2\pi J_{12}\hat{I}_{1z}\hat{I}_{2z}. \qquad (7.14)$$

Ω_1 and Ω_2 are the offsets of spin one and two, respectively. J_{12} is the scalar coupling constant between spins one and two, and is given in Hz. However, as we are writing the Hamiltonian in angular frequency units, we have to multiply J_{12} by 2π in order to put everything into rad s^{-1}. This is a bit awkward, but the well-established convention is always to write scalar coupling constants in Hz, so we will have to put up with the factor of 2π.

There are three terms in this Hamiltonian. These describe, in turn: the evolution of the offset of spin one; the offset of spin two; and the scalar coupling between the two spins. It turns out that we can consider the effect of these three terms one at a time, and in any order (this is because the operators commute with one another). So, in the arrow notation, the result of a period of free precession

$$\hat{\rho}(0) \xrightarrow{\hat{H}_{\text{two spins}}t} \hat{\rho}(t)$$

can be worked out by three successive transformations (in any order):

$$\hat{\rho}(0) \xrightarrow{\Omega_1 t \hat{I}_{1z}} \xrightarrow{\Omega_2 t \hat{I}_{2z}} \xrightarrow{2\pi J_{12} t \hat{I}_{1z} \hat{I}_{2z}} \hat{\rho}(t).$$

The first two of these transformations are simply rotations about z, and we can work out the effect of these using the approach which has already been described. The new thing we need to deal with is the effect of the term $2\pi J_{12} t \hat{I}_{1z} \hat{I}_{2z}$.

The effect of this term on the operators $\hat{I}_{1x}$ and $\hat{I}_{1y}$ can be deduced from the following identities:

$$\exp\left(-i\theta \hat{I}_{1z} \hat{I}_{2z}\right) \hat{I}_{1x} \exp\left(i\theta \hat{I}_{1z} \hat{I}_{2z}\right) \equiv \cos\left(\tfrac{1}{2}\theta\right) \hat{I}_{1x} + \sin\left(\tfrac{1}{2}\theta\right) 2\hat{I}_{1y}\hat{I}_{2z}, \quad (7.15)$$

$$\exp\left(-i\theta \hat{I}_{1z} \hat{I}_{2z}\right) \hat{I}_{1y} \exp\left(i\theta \hat{I}_{1z} \hat{I}_{2z}\right) \equiv \cos\left(\tfrac{1}{2}\theta\right) \hat{I}_{1y} - \sin\left(\tfrac{1}{2}\theta\right) 2\hat{I}_{1x}\hat{I}_{2z}. \quad (7.16)$$

The important thing to notice here is that the sine and cosine terms are of *half* the angle θ. This is in contrast to all the identities we have come across before, which depend on the full angle θ. These identities tell us that the effect of coupling is to cause the in-phase terms $\hat{I}_{1x}$ and $\hat{I}_{1y}$ to evolve into the anti-phase terms $2\hat{I}_{1y}\hat{I}_{2z}$ and $2\hat{I}_{1x}\hat{I}_{2z}$. Note the shift of axis: an in-phase term along x becomes an anti-phase term along y.

For example, suppose we start with the term $\hat{I}_{1x}$ and allow it to evolve under coupling for a time τ. In the arrow notation the transformation is

$$\hat{I}_{1x} \xrightarrow{2\pi J_{12} \tau \hat{I}_{1z} \hat{I}_{2z}} \hat{\rho}(\tau).$$

To work out the effect of this we need the identity of Eq. 7.15 with $\theta = 2\pi J_{12}\tau$. Noting that $\tfrac{1}{2}\theta = \pi J_{12}\tau$, the evolution is

$$\hat{I}_{1x} \xrightarrow{2\pi J_{12} \tau \hat{I}_{1z} \hat{I}_{2z}} \cos\left(\pi J_{12}\tau\right) \hat{I}_{1x} + \sin\left(\pi J_{12}\tau\right) 2\hat{I}_{1y}\hat{I}_{2z}.$$

There will be complete conversion to the anti-phase term when the $\sin\left(\pi J_{12}\tau\right) = 1$, which is when $\left(\pi J_{12}\tau\right) = \pi/2$ i.e. $\tau = 1/(2J_{12})$.

In a similar way, anti-phase terms become in-phase according to the following identities:

$$\exp\left(-i\theta \hat{I}_{1z} \hat{I}_{2z}\right) 2\hat{I}_{1x}\hat{I}_{2z} \exp\left(i\theta \hat{I}_{1z} \hat{I}_{2z}\right) \equiv \cos\left(\tfrac{1}{2}\theta\right) 2\hat{I}_{1x}\hat{I}_{2z} + \sin\left(\tfrac{1}{2}\theta\right) \hat{I}_{1y}, \quad (7.17)$$

$$\exp\left(-i\theta \hat{I}_{1z} \hat{I}_{2z}\right) 2\hat{I}_{1y}\hat{I}_{2z} \exp\left(i\theta \hat{I}_{1z} \hat{I}_{2z}\right) \equiv \cos\left(\tfrac{1}{2}\theta\right) 2\hat{I}_{1y}\hat{I}_{2z} - \sin\left(\tfrac{1}{2}\theta\right) \hat{I}_{1x}. \quad (7.18)$$

For example, by using the second identity we can see that

$$2\hat{I}_{1y}\hat{I}_{2z} \xrightarrow{2\pi J_{12} \tau \hat{I}_{1z} \hat{I}_{2z}} \cos\left(\pi J_{12}\tau\right) 2\hat{I}_{1y}\hat{I}_{2z} - \sin\left(\pi J_{12}\tau\right) \hat{I}_{1x}.$$

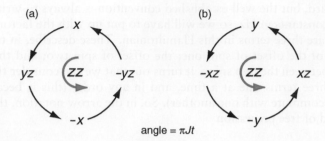

Fig. 7.6 Diagrams, similar to those of Fig. 7.4 on page 148, for working out the effect of the evolution of scalar coupling on in-phase and anti-phase terms. The letters x and y represent the in-phase terms about the corresponding axes, whereas xz and yz represent anti-phase terms about these axes. To work out the evolution of a term, we simply locate it in (a) or (b); this term will evolve into the term indicated by the arrow, and with the coefficient $\sin(\pi Jt)$. For example, $-2\hat{I}_{1x}\hat{I}_{2z}$ is located in (b), and the arrow takes us to the term $-\hat{I}_{1y}$; the result of evolution of the coupling is therefore $\cos(\pi Jt)$ times the original term plus $\sin(\pi Jt)$ times the new term, i.e. $-\cos(\pi Jt)2\hat{I}_{1x}\hat{I}_{2z} - \sin(\pi Jt)\hat{I}_{1y}$.

There is thus complete conversion of the anti-phase to the in-phase operator when $\tau = 1/(2J_{12})$. We saw above that the same value of the delay gives complete conversion of in-phase to anti-phase.

This interconversion of in-phase and anti-phase magnetization can be summarized in the diagrams shown in Fig. 7.6, which are similar to those of Fig. 7.4 on page 148.

The evolution of the in-phase and anti-phase terms of spin two follows the same pattern as that for spin one: all we have to do is swap the indices 1 and 2. So, for example, Eq. 7.17 on the preceding page

$$\exp(-i\theta\hat{I}_{1z}\hat{I}_{2z})\,2\hat{I}_{1x}\hat{I}_{2z}\,\exp(i\theta\hat{I}_{1z}\hat{I}_{2z}) \equiv \cos(\tfrac{1}{2}\theta)\,2\hat{I}_{1x}\hat{I}_{2z} + \sin(\tfrac{1}{2}\theta)\,\hat{I}_{1y},$$

becomes

$$\exp(-i\theta\hat{I}_{1z}\hat{I}_{2z})\,2\hat{I}_{1z}\hat{I}_{2x}\,\exp(i\theta\hat{I}_{1z}\hat{I}_{2z}) \equiv \cos(\tfrac{1}{2}\theta)\,2\hat{I}_{1z}\hat{I}_{2x} + \sin(\tfrac{1}{2}\theta)\,\hat{I}_{2y}.$$

We can also use Fig. 7.6 provided we interpret x and y as referring to $\hat{I}_{2x}$ and $\hat{I}_{2y}$, and xz and yz as referring to $2\hat{I}_{1z}\hat{I}_{2x}$ and $2\hat{I}_{1z}\hat{I}_{2y}$.

For example, the evolution of the term $-\hat{I}_{2x}$ is found from (a), where we see that this term evolves to $-2\hat{I}_{1z}\hat{I}_{2y}$. In the arrow notation this is:

$$-\hat{I}_{2x} \xrightarrow{\,2\pi J_{12}\tau\hat{I}_{1z}\hat{I}_{2z}\,} -\cos(\pi J_{12}\tau)\,\hat{I}_{2x} - \sin(\pi J_{12}\tau)\,2\hat{I}_{1z}\hat{I}_{2y}.$$

We are now in a position to explore more closely why $\hat{I}_{1x}$ is called an in-phase term and $2\hat{I}_{1x}\hat{I}_{2z}$ an anti-phase term.

7.5 In-phase and anti-phase terms

We mentioned above that operators like $\hat{I}_{1x}$ are called in-phase terms and products such as $2\hat{I}_{1x}\hat{I}_{2z}$ are called anti-phase terms. Now that we understand how to deal with the evolution due to coupling, we are in a position to explain just precisely what in- and anti-phase mean. As our discussion progresses, we will see that these anti-phase terms play a pivotal role in multiple-pulse NMR experiments.

7.5.1   In-phase terms

Let us imagine that at time zero we just have the operator $\hat{I}_{1x}$. We then allow this to evolve freely for a time t, all the time observing the x- and y-magnetizations. What we are going to do is work out how these magnetizations vary with time, and hence the form of the time-domain signal and the corresponding spectrum.

The evolution is controlled by the Hamiltonian we introduced before (Eq. 7.14 on page 150):

$$\hat{H}_{\text{two spins}} = \Omega_1 \hat{I}_{1z} + \Omega_2 \hat{I}_{2z} + 2\pi J_{12} \hat{I}_{1z} \hat{I}_{2z}.$$

We need to work out the effect of the three terms in turn. The first is simply a z-rotation due to the offset term for spin one, $\Omega_1 \hat{I}_{1z}$, which we see from (c) in Fig. 7.4 on page 148 takes $\hat{I}_{1x}$ to the new operator $\hat{I}_{1y}$:

$$\hat{I}_{1x} \xrightarrow{\ \Omega_1 t \hat{I}_{1z}\ } \cos(\Omega_1 t)\, \hat{I}_{1x} + \sin(\Omega_1 t)\, \hat{I}_{1y}. \tag{7.19}$$

The next term we need to consider is the offset term for spin two, $\Omega_2 \hat{I}_{2z}$; as the operator here refers to spin *two* it has *no effect* on any spin *one* operator. This is a general principle which we will use often.

Finally, we need to consider the effect of the scalar coupling, $2\pi J_{12} \hat{I}_{1z} \hat{I}_{2z}$, on each term on the right of Eq. 7.19. For the term $\hat{I}_{1x}$ we need (a) from Fig. 7.6 on the preceding page to see that the new operator is $2\hat{I}_{1y}\hat{I}_{2z}$; for the term $\hat{I}_{1y}$ we need (b) to see that the new operator is $-2\hat{I}_{1x}\hat{I}_{2z}$. The overall result is

$$\cos(\Omega_1 t)\, \hat{I}_{1x} + \sin(\Omega_1 t)\, \hat{I}_{1y} \xrightarrow{\ 2\pi J_{12} t \hat{I}_{1z} \hat{I}_{2z}\ }$$

$$\underbrace{\cos(\pi J_{12} t) \cos(\Omega_1 t)\, \hat{I}_{1x}}_{x\text{-magnetization}} + \sin(\pi J_{12} t) \cos(\Omega_1 t)\, 2\hat{I}_{1y}\hat{I}_{2z}$$

$$\underbrace{+ \cos(\pi J_{12} t) \sin(\Omega_1 t)\, \hat{I}_{1y}}_{y\text{-magnetization}} - \sin(\pi J_{12} t) \sin(\Omega_1 t)\, 2\hat{I}_{1x}\hat{I}_{2z}.$$

At time t the observable x-magnetization (on spin one) is given by the coefficient multiplying the operator $\hat{I}_{1x}$. Similarly, the y-magnetization is given by the coefficient multiplying the operator $\hat{I}_{1y}$.

As explained in section 5.2 on page 82, we usually represent the NMR signal as a complex number, the real part being proportional to the x-magnetization and the imaginary part proportional to the y-magnetization. Thus, the signal at time t, $S(t)$, is

$$
\begin{aligned}
S(t) &= \cos(\pi J_{12} t) \cos(\Omega_1 t) + \mathrm{i} \cos(\pi J_{12} t) \sin(\Omega_1 t) \\
&= \cos(\pi J_{12} t) \exp(\mathrm{i}\Omega_1 t) \\
&= \tfrac{1}{2}\left[\exp(\mathrm{i}\pi J_{12} t) + \exp(-\mathrm{i}\pi J_{12} t)\right] \exp(\mathrm{i}\Omega_1 t) \\
&= \tfrac{1}{2}\exp(\mathrm{i}[\Omega_1 + \pi J_{12}]t) + \tfrac{1}{2}\exp(\mathrm{i}[\Omega_1 - \pi J_{12}]t).
\end{aligned}
$$

We have used quite a lot of manipulations here. To go to the second line we have used the identity $\cos\theta + i\sin\theta \equiv \exp(i\theta)$, and to go to the third line we have rewritten the cosine in terms of complex exponentials using the identity $\cos\theta \equiv \frac{1}{2}[\exp(i\theta) + \exp(-i\theta)]$. Finally, to go to the last line we have multiplied out the square brackets and used $\exp(A)\exp(B) \equiv \exp(A + B)$.

The time-domain signal is the sum of two exponential terms, one oscillating at frequency $(\Omega_1 + \pi J_{12})$ and one at $(\Omega_1 - \pi J_{12})$; these two terms are both positive and have the same size. In practice, these signals will not only oscillate but will also decay over time due to relaxation. If we assume, for simplicity, that the decay is exponential, then the signal is of the form

$$S(t) = \tfrac{1}{2}\exp(i[\Omega_1 + \pi J_{12}]t)\exp(-Rt) + \tfrac{1}{2}\exp(i[\Omega_1 - \pi J_{12}]t)\exp(-Rt).$$

As we saw in section 5.3 on page 83, Fourier transformation of such a signal will give rise (in the real part of the spectrum) to two absorption mode lines of the same height, one centred at $(\Omega_1 + \pi J_{12}t)$, and one centred at $(\Omega_1 - \pi J_{12}t)$. These are, of course, the two lines of the spin-one doublet; the spectrum is illustrated in Fig. 7.7. Remember that we are working here in angular frequency units, so the lines are separated by $2\pi J_{12}$ rad s^{-1}, which is J_{12} Hz.

We can now see why the term $\hat{I}_{1x}$ is described as 'in-phase', since if we observe its evolution it gives rise to a spectrum consisting of the spin one doublet, with both lines positive and of the same intensity. A calculation along the same lines starting with $\hat{I}_{1y}$ gives the result

$$S(t) = \tfrac{1}{2}i\exp(i[\Omega_1 + \pi J_{12}]t) + \tfrac{1}{2}i\exp(i[\Omega_1 - \pi J_{12}]t).$$

This is the same as for $\hat{I}_{1x}$, apart from the factor i, which is simply a phase factor, corresponding to a phase shift of $\pi/2$ radians or 90° (see section 5.3.2 on page 85). In practice this means that if the spectrum obtained from the evolution of $\hat{I}_{1x}$ is phased to give absorption mode lines in the real part, that from $\hat{I}_{1y}$ will give dispersion mode lines. Remember that the relative phase is arbitrary, so we could just as well phase the spectrum so that the doublet from $\hat{I}_{1x}$ is dispersive, and that from $\hat{I}_{1y}$ is absorptive. The important thing is that $\hat{I}_{1x}$ and $\hat{I}_{1y}$ both give rise to in-phase doublets.

Similar calculations for $\hat{I}_{2x}$ and $\hat{I}_{2y}$ show that these two operators give rise to in-phase doublets centred on the offset of spin two; both lines in the doublet have the same sign. If one doublet is phased to absorption, the other will be in dispersion.

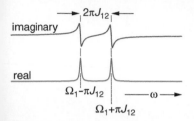

Fig. 7.7 The term $\hat{I}_{1x}$ evolves over time to give an observable signal whose Fourier transform consists of the two lines of the spin-one doublet i.e. a line at $(\Omega_1 + \pi J_{12})$ and a line at $(\Omega_1 - \pi J_{12})$. In the real part of the spectrum the lines have the absorption lineshape, whereas in the imaginary part the lineshape is dispersive. The important thing is that the two lines of the doublet both have the same sign; this is why the term $\hat{I}_{1x}$ is described as 'in-phase'.

7.5.2   Anti-phase terms

We now turn to the anti-phase operators. Strictly speaking, these are not observable in the sense that they do not give rise to transverse magnetization. However, we will show in this section that over time anti-phase operators evolve into in-phase operators which are observable.

Imagine that at time zero we have just $2\hat{I}_{1x}\hat{I}_{2z}$, and that we allow this to evolve for time t, observing the x- and y-magnetizations as time proceeds. Once again, we need to consider the effect of the offset of spin one, the offset of spin two, and the coupling.

The term in the Hamiltonian which represents the offset of spin one is $\Omega_1\hat{I}_{1z}$. This spin-one operator has no effect on any of the spin-two operators, so its effect on $2\hat{I}_{1x}\hat{I}_{2z}$ is the same as its effect on $\hat{I}_{1x}$. In the following, we emphasize this by placing curly braces around $\hat{I}_{2z}$ in the following:

$$2\hat{I}_{1x}\{\hat{I}_{2z}\} \xrightarrow{\Omega_1 t \hat{I}_{1z}} \cos(\Omega_1 t)\, 2\hat{I}_{1x}\{\hat{I}_{2z}\} + \sin(\Omega_1 t)\, 2\hat{I}_{1y}\{\hat{I}_{2z}\};$$

essentially, the operator $\hat{I}_{2z}$ carries through as a harmless factor, and $\hat{I}_{1x}$ evolves into $\hat{I}_{1y}$.

The next term to consider is that for the offset of spin two: $\Omega_2 \hat{I}_{2z}$. This has no effect on the spin-one operators and, as the spin-two operator is $\hat{I}_{2z}$, this is also unaffected as a z-operator is unaffected by a z-rotation.

Finally, we have to consider the coupling, which gives the following result (check it for yourself using Fig. 7.6 on page 152):

$$\cos(\Omega_1 t)\, 2\hat{I}_{1x}\hat{I}_{2z} + \sin(\Omega_1 t)\, 2\hat{I}_{1y}\hat{I}_{2z} \xrightarrow{2\pi J_{12} t \hat{I}_{1z}\hat{I}_{2z}}$$

$$\cos(\pi J_{12} t)\cos(\Omega_1 t)\, 2\hat{I}_{1x}\hat{I}_{2z} + \underbrace{\sin(\pi J_{12} t)\cos(\Omega_1 t)\,\hat{I}_{1y}}_{y\text{-magnetization}}$$

$$+ \cos(\pi J_{12} t)\sin(\Omega_1 t)\, 2\hat{I}_{1y}\hat{I}_{2z} - \underbrace{\sin(\pi J_{12} t)\sin(\Omega_1 t)\,\hat{I}_{1x}}_{x\text{-magnetization}}.$$

As before, M_x is given by the coefficient of $\hat{I}_{1x}$ and M_y by the coefficient of $\hat{I}_{1y}$; so, $M_x + iM_y$ is:

$$
\begin{aligned}
S(t) &= -\sin(\pi J_{12} t)\sin(\Omega_1 t) + i\sin(\pi J_{12} t)\cos(\Omega_1 t) \\
&= i\sin(\pi J_{12} t)\left[\cos(\Omega_1 t) + i\sin(\Omega_1 t)\right] \\
&= i\sin(\pi J_{12} t)\exp(i\Omega_1 t) \\
&= i\frac{1}{2i}\left[\exp(i\pi J_{12} t) - \exp(-i\pi J_{12} t)\right]\exp(i\Omega_1 t) \\
&= \frac{1}{2}\exp(i[\Omega_1 + \pi J_{12}]t) - \frac{1}{2}\exp(i[\Omega_1 - \pi J_{12}]t).
\end{aligned}
$$

We have used the same manipulations as in the case of the evolution of the in-phase term, except that to go to the fourth line the identity $\sin\theta \equiv (1/2i)[\exp(i\theta) - \exp(-i\theta)]$ has been used.

As with the term $\hat{I}_{1x}$, we find a term oscillating at $(\Omega_1 + \pi J_{12})$ and one at $(\Omega_1 - \pi J_{12})$. These are the two lines of the spin-one doublet. However, the crucial thing here is that one of the terms is multiplied by a minus sign. So, on Fourier transformation, one peak will be positive and one will be negative, as shown in Fig. 7.8. This is why the operator $2\hat{I}_{1x}\hat{I}_{2z}$ is referred to as an *anti-phase* operator.

In section 3.6 on page 38 it was explained how the two lines of the spin-one doublet could be associated with spin two being in the α or β state. What we see in the case of the anti-phase term is that the *sign* of the peak (i.e. whether it is positive or negative) also depends on the spin state of spin two.

A similar calculation for $2\hat{I}_{1y}\hat{I}_{2z}$ shows that this is also an anti-phase doublet on spin one, but with the opposite lineshape to that for $2\hat{I}_{1x}\hat{I}_{2z}$. Similarly, $2\hat{I}_{1z}\hat{I}_{2x}$ and $2\hat{I}_{1z}\hat{I}_{2y}$ correspond to anti-phase doublets on spin two.

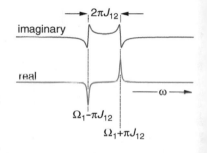

Fig. 7.8 The term $2\hat{I}_{1x}\hat{I}_{2z}$ evolves over time to give an observable signal whose Fourier transform consists of the two lines of the spin-one doublet. However, in contrast to the in-phase term $\hat{I}_{1x}$, the two lines from $2\hat{I}_{1x}\hat{I}_{2z}$ have opposite signs – hence the description of this term as 'anti-phase'.

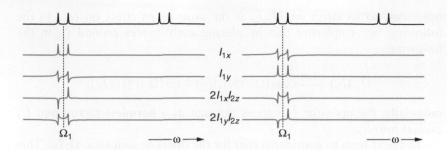

Fig. 7.9 Spectra resulting from the four observable operators of spin one. The complete spectrum is shown in black at the top of the diagram; on the left, the spectra are phased such that x-magnetization gives rise to absorption mode lines, whereas on the right the phase is such that y-magnetization gives absorption mode lines i.e. there is a phase shift of 90° between the two sets of spectra.

Fig. 7.10 Spectra resulting from the four observable operators of spin two; the format is the same as for Fig. 7.9.

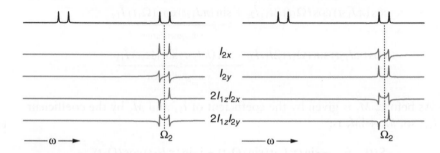

7.5.3   Observable terms

Strictly speaking, the only terms which give rise to observable magnetization are $\hat{I}_{1x}$, $\hat{I}_{1y}$, $\hat{I}_{2x}$ and $\hat{I}_{2y}$. However we have shown that if we start with the operator $2\hat{I}_{1x}\hat{I}_{2z}$ at time zero, this evolves in such a way as to give observable signals. So, it is usual to 'pretend' that the anti-phase terms such as $2\hat{I}_{1x}\hat{I}_{2z}$ are observable, in the sense that over time they will evolve into observable signals.

So, when we make a calculation on a pulse sequence we only need to carry it on up to the very beginning of data acquisition. At this point we simply inspect the terms we have, pick out those which are observable and simply deduce the form of the spectrum by realizing which spin they are on, and whether or not they are in-phase or anti-phase.

We also need to take into account the axes along which the operators are aligned, as this affects the lineshape in the spectrum. So, for example, if we have adjusted the phase in the spectrum so that $\hat{I}_{1x}$ gives an absorption mode in-phase doublet, then $2\hat{I}_{1x}\hat{I}_{2z}$ will give an absorption mode anti-phase doublet, whereas $2\hat{I}_{1y}\hat{I}_{2z}$ will give a dispersion mode anti-phase doublet. On the other hand, we could just as well phase the spectrum so that $\hat{I}_{1y}$ gives an absorption mode in-phase doublet, in which case $2\hat{I}_{1x}\hat{I}_{2z}$ will give a dispersion mode anti-phase doublet.

Figures 7.9 and 7.10 show the spectra arising from the observable operators on spin one and spin two, respectively. Each set of spectra appear twice: once phased such that x-magnetization gives rise to absorption mode lineshapes, and once such that y-magnetization gives rise to the absorption mode.

7.6   Hamiltonians for two spins

We have already introduced and discussed the free-precession Hamiltonian for two spins (Eq. 7.14 on page 150):

$$\hat{H}_{\text{two spins}} = \Omega_1 \hat{I}_{1z} + \Omega_2 \hat{I}_{2z} + 2\pi J_{12} \hat{I}_{1z} \hat{I}_{2z}.$$

It was also described how the evolution caused by this Hamiltonian can be worked out by considering the effect of the three terms in turn, and in any order. We have already seen that, as spin-one operators are unaffected by rotations due to spin-two operators (and vice versa), it is often the case that one of these three terms has no effect on the evolution.

For hard RF pulses, the Hamiltonian is analogous to that for a single spin (Eq. 7.6 on page 141), except that there is a term for spin one and a term for spin two:

$$\hat{H}_x = \omega_1 \hat{I}_{1x} + \omega_1 \hat{I}_{2x}.$$

Once more, these two terms commute, so that the evolution caused by this Hamiltonian can be worked out by considering each term in turn. In the arrow notation this evolution for time t_p is represented:

$$\hat{\rho}(0) \xrightarrow{\omega_1 t_p \hat{I}_{1x}} \xrightarrow{\omega_1 t_p \hat{I}_{2x}} \hat{\rho}(t_p).$$

Noting that the flip angle α is given by $\omega_1 t_p$ means that we can write these two arrows as

$$\hat{\rho}(0) \xrightarrow{\alpha \hat{I}_{1x}} \xrightarrow{\alpha \hat{I}_{2x}} \hat{\rho}(t_p).$$

If the pulse is applied only to one of the spins, then only the operator for that spin is present in the Hamiltonian. For example, a pulse to spin one has the Hamiltonian

$$\hat{H}_{x,\text{ spin }1} = \omega_1 \hat{I}_{1x}.$$

As before, for a pulse about y, the operators in the Hamiltonian are changed to $\hat{I}_y$:

$$\hat{H}_y = \omega_1 \hat{I}_{1y} + \omega_1 \hat{I}_{2y}.$$

7.7   Notation for heteronuclear spin systems

As far as the product operator approach is concerned, it makes no difference whether the two spins are of the same type (e.g. both protons), or of different types (e.g. one proton and one ^{13}C). However, when we are analysing heteronuclear pulse sequences it is sometimes useful to modify the operator notation somewhat so as to create a stronger distinction between the spin-one and spin-two operators.

The usual way to do this is to call one of the spins the 'I spin' and the other the 'S spin'. The operators for the I spin are

$$\hat{I}_x \quad \hat{I}_y \quad \hat{I}_z,$$

and those for the S spin are

$$\hat{S}_x \quad \hat{S}_y \quad \hat{S}_z.$$

This is only a change of notation, so everything we have done up to now still holds. All we have to do is replace the operators $\hat{I}_{1\gamma}$ with $\hat{I}_{\gamma}$, where $\gamma = x, y, z$, and similarly replace $\hat{I}_{2\gamma}$ with $\hat{S}_{\gamma}$.

In this notation, instead of writing the offsets of spins one and two as Ω_1 and Ω_2, we write the offsets of the I and S spin as Ω_I and Ω_S. The coupling between the two spins, which was J_{12}, becomes J_{IS}. Using this notation, the free precession Hamiltonian is

$$\hat{H}_{\text{two spins}} = \Omega_I \hat{I}_z + \Omega_S \hat{S}_z + 2\pi J_{IS} \hat{I}_z \hat{S}_z.$$

For an x-pulse to the I spin the Hamiltonian is

$$\hat{H}_{x,I} = \omega_1 \hat{I}_x,$$

whereas for a y-pulse to the S spin the Hamiltonian is

$$\hat{H}_{y,S} = \omega_1 \hat{S}_y.$$

7.8   Spin echoes and J-modulation

In section 4.9 on page 63 we used the vector model to explain how it is that the spin echo refocuses the evolution of the offset (chemical shift), and earlier in this chapter (section 7.2.2 on page 145) we repeated the analysis of the spin echo using operators. However, in both cases we only considered the evolution of the offset; we were not in a position to consider what happens when a scalar coupling is present – which is precisely what we are going to do now.

What we will find is that, in contrast to what happens to the offset, the evolution of the scalar coupling is *not* refocused in a spin echo. In fact, it appears that the coupling evolves throughout the entire duration of the spin echo, just as if the refocusing pulse were not there. This feature of the spin echo turns out to be absolutely crucial in multiple-pulse NMR experiments.

To start with, we will consider the case where the two spins which are coupled are of the same type e.g. two protons: this is called a *homonuclear* spin system. Then, we will go on to consider the case where the two spins are different e.g. ^{13}C and proton: this is called a *heteronuclear* spin system. The key difference in this second case is that we can choose whether to apply the 180° pulse to either one of the spins, or to both. We shall see that this flexibility allows us to choose whether the coupling is refocused or not, a property which turns out to be very important in heteronuclear multiple-pulse experiments.

So far, we have used the term spin echo as the name for the whole sequence (90° – delay – 180°– delay –). However, the initial 90° pulse is just there to excite transverse magnetization, the behaviour of which during the spin echo is what we are really interested in. So, from now on we will use the term spin echo for the element (delay – 180°– delay), and simply imagine that some transverse magnetization is present at its start.

7.8.1   Spin echo in a homonuclear spin system

First, we are going to analyse the evolution during a spin echo applied to a homonuclear two-spin system. The pulse sequence is shown in Fig. 7.11. As the spin system is homonuclear, the π pulse is applied to both spins, and to start with we will assume that this pulse is applied along the x-axis.

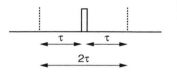

Fig. 7.11 Pulse sequence for the spin echo. From now on, we will regard the (delay – 180° – delay) sequence i.e. that part between the dashed lines, as the spin echo. The initial 90° pulse, included in Fig. 7.2 on page 145, is not considered to be part of the spin echo. The duration of the 180° pulse, indicated by the open rectangle, is negligible compared with the delays τ.

During the delays τ, both offsets and the scalar coupling affect the evolution. We have already shown that the offset is refocused, so to simplify the calculation here we will simply ignore the offset entirely, safe in the knowledge that it has no overall effect. You might legitimately object to this simplification, as it could be that the presence of the coupling somehow interferes with the refocusing of the offset. Rest assured that this is in fact not the case. The technical reason for this is that the terms in the Hamiltonian which describe offsets and coupling commute with one another, and so act entirely independently.

Let us imagine that at the start of the echo sequence we have in-phase x-magnetization of spin one: $\hat{I}_{1x}$. The first thing to consider is the evolution of the coupling during the delay τ. To work out what happens, all we need to do is refer to Fig. 7.6 (a) on page 152, and note that $\hat{I}_{1x}$ evolves into $2\hat{I}_{1y}\hat{I}_{2z}$, with the angle being $\pi J_{12}\tau$:

$$\hat{I}_{1x} \xrightarrow{\; 2\pi J_{12}\tau\hat{I}_{1z}\hat{I}_{2z} \;} \cos\left(\pi J_{12}\tau\right)\hat{I}_{1x} + \sin\left(\pi J_{12}\tau\right)2\hat{I}_{1y}\hat{I}_{2z}. \tag{7.20}$$

Next comes the π pulse, applied about the x-axis. The effect of this pulse is determined by treating it as a π rotation of spin one and then of spin two (section 7.6 on page 157). As was described in section 7.3.1 on page 146, such π pulses simply invert the sign of some terms. The term $\hat{I}_{1x}$ is unaffected, as an x-pulse does not affect x-operators. In the product $2\hat{I}_{1y}\hat{I}_{2z}$, $\hat{I}_{1y}$ changes sign to $-\hat{I}_{1y}$, and the same is true for $\hat{I}_{2z}$ which goes to $-\hat{I}_{2z}$. Since $-1 \times -1 = +1$, the net result is that the product $2\hat{I}_{1y}\hat{I}_{2z}$ is unaffected by the π pulse. Overall, therefore, nothing happens:

$$\cos\left(\pi J_{12}\tau\right)\hat{I}_{1x} + \sin\left(\pi J_{12}\tau\right)2\hat{I}_{1y}\hat{I}_{2z} \xrightarrow{\; \pi\hat{I}_{1x}+\pi\hat{I}_{2x} \;} \cos\left(\pi J_{12}\tau\right)\hat{I}_{1x} + \sin\left(\pi J_{12}\tau\right)2\hat{I}_{1y}\hat{I}_{2z}.$$

Finally, we have to consider the evolution during the second delay τ. Taking the terms one at a time, the evolution of $\hat{I}_{1x}$ is just as before:

$$\cos\left(\pi J_{12}\tau\right)\hat{I}_{1x} \xrightarrow{\; 2\pi J_{12}\tau\hat{I}_{1z}\hat{I}_{2z} \;}$$
$$\cos\left(\pi J_{12}\tau\right)\cos\left(\pi J_{12}\tau\right)\hat{I}_{1x} + \sin\left(\pi J_{12}\tau\right)\cos\left(\pi J_{12}\tau\right)2\hat{I}_{1y}\hat{I}_{2z}.$$

For the evolution of the term $2\hat{I}_{1y}\hat{I}_{2z}$ we need Fig. 7.6 (a) to see that it evolves to $-\hat{I}_{1x}$:

$$\sin\left(\pi J_{12}\tau\right)2\hat{I}_{1y}\hat{I}_{2z} \xrightarrow{\; 2\pi J_{12}\tau\hat{I}_{1z}\hat{I}_{2z} \;}$$
$$\cos\left(\pi J_{12}\tau\right)\sin\left(\pi J_{12}\tau\right)2\hat{I}_{1y}\hat{I}_{2z} - \sin\left(\pi J_{12}\tau\right)\sin\left(\pi J_{12}\tau\right)\hat{I}_{1x}.$$

The result of this evolution is therefore

$$\left[\cos\left(\pi J_{12}\tau\right)\cos\left(\pi J_{12}\tau\right) - \sin\left(\pi J_{12}\tau\right)\sin\left(\pi J_{12}\tau\right)\right]\hat{I}_{1x}$$
$$+ \left[\sin\left(\pi J_{12}\tau\right)\cos\left(\pi J_{12}\tau\right) + \cos\left(\pi J_{12}\tau\right)\sin\left(\pi J_{12}\tau\right)\right]2\hat{I}_{1y}\hat{I}_{2z}.$$

This looks a little complicated until we spot that the expression multiplying $\hat{I}_{1x}$ is of the form $\cos^2\theta - \sin^2\theta$; the identity $\cos(2\theta) \equiv \cos^2\theta - \sin^2\theta$ can therefore be used to simplify it to $\cos(2\pi J_{12}\tau)$. Similarly, the expression multiplying $2\hat{I}_{1y}\hat{I}_{2z}$ is of the form $2\sin\theta\cos\theta$; the identity $\sin(2\theta) \equiv 2\sin\theta\cos\theta$ simplifies the expression to $\sin(2\pi J_{12}\tau)$. Using these simplifications gives us the the final result:

$$\cos(2\pi J_{12}\tau)\hat{I}_{1x} + \sin(2\pi J_{12}\tau)2\hat{I}_{1y}\hat{I}_{2z}. \tag{7.21}$$

Compare this final result with Eq. 7.20 on the previous page which gives the result of evolution of the coupling for time τ. If you replace τ in Eq. 7.20 by 2τ you obtain Eq. 7.21. What we see is that the overall effect of the spin echo is the *same* as allowing the coupling to evolve for time 2τ. The evolution of the coupling is *not* affected by the sequence – in contrast to the offset which is refocused.

You will remember that, arbitrarily, we assumed that there was just in-phase x-magnetization present at the start of the spin echo. We need to check that our conclusion that the coupling is not refocused applies generally, and not just to this particular starting case. First, let us consider in-phase y-magnetization: the calculation proceeds as before, although this time we need Fig. 7.6 (b) on page 152 to see that $\hat{I}_{1y}$ evolves to $-2\hat{I}_{1x}\hat{I}_{2z}$:

$$\hat{I}_{1y} \xrightarrow{2\pi J_{12}\tau\hat{I}_{1z}\hat{I}_{2z}} \cos(\pi J_{12}\tau)\,\hat{I}_{1y} - \sin(\pi J_{12}\tau)\,2\hat{I}_{1x}\hat{I}_{2z}. \tag{7.22}$$

This time the π pulse inverts the first term $\hat{I}_{1y}$, and also inverts $\hat{I}_{2z}$ from the product $2\hat{I}_{1x}\hat{I}_{2z}$; overall, both terms are multiplied by -1:

$$\cos(\pi J_{12}\tau)\,\hat{I}_{1y} - \sin(\pi J_{12}\tau)\,2\hat{I}_{1x}\hat{I}_{2z} \xrightarrow{\pi\hat{I}_{1x}+\pi\hat{I}_{2x}}$$
$$-\cos(\pi J_{12}\tau)\,\hat{I}_{1y} + \sin(\pi J_{12}\tau)\,2\hat{I}_{1x}\hat{I}_{2z}.$$

For the evolution during the second delay τ, each term is considered separately:

$$-\cos(\pi J_{12}\tau)\,\hat{I}_{1y} \xrightarrow{2\pi J_{12}\tau\hat{I}_{1z}\hat{I}_{2z}}$$
$$-\cos(\pi J_{12}\tau)\cos(\pi J_{12}\tau)\,\hat{I}_{1y} + \sin(\pi J_{12}\tau)\cos(\pi J_{12}\tau)\,2\hat{I}_{1x}\hat{I}_{2z};$$

where we have used Fig. 7.6 (b) to see that $-\hat{I}_{1y}$ evolves to $2\hat{I}_{1x}\hat{I}_{2z}$. Using the same diagram, we see that $2\hat{I}_{1x}\hat{I}_{2z}$ evolves to $\hat{I}_{1y}$, giving

$$\sin(\pi J_{12}\tau)\,2\hat{I}_{1x}\hat{I}_{2z} \xrightarrow{2\pi J_{12}\tau\hat{I}_{1z}\hat{I}_{2z}}$$
$$\cos(\pi J_{12}\tau)\sin(\pi J_{12}\tau)\,2\hat{I}_{1x}\hat{I}_{2z} + \sin(\pi J_{12}\tau)\sin(\pi J_{12}\tau)\,\hat{I}_{1y}.$$

So, the final result is:

$$-[\cos(\pi J_{12}\tau)\cos(\pi J_{12}\tau) - \sin(\pi J_{12}\tau)\sin(\pi J_{12}\tau)]\,\hat{I}_{1y}$$
$$+[\sin(\pi J_{12}\tau)\cos(\pi J_{12}\tau) + \cos(\pi J_{12}\tau)\sin(\pi J_{12}\tau)]\,2\hat{I}_{1x}\hat{I}_{2z}.$$

Applying the identities for $\cos(2\theta)$ and $\sin(2\theta)$ we used before, this simplifies to

$$-\cos(2\pi J_{12}\tau)\hat{I}_{1y} + \sin(2\pi J_{12}\tau)2\hat{I}_{1x}\hat{I}_{2z}. \tag{7.23}$$

Once again, we see that the evolution of the coupling has not been refocused. Comparing Eq. 7.23 with Eq. 7.22, the result of evolution for time τ, we see that the result of the spin echo is the same as evolution for 2τ together with an overall change of sign. We will have more to say about this sign change in a moment.

Finally, we should consider the effect of the spin echo on the anti-phase terms $2\hat{I}_{1x}\hat{I}_{2z}$ and $2\hat{I}_{1y}\hat{I}_{2z}$. We will not go through the calculation step-by-step (this is something you can do for yourself), but simply quote the results; for convenience, they are summarized in the following table. Note that each initial state evolves into two terms, one multiplied by $\cos(2\pi J_{12}\tau)$ and one by $\sin(2\pi J_{12}\tau)$.

	final state	
initial state	$\times \cos(2\pi J_{12}\tau)$	$\times \sin(2\pi J_{12}\tau)$
$\hat{I}_{1x}$	$\hat{I}_{1x}$	$2\hat{I}_{1y}\hat{I}_{2z}$
$\hat{I}_{1y}$	$-\hat{I}_{1y}$	$2\hat{I}_{1x}\hat{I}_{2z}$
$2\hat{I}_{1x}\hat{I}_{2z}$	$-2\hat{I}_{1x}\hat{I}_{2z}$	$-\hat{I}_{1y}$
$2\hat{I}_{1y}\hat{I}_{2z}$	$2\hat{I}_{1y}\hat{I}_{2z}$	$-\hat{I}_{1x}$

For all initial states, the coupling evolves for time 2τ. Careful inspection of this table will show that the result of the spin echo is equivalent to (in either order)

- evolution of the coupling for time 2τ

- a 180° pulse (here about *x*).

This is a very handy way of dealing with the evolution due to spin echoes – we will use it often.

For example, consider starting with $\hat{I}_{1x}$: evolution for 2τ gives

$$\hat{I}_{1x} \xrightarrow{\;2\pi J_{12}(2\tau)\hat{I}_{1z}\hat{I}_{2z}\;} \cos(2\pi J_{12}\tau)\,\hat{I}_{1x} + \sin(2\pi J_{12}\tau)\,2\hat{I}_{1y}\hat{I}_{2z}.$$

A 180° pulse about *x* leaves the terms unaffected

$$\cos(2\pi J_{12}\tau)\,\hat{I}_{1x} + \sin(2\pi J_{12}\tau)\,2\hat{I}_{1y}\hat{I}_{2z} \xrightarrow{\;\pi\hat{I}_{1x}+\pi\hat{I}_{2x}\;}$$
$$\cos(2\pi J_{12}\tau)\,\hat{I}_{1x} + \sin(2\pi J_{12}\tau)\,2\hat{I}_{1y}\hat{I}_{2z};$$

this is exactly the result we obtained before, and is given on the first line of the table.

As another example, consider starting with $2\hat{I}_{1x}\hat{I}_{2z}$: evolution for 2τ gives

$$2\hat{I}_{1x}\hat{I}_{2z} \xrightarrow{\;2\pi J_{12}(2\tau)\hat{I}_{1z}\hat{I}_{2z}\;} \cos(2\pi J_{12}\tau)\,2\hat{I}_{1x}\hat{I}_{2z} + \sin(2\pi J_{12}\tau)\,\hat{I}_{1y}.$$

A 180° pulse about *x* changes the signs of both terms

$$\cos(2\pi J_{12}\tau)\,2\hat{I}_{1x}\hat{I}_{2z} + \sin(2\pi J_{12}\tau)\,\hat{I}_{1y} \xrightarrow{\;\pi\hat{I}_{1x}+\pi\hat{I}_{2x}\;}$$
$$-\cos(2\pi J_{12}\tau)\,2\hat{I}_{1x}\hat{I}_{2z} - \sin(2\pi J_{12}\tau)\,\hat{I}_{1y},$$

which is the result on line three of the table. If we change the phase of the 180° pulse from x to y, the same principle applies: the result of the spin echo is evolution for time 2τ, followed by a 180° pulse, this time about y.

Of course, we might just as well start with magnetization on spin two rather than spin one. The general result we have found above still applies e.g.

$$\hat{I}_{2y} \xrightarrow{\ \tau - 180^\circ - \tau\ } -\cos\left(2\pi J_{12}\tau\right)\hat{I}_{2y} + \sin\left(2\pi J_{12}\tau\right)2\hat{I}_{1z}\hat{I}_{2x}.$$

7.8.2   Summary

It is useful at this point to summarize the properties of the spin echo when applied to homonuclear spin systems:

- The offset is refocused i.e. it can be ignored.

- The coupling is not refocused.

- The overall result of the spin echo is equivalent to evolution of the coupling for time 2τ followed by a 180° pulse (of the appropriate phase).

The homonuclear spin echo is said to be *modulated* by the coupling, in the sense that the result depends, in an oscillatory way, on the coupling. You will also encounter the term *J-modulated spin echo* as a description of this effect.

7.8.3   Spectra from a *J*-modulated spin echo

A good way of seeing what this *J*-modulation means in practical terms is to consider an experiment where we start with in-phase x-magnetization, $\hat{I}_{1x}$, apply a spin echo sequence and then observe the result straight away. A suitable pulse sequence is shown in Fig. 7.12. Just after the spin echo, and right at the start of acquisition, we have shown in the previous section that the terms present are

$$\cos\left(2\pi J_{12}\tau\right)\hat{I}_{1x} + \sin\left(2\pi J_{12}\tau\right)2\hat{I}_{1y}\hat{I}_{2z}.$$

From the discussion in section 7.5 on page 152, we know that the term $\hat{I}_{1x}$ will give rise to the spin-one doublet in which both lines have the same sign and amplitude i.e. they are in-phase. The term $2\hat{I}_{1y}\hat{I}_{2z}$ will give the same doublet, but this time with one line positive and one line negative i.e. anti-phase. In addition, if we phase the spectrum such that magnetization initially along x gives absorption mode lineshapes, the in-phase doublet will be in absorption but the anti-phase doublet will be in dispersion.

The spectrum will therefore consist of a superposition of two doublets: there will be an in-phase absorption doublet, with intensity proportional to $\cos\left(2\pi J_{12}\tau\right)$, and an anti-phase dispersion doublet with intensity proportional to $\sin\left(2\pi J_{12}\tau\right)$. Clearly, as the delay τ changes, the ratio between the in-phase and anti-phase contributions changes.

This effect of changing the spin-echo delay τ is shown in Fig. 7.13 on the facing page. At the top, we have $\tau = 0$, and so see just the in-phase absorption multiplet. As τ increases, the proportion of the in-phase

Fig. 7.12 A simple sequence for observing *J*-modulated spin echoes. The initial 90° pulse about y generates in-phase x-magnetization; a spin echo of total duration 2τ follows, and then the signal is recorded.

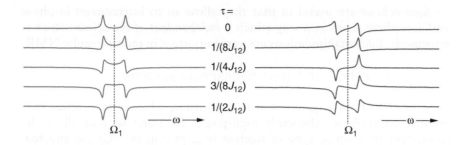

Fig. 7.13 Illustration of the result expected for the spin echo sequence of Fig. 7.12 on the preceding page. Different values of the delay τ are shown, starting with $\tau = 0$ at the top. On the left the spectra are phased so that x-magnetization gives rise to absorption mode lineshapes, whereas on the right the phase is such that y-magnetization gives rise to absorption mode lineshapes. For $\tau = 0$ we start with in-phase magnetization along x, $\hat{I}_{1x}$; as τ increases, the amount of anti-phase magnetization increases, and when $\tau = 1/(4J_{12})$ there is complete conversion to anti-phase magnetization along y, $2\hat{I}_{1y}\hat{I}_{2z}$. As τ increases beyond this point, the amount of in-phase magnetization increases, but in the negative sense; when $\tau = 1/(2J_{12})$ there is complete conversion to negative in-phase magnetization, $-\hat{I}_{1x}$.

contribution decreases and that of the anti-phase term increases. Note that in-phase magnetization along x becomes anti-phase magnetization along y, so there is a change in lineshape from absorption to dispersion (or *vice versa*).

We can work out the value of the delay which will give complete conversion to anti-phase by noting that this will be when

$$\sin(2\pi J_{12}\tau_{\text{opt}}) = 1,$$

which is when $(2\pi J_{12}\tau_{\text{opt}}) = \pi/2$ i.e. $\tau_{\text{opt}} = 1/(4J_{12})$. For this value of the delay, the amount of in-phase magnetization is $\cos(2\pi J_{12}/[4J_{12}]) = \cos(\pi/2) = 0$, so there is indeed complete conversion to anti-phase, as can be seen in Fig. 7.13.

As τ increases beyond $1/(4J_{12})$, the amount of anti-phase decreases and the amount of in-phase increases, but this time the in-phase doublet is negative. Complete conversion to the inverted doublet occurs when $\cos(2\pi J_{12}\tau) = -1$, which is when $\tau = 1/(2J_{12})$.

If we imagine starting the experiment not with in-phase magnetization, but with the anti-phase state $2\hat{I}_{1x}\hat{I}_{2z}$, then the effect of the spin echo is to give

$$-\cos(2\pi J_{12}\tau)2\hat{I}_{1x}\hat{I}_{2z} - \sin(2\pi J_{12}\tau)\hat{I}_{1y}.$$

This time, anti-phase magnetization evolves into in-phase and, as before, complete conversion to in-phase requires $\tau = 1/(4J_{12})$. In summary:

- The *J*-modulation during a spin echo causes an oscillatory interchange between in-phase and anti-phase terms.

- The in-phase and anti-phase terms are along orthogonal axes e.g. x and y.

- Complete conversion of in-phase to anti-phase, or *vice versa*, requires a spin-echo delay τ of $1/(4J_{12})$, which is a total delay of $1/(2J_{12})$.

Spin echoes are useful in that they allow us to interconvert in-phase and anti-phase terms in a way which is *independent of the offset* (which is refocused). Such manipulations are very important in multiple-pulse NMR.

7.8.4   Spin echoes in heteronuclear spin systems

In section 4.11 on page 67 it was noted that, whereas in practice it is possible to apply a sufficiently high-power RF pulse so that all of the resonances of a single type of nucleus (e.g. proton or ^{13}C) are affected, such a pulse applied to one type of nucleus will not affect another. So, for example, a hard pulse set to cover the range of ^{13}C chemical shifts will have no effect on protons, and vice versa.

Therefore, when it comes to forming a spin echo in a heteronuclear spin system, we can choose to which type of nucleus the 180° pulse is applied. In a two-spin system there are three possibilities, shown in Fig. 7.14. For the discussion in this section we will switch to calling the two spins I and S, rather than one and two, as this is the usual notation for heteronuclear spin systems (see section 7.7 on page 157). Typically I will be proton, and S will be a heteronucleus, such as ^{13}C, ^{31}P or ^{15}N.

In sequence (a) 180° pulses are applied to both spins. This is exactly the same as in the homonuclear spin system, so we can immediately deduce that the offsets of both spins are refocused, whereas the coupling evolves for time 2τ. In (b) and (c) there is a single 180° pulse applied to either the I spin or the S spin. We will now work out what happens in these two sequences.

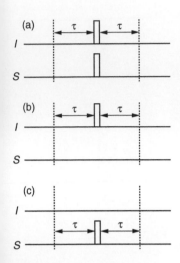

Fig. 7.14 Three different spin echo pulse sequences which can be applied in heteronuclear spin systems where we have the option of applying 180° pulses to: (a) both spins; (b) only the I spin; and (c) only the S spin. Sequence (a) gives an identical result to a homonuclear spin echo i.e. the offset is refocused and the coupling is not. Sequence (b) refocuses the offset of the I spin (but not of the S spin), and also refocuses the coupling. Sequence (c) refocuses the coupling, leaves the offset of the I spin unaffected, but refocuses the offset of the S spin. In all cases the start and end of the echo sequence is indicated by the dashed lines.

180° pulse to the I spin only – sequence (b)

From all we have done so far, it is clear that in this sequence the offset of the I spin is refocused by the action of the 180° pulse applied to that spin. So, if there are terms such as $\hat{I}_x$ or $2\hat{I}_y\hat{S}_z$ present at the start of the sequence, we expect that at the end of the sequence they will be unaffected by the offset of the I spin.

On the other hand, we do *not* expect the offset of the S spin to be refocused as there is no 180° pulse applied to this spin. Thus terms such as $\hat{S}_x$ or $2\hat{I}_z\hat{S}_y$ will simply continue to evolve under the influence of the offset throughout the whole period 2τ.

To work out what happens to the coupling requires an explicit calculation. Let us start with the term $\hat{I}_x$, and follow its fate through the sequence. The evolution during the first delay τ is exactly as before:

$$\hat{I}_x \xrightarrow{\ 2\pi J_{IS}\tau\hat{I}_z\hat{S}_z\ } \cos{(\pi J_{IS}\tau)}\,\hat{I}_x + \sin{(\pi J_{IS}\tau)}\,2\hat{I}_y\hat{S}_z.$$

The 180° pulse is only applied to the I spin, so it is only the I-spin operators whose sign can be changed. As before, the x-pulse does not affect $\hat{I}_x$ but the operator $\hat{I}_y$ in the product $2\hat{I}_y\hat{S}_z$ is inverted. The overall result is

$$\cos{(\pi J_{IS}\tau)}\,\hat{I}_x + \sin{(\pi J_{IS}\tau)}\,2\hat{I}_y\hat{S}_z \xrightarrow{\ \pi\hat{I}_x\ }$$

$$\cos{(\pi J_{IS}\tau)}\,\hat{I}_x - \sin{(\pi J_{IS}\tau)}\,2\hat{I}_y\hat{S}_z. \tag{7.24}$$

We now allow each term to evolve once more for delay τ under the coupling.

$$\cos(\pi J_{IS}\tau)\,\hat{I}_x \xrightarrow{2\pi J_{IS}\tau\hat{I}_z\hat{S}_z}$$
$$\cos(\pi J_{IS}\tau)\cos(\pi J_{IS}\tau)\,\hat{I}_x + \sin(\pi J_{IS}\tau)\cos(\pi J_{IS}\tau)\,2\hat{I}_y\hat{S}_z.$$

$$-\sin(\pi J_{IS}\tau)\,2\hat{I}_y\hat{S}_z \xrightarrow{2\pi J_{IS}\tau\hat{I}_z\hat{S}_z}$$
$$-\cos(\pi J_{IS}\tau)\sin(\pi J_{IS}\tau)\,2\hat{I}_y\hat{S}_z + \sin(\pi J_{IS}\tau)\sin(\pi J_{IS}\tau)\,\hat{I}_x.$$

Collecting the terms together we see that the two anti-phase terms cancel one another out, whereas the in-phase terms are

$$[\cos(\pi J_{IS}\tau)\cos(\pi J_{IS}\tau) + \sin(\pi J_{IS}\tau)\sin(\pi J_{IS}\tau)]\,\hat{I}_x.$$

Using the identity that $\cos^2\theta + \sin^2\theta \equiv 1$, we see that the result is simply $\hat{I}_x$. In other words, at the end of the sequence it appears that the coupling *does not* affect the evolution of $\hat{I}_x$ i.e. the coupling is refocused.

Repeating the calculation with different initial states shows the following overall results:

$$\hat{I}_x \xrightarrow{\tau-\pi_x(I\text{ spin})-\tau} \hat{I}_x$$
$$\hat{I}_y \xrightarrow{\tau-\pi_x(I\text{ spin})-\tau} -\hat{I}_y$$
$$2\hat{I}_x\hat{S}_z \xrightarrow{\tau-\pi_x(I\text{ spin})-\tau} 2\hat{I}_x\hat{S}_z$$
$$2\hat{I}_y\hat{S}_z \xrightarrow{\tau-\pi_x(I\text{ spin})-\tau} -2\hat{I}_y\hat{S}_z.$$

It is clear from these that, for transverse *I*-spin operators, the overall outcome is the same as a 180° pulse about the *x*-axis applied to the *I* spin.

We now turn to transverse *S* spin operators; first we will consider $\hat{S}_x$:

$$\hat{S}_x \xrightarrow{2\pi J_{IS}\tau\hat{I}_z\hat{S}_z} \cos(\pi J_{IS}\tau)\,\hat{S}_x + \sin(\pi J_{IS}\tau)\,2\hat{I}_z\hat{S}_y.$$

The 180°(*x*) pulse applied to the *I* spin inverts only the operator $\hat{I}_z$ in the product $2\hat{I}_z\hat{S}_y$

$$\cos(\pi J_{IS}\tau)\,\hat{S}_x + \sin(\pi J_{IS}\tau)\,2\hat{I}_z\hat{S}_y \xrightarrow{\pi\hat{I}_x} \cos(\pi J_{IS}\tau)\,\hat{S}_x - \sin(\pi J_{IS}\tau)\,2\hat{I}_z\hat{S}_y.$$

Comparing this result with Eq. 7.24 on the preceding page, we see that the terms on the right are the same apart from the fact that the *I* and *S* operators have been swapped. We do not therefore need to complete the rest of the calculation as it will be the same as before, giving the final result $\hat{S}_x$. The coupling is refocused.

Working through all of the other *S*-spin transverse operators shows us that the coupling is refocused in each case. Do not forget, however, that these transverse *S*-spin terms will still evolve under the offset of that spin for the entire duration 2τ.

It is interesting to note that, as far as transverse operators of the *S* spin are concerned, the only effect of the 180° pulse to the *I* spin is to invert the operator $\hat{I}_z$ when it appears in anti-phase terms. For this reason, when thinking about the *S*-spin operators, the 180° pulse to the *I* spin is often called an *inversion* pulse, rather than a refocusing pulse.

180° pulse to the S spin only – sequence (c)

Working out what happens here need not detain us for long. We can re-use all of the calculations for the echo with the 180° pulse applied to the *I* spin simply by swapping the *I* and *S* operators. So, we expect that transverse operators of the *S* spin will be affected neither by the coupling nor the offset – both are refocused. For these operators, the overall effect of the sequence is just the 180° pulse to the *S* spin. Transverse operators on the *I* spin will evolve under the offset of the *I* spin, but the coupling is once again refocused.

7.8.5   Summary

In summary, the echo sequence shown in Fig. 7.14 (a) on page 164 refocuses the offset, but allows the coupling to evolve. The effects of sequences (b) and (c) on transverse operators for different spins are summarized in the following table:

transverse operator on	sequence (b)		sequence (c)	
	offset	coupling	offset	coupling
I spin	refoc.	refoc.	not refoc.	refoc.
S spin	not refoc.	refoc.	refoc.	refoc.

It is clear that in a heteronuclear spin system we will be able to control the evolution of the coupling and offset separately by choosing the appropriate kind of spin echo. This freedom is exceptionally important in the operation of heteronuclear multiple-pulse experiments.

7.9   Coherence transfer

We are now in a position to understand one of the key building blocks of multiple-pulse NMR experiments – coherence transfer. The idea is remarkably simple: suppose that we have generated some anti-phase magnetization of spin one, aligned along the *y*-axis, $2\hat{I}_{1y}\hat{I}_{2z}$. From what we have seen so far, such a state can easily be generated by allowing the scalar coupling to evolve for an appropriate time. We now apply a 90° pulse (to both spins), about the *x*-axis. The result is

$$\underbrace{2\hat{I}_{1y}\hat{I}_{2z}}_{\text{on spin 1}} \xrightarrow{(\pi/2)\hat{I}_{1x}} 2\hat{I}_{1z}\hat{I}_{2z} \xrightarrow{(\pi/2)\hat{I}_{2x}} \underbrace{-2\hat{I}_{1z}\hat{I}_{2y}}_{\text{on spin 2}}.$$

The key point here is that the term which started out as transverse magnetization on spin one, $2\hat{I}_{1y}\hat{I}_{2z}$, ends up transverse on spin two, $2\hat{I}_{1z}\hat{I}_{2y}$. This process is called *coherence transfer* as transverse magnetization, which is a coherence, is transferred from one spin to another.

An in-phase term cannot be transferred from one spin to another by the action of a pulse – such a transfer is a unique feature of anti-phase terms. As we have seen, anti-phase terms arise because of the evolution of a scalar coupling, so it is *only* if such a coupling exists between two spins that we can have coherence transfer via the anti-phase state. Therefore, the existence of coherence transfer between two spins is indicative of a coupling between those spins; this connection is very important in two-dimensional spectroscopy.

7.10 The INEPT experiment

The INEPT experiment is an excellent demonstration of how coherence transfer using anti-phase states can be used to great advantage. In addition, the basic idea used in this experiment is used over and over again in many more complex heteronuclear multiple-pulse experiments. However, to understand the motivation for developing the INEPT experiment, we need to back-track somewhat and discuss the factors which influence the size of the equilibrium magnetization.

NMR spectroscopists love to give their new experiments names or acronyms. The more whimsical or ironic the name, the better. INEPT stands for Insensitive Nuclei Enhanced by Polarization Transfer. You can judge for yourself whether whimsy or accuracy was what led to the choice of this acronym!

7.10.1 Why the experiment was developed

In section 4.1.3 on page 49 we described how, at equilibrium, the bulk z-magnetization arose due to the preferential population of the lower energy orientations of the individual magnetic moments. The closer the magnetic moment lies to the applied field (which is along the $+z$-axis), the lower the energy, so preferential population of lower energy orientations leads to bulk z magnetization.

The energy of interaction between an individual magnetic moment and the applied field is proportional to the gyromagnetic ratio of the nucleus and the applied magnetic field strength. Increasing either of these factors increases the energetic preference for orientations lying close to the field direction, and hence increases the size of the equilibrium magnetization.

The size of the signal we detect at the end of an experiment ultimately depends on the size of the equilibrium magnetization from which we start: the larger the equilibrium magnetization, and the larger the observed signal. This is one of the reasons why so much effort has been put into increasing the strength of the applied magnetic field, as doing so increases the equilibrium magnetization and hence the strength of the signals. As a result, the sensitivity is improved.

For a fixed magnetic field strength, nuclei with higher gyromagnetic ratios will have larger equilibrium magnetizations, and hence – all other things being equal – have higher sensitivity. The INEPT experiment was conceived as a way of enhancing the signals observed from a low gyromagnetic ratio nucleus by transferring to it the larger equilibrium magnetization of a higher gyromagnetic ratio nucleus. Typically, the high gyromagnetic ratio nucleus is a proton, whose magnetization is transferred to nuclei such as ^{13}C or ^{15}N.

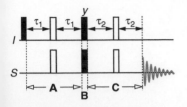

Fig. 7.15 Pulse sequence for the INEPT experiment; note that the second 90° pulse to the I spin is of phase y. The experiment results in an observable signal on the S spin whose size depends on the equilibrium magnetization of the I spin. By choosing the I spin to have a higher gyromagnetic ratio than the S spin, the signal observed on S will be larger than for a simple pulse–acquire experiment on that spin. The sequence works by generating an anti-phase state on the I spin during period **A**, and then transferring it to the S spin using the two pulses in period **B**. During period **C** the anti-phase state evolves into an in-phase state; it is then observed. The optimum value for both delays τ_1 and τ_2 is $1/(4J_{IS})$.

7.10.2   Analysis of the INEPT experiment

The pulse sequence for INEPT is shown in Fig. 7.15. Broadly speaking, what happens is that during period **A** an anti-phase state is generated on the I spin. The two pulses in period **B** transfer this anti-phase state to S, and then during period **C** the anti-phase state evolves back into an in-phase state; it is then observed on the S spin. Let us now see, in detail, how this all works.

As the two spins will have different gyromagnetic ratios, we want to keep track of the fact that their equilibrium magnetizations are different. So, rather than writing the equilibrium magnetization on spin one as $\hat{I}_z$, we will write it as $k_I \hat{I}_z$, where k_I is a parameter which gives the overall size of the equilibrium magnetization. Similarly, the equilibrium magnetization on spin two will be written $k_S \hat{S}_z$.

To start with, we are going to consider the fate of the equilibrium magnetization on spin one. The initial $90°(x)$ pulse rotates this to $-k_I \hat{I}_y$. Looking at the pulse sequence, we recognize that period **A** is a spin echo; as both spins experience a 180° pulse, it follows that the offsets are refocused but the coupling continues to evolve for the whole of the period $2\tau_1$. It was shown in section 7.8.1 on page 159 that the overall effect of such a spin echo is evolution of the coupling for $2\tau_1$, followed by a 180° pulse to both spins. So, at the end of period **A** the state of the spin system is

$$k_I \cos\left(2\pi J_{IS}\tau_1\right)\hat{I}_y - k_I \sin\left(2\pi J_{IS}\tau_1\right)2\hat{I}_x\hat{S}_z.$$

Period **B** consists of the two 90° pulses, but note that the pulse to the I spin is about the y-axis; these are the pulses which cause the coherence transfer. The order in which they act is not important as they are on different spins.

The operator $\hat{I}_y$ will not be affected by either the y-pulse to I, or the pulse to S. We can therefore discard this term right away as there are no more coherence transfer steps and therefore it will not contribute to any observable signal on the S spin. Remember that in this heteronuclear experiment we are only going to observe the signals from S.

The term $2\hat{I}_x\hat{S}_z$ is affected by the two pulses as follows:

$$- k_I \sin\left(2\pi J_{IS}\tau_1\right)2\hat{I}_x\hat{S}_z \xrightarrow{(\pi/2)\hat{I}_y} k_I \sin\left(2\pi J_{IS}\tau_1\right)2\hat{I}_z\hat{S}_z$$

$$\xrightarrow{(\pi/2)\hat{S}_x} -k_I \sin\left(2\pi J_{IS}\tau_1\right)2\hat{I}_z\hat{S}_y.$$

These two pulses have transferred our anti-phase state from the I spin to the S spin. We could observe it immediately, resulting in a spectrum with anti-phase doublets; the resulting simplified INEPT pulse sequence is shown in Fig. 7.16 (a) on the facing page. However, it is more common to allow the anti-phase state to evolve into an in-phase state before it is observed.

Continuing with the INEPT pulse sequence of Fig. 7.15, we recognize that, like period **A**, period **C** is a spin echo, during which offsets are refocused, and the coupling continues to evolve. As before, we can compute the overall effect of this spin echo by allowing the coupling to evolve for $2\tau_2$, and then applying a 180° pulse to both spins. Thus, at the end of this period we have

$$- k_I \cos(2\pi J_{IS}\tau_2) \sin(2\pi J_{IS}\tau_1) 2\hat{I}_z\hat{S}_y$$
$$+ k_I \sin(2\pi J_{IS}\tau_2) \sin(2\pi J_{IS}\tau_1) \hat{S}_x. \tag{7.25}$$

The in-phase signal on the S spin, $\hat{S}_x$, is largest when both of the sine terms multiplying it are $= 1$, which is when the argument of the sine is $\pi/2$:

$$2\pi J_{IS}\tau_{1,\mathrm{opt}} = \pi/2.$$

This tells us that $\tau_{1,\mathrm{opt}} = 1/(4J_{IS})$ gives us the strongest signal; the optimum value for τ_2 is the same. These delays are those which result in complete interconversion of in- and anti-phase magnetization during the two spin echoes.

With this optimum value for the delays τ_1 and τ_2, the final observable term is

$$k_I\hat{S}_x;$$

compare this with the result of a simple pulse–acquire experiment directly on the S spin, which would give the term

$$-k_S\hat{S}_y.$$

Apart from the trivial difference in phase (and sign) between these two signals, the key thing is that in the INEPT experiment the signal is proportional to k_I, whereas in the pulse–acquire experiment the signal is proportional to k_S. If the I spin has a greater gyromagnetic ratio than the S spin, the result will be a stronger signal from the INEPT experiment by a factor k_I/k_S.

Note that because offsets are refocused throughout the pulse sequence, the enhancement produced by the INEPT experiment is independent of the offset of either spin. So, for example, in a molecule *all* of the ^{13}C nuclei bearing an attached proton can have their signals enhanced at the same time by the transfer of magnetization from the attached protons.

The final thing to note is that the transverse magnetization present during the two periods **A** and **C** will decay due to relaxation. As a result, not all of the equilibrium magnetization will be transferred from I to S i.e. the enhancement will be less than the theoretical maximum.

7.10.3 Decoupling in the INEPT experiment

As we described in section 2.4.1 on page 14, when observing heteronuclei such as ^{13}C, it is usual to employ broadband decoupling of the protons so as to remove any splittings from the ^{13}C spectrum due to ^{13}C–^{1}H couplings. There are two reasons for doing this: first, the spectrum is simplified by collapsing the multiplets to single lines; secondly, the sensitivity is improved as all of the intensity which was spread across a multiplet is now concentrated in one line.

If we have an in-phase state on the S spin, then applying broadband decoupling to the I spin will cause the in-phase doublet to collapse to a single line. However, if we have an anti-phase state, decoupling I will simply make the multiplet disappear; these different outcomes are

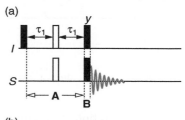

(a)

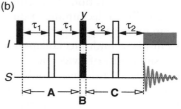

(b)

Fig. 7.16 Two modified INEPT experiments. In sequence (a) the S-spin signal is observed immediately after it has been transferred from I; as a result, the S-spin multiplet will appear in anti-phase. Sequence (b) is identical to that shown in Fig. 7.15 on the preceding page, except that broadband decoupling of the I spin (denoted by the blue rectangle) is applied during acquisition of the signal on S. As a result, the S-spin multiplet collapses to a single line.

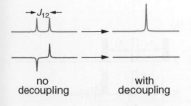

Fig. 7.17 Illustration of the effect of broadband decoupling of the I spin on the S spin multiplet. An in-phase multiplet collapses to a single line of twice the intensity; an anti-phase doublet collapses to nothing. This can be thought of as a result of setting the coupling to zero.

illustrated in Fig. 7.17. We can think of this as a result of the decoupling effectively setting the coupling to zero, so that the positive and negative lines of the anti-phase multiplet cancel one another.

In the INEPT experiment, if we wish to observe the S-spin signal in the presence of broadband decoupling of the I spin, then it is essential to allow the anti-phase term which appears at the end of period **B** to evolve into an in-phase term. This is achieved during the second spin echo, period **C**. At the end of this period, it is possible to turn on the broadband decoupling and observe the S-spin signal, see Fig. 7.16 (b) on the preceding page.

Assuming that only the in-phase signal will be observed during decoupled acquisition, the only term at the end of period **C**, given by Eq. 7.25 on the previous page, which is important is

$$k_I \sin\left(2\pi J_{IS}\tau_2\right)\sin\left(2\pi J_{IS}\tau_1\right)\hat{S}_x.$$

It was noted above that the optimum values of both of the τ delays is $1/(4J_{IS})$; any value other than this will result in a reduction in the intensity of the signal, and so reduce the advantage of the INEPT experiment.

In a real molecule not all of the couplings have the same value, so a compromise has to be made when it comes to choosing the values of the delays τ. As a result, not all of the spins will experience the same enhancement.

Generally, the INEPT experiment is most successful when large one-bond heteronuclear couplings are used to transfer the magnetization from one spin to another. Not only do such large couplings keep the delays τ short, thus reducing any losses due to relaxation, but they also tend to have a limited range of values, making it possible to choose a good compromise value for τ.

7.10.4 Suppressing the signal from the equilibrium magnetization on the S spin

So far we have not considered the fate of the equilibrium magnetization on the S spin, $k_S \hat{S}_z$. This is inverted by the first 180° pulse to the S spin and then rotated onto the $+y$-axis by the 90° pulse, to give $\hat{S}_y$. During the spin echo, period **C**, this in-phase term evolves to give:

$$-k_S \cos\left(2\pi J_{IS}\tau_2\right)\hat{S}_y + k_S \sin\left(2\pi J_{IS}\tau_2\right) 2\hat{I}_z\hat{S}_x.$$

As usual, we have worked this out by allowing the coupling to evolve for 2τ and then applying a 180° pulse to both spins. If we assume that we are using broadband decoupling of the I spin during acquisition, then only the first term is observable. Furthermore, if τ_2 is set to the optimum value of $1/(4J_{IS})$, the cosine term will be zero, so this in-phase term disappears i.e. there is no visible contribution due to the S-spin equilibrium magnetization.

However, we may not be able to use the optimum value of τ_2, so there is the possibility of some of this term in $\hat{S}_y$ being present at the start of acquisition. The problem is that this term is along y, whereas the term transferred from the I spin is along x. The result will therefore be a phase distortion of the spectrum. It is therefore desirable to remove the contribution from equilibrium magnetization on spin two, so as to remove the phase distortion.

The way this is done is to repeat the experiment twice. In the first experiment, everything is as we have described, so the two in-phase signals are

Experiment 1: $\underbrace{-k_S \cos\left(2\pi J_{12}\tau_2\right)\hat{S}_y}_{\text{from equil. mag. on the } S \text{ spin}}$ $\underbrace{+k_I \sin\left(2\pi J_{12}\tau_2\right)\sin\left(2\pi J_{12}\tau_1\right)\hat{S}_x}_{\text{from equil. mag. on the } I \text{ spin}}$.

In the second experiment, we change the phase of the very first pulse from x to $-x$. If you work through the calculations again, you will find that this alters the sign of the terms which arise from the equilibrium magnetization of the I spin. However, the terms which arise from the equilibrium magnetization of the S spin are not affected, simply because this first pulse has no effect on the S-spin terms. So, the outcome of the second experiment is

Experiment 2: $\underbrace{-k_S \cos\left(2\pi J_{IS}\tau_2\right)\hat{S}_y}_{\text{from equil. mag. on the } S \text{ spin}}$ $\underbrace{-k_I \sin\left(2\pi J_{IS}\tau_2\right)\sin\left(2\pi J_{IS}\tau_1\right)\hat{S}_x}_{\text{from equil. mag. on the } I \text{ spin}}$.

All we have to do is to subtract the data from these two experiments; the terms arising from the S-spin equilibrium magnetization will cancel, whilst those arising from the I-spin equilibrium magnetization will add:

Experiment 1 $-$ Experiment 2 $= 2k_I \sin\left(2\pi J_{IS}\tau_2\right)\sin\left(2\pi J_{IS}\tau_1\right)\hat{S}_x$.

This procedure is an example of an idea called *difference spectroscopy*, in which we separate out a wanted from an unwanted term by shifting the phase of a pulse in such a way that one term changes sign and the other does not. If we understand how the pulse sequence works, we can work out which pulse phase we need to alter.

The INEPT pulse sequence combines all of the key ideas we have introduced in this chapter: spin echoes are used to interconvert in- and anti-phase terms, independent of offset, and pulses are used to cause anti-phase terms to be transferred from one spin to another.

7.11 Selective COSY

In this section we are going to describe a simple experiment which is an analogue of the very important two-dimensional COSY experiment; both experiments enable us to identify which spins are coupled to one another. The experiment we are going to describe uses a selective pulse, which is a pulse with such a weak RF field that, even in a homonuclear spin system, only one multiplet is affected. More details about such pulses can be found in section 4.11.2 on page 69.

The pulse sequence is show in Fig. 7.18 on the next page. To start with we will consider only the fate of the equilibrium magnetization of spin one. The first pulse is made selective so that only spin one is excited; we will also put the transmitter on resonance with this spin, so that its offset is zero. The effect of this selective $90°$ pulse is simply to generate $-\hat{I}_{1y}$.

During the delay τ the coupling evolves but, as we have assumed that spin one is on resonance ($\Omega_1 = 0$), there is no evolution of the offset. So, at

COSY stands for COrrelation SpectroscopY. It was originally conceived as a two-dimensional experiment, and this version will be discussed in detail in Chapter 8.

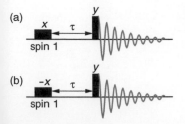

Fig. 7.18 Pulse sequence for the selective COSY experiment. The first pulse, indicated by a small filled-in rectangle, is made selective so that it only affects one multiplet, here spin one. After a delay τ, in which anti-phase magnetization develops, a non-selective 90° pulse is applied, followed by acquisition. The experiment is repeated twice, once with the first pulse phase x, sequence (a), and once with this pulse phase $-x$, sequence (b). Subtracting the two sets of data gives a spectrum in which only the multiplet from spin one, and multiplets from any spin coupled to it, are present.

the end of the delay we have a mixture of in- and anti-phase magnetization:

$$- \cos\left(\pi J_{12}\tau\right)\hat{I}_{1y} + \sin\left(\pi J_{12}\tau\right) 2\hat{I}_{1x}\hat{I}_{2z}.$$

The second pulse is non-selective, and so affects both spin one and spin two; note that the pulse is about the y-axis. After this pulse we have

$$- \cos\left(\pi J_{12}\tau\right)\hat{I}_{1y} - \sin\left(\pi J_{12}\tau\right) 2\hat{I}_{1z}\hat{I}_{2x}.$$

As expected, the in-phase term is unaffected, but the anti-phase term has been transferred to spin two.

We must also consider the fate of the equilibrium magnetization of spin two. This is unaffected by the first pulse, and simply rotated to x by the second. In summary, at the start of acquisition we have:

$$\begin{aligned} \text{from spin one:} \quad &- \cos\left(\pi J_{12}\tau\right)\hat{I}_{1y} - \sin\left(\pi J_{12}\tau\right) 2\hat{I}_{1z}\hat{I}_{2x} \\ \text{from spin two:} \quad &\hat{I}_{2x}. \end{aligned} \quad (7.26)$$

The term $2\hat{I}_{1z}\hat{I}_{2x}$ arises from coherence transfer from spin one to spin two. However, the term $\hat{I}_{2x}$ simply comes from the equilibrium magnetization of spin two, and its presence (on top of the anti-phase term) serves only to confuse the spectrum. This is illustrated in Fig. 7.19 (a) on the facing page, where we see the superposition of the operators $\hat{I}_{2x}$ and $2\hat{I}_{1z}\hat{I}_{2x}$ leads to an odd-looking multiplet on spin two.

As with INEPT, this unwanted signal is suppressed simply by repeating the experiment using the sequence of Fig. 7.18 (b), in which the first pulse has phase $-x$. Working through the calculation shows that this changes the sign of both of the terms which arise from spin one, but leaves the sign of the term arising from spin two unaffected. This is, of course, because spin two is unaffected by the initial selective pulse.

The outcome of the second experiment is therefore

$$\begin{aligned} \text{from spin one:} \quad &+ \cos\left(\pi J_{12}\tau\right)\hat{I}_{1y} + \sin\left(\pi J_{12}\tau\right) 2\hat{I}_{1z}\hat{I}_{2x} \\ \text{from spin two:} \quad &\hat{I}_{2x}; \end{aligned} \quad (7.27)$$

this is shown in Fig. 7.19 (b). Subtracting the two experiments, Eq. 7.27 − Eq. 7.26, gives us just the required signals which arise from spin one:

$$2 \cos\left(\pi J_{12}\tau\right)\hat{I}_{1y} + 2 \sin\left(\pi J_{12}\tau\right) 2\hat{I}_{1z}\hat{I}_{2x}.$$

This difference spectrum is shown in Fig. 7.19 (c).

The difference spectrum shows an in-phase multiplet from the spin we originally excited, here spin one. In contrast, the multiplet from the coupled spin, here spin two, appears in anti-phase and shifted in phase by 90°.

The presence of this anti-phase multiplet in the spectrum shows that there is a coupling between the spin associated with this multiplet and spin one. We can see this as if $J_{12} = 0$ the intensity of the anti-phase term, $\sin\left(\pi J_{12}\tau\right)$, goes to zero. More generally, the intensity of the anti-phase term depends on the choice of τ and the coupling present in the system; typically one chooses a value of τ of the order of $1/(2J_{12})$ for the largest couplings expected.

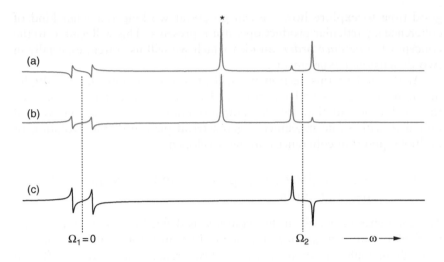

Fig. 7.19 Spectrum (a) is the outcome of the selective COSY experiment of Fig. 7.18 (a) on the preceding page in which spin one experiences the selective pulse; the spectrum is phased such that x-magnetization gives absorption lines. On spin one we see a dispersive in-phase doublet arising from the term $\hat{I}_{1y}$. Two operators contribute to the spin two multiplet: the first is $\sin{(\pi J_{12}\tau)}\,2\hat{I}_{1z}\hat{I}_{2x}$, which arises from coherence transfer from spin one, the second is $\hat{I}_{2x}$ which simply arises from the equilibrium magnetization on spin two. The superposition of these two operators results in the unsymmetrical spin two multiplet. Spectrum (b) was recorded using sequence (b) in Fig. 7.18; in this spectrum, the signals derived from the equilibrium magnetization of spin one are inverted, but those arising from equilibrium magnetization on spin two are unaffected. Taking the difference (b) − (a) gives spectrum (c), which contains only signals which derive from the equilibrium magnetization of spin one: we see a nice anti-phase doublet on spin two. The peak marked with a * is from another spin, not coupled to spins one or two; this peak is also eliminated from the difference spectrum, (c). In summary, (c) contains signals *only* from spin one or those spins coupled to it.

The difference procedure also makes sure that signals from any spins not coupled to spin one do not appear in the final spectrum. Thus what we have in the difference spectrum is just the multiplet of the initially excited spin (spin one), and the multiplets from *any* spins coupled to spin one. You can see how this could be useful for tracing out the network of couplings in a molecule.

This experiment once more demonstrates coherence transfer via an anti-phase state, and also the use of difference spectroscopy to suppress unwanted signals.

7.12 Coherence order and multiple-quantum coherences

When we introduced the product operators for two spins we noted that operators such as $\hat{I}_{1x}$ and $\hat{I}_{2y}$ corresponded to transverse magnetization (or single-quantum coherence), whereas operator products such as $2\hat{I}_{1x}\hat{I}_{2y}$ corresponded to unobservable multiple-quantum coherence. Now is a good time to explore how we can go about working out what kind of coherence a particular product operator represents. This will lead us to the concept of *coherence order*, an idea which we will use often, especially in two-dimensional experiments.

At the end of this section we will also look at how these multiple-quantum coherences evolve over time. There are some differences between this evolution and that for the simple operators $\hat{I}_x$ and $\hat{I}_y$, but we will find that with some ingenuity a geometrical picture of the evolution of multiple-quantum coherences can be developed.

7.12.1   Raising and lowering operators: the classification of coherence order

The coherence order, given the symbol p, is defined by what happens to a particular state (e.g. an operator or product operator) when a z-rotation through an angle ϕ is applied. A state of coherence order zero is unaffected by the rotation. From what we know already we can deduce that the operator $\hat{I}_z$ must be of coherence order zero as it is not affected by a z-rotation.

A state of coherence order $+1$ rotates through an angle $-\phi$ under this z-rotation, whereas a state of coherence order $+2$ rotates through an angle -2ϕ. Coherence order is a signed quantity, so there also exist states of order -1 and -2. The z-rotation rotates these states through angles of $+\phi$ and $+2\phi$, respectively.

In order to work out the coherence order of a particular operator or product operator, we need to introduce the *raising operator* $\hat{I}_+$, and the *lowering operator* $\hat{I}_-$. These are defined as follows:

$$\hat{I}_+ \equiv \hat{I}_x + i\hat{I}_y \qquad \hat{I}_- \equiv \hat{I}_x - i\hat{I}_y. \qquad (7.28)$$

The point of introducing these operators is that it turns out that, more or less by definition, $\hat{I}_+$ has coherence order $+1$ and $\hat{I}_-$ has coherence order -1. We can work out the coherence order of other operators by expressing them in terms of the raising and lowering operators.

If we add together the definitions in Eq. 7.28, the term in $\hat{I}_y$ cancels; similarly, if we subtract the two equations the term in $\hat{I}_x$ cancels. So, we can write:

$$\hat{I}_x \equiv \tfrac{1}{2}\left(\hat{I}_+ + \hat{I}_-\right) \qquad \hat{I}_y \equiv \tfrac{1}{2i}\left(\hat{I}_+ - \hat{I}_-\right). \qquad (7.29)$$

Therefore, both $\hat{I}_x$ and $\hat{I}_y$ are equal mixtures of coherence orders $+1$ and -1; these operators therefore represent *single-quantum coherence*. Recalling that $\hat{I}_x$ and $\hat{I}_y$ also represent observable transverse magnetization, we can see that such magnetization has coherence order ± 1.

To classify the product operators for two spins we simply introduce raising and lowering operators for each spin, defined in an identical way to Eq. 7.28:

$$\hat{I}_{1+} \equiv \hat{I}_{1x} + i\hat{I}_{1y} \qquad \hat{I}_{1-} \equiv \hat{I}_{1x} - i\hat{I}_{1y}$$
$$\hat{I}_{2+} \equiv \hat{I}_{2x} + i\hat{I}_{2y} \qquad \hat{I}_{2-} \equiv \hat{I}_{2x} - i\hat{I}_{2y}.$$

These can be rearranged to give analogous expressions to those of Eq. 7.29 on the facing page:

$$\hat{I}_{1x} \equiv \tfrac{1}{2}\left(\hat{I}_{1+} + \hat{I}_{1-}\right) \qquad \hat{I}_{1y} \equiv \tfrac{1}{2i}\left(\hat{I}_{1+} - \hat{I}_{1-}\right)$$
$$\hat{I}_{2x} \equiv \tfrac{1}{2}\left(\hat{I}_{2+} + \hat{I}_{2-}\right) \qquad \hat{I}_{2y} \equiv \tfrac{1}{2i}\left(\hat{I}_{2+} - \hat{I}_{2-}\right). \tag{7.30}$$

Using Eq. 7.30, a product such as $2\hat{I}_{1x}\hat{I}_{2z}$ can be written

$$2\hat{I}_{1x}\hat{I}_{2z} \equiv \left(\hat{I}_{1+} + \hat{I}_{1-}\right)\hat{I}_{2z}.$$

To work out the overall coherence order, we sum the coherence order for spin one and that for spin two. As $\hat{I}_{2z}$ has coherence order zero, the coherence order of $2\hat{I}_{1x}\hat{I}_{2z}$ is therefore an equal mixture of $+1$ and -1.

The product $2\hat{I}_{1x}\hat{I}_{2x}$ is more interesting:

$$\begin{aligned} 2\hat{I}_{1x}\hat{I}_{2x} &\equiv 2 \times \tfrac{1}{2}\left(\hat{I}_{1+} + \hat{I}_{1-}\right) \times \tfrac{1}{2}\left(\hat{I}_{2+} + \hat{I}_{2-}\right) \\ &= \tfrac{1}{2}(\underbrace{\hat{I}_{1+}\hat{I}_{2+}}_{p=+2} + \underbrace{\hat{I}_{1-}\hat{I}_{2-}}_{p=-2} + \underbrace{\hat{I}_{1+}\hat{I}_{2-}}_{p=0} + \underbrace{\hat{I}_{1-}\hat{I}_{2+}}_{p=0}). \end{aligned}$$

The term $\hat{I}_{1+}\hat{I}_{2+}$ has coherence order $+1$ for spin one, and also $+1$ for spin two; so, the overall coherence order is $+1 + 1 = +2$. The coherence orders p of the other terms are given underneath each. We see that $2\hat{I}_{1x}\hat{I}_{2x}$ is an equal mixture of coherence orders $+2$ and -2, *double-quantum coherence*, and coherence order 0, *zero-quantum coherence*.

Similar calculations show that all of the products involving two transverse operators are likewise mixtures of double- and zero-quantum coherence. The results are summarized in the following table:

product operator	double-quantum part	zero-quantum part
$2\hat{I}_{1x}\hat{I}_{2x}$	$\tfrac{1}{2}(\hat{I}_{1+}\hat{I}_{2+} + \hat{I}_{1-}\hat{I}_{2-})$	$\tfrac{1}{2}(\hat{I}_{1+}\hat{I}_{2-} + \hat{I}_{1-}\hat{I}_{2+})$
$2\hat{I}_{1x}\hat{I}_{2y}$	$\tfrac{1}{2i}(\hat{I}_{1+}\hat{I}_{2+} - \hat{I}_{1-}\hat{I}_{2-})$	$\tfrac{1}{2i}(-\hat{I}_{1+}\hat{I}_{2-} + \hat{I}_{1-}\hat{I}_{2+})$
$2\hat{I}_{1y}\hat{I}_{2x}$	$\tfrac{1}{2i}(\hat{I}_{1+}\hat{I}_{2+} - \hat{I}_{1-}\hat{I}_{2-})$	$\tfrac{1}{2i}(\hat{I}_{1+}\hat{I}_{2-} - \hat{I}_{1-}\hat{I}_{2+})$
$2\hat{I}_{1y}\hat{I}_{2y}$	$-\tfrac{1}{2}(\hat{I}_{1+}\hat{I}_{2+} + \hat{I}_{1-}\hat{I}_{2-})$	$\tfrac{1}{2}(\hat{I}_{1+}\hat{I}_{2-} + \hat{I}_{1-}\hat{I}_{2+})$

We can look at the contents of this table in another way. Suppose we take the sum $(2\hat{I}_{1x}\hat{I}_{2x} + 2\hat{I}_{1y}\hat{I}_{2y})$; we can see that the double-quantum parts will cancel, and the zero-quantum parts will add, so $(2\hat{I}_{1x}\hat{I}_{2x} + 2\hat{I}_{1y}\hat{I}_{2y})$ is *pure* zero-quantum. Similarly, for the difference $(2\hat{I}_{1x}\hat{I}_{2x} - 2\hat{I}_{1y}\hat{I}_{2y})$ the zero-quantum parts cancel and the double-quantum parts add, so $(2\hat{I}_{1x}\hat{I}_{2x} - 2\hat{I}_{1y}\hat{I}_{2y})$ is pure double quantum.

By adding and subtracting the rows in the table we can construct two pure double-quantum operators, denoted $\hat{DQ}_x$ and $\hat{DQ}_y$, and two pure zero-quantum operators, denoted $\hat{ZQ}_x$ and $\hat{ZQ}_y$. The definitions are given in the table on the following page.

operator	definition
$\hat{D}Q_x$	$(2\hat{I}_{1x}\hat{I}_{2x} - 2\hat{I}_{1y}\hat{I}_{2y})$
$\hat{D}Q_y$	$(2\hat{I}_{1x}\hat{I}_{2y} + 2\hat{I}_{1y}\hat{I}_{2x})$
$\hat{Z}Q_x$	$(2\hat{I}_{1x}\hat{I}_{2x} + 2\hat{I}_{1y}\hat{I}_{2y})$
$\hat{Z}Q_y$	$(2\hat{I}_{1y}\hat{I}_{2x} - 2\hat{I}_{1x}\hat{I}_{2y})$

The designations $\hat{D}Q_x$ and $\hat{D}Q_y$ are arbitrary, but as we shall see soon, quite useful.

7.12.2 Generation of multiple-quantum coherence

Multiple-quantum coherence is generated by applying a pulse to an anti-phase state. For example, if we have the state $2\hat{I}_{1x}\hat{I}_{2z}$ and apply a non-selective $90°(x)$ pulse the result is the generation of a mixture of zero- and double-quantum coherence:

$$2\hat{I}_{1x}\hat{I}_{2z} \xrightarrow{(\pi/2)\hat{I}_{1x}+(\pi/2)\hat{I}_{1x}} -2\hat{I}_{1x}\hat{I}_{2y}.$$

It is interesting to note that the same anti-phase term undergoes coherence transfer if the $90°$ pulse is applied about the y-axis:

$$2\hat{I}_{1x}\hat{I}_{2z} \xrightarrow{(\pi/2)\hat{I}_{1y}+(\pi/2)\hat{I}_{1y}} -2\hat{I}_{1z}\hat{I}_{2x}.$$

Anti-phase states are thus crucial in both coherence transfer and the generation of multiple-quantum coherence.

7.12.3 Evolution of multiple-quantum coherence

Having generated a multiple-quantum state, we now need to know how it will evolve under the free precession Hamiltonian, Eq. 7.14 on page 150,

$$\hat{H}_{\text{two spins}} = \Omega_1\hat{I}_{1z} + \Omega_2\hat{I}_{2z} + 2\pi J_{12}\hat{I}_{1z}\hat{I}_{2z}.$$

It turns out that both double- and zero-quantum states are *unaffected* by the coupling term between the two spins involved in the coherence. In the case of two spins, it means that the coupling term $2\pi J_{12}\hat{I}_{1z}\hat{I}_{2z}$ does not affect the evolution. This leaves just the offset terms which can, as usual, be treated sequentially.

As an example, we will consider the evolution of the pure double-quantum term $\hat{D}Q_x$, which is $(2\hat{I}_{1x}\hat{I}_{2x} - 2\hat{I}_{1y}\hat{I}_{2y})$. First, we will consider the $2\hat{I}_{1x}\hat{I}_{2x}$ term:

$$2\hat{I}_{1x}\hat{I}_{2x} \xrightarrow{\Omega_1 t\hat{I}_{1z}+\Omega_2 t\hat{I}_{2z}}$$
$$2\left[\cos\left(\Omega_1 t\right)\hat{I}_{1x} + \sin\left(\Omega_1 t\right)\hat{I}_{1y}\right]\left[\cos\left(\Omega_2 t\right)\hat{I}_{2x} + \sin\left(\Omega_2 t\right)\hat{I}_{2y}\right].$$

Next, the $-2\hat{I}_{1y}\hat{I}_{2y}$ term:

$$-2\hat{I}_{1y}\hat{I}_{2y} \xrightarrow{\Omega_1 t\hat{I}_{1z}+\Omega_2 t\hat{I}_{2z}}$$
$$-2\left[\cos\left(\Omega_1 t\right)\hat{I}_{1y} - \sin\left(\Omega_1 t\right)\hat{I}_{1x}\right]\left[\cos\left(\Omega_2 t\right)\hat{I}_{2y} - \sin\left(\Omega_2 t\right)\hat{I}_{2x}\right].$$

Collecting together the terms we have:

$$[\cos{(\Omega_1 t)}\cos{(\Omega_2 t)} - \sin{(\Omega_1 t)}\sin{(\Omega_2 t)}]\,(2\hat{I}_{1x}\hat{I}_{2x} - 2\hat{I}_{1y}\hat{I}_{2y})$$
$$+ [\cos{(\Omega_1 t)}\sin{(\Omega_2 t)} + \sin{(\Omega_1 t)}\cos{(\Omega_2 t)}]\,(2\hat{I}_{1x}\hat{I}_{2y} + 2\hat{I}_{1y}\hat{I}_{2x}).$$

Using the identities:

$$\cos{(A + B)} \equiv \cos A \cos B - \sin A \sin B$$
$$\sin{(A + B)} \equiv \cos A \sin B + \sin A \cos B$$

we can rewrite the result as

$$\cos{([\Omega_1 + \Omega_2]t)}\,(2\hat{I}_{1x}\hat{I}_{2x} - 2\hat{I}_{1y}\hat{I}_{2y}) + \sin{([\Omega_1 + \Omega_2]t)}\,(2\hat{I}_{1x}\hat{I}_{2y} + 2\hat{I}_{1y}\hat{I}_{2x}).$$

This can be further tidied up by inserting the definitions of $\hat{DQ}_x$ and $\hat{DQ}_y$ from the table on the previous page to give:

$$\cos{([\Omega_1 + \Omega_2]t)}\,\hat{DQ}_x + \sin{([\Omega_1 + \Omega_2]t)}\,\hat{DQ}_y.$$

So, overall the evolution of $\hat{DQ}_x$ is

$$\hat{DQ}_x \xrightarrow{\Omega_1 t \hat{I}_{1z} + \Omega_2 t \hat{I}_{2z}} \cos{([\Omega_1 + \Omega_2]t)}\,\hat{DQ}_x + \sin{([\Omega_1 + \Omega_2]t)}\,\hat{DQ}_y.$$

Overall, $\hat{DQ}_x$ evolves into $\hat{DQ}_y$ at a rate determined by the frequency $[\Omega_1 + \Omega_2]$. We call this sum of the offsets the double quantum precession frequency, Ω_{DQ}:

$$\Omega_{DQ} = \Omega_1 + \Omega_2.$$

Using this, the evolution of $\hat{DQ}_x$ becomes:

$$\hat{DQ}_x \xrightarrow{\Omega_1 t \hat{I}_{1z} + \Omega_2 t \hat{I}_{2z}} \cos{(\Omega_{DQ}t)}\,\hat{DQ}_x + \sin{(\Omega_{DQ}t)}\,\hat{DQ}_y.$$

There is a complete analogy between this evolution of the double-quantum term and that of a simple operator such as $\hat{I}_x$:

$$\hat{I}_x \xrightarrow{\Omega t \hat{I}_z} \cos{(\Omega t)}\,\hat{I}_x + \sin{(\Omega t)}\,\hat{I}_y.$$

This is why we chose the symbols $\hat{DQ}_x$ and $\hat{DQ}_y$. Similarly, we can use diagrams such a those in Fig. 7.4 on page 148 to determine the way in which $\hat{DQ}_x$ and $\hat{DQ}_y$ evolve into one another; a suitable diagram is shown in Fig. 7.20 (a) on the following page. Using this we can determine very simply that:

$$-\hat{DQ}_y \xrightarrow{\Omega_1 t \hat{I}_{1z} + \Omega_2 t \hat{I}_{2z}} -\cos{(\Omega_{DQ}t)}\,\hat{DQ}_y + \sin{(\Omega_{DQ}t)}\,\hat{DQ}_x,$$

and so on.

In section 3.6.1 on page 39 we found that, in a two-spin system, the frequency of the double-quantum transition between levels $\alpha\alpha$ and $\beta\beta$ was the sum of the Larmor frequencies of the two spins. Here, we have found that double-quantum coherence evolves at the sum of the *offsets* of the two spins. The difference comes about as for the present discussion we are

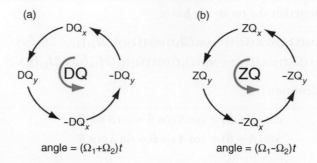

Fig. 7.20 Diagrams, analogous to those of Fig. 7.4 on page 148, for determining the evolution of (a) double quantum, and (b) zero quantum, during a delay. As before, having located the old operator, we find the new operator as the next one round the circle, indicated by the arrows. In the case of the double-quantum coherences, diagram (a), the old operator is multiplied by the term $\cos{(\Omega_{DQ}t)}$ and the new by $\sin{(\Omega_{DQ}t)}$, where $\Omega_{DQ} = (\Omega_1 + \Omega_2)$. For the zero-quantum coherences, (b), the terms are multiplied by $\cos{(\Omega_{ZQ}t)}$ and $\sin{(\Omega_{ZQ}t)}$, where $\Omega_{ZQ} = (\Omega_1 - \Omega_2)$.

working in the rotating frame, rather than the laboratory frame used in Chapter 3, so the Larmor frequencies are replaced by the offsets.

The zero-quantum terms evolve in an analogous way, except this time the frequency is the *difference* of the offsets: $\Omega_{ZQ} = [\Omega_1 - \Omega_2]$. Figure 7.20 (b) shows how the operators evolve into one another, for example:

$$-\hat{Z}Q_x \xrightarrow{\Omega_1 t \hat{I}_{1z} + \Omega_2 t \hat{I}_{2z}} -\cos{(\Omega_{ZQ}t)}\,\hat{Z}Q_x - \sin{(\Omega_{ZQ}t)}\,\hat{Z}Q_y.$$

7.13 Summary

We have covered a great deal of ground in this chapter, but in doing so we have laid the basis for understanding just about any multiple-pulse NMR experiment. Straight away in the next chapter we will use all that we have developed here to help us to understand how two-dimensional NMR works. We will find that we can make fast progress with understanding such experiments now that we have the product operator method 'under our belts'.

The key points of the method are summarized here:

- There are fifteen product operators needed to describe a two-spin system: they are listed, along with their interpretation, in the table on page 150.

- The way in which the operators evolve under pulses, offsets and coupling can be deduced from Fig. 7.4 on page 148 and Fig. 7.6 on page 152.

- The distinction between in-phase and anti-phase operators is particularly important; Figs 7.9 and 7.10 on page 156 illustrate these.

- The evolution of coupling interconverts in- and anti-phase terms. Spin echoes are a convenient way of achieving such interconversions independent of the offset.

- Anti-phase terms can be transferred to other spins or to multiple-quantum coherences by the action of pulses.

7.14   Further reading

Product operators for two spins:

Chapters 3 and 4 from P. J. Hore, J. A. Jones and S. Wimperis, *NMR: The Toolkit* (Oxford University Press, 2000).

Chapter 3 from R. Freeman, *Spin Choreography* (Spektrum, 1997).

The quantum mechanics of two coupled spins, including product operators:

Chapters 15 and 16 from M. H. Levitt, *Spin Dynamics* (2nd edition, John Wiley & Sons, Ltd, 2008).

Chapter 2 from F. J. M. van de Ven, *Multidimensional NMR in Liquids* (VCH, 1995).

A full account of the product operator method:

O. W. Sørensen, G. W. Eich, M. H. Levitt, G. Bodenhausen and R. R. Ernst, *Progress in Nuclear Magnetic Resonance Spectroscopy*, **16**, 163–192 (1983).

7.15   Exercises

7.1   Using Fig. 7.4 on page 148, determine the result of the following rotations:

$$\exp\left(-i\theta\hat{I}_x\right)\hat{I}_y\exp\left(i\theta\hat{I}_x\right) \qquad\qquad \exp\left(-i\theta\hat{I}_z\right)\left(-\hat{I}_y\right)\exp\left(i\theta\hat{I}_z\right)$$

$$\exp\left(-i\theta\hat{S}_y\right)\hat{S}_z\exp\left(i\theta\hat{S}_y\right) \qquad\qquad \exp\left(-i(\theta/2)\hat{I}_y\right)\hat{I}_x\exp\left(i(\theta/2)\hat{I}_y\right)$$

$$\exp\left(-i\theta\hat{I}_x\right)\hat{I}_x\exp\left(i\theta\hat{I}_x\right) \qquad\qquad \exp\left(i\theta\hat{I}_z\right)\left(-\hat{I}_z\right)\exp\left(-i\theta\hat{I}_z\right).$$

Write each of these transformations using the arrow notation introduced in section 7.1.2 on page 141.

7.2   Following the same approach as in section 7.2.2 on page 145, show that (for a single spin) a spin-echo sequence in which the 180° pulse is about y

$$90°(x) - \tau - 180°(y) - \tau-$$

results in the magnetization being refocused on the $-y$-axis.

7.3   Determine the outcome of the following rotations:

$$\hat{I}_y \xrightarrow{(\pi/2)\hat{I}_y} \qquad\qquad \hat{I}_z \xrightarrow{(\pi/2)\hat{I}_y}$$

$$\hat{I}_y \xrightarrow{-(\pi/2)\hat{I}_y} \qquad\qquad \hat{I}_z \xrightarrow{-(\pi/2)\hat{I}_y}$$

$$\hat{S}_y \xrightarrow{\pi\hat{S}_y} \qquad\qquad \hat{S}_z \xrightarrow{\pi\hat{S}_y}$$

$$\hat{I}_x \xrightarrow{-\pi\hat{I}_y} \qquad\qquad \hat{I}_z \xrightarrow{-\pi\hat{I}_y}.$$

7.4   A variant on the $1 - \bar{1}$ sequence, described in section 7.3.3 on page 148, is the sequence:

$$90°(x) - \tau - 90°(y).$$

Show that, for a one-spin system, this sequence gives rise to transverse magnetization which varies as $\cos\left(\Omega\tau\right)$.

Hence show that there is a null in the excitation at $\Omega = \pi/(2\tau)$; give the position of this null in frequency units (Hz). At what offsets is the excitation a maximum?

7.5   Using Fig. 7.6 on page 152, determine the outcome of the following, all of which involve the evolution of coupling:

$$-\hat{I}_{1y} \xrightarrow{2\pi J_{12}\tau\hat{I}_{1z}\hat{I}_{2z}} \qquad\qquad \hat{I}_{2y} \xrightarrow{2\pi J_{12}\tau\hat{I}_{1z}\hat{I}_{2z}}$$

$$-2\hat{I}_{1x}\hat{I}_{2z} \xrightarrow{2\pi J_{12}\tau\hat{I}_{1z}\hat{I}_{2z}} \qquad\qquad 2\hat{I}_{1z}\hat{I}_{2y} \xrightarrow{2\pi J_{12}\tau\hat{I}_{1z}\hat{I}_{2z}}$$

$$\hat{S}_x \xrightarrow{2\pi J_{IS}(\tau/2)\hat{I}_z\hat{S}_z} \qquad\qquad \hat{I}_{2z} \xrightarrow{2\pi J_{12}\tau\hat{I}_{1z}\hat{I}_{2z}}.$$

7.6   Following the approach of section 7.5 on page 152, show that the observable signal arising from $\hat{I}_{1y}$ is of the form

$$S(t) = \tfrac{1}{2}i\exp\left(i[\Omega_1 + \pi J_{12}]t\right) + \tfrac{1}{2}i\exp\left(i[\Omega_1 - \pi J_{12}]t\right).$$

Hence describe the spectrum you expect to see.

Similarly, determine the observable signal arising from $2\hat{I}_{1y}\hat{I}_{2z}$, and hence the form of the spectrum.

7.7 Assuming that magnetization along the y-axis gives rise to an absorption mode lineshape, draw sketches of the spectra which arise from the following operators:

$$\hat{I}_{1y} \qquad \hat{I}_{2x} \qquad 2\hat{I}_{1y}\hat{I}_{2z} \qquad 2\hat{I}_{1z}\hat{I}_{2x}.$$

Describe each spectrum in words.

7.8 Give the outcome of the following evolution due to pulses or delays. In each case, describe the overall transformation in words.

$$\hat{I}_{1x} \xrightarrow{\omega_1 t_p \hat{I}_{1y}} \qquad\qquad 2\hat{I}_{1x}\hat{I}_{2z} \xrightarrow{(\pi/2)(\hat{I}_{1y}+\hat{I}_{2y})}$$

$$2\hat{I}_{1x}\hat{I}_{2z} \xrightarrow{-\pi\hat{I}_{2y}} \qquad\qquad \hat{S}_x \xrightarrow{\Omega_I t \hat{I}_z} \xrightarrow{\Omega_S t \hat{S}_z}$$

$$-\hat{I}_{1x} \xrightarrow{2\pi J_{12} t \hat{I}_{1z}\hat{I}_{2z}} \qquad\qquad -2\hat{I}_{1z}\hat{I}_{2y} \xrightarrow{2\pi J_{12} t \hat{I}_{1z}\hat{I}_{2z}}.$$

7.9 Using the same approach as in section 7.8.1 on page 159, show that the effect of a spin echo (in a homonuclear system) on the terms $2\hat{I}_{1x}\hat{I}_{2z}$ and $2\hat{I}_{1y}\hat{I}_{2z}$ is as follows:

$$2\hat{I}_{1x}\hat{I}_{2z} \xrightarrow{\tau-\pi_x-\tau} -\cos\left(2\pi J_{12}\tau\right) 2\hat{I}_{1x}\hat{I}_{2z} - \sin\left(2\pi J_{12}\tau\right)\hat{I}_{1y}$$

$$2\hat{I}_{1y}\hat{I}_{2z} \xrightarrow{\tau-\pi_x-\tau} \cos\left(2\pi J_{12}\tau\right) 2\hat{I}_{1y}\hat{I}_{2z} - \sin\left(2\pi J_{12}\tau\right)\hat{I}_{1x}.$$

Using the idea that a spin echo is equivalent to evolution of the coupling for time 2τ, followed by a $180°$ pulse, draw up a table similar to that on page 161 for a spin echo in which the $180°$ pulse is applied about the y-axis.

Extend both your table and that on page 161 to include in- and anti-phase operators on spin two.

7.10 For a homonuclear two-spin system, what delay τ in a spin-echo sequence would you use to achieve the following overall transformations? (Apart from in the last transformation, do not worry about the sign of any term.)

$$\hat{I}_{2y} \longrightarrow 2\hat{I}_{1z}\hat{I}_{2x} \qquad 2\hat{I}_{1z}\hat{I}_{2x} \longrightarrow \hat{I}_{2y}$$

$$\hat{I}_{1x} \longrightarrow \tfrac{1}{\sqrt{2}}\hat{I}_{1x} + \tfrac{1}{\sqrt{2}}2\hat{I}_{1y}\hat{I}_{2z} \qquad \hat{I}_{1x} \longrightarrow -\hat{I}_{1x}.$$

7.11 Consider the spin-echo pulse sequence shown in Fig. 7.14 (c) on page 164. By considering the evolution of the operators $\hat{S}_x$ and $2\hat{I}_z\hat{S}_x$, show that the coupling is refocused, and that, from the point of view of the evolution of these operators, the sequence is equivalent to a $180°$ pulse to the S spin.

Without detailed calculations, state what effect you expect sequence (c) to have on the evolution of the operators $\hat{I}_x$ and $2\hat{I}_x\hat{S}_z$.

7.12   Why does the second 90° pulse to spin one in the INEPT experiment (Fig. 7.15 on page 168) have to be about the y-axis?

Show that changing the phase of the first 90° pulse from x to $-x$ results in the following observables on the S spin at the start of acquisition:

$$-k_S \cos\left(2\pi J_{IS}\tau_2\right)\hat{S}_y - k_I \sin\left(2\pi J_{IS}\tau_2\right)\sin\left(2\pi J_{IS}\tau_1\right)\hat{S}_x.$$

7.13   Specify the coherence order (or orders) of the following operators:

$$\hat{I}_+ \qquad \hat{I}_z \qquad \hat{I}_- \qquad \hat{I}_{1x} \qquad 2\hat{I}_{1z}\hat{I}_{2y} \qquad 2\hat{I}_{1z}\hat{I}_{2z} \qquad \hat{I}_{1+}\hat{I}_{2-} \qquad \hat{I}_{1x}\hat{I}_{2y}.$$

7.14   By expressing $\hat{I}_x$ and $\hat{I}_y$ in terms of $\hat{I}_+$ and $\hat{I}_-$, verify the relationships given in the table on page 175.

7.15   Consider the pulse sequence shown below.

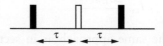

Starting with equilibrium magnetization on spin one, $\hat{I}_{1z}$, show that the sequence generates a mixture of double- and zero-quantum coherence. Find the value of τ which gives the maximum amount of multiple-quantum coherence. [Hint: can you spot the spin echo? If so, the calculation is much simpler.]

Show that if we start with equilibrium magnetization on both spins one and two, i.e. $\hat{I}_{1z} + \hat{I}_{2z}$, the sequence generates only double-quantum coherence.

7.16   Show that

$$\hat{ZQ}_x \xrightarrow{\ \Omega_1 t \hat{I}_{1z} + \Omega_2 t \hat{I}_{2z}\ } \cos\left(\left[\Omega_1 - \Omega_2\right]t\right)\hat{ZQ}_x + \sin\left(\left[\Omega_1 - \Omega_2\right]t\right)\hat{ZQ}_y.$$

8

Two-dimensional NMR

There can be little doubt that the introduction of two-dimensional NMR has made structure determination by NMR much easier, and has also greatly increased the complexity of problems which can be tackled. Two-dimensional NMR has now become so routine that we think nothing of requesting a two-dimensional COSY or HMQC experiment in order to help us unravel a problem. Such experiments are straightforward to interpret, and have proved to be very reliable, which accounts for their popularity.

Two-dimensional spectroscopy has also made it possible to use NMR to determine the structures of biomolecules, such as proteins, DNA and RNA – tackling molecules of this size would have been quite unthinkable before the advent of two-dimensional techniques. Once the idea of two dimensions was firmly established, the extension to three or even four dimensions followed on quite naturally, and such experiments open up the possibility of studying even larger biomolecules.

In this chapter we are going to be concerned with the simplest and most frequently used two-dimensional experiments. These will serve as our introduction to the key ideas in two-dimensional NMR which are the basis of more elaborate experiments.

The basic idea behind two-dimensional NMR is quite simple, but it is one of those simple ideas capable of great elaboration. As shown in Fig. 8.1, in conventional (one-dimensional) NMR we have a plot of intensity against frequency, whereas in two-dimensional NMR we plot intensity against *two* frequency axes; each peak in a two-dimensional spectrum thus has an intensity and two frequency co-ordinates. What these two co-ordinates represent depends on the experiment in question.

Probably the most useful two-dimensional experiments are those in which the position of the peak shows a correlation between two quantities. For example, in the COSY experiment the frequency co-ordinates of the peaks give the chemical shifts of coupled spins. Another example is the HMQC experiment, in which one frequency co-ordinate gives the ^{13}C chemical shift, while the other gives the chemical shift of the attached proton.

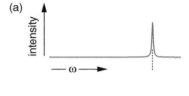

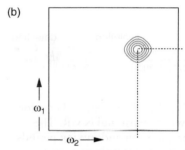

Fig. 8.1 In conventional (one-dimensional) NMR we plot the intensity of absorption against frequency, as shown in (a); each peak has a single frequency coordinate. In two-dimensional NMR, (b), each peak has two frequency coordinates, measured along the ω_1 and ω_2 axes. It is usual to present two-dimensional spectra as contour plots in which points of equal intensity are joined by lines, just as in a topographic map.

Understanding NMR Spectroscopy, Second Edition James Keeler
© 2010 John Wiley & Sons, Ltd

In this chapter we will look in detail at a number of important two-dimensional experiments, and will analyse them using the product operator method introduced in the previous chapter. At this stage, we will restrict the discussion to two-dimensional experiments which involve transfer of magnetization through scalar couplings. Discussion of the important NOESY experiment, in which the magnetization transfer arises due to relaxation effects, is delayed until the next chapter which is devoted to relaxation.

For the whole of this chapter we will restrict ourselves to discussing just two coupled spins. Although this is the simplest system in which magnetization transfer through scalar couplings can be seen, all of the important ideas about how two-dimensional NMR experiments work can be understood by considering this simple spin system. There are, however, some additional features which can only be seen if we have three or more spins; for a selected set of experiments, these are discussed in Chapter 10.

Before embarking on our discussion of specific experiments, we will consider the general scheme for two-dimensional NMR, and discuss the important topic of lineshapes.

8.1   The general scheme for two-dimensional NMR

A general way of representing just about all two-dimensional experiments is shown in Fig. 8.2. We start with the *preparation period*, during which the equilibrium magnetization is transformed into some kind of coherence which then evolves for the *evolution period*, t_1.

The preparation period might be something as simple as a 90° pulse, which would generate transverse magnetization (single-quantum coherence). However, this period could be a more complicated set of pulses and delays. For example, it might be a sequence designed to generate multiple-quantum coherence, or an INEPT-style sequence designed to transfer magnetization from another type of nucleus.

The evolution period, t_1, is not a fixed time; rather, t_1 is incremented systematically in a series of separate experiments. We will have more to say about how this is done in the next section. The second important thing about the evolution period is that no observations are made during it. So, the coherence which evolves during t_1 need not be observable e.g. it could be multiple-quantum coherence. This ability to follow the evolution of unobservable coherences is an important feature of two-dimensional NMR.

Next comes the *mixing period*, during which the coherence present at the end of t_1 is manipulated into an observable signal which can be recorded during the *detection period*, t_2. For example, if multiple-quantum coherence is present during t_1, then the mixing period needs to be devised in such a way as to transform the multiple-quantum coherence into an observable signal. During the mixing time it is also common for magnetization to be transferred from one spin to another, for example through a scalar coupling. Ultimately, it is the form of the mixing period which determines the information content of the spectrum.

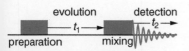

Fig. 8.2 The general scheme for two-dimensional NMR. The preparation and mixing periods may be as simple as a single pulse, or may consist of much more complex arrangements of pulses and delays. The coherence generated during the preparation period evolves for time t_1, but there is no detection during this time. After the mixing period, the signal is detected during time t_2.

8.1.1   How two-dimensional spectra are recorded

At the start of Chapter 5, we described how the FID is recorded at regular time intervals, leading to a series of data points which represent the time-domain function. In a two-dimensional experiment, the same approach is used, with data being recorded at regular intervals in both t_1 and t_2.

The process of acquiring a two-dimensional time-domain data set is illustrated in Fig. 8.3. First, t_1 is set to zero, the pulse sequence is executed and the FID recorded as a series of data points in the usual way. The resulting set of data is stored away (in memory or on computer disc). This set of data is called the first t_1 increment.

Next, t_1 is set to Δ_1, the sampling interval in that dimension. Once again, the sequence is executed, the data recorded and stored away separately. These data form the second t_1 increment.

The process is repeated with $t_1 = 2\Delta_1$, $t_1 = 3\Delta_1$ ... until sufficient data in the t_1 dimension have been built up. We can imagine all these data as forming a *matrix*. The first row is the t_2 data for $t_1 = 0$, the second row is the t_2 data for $t_1 = \Delta_1$, the third for $t_1 = 2\Delta_1$ and so on. This two-dimensional data set can be represented as the time-domain function $S(t_1, t_2)$.

The t_2 data are recorded in real time, just as in a conventional experiment. It is therefore not time consuming to record thousands of data points, should we wish to. However, in t_1 it is rather a different story, as for each data point (each increment) we have to execute the whole pulse sequence. So, data as a function of t_1 is rather more time consuming to record. It is therefore uncommon to record more than a few hundred increments of t_1.

8.1.2   How the data are processed

In conventional (one-dimensional) NMR we take the time-domain function $S(t)$, and subject it to a Fourier transform to give the frequency-domain function, or spectrum, $S(\omega)$. In two-dimensional NMR we have a time-domain function which depends on t_1 and t_2, so in order to arrive at the spectrum, we need to Fourier transform with respect to both times.

The way in which this is done is visualized in Fig. 8.4 on the following page. We start with the time-domain data which can be thought of as a matrix. A row in the matrix corresponds to a particular value of t_1, whereas a column corresponds to a particular value of t_2. Remember that the data are sampled at regular intervals, so that the first row corresponds to $t_1 = 0$, the second to $t_1 = \Delta_1$, the third to $t_1 = 2\Delta_1$ and so on. Similarly, the data in the first column correspond to $t_2 = 0$, the second to $t_2 = \Delta_2$ and so on (Δ_2 is the sampling interval in t_2).

The first step is to extract each row from the matrix in turn, subject it to the usual Fourier transform, and then construct a new matrix out of these transformed rows. What this process gives us is a series of spectra, with the running frequency variable ω_2 (in angular frequency units). Each row in the new matrix corresponds to a different t_1 value.

We now take each column in turn from this new matrix. A column corresponds to a particular ω_2 frequency, and the data points in the column correspond to increasing values of t_1; these time-domain data are often

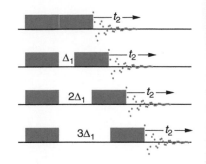

Fig. 8.3 Illustration of how a two-dimensional data set is recorded for the general sequence of Fig. 8.2 on the facing page. First, t_1 is set to zero, the sequence is executed and the FID is digitized at regular intervals as a function of t_2; the resulting time-domain signal is stored away. Next, t_1 is set to the sampling interval Δ_1. The sequence is then executed and the data (as a function of t_2) are recorded and stored away. The whole process is repeated for the next increment of t_1, in which $t_1 = 2\Delta_1$ and so on for as many increments of t_1 as are required.

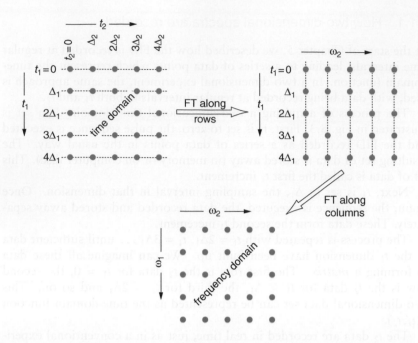

Fig. 8.4 Visualization of how a two-dimensional time-domain matrix is converted to a two-dimensional spectrum. The original data, top left, are arranged in a matrix with successive rows corresponding to longer t_1 values, and successive columns corresponding to longer t_2 values. The data along t_2 are sampled at intervals of Δ_2, whereas those along t_1 are sampled at intervals of Δ_1. The first step is to take the rows, Fourier transform them and then construct a new matrix (top right). In this matrix, successive rows still correspond to increasing values of t_1, but the columns now correspond to different ω_2 frequencies, resulting from the Fourier transformation with respect to t_2. In the second step, columns from the top right matrix are subjected to a Fourier transform. The data in these columns correspond to increasing values of t_1, and so Fourier transformation gives a spectrum with running frequency variable ω_1. These transformed columns are used to construct the bottom matrix, which is the two-dimensional frequency-domain spectrum.

called *interferograms*. Each column is subject to a Fourier transform and then used to construct the final matrix. The columns of this matrix have the running frequency variable ω_1 (in angular units). Figure 8.5 on the next page shows an example of the result of these two separate transforms on a simple time-domain data set.

This final matrix is our two-dimensional spectrum, with frequency axes ω_1, corresponding to the evolution in t_1, and ω_2, corresponding to the evolution in t_2. Just as for one-dimensional spectra, the time-domain data can be manipulated using weighting functions, and the final spectrum subject to phase correction. The only difference here is that we need separate weighting and phasing in each dimension.

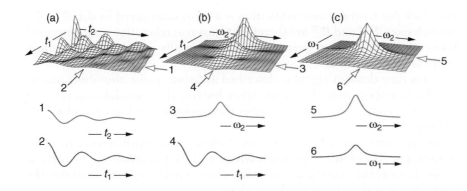

Fig. 8.5 Illustration of the process of double Fourier transformation on a simple data set. The original time-domain data, shown in (a), consists of a damped cosine wave in each dimension. Fourier transformation with respect to t_2 gives the data shown in (b), and then the second Fourier transform with respect to t_1 gives the two-dimensional spectrum shown in (c); as expected, the spectrum shows a single line. Two typical cross-sections, taken through each data set at the positions indicated by the numbered arrows, are shown beneath each data set. Cross-sections 1 and 2 are taken parallel to t_2 and t_1, respectively, and show damped cosine waves. In data set (b), cross-section 3, which is taken parallel to ω_2, shows a peak. However, cross-section 4, which is taken parallel to t_1, still shows a cosine wave. The peak is said to be modulated in t_1 by the cosine wave. In the final spectrum (c), cross-sections 5 and 6 both show a single peak. The intensity in any particular cross-section depends on the coordinate at which it is taken.

8.2 Modulation and lineshapes

In section 5.3 on page 83 we saw that Fourier transformation of an exponentially damped time-domain function gives a spectrum with the absorption mode Lorentzian lineshape in its real part (Eq. 5.7 on page 85):

$$S_0 \exp\left(i\Omega t\right) \exp\left(-Rt\right) \xrightarrow{\text{FT}} S_0 \left[A(\omega) + iD(\omega)\right],$$

where $A(\omega)$ is the absorption mode Lorentzian and $D(\omega)$ is the corresponding dispersion mode lineshape. We now need to work out what to expect when we subject a two-dimensional time-domain data set to a two-dimensional Fourier transformation.

8.2.1 Cosine amplitude modulated data

As we will see when we analyse some particular experiments, a typical two-dimensional time-domain function is of the form:

$$S(t_1, t_2) = S_0 \underbrace{\left[\cos\left(\Omega_A t_1\right)\exp\left(-R^{(1)}t_1\right)\right]}_{\text{modulation in } t_1} \times \left[\exp\left(i\Omega_B t_2\right)\exp\left(-R^{(2)}t_2\right)\right].$$

In this expression, S_0 gives the overall amplitude of the signal, Ω_A is the modulating frequency in t_1, and $R^{(1)}$ is the decay constant in this dimension. Similarly, Ω_B and $R^{(2)}$ are the frequency and the decay constant in t_2. The decay of the signal represented by the terms $\exp\left(-R^{(1)}t_1\right)$ and $\exp\left(-R^{(2)}t_2\right)$

is in fact due to transverse relaxation – a topic considered in detail in the next chapter. $R^{(1)}$ and $R^{(2)}$ are therefore transverse relaxation rate constants; however, for the present purposes it does not really matter what the origin of these decay terms is.

This time-domain signal is described as being *cosine amplitude modulated* with respect to t_1. The name arises because the t_1 modulation is of the form of a cosine, which simply varies the amplitude, but not the phase, of the signal.

If we Fourier transform this time-domain signal with respect to t_2 we obtain, just as before, a spectrum whose real part contains an absorption mode Lorentzian centred at Ω_B, and whose imaginary part contains the corresponding dispersion mode lineshape.

$$S(t_1, t_2) \xrightarrow{\text{FT along } t_2} S(t_1, \omega_2)$$
$$\xrightarrow{\text{FT along } t_2} S_0 \left[\cos(\Omega_A t_1) \exp\left(-R^{(1)} t_1\right) \right] \times [A_2(\Omega_B) + i\, D_2(\Omega_B)].$$

The notation here is that $A_2(\Omega)$ represents an absorption Lorentzian in the ω_2 dimension, centred at frequency Ω; similarly, $D_2(\Omega)$ represents the corresponding dispersion lineshape.

The next step is to Fourier transform $S(t_1, \omega_2)$ with respect to t_1. In contrast to the modulation in t_2, which is of the form of a complex exponential $\exp(i\Omega t)$, the modulation in t_1 is simply a cosine wave. As a consequence, to generate the spectrum we need to use a slightly different kind of Fourier transform, called a *cosine Fourier transform*.

The cosine Fourier transform of a damped cosine wave gives the absorption mode Lorentzian:

$$S_0 \cos(\Omega t) \exp(-Rt) \xrightarrow{\cos \text{FT}} S_0 A(\omega).$$

Note that, in contrast to the transform of the complex exponential, the resulting spectrum is real.

Using this, we can determine the result of the cosine transform with respect to t_1:

$$S(t_1, \omega_2) \xrightarrow{\cos \text{FT along } t_1} S(\omega_1, \omega_2)$$
$$\xrightarrow{\cos \text{FT along } t_1} S_0 [A_1(\Omega_A)] \times [A_2(\Omega_B) + i\, D_2(\Omega_B)]$$
$$\xrightarrow{\cos \text{FT along } t_1} S_0 [A_1(\Omega_A) A_2(\Omega_B) + i\, A_1(\Omega_A) D_2(\Omega_B)],$$

where $A_1(\Omega)$ represents an absorption mode Lorentzian centred at Ω in the ω_1 dimension.

The real part of the spectrum $S(\omega_1, \omega_2)$ is $S_0 A_1(\Omega_A) A_2(\Omega_B)$, which is a peak at frequency $\{\omega_1 = \Omega_A,\ \omega_2 = \Omega_B\}$ with the absorption lineshape in each dimension. This lineshape is called a *double absorption* Lorentzian, and is illustrated in Fig. 8.6 on the facing page. The imaginary part has a lineshape which is dispersive in the ω_2 dimension. Such a lineshape is not suitable for high-resolution spectra, so we simply choose to display the real part.

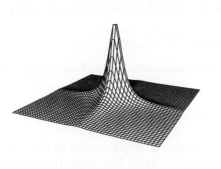 

Fig. 8.6 Two views of the double absorption mode Lorentzian lineshape, commonly encountered in two-dimensional spectra. On the left is shown a perspective view, and on the right is shown a contour plot. A cross-section through this lineshape taken parallel to either axis shows an absorption mode line whose intensity depends on where the cross-section is taken.

8.2.2 Sine amplitude modulated data

Another commonly encountered two-dimensional time-domain function has sine, rather than cosine, amplitude modulation:

$$S(t_1, t_2) = S_0 \underbrace{\left[\sin(\Omega_A t_1) \exp\left(-R^{(1)} t_1\right)\right]}_{\text{modulation in } t_1} \times \left[\exp(i\Omega_B t_2) \exp\left(-R^{(2)} t_2\right)\right].$$

The transform with respect to t_2 gives the absorption and dispersion mode lineshape, just as in the case of the cosine modulated data:

$$S(t_1, \omega_2) = S_0 \left[\sin(\Omega_A t_1) \exp\left(-R^{(1)} t_1\right)\right] \times [A_2(\Omega_B) + i D_2(\Omega_B)].$$

This time, the modulation with respect to t_1 is of the form of sinc, so to transform the data in t_1 we need a *sine Fourier transform*. Such a transform of a damped sine wave again gives the absorption mode Lorentzian:

$$S_0 \sin(\Omega t) \exp(-Rt) \xrightarrow{\text{sin FT}} S_0 A(\omega).$$

Using this for the transform with respect to t_1, we can work out the form of the resulting spectrum:

$$S(t_1, \omega_2) \xrightarrow{\text{sin FT along } t_1} S(\omega_1, \omega_2)$$

$$\xrightarrow{\text{sin FT along } t_1} S_0 [A_1(\Omega_A)] \times [A_2(\Omega_B) + i D_2(\Omega_B)]$$

$$\xrightarrow{\text{sin FT along } t_1} S_0 [A_1(\Omega_A) A_2(\Omega_B) + i A_1(\Omega_A) D_2(\Omega_B)].$$

As before, the real part of the spectrum contains the required double absorption lineshape.

8.2.3 Mixed cosine and sine modulation

If the data are either cosine or sine modulated, we can obtain the desired double absorption lineshape by selecting the appropriate type of transform in the t_1 dimension. Unfortunately, there are cases where, in a single experiment, some of the data are cosine and some are sine modulated, so it is not possible to choose the appropriate transform for all of the data.

A *cosine* Fourier transform of a *sine* modulated signal gives the dispersion lineshape:

$$S_0 \sin(\Omega t) \exp(-Rt) \xrightarrow{\cos \text{FT}} S_0 D(\omega).$$

A *sine* Fourier transform of a *cosine* modulated signal also gives the dispersion lineshape:

$$S_0 \cos(\Omega t) \exp(-Rt) \xrightarrow{\sin \text{FT}} S_0 D(\omega).$$

So, if we select the kind of Fourier transform which results in the cosine modulated data giving the desired absorption mode lineshape, then the sine modulated data will give peaks with the undesirable dispersion lineshape. Similarly, selecting the correct transform for the sine modulated data will result in dispersion mode lineshapes for the cosine modulated data. There is no way round this problem, although we will see in due course that it is sometimes possible to modify an experiment so as to obtain one kind of modulation.

8.2.4 Labelling the axes of two-dimensional spectra

In all of the theoretical approaches we use to analyse NMR experiments, it is generally more convenient to express any frequencies in rad s^{-1}, rather than in Hz. This is why we so far have labelled the frequency axes of our two-dimensional spectra as ω_1 and ω_2, so as to indicate that they are in angular frequency units.

However, when we are interpreting and working with practical spectra, we are certainly going to be using Hz or ppm, and not rad s^{-1}. Of course, ppm is not really a frequency scale at all, but such values can readily be converted into frequencies.

Since this chapter is concerned with using theoretical methods to predict and understand the form of two-dimensional spectra, we will label the axes ω_1 and ω_2 i.e. implying angular frequency units. This gives us a common unit for the axes and quantities such as the offsets of the two spins, Ω_1 and Ω_2. However, when we give experimental examples of two-dimensional spectra we will label the scales in ppm, as is clearly most natural.

Often, we will want to indicate that the splitting between two peaks in a two-dimensional spectrum is given by the coupling between the two spins, J_{12} (Hz). As the frequency scale is in rad s^{-1}, the splitting should be labelled $2\pi J_{12}$, so that it too will be in rad s^{-1}. Although this is technically correct, we will not do it as the result would be cumbersome and fussy. So, even though the scale is in rad s^{-1}, we will mark the splitting as J_{12}; from the context, it will always be clear what is going on.

8.3   COSY

COSY: COrrelation SpectroscopY

The COSY experiment, and its variants, is one of the most popular and useful of all two-dimensional experiments. It is a homonuclear experiment, mostly used for analysing proton spectra. From a COSY spectrum it is possible to identify the chemical shifts of spins which are scalar coupled to one another, thus enabling us to trace out the *J*-coupling network in the molecule.

Figure 8.7 shows a schematic COSY spectrum; broadly speaking it contains two kinds of peaks: *cross peaks*, here shown in blue, and *diagonal peaks* show in dark grey. Cross peaks have *different* frequency coordinates in ω_1 and ω_2. Such a peak appearing at frequency $\omega_1 = \Omega_A$, $\omega_2 = \Omega_B$ shows that a spin at offset (chemical shift) Ω_A is coupled to another spin at offset (chemical shift) Ω_B. Thus, the spectrum in Fig. 8.7 shows the presence of the following couplings: *A–B*, *B–D* and *C–E*. Diagonal peaks have the *same* frequency coordinates in ω_1 and ω_2, and are centred at the offset (chemical shift) of each spin. These peaks do not convey any particular information about the connectivity of the spins, but serve to locate the shifts in the spectrum.

We will see shortly that the 'blobs' used in Fig. 8.7 to represent cross and diagonal peaks are not single lines but collections of peaks which form a two-dimensional multiplet. So, to be more precise we should refer to *cross-peak multiplets* and *diagonal-peak multiplets*.

8.3.1 Overall form of the COSY spectrum

The pulse sequence for the COSY experiment in shown in Fig. 8.8 on the following page. Although the pulse sequence is very simple, working out the detailed form of the spectrum is quite an involved process, so we will go through the calculation slowly.

We will start with equilibrium magnetization on spin one, $\hat{I}_{1z}$, which is rotated to $-\hat{I}_{1y}$ by the first 90° pulse. During t_1 this magnetization evolves under the offset of spin one and the coupling between the two spins. Evolution of the offset gives:

$$-\hat{I}_{1y} \xrightarrow{\Omega_1 t_1 \hat{I}_{1z}} -\cos(\Omega_1 t_1)\,\hat{I}_{1y} + \sin(\Omega_1 t_1)\,\hat{I}_{1x}.$$

Each term evolves under the coupling to give :

$$-\cos(\Omega_1 t_1)\,\hat{I}_{1y} \xrightarrow{2\pi J_{12} t_1 \hat{I}_{1z}\hat{I}_{2z}}$$
$$-\cos(\pi J_{12} t_1)\cos(\Omega_1 t_1)\,\hat{I}_{1y} + \sin(\pi J_{12} t_1)\cos(\Omega_1 t_1)\,2\hat{I}_{1x}\hat{I}_{2z};$$

$$\sin(\Omega_1 t_1)\,\hat{I}_{1x} \xrightarrow{2\pi J_{12} t_1 \hat{I}_{1z}\hat{I}_{2z}}$$
$$\cos(\pi J_{12} t_1)\sin(\Omega_1 t_1)\,\hat{I}_{1x} + \sin(\pi J_{12} t_1)\sin(\Omega_1 t_1)\,2\hat{I}_{1y}\hat{I}_{2z}.$$

Finally, each of these four terms is rotated by the second 90° pulse:

$$-\cos(\pi J_{12} t_1)\cos(\Omega_1 t_1)\,\hat{I}_{1y} \xrightarrow{(\pi/2)(\hat{I}_{1x}+\hat{I}_{2x})} -\cos(\pi J_{12} t_1)\cos(\Omega_1 t_1)\,\hat{I}_{1z} \quad [1]$$

$$\sin(\pi J_{12} t_1)\cos(\Omega_1 t_1)\,2\hat{I}_{1x}\hat{I}_{2z} \xrightarrow{(\pi/2)(\hat{I}_{1x}+\hat{I}_{2x})} -\sin(\pi J_{12} t_1)\cos(\Omega_1 t_1)\,2\hat{I}_{1x}\hat{I}_{2y} \quad [2]$$

$$\cos(\pi J_{12} t_1)\sin(\Omega_1 t_1)\,\hat{I}_{1x} \xrightarrow{(\pi/2)(\hat{I}_{1x}+\hat{I}_{2x})} \cos(\pi J_{12} t_1)\sin(\Omega_1 t_1)\,\hat{I}_{1x} \quad [3]$$

$$\sin(\pi J_{12} t_1)\sin(\Omega_1 t_1)\,2\hat{I}_{1y}\hat{I}_{2z} \xrightarrow{(\pi/2)(\hat{I}_{1x}+\hat{I}_{2x})} -\sin(\pi J_{12} t_1)\sin(\Omega_1 t_1)\,2\hat{I}_{1z}\hat{I}_{2y}. \quad [4]$$

This brings us to the start of t_2, so from now on we need only consider the observable terms, which are [3] and [4]. Term [1] represents

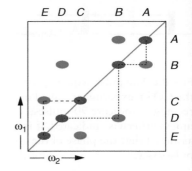

Fig. 8.7 A schematic COSY spectrum, indicating how it is used to determine the chemical shifts of coupled spins. In this example there are five spins, *A–E*, with the offsets indicated. Two kinds of peaks appear in the spectrum: *diagonal peaks*, shown in dark grey, and *cross peaks*, shown in blue. Diagonal peaks have the same frequency coordinates (chemical shifts) in each dimension, whereas for cross peaks the coordinates are different. The appearance of the cross peak at the ω_1 frequency of *B* and the ω_2 frequency of *A* indicates that *A* and *B* are coupled. Using the same interpretation for the other cross peaks, we find that *B* is further coupled to *D* (connections indicated by the dashed lines in the lower triangle). Similarly, *C* is coupled to *E*, but not to any of the other spins (the dashed lines in the upper triangle show this connection). Overall, the COSY spectrum allows us to trace out the network of coupled spins in the molecule. Note that the spectrum has symmetry about the diagonal, shown by the blue line.

Fig. 8.8 The pulse sequence for the COSY experiment. Filled in rectangles indicate 90° pulses; unless otherwise indicated, it is assumed that the phase of the pulses is *x*.

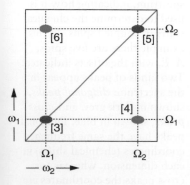

Fig. 8.9 Schematic COSY spectrum for two-spin system. There are two diagonal-peak multiplets, shown in dark grey, centred at $\{\omega_1, \omega_2\} = \{\Omega_1, \Omega_1\}$ and $\{\Omega_2, \Omega_2\}$. In addition, there are two cross-peak multiplets, shown in blue, centred at $\{\Omega_1, \Omega_2\}$ and $\{\Omega_2, \Omega_1\}$ (the internal structure of the multiplets is not shown). The numbers in square braces refer to the terms in the calculation which give rise to each feature. If the coupling between the spins goes to zero, the cross-peak multiplets disappear.

z-magnetization, and term [2] is multiple-quantum coherence, neither of which is observable.

The operator in term [3] is $\hat{I}_{1x}$, which will give rise to a doublet on spin one in the ω_2 dimension. This term is modulated in t_1 by $\sin(\Omega_1 t_1)$ i.e. it is modulated at the offset of spin one, Ω_1. Thus, in the two-dimensional spectrum, term [3] gives rise to a feature centred at the offset of spin one in ω_2, and the offset of spin one in ω_1: in other words, a diagonal peak (or more precisely a diagonal-peak multiplet). The position of the peak is shown in Fig. 8.9.

In contrast, the operator in term [4] is $2\hat{I}_{1z}\hat{I}_{2y}$; this gives rise to an anti-phase doublet centred at the shift of spin *two* in the ω_2 dimension. Like term [3], [4] is also modulated in t_1 according to $\sin(\Omega_1 t_1)$. So, overall, term [4] gives rise to a feature centred at Ω_1 in ω_1 and Ω_2 in ω_2; this is a cross-peak multiplet. Once again, the position of the peak is shown in Fig. 8.9.

We started the calculation with equilibrium magnetization on spin one, $\hat{I}_{1z}$. If we repeat the calculation starting with equilibrium magnetization on spin two, $\hat{I}_{2z}$, the resulting observable terms are

$$\cos(\pi J_{12}t_1)\sin(\Omega_2 t_1)\,\hat{I}_{2x} \qquad [5]$$

$$-\sin(\pi J_{12}t_1)\sin(\Omega_2 t_1)\,2\hat{I}_{1y}\hat{I}_{2z}. \qquad [6]$$

Using the same interpretation as above, term [5] is the diagonal-peak multiplet centred at Ω_2 in each dimension, and [6] is the cross-peak multiplet centred at Ω_2 in ω_1 and Ω_1 in ω_2. So, the complete COSY spectrum consists of two diagonal-peak multiplets and two cross-peak multiplets, as shown schematically in Fig. 8.9.

Looking back through the calculation we can see that the cross peak, term [4], arises from magnetization on spin one which went anti-phase during t_1 and was then transferred to spin two by the second 90° pulse. In other words, the cross peaks arise due to coherence transfer via the coupling.

If the coupling J_{12} is zero, no such anti-phase magnetization is generated, and so there is no cross peak. We can see this from the calculation as term [4],

$$-\sin(\pi J_{12}t_1)\sin(\Omega_1 t_1)\,2\hat{I}_{1z}\hat{I}_{2y}, \qquad [4]$$

will be zero if $J_{12} = 0$ as this makes $\sin(\pi J_{12}t_1) = 0$.

The next task is to work out the detailed form of the two-dimensional multiplets. We will find that each consists of four separate peaks, but that the phase and sign of the four peaks is significantly different between the cross- and diagonal-peak multiplets.

8.3.2   Detailed form of the two-dimensional multiplets

The cross-peak multiplet

The cross-peak multiplet arises from term [4]:

$$-\sin(\pi J_{12}t_1)\sin(\Omega_1 t_1)\,2\hat{I}_{1z}\hat{I}_{2y}. \qquad [4]$$

As was shown in section 7.5.2 on page 154, evolution of the term $2\hat{I}_{1z}\hat{I}_{2y}$ during t_2 gives rise to a time-domain signal of the form

$$\tfrac{1}{2}i\exp\left(i[\Omega_2 t_2 + \pi J_{12}t_2]\right) - \tfrac{1}{2}i\exp\left(i[\Omega_2 t_2 - \pi J_{12}t_2]\right).$$

As expected, we have terms oscillating at $(\Omega_2 + \pi J_{12})$ and $(\Omega_2 - \pi J_{12})$, which are the frequencies of the two lines of the spin-two multiplet.

If we impose an exponential decay on this time-domain signal, we obtain

$$\tfrac{1}{2}i\exp\left(i[\Omega_2 t_2 + \pi J_{12}t_2]\right)\exp\left(-R^{(2)}t_2\right)$$
$$-\tfrac{1}{2}i\exp\left(i[\Omega_2 t_2 - \pi J_{12}t_2]\right)\exp\left(-R^{(2)}t_2\right). \tag{8.1}$$

The Fourier transform of an exponentially decaying oscillation gives the usual spectrum, with the absorption mode in the real part:

$$S_0\exp\left(i\Omega t_2\right)\exp\left(-R^{(2)}t_2\right) \xrightarrow{\text{FT}} S_0\left[A_2(\Omega) + iD_2(\Omega)\right],$$

so Fourier transformation of the time-domain signal in Eq. 8.1 gives:

$$\tfrac{1}{2}i\left[A_2(\Omega_2 + \pi J_{12}) + iD_2(\Omega_2 + \pi J_{12})\right]$$
$$-\tfrac{1}{2}i\left[A_2(\Omega_2 - \pi J_{12}) + iD_2(\Omega_2 - \pi J_{12})\right]. \tag{8.2}$$

As expected, there are two peaks, one at $(\Omega_2 + \pi J_{12})$ and one at $(\Omega_2 - \pi J_{12})$. Most importantly, the peak at $(\Omega_2 + \pi J_{12})$ is positive whereas that at $(\Omega_2 - \pi J_{12})$ is negative, so what we have is an anti-phase doublet, just as expected for the term $2\hat{I}_{1z}\hat{I}_{2y}$.

Due to the factor of $\tfrac{1}{2}i$ which is multiplying the whole expression in Eq. 8.2, the desirable absorption mode lineshape appears in the *imaginary* part of the spectrum. The normal practice is to adjust the phase of the spectrum so that the absorption mode lineshape appears in the real part. In this case the required phase correction is $-90°$ or $(-\pi/2)$ radians; such a correction is achieved by multiplying by $\exp\left(-i[\pi/2]\right) \equiv -i$. Noting that $i \times (-i) = 1$, the result of applying this phase correction to Eq. 8.2 is

$$\tfrac{1}{2}\left[A_2(\Omega_2 + \pi J_{12}) + iD_2(\Omega_2 + \pi J_{12})\right]$$
$$-\tfrac{1}{2}\left[A_2(\Omega_2 - \pi J_{12}) + iD_2(\Omega_2 - \pi J_{12})\right]. \tag{8.3}$$

Now the absorption mode lineshape appears in the real part of the spectrum.

For term [4], the modulation with respect to t_1 is of the form $-\sin(\pi J_{12}t_1)\sin(\Omega_1 t_1)$. This product of two sine terms can be expanded using the trigonometric identity

$$\sin A \sin B \equiv \tfrac{1}{2}\left[\cos(A - B) - \cos(A + B)\right]$$

to give

$$\tfrac{1}{2}\left[\cos(\Omega_1 t_1 + \pi J_{12}t_1) - \cos(\Omega_1 t_1 - \pi J_{12}t_1)\right].$$

As before, we impose an exponential decay to give

$$\tfrac{1}{2}\left[\cos(\Omega_1 t_1 + \pi J_{12}t_1)\exp\left(-R^{(1)}t_1\right) - \cos(\Omega_1 t_1 - \pi J_{12}t_1)\exp\left(-R^{(1)}t_1\right)\right].$$

Fig. 8.10 Contour plot of the cross-peak multiplet centred at $\omega_1 = \Omega_1$ and $\omega_2 = \Omega_2$. Positive contours are shown in blue and negative ones are shown in dark grey. Alongside each axis is plotted an anti-phase doublet, and the four peaks in the two-dimensional spectrum can be constructed by 'multiplying together' these two anti-phase doublets. For example, the top right-hand peak is positive, as the peaks along ω_1 and ω_2 from which it is derived are both positive. In contrast, the top left-hand peak is negative, as it is constructed from the product of a positive peak (along ω_1) and a negative peak (along ω_2).

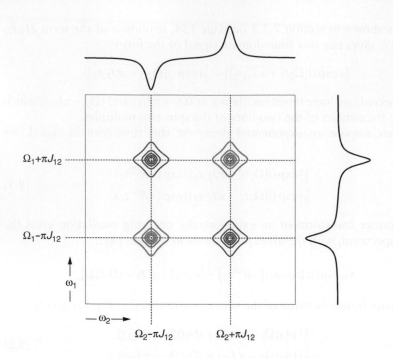

The cosine Fourier transformation of this with respect to t_1 gives two absorption mode peaks:

$$\tfrac{1}{2}\left[A_1(\Omega_1 + \pi J_{12}) - A_1(\Omega_1 - \pi J_{12})\right]. \tag{8.4}$$

What we have here are the two lines of the spin-one doublet; one line is positive and one is negative i.e. the doublet is in anti-phase.

Equation 8.3 on the preceding page gives the spectrum in the ω_2 dimension, and Eq. 8.4 gives the spectrum in the ω_1 dimension. Multiplying the two together will give us the overall form of the two-dimensional spectrum. If we take just the real part of the ω_2 spectrum (as this has the absorption mode lineshape), the result is

$$\underbrace{\tfrac{1}{2}\left[A_1(\Omega_1 + \pi J_{12}) - A_1(\Omega_1 - \pi J_{12})\right]}_{\omega_1 \text{ spectrum}} \times \underbrace{\tfrac{1}{2}\left[A_2(\Omega_2 + \pi J_{12}) - A_2(\Omega_2 - \pi J_{12})\right]}_{\omega_2 \text{ spectrum}}.$$

Multiplying this out gives us four lines, each with the double absorption lineshape:

$$+\tfrac{1}{4}A_1(\Omega_1 + \pi J_{12})\, A_2(\Omega_2 + \pi J_{12})$$
$$-\tfrac{1}{4}A_1(\Omega_1 + \pi J_{12})\, A_2(\Omega_2 - \pi J_{12})$$
$$-\tfrac{1}{4}A_1(\Omega_1 - \pi J_{12})\, A_2(\Omega_2 + \pi J_{12})$$
$$+\tfrac{1}{4}A_1(\Omega_1 - \pi J_{12})\, A_2(\Omega_2 - \pi J_{12}).$$

Figure 8.10 shows a contour plot of these four peaks. This pattern is called an *anti-phase square array*, on account of the sign alternation of the peaks in each dimension. The array of peaks is centred at $\omega_1 = \Omega_1$ and $\omega_2 = \Omega_2$, and is split by $2\pi J_{12}$ in each dimension.

The process of finding the frequencies and signs of the four peaks is, to say the least, rather convoluted, but can be speeded up by using the following approach.

In term [4] the operator is $2\hat{I}_{1z}\hat{I}_{2y}$, and we already know that this gives rise, with suitable phasing, to an anti-phase doublet on spin two. In t_1, we saw that the modulation of the signal, $-\sin(\pi J_{12}t_1)\sin(\Omega_1 t_1)$, could be expressed as

$$\tfrac{1}{2}\left[\cos(\Omega_1 t_1 + \pi J_{12}t_1) - \cos(\Omega_1 t_1 - \pi J_{12}t_1)\right].$$

The Fourier transform of this gives an anti-phase doublet on spin one.

The two-dimensional multiplet can be found by imagining these two anti-phase doublets along the ω_1 and ω_2 axes, as show in Fig. 8.10 on the facing page, and then multiplying them so as to create the two-dimensional multiplet. To describe this process in words makes it sound very complicated, but if you look at Fig. 8.10 it should be clear what the process is.

If the coupling J_{12} goes to zero, then the four peaks in the anti-phase square array will fall on top of one another and cancel completely. So, if the coupling is zero, there is no cross-peak multiplet.

The diagonal-peak multiplet

The diagonal-peak multiplet is represented by term [3] from page 191:

$$\cos(\pi J_{12}t_1)\sin(\Omega_1 t_1)\,\hat{I}_{1x}. \qquad [3]$$

From section 7.5.1 on page 153 we know that evolution of the term $\hat{I}_{1x}$ during t_2 will give the following signal

$$\tfrac{1}{2}\exp(\mathrm{i}[\Omega_1 t_2 + \pi J_{12}t_2]) + \tfrac{1}{2}\exp(\mathrm{i}[\Omega_1 t_2 - \pi J_{12}t_2]).$$

If we assume that this is decaying exponentially, then Fourier transformation with respect to t_2 gives the following frequency-domain signal:

$$\begin{aligned} &\tfrac{1}{2}\left[A_2(\Omega_1 + \pi J_{12}) + \mathrm{i}\,D_2(\Omega_1 + \pi J_{12})\right] \\ &+ \tfrac{1}{2}\left[A_2(\Omega_1 - \pi J_{12}) + \mathrm{i}\,D_2(\Omega_1 - \pi J_{12})\right]. \end{aligned} \qquad (8.5)$$

This time, the real part contains the required double absorption lineshape, so no phase correction is required. As expected for the term $\hat{I}_{1x}$, the spectrum is an in-phase doublet of spin one.

The modulation of the signal with respect to t_1 is the product of a cosine and a sine term. This can be expanded using the trigonometric identity

$$\sin A \cos B \equiv \tfrac{1}{2}\left[\sin(A+B) + \sin(A-B)\right]$$

to give

$$\tfrac{1}{2}\left[\sin(\Omega_1 t_1 + \pi J_{12}t_1) + \sin(\Omega_1 t_1 - \pi J_{12}t_1)\right].$$

Again assuming an exponential decay, a sine Fourier transform with respect to t_1 gives two absorption mode peaks of the same sign:

$$\tfrac{1}{2}\left[A_1(\Omega_1 + \pi J_{12}) + A_1(\Omega_1 - \pi J_{12})\right]. \qquad (8.6)$$

Fig. 8.11 Contour plot of the diagonal-peak multiplet centred at the offset of spin one in both dimensions; positive contours are indicated by blue lines. Note that, in contrast to the cross-peak multiplet shown in Fig. 8.10 on page 194, all four peaks have the same sign. Alongside each axis is plotted an in-phase doublet; the two-dimensional multiplet can be constructed by multiplying these together.

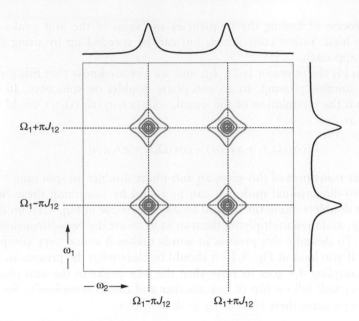

These two lines form the in-phase doublet on spin one.

The two-dimensional spectrum is the product of the ω_2 part, Eq. 8.5 on the preceding page, and the ω_1 part, Eq. 8.6 on the previous page. Taking just the real part of Eq. 8.5 on the preceding page we find:

$$\underbrace{\tfrac{1}{2}\left[A_1(\Omega_1 + \pi J_{12}) + A_1(\Omega_1 - \pi J_{12})\right]}_{\omega_1 \text{ spectrum}} \times \underbrace{\tfrac{1}{2}\left[A_2(\Omega_1 + \pi J_{12}) + A_2(\Omega_1 - \pi J_{12})\right]}_{\omega_2 \text{ spectrum}}.$$

Multiplying this out gives four double absorption lines, which are all positive.

$$+\tfrac{1}{4}A_1(\Omega_1 + \pi J_{12})\,A_2(\Omega_1 + \pi J_{12})$$
$$+\tfrac{1}{4}A_1(\Omega_1 + \pi J_{12})\,A_2(\Omega_1 - \pi J_{12})$$
$$+\tfrac{1}{4}A_1(\Omega_1 - \pi J_{12})\,A_2(\Omega_1 + \pi J_{12})$$
$$+\tfrac{1}{4}A_1(\Omega_1 - \pi J_{12})\,A_2(\Omega_1 - \pi J_{12}).$$

Figure 8.11 shows a schematic contour plot of these four peaks. The array of peaks is centred at $\omega_1 = \Omega_1$ and $\omega_2 = \Omega_1$, and is split by $2\pi J_{12}$ in each dimension. Note that, in contrast to the cross-peak multiplet, all of the lines in the diagonal-peak multiplet have the same sign. If the coupling goes to zero, these four peaks fall on top of one another, but in contrast to the cross-peak multiplet, the four lines reinforce one another.

As before, we can speed things up by realizing that the spectrum in ω_2 is an in-phase doublet on spin one. The form of the modulation in t_1

$$\tfrac{1}{2}\left[\sin\left(\Omega_1 t_1 + \pi J_{12} t_1\right) + \sin\left(\Omega_1 t_1 - \pi J_{12} t_1\right)\right],$$

tells us that the spectrum in ω_1 is also an in-phase doublet on spin one. Multiplying these two together, in the manner shown in Fig. 8.11, gives the four positive lines of the diagonal-peak multiplet.

 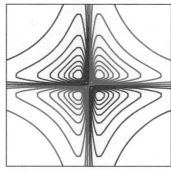

Fig. 8.12 Two views of the double dispersion mode Lorentzian lineshape: on the left is shown a perspective view, and on the right is shown a contour plot (positive contours are blue, negative contours are dark grey). A cross-section through this lineshape taken parallel to either axis shows a dispersion mode line.

8.3.3 Phase properties of the COSY spectrum

Looking back through the previous section, you will see that we used rather different processing to obtain the cross- and diagonal-peak multiplets shown in Figs 8.10 and 8.11.

- For the cross peak, we applied a 90° phase correction in the ω_2 dimension and used a cosine Fourier transform in t_1.

- For the diagonal peak, no phase correction was used in ω_2, and a sine Fourier transform was used in t_1.

We can, of course, choose to process the data any way we like, but the *whole* spectrum is processed at the same time: we cannot choose one kind of processing for the cross peaks, and a different one for the diagonal peaks. It turns out that, if we choose the processing which results in the cross peaks appearing in double absorption, then the diagonal peaks will appear in *double dispersion*.

The reason for this can be seen by looking at the terms which give rise to the cross and diagonal peaks. The diagonal-peak term is

$$\cos\left(\pi J_{12}t_1\right)\sin\left(\Omega_1 t_1\right)\hat{I}_{1x}; \qquad [3]$$

expanding the trigonometric terms as we did before gives

$$\tfrac{1}{2}\left[\sin\left(\Omega_1 t_1 + \pi J_{12}t_1\right) + \sin\left(\Omega_1 t_1 - \pi J_{12}t_1\right)\right]\hat{I}_{1x}. \qquad (8.7)$$

The cross-peak term is

$$-\sin\left(\pi J_{12}t_1\right)\sin\left(\Omega_1 t_1\right)2\hat{I}_{1z}\hat{I}_{2y}, \qquad [4]$$

which expands to

$$\tfrac{1}{2}\left[\cos\left(\Omega_1 t_1 + \pi J_{12}t_1\right) - \cos\left(\Omega_1 t_1 - \pi J_{12}t_1\right)\right]2\hat{I}_{1z}\hat{I}_{2y}. \qquad (8.8)$$

Comparing Eqs 8.7 and 8.8, we see that in the first the magnetization observed during t_2 appears along the x-axis, whereas in the second it appears along the y-axis; this accounts for the 90° phase shift in the ω_2 dimension. Similarly, the t_1 modulation in the first appears as a sine, whereas in the second it appears as a cosine; this accounts for the change in lineshape in the ω_1 dimension.

Fig. 8.13 Schematic COSY spectra of a two-spin system. In (a) the processing has been chosen so that the cross peaks have the double absorption lineshape and the diagonal peaks have the double dispersion lineshape; in (b), the processing has been chosen so that the lineshapes are the other way round. The anti-phase square arrays are clearly visible in (a), but harder to spot in (b). On account of their dispersion lineshape, the diagonal peaks in (a) spread much more than in (b).

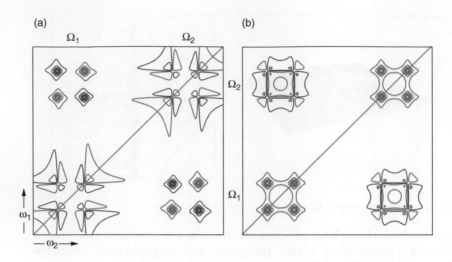

The double dispersion lineshape is illustrated in Fig. 8.12 on the previous page. As the dispersive line has positive and negative parts, the two-dimensional lineshape alternates sign in a four-fold pattern; furthermore, compared with the double absorption lineshape, the peak height is reduced by a factor of $\frac{1}{4}$. The combination of the broad wings of the dispersion lineshape, and the alternating signs, makes this lineshape very undesirable for high-resolution work. The double absorption lineshape, illustrated in Fig. 8.6 on page 189, is much preferred.

Figure 8.13 shows schematic COSY spectra processed in such a way as to have either the cross peaks or the diagonal peaks in double absorption. It is clear from these plots that the double dispersion lineshape, whether it appears on the diagonal or the cross peaks, is simply not desirable.

8.3.4   How small a coupling can we detect with COSY?

We commented above that, if the coupling goes to zero the cross peak disappears due to the cancellation of the anti-phase lines. However, what happens if the coupling is not zero, but just small: will the cross peak be detectable?

The answer to this question is illustrated in Fig. 8.14 on the facing page, which shows cross-sections through a series of cross-peak multiplets in which the coupling constant is successively halved. Thus if the coupling constant for the left-most doublet is J_{max}, for the next it is $J_{max}/2$, for the next it is $J_{max}/4$ and so on; the linewidth has been kept constant, and is about one-fifth of J_{max}.

As the coupling constant becomes smaller and smaller, the two lines begin to overlap and, since they are of opposite sign, they begin to cancel one another out. So, as we go from left to right in the diagram, the overall intensity of the cross peaks gets smaller and smaller. However, in (a) the anti-phase multiplet is still clearly visible even on the far right-hand side where the coupling has been reduced by a factor of 1/64.

The spectra shown in (a) are unrealistic, though, as unlike experimental spectra they contain no noise. The series of spectra (b) are the same as (a)

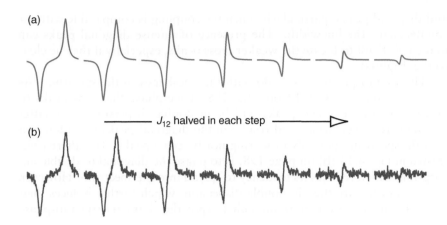

Fig. 8.14 Both (a) and (b) show cross-sections taken through a series of cross-peak multiplets in which the coupling constant has been halved in each successive step. Since the two lines are in anti-phase, once the coupling constant becomes comparable with the linewidth, cancellation starts to occur, leading to an overall reduction in intensity. The series of spectra in (b) differ in that noise has been added to them. In these spectra we see that as the coupling constant decreases, the signal-to-noise ratio decreases, indeed the effect is sufficient to make it impossible to discern the anti-phase doublet in the right-most spectrum. The smallest coupling which can be detected thus depends on the linewidth and the noise level in the spectrum.

except that noise has been added. Now, as the overall intensity decreases, we see that the signal-to-noise ratio also decreases. In fact, for the right-most spectrum, the anti-phase doublet is not really visible.

The amount of cancellation in an anti-phase doublet depends on the size of the coupling relative to the linewidth. If the coupling is much larger than the linewidth, there will be no cancellation, but as the two become comparable a significant amount of cancellation occurs. Once the linewidth becomes larger than the coupling, the amount of cancellation will be very significant.

For a given combination of linewidth and coupling constant, whether or not a cross peak is visible above the noise depends on the signal-to-noise ratio. Therefore, to detect cross peaks due to the smallest coupling constants we need to make sure that the linewidth is minimized and the signal-to-noise ratio maximized.

8.3.5 The problem with COSY

The basic COSY experiment is undoubtedly extraordinarily useful, but it does suffer from two drawbacks, both of which are associated with the detailed form of the cross- and diagonal-peak multiplets.

The first problem is a consequence of the anti-phase structure of the cross-peak multiplet, which contrasts with the in-phase structure of the diagonal-peak multiplet. In the cross-peak multiplet the lines tend to cancel one another out, leading to a reduction in intensity, whereas in the diagonal-peak multiplet the lines tend to reinforce one another. As a result, there can be a considerable difference in the overall intensity of the cross and diagonal peaks, particularly when the coupling is comparable with, or smaller than, the linewidth. The presence of intense diagonal peaks can make it difficult to locate the weaker cross peaks, especially if they lie close to the diagonal.

The second problem is to do with the lineshapes in the spectrum. As was illustrated in Fig. 8.13 on page 198, if we phase the cross peaks to absorption, the diagonal peaks will be in double dispersion. The latter lineshape is rather broad and results in the diagonal peaks spreading out into the spectrum, possibly obscuring nearby cross peaks. The alternative, shown in Fig. 8.13 (b) on page 198, is to phase the diagonal to double absorption, which reduces its tendency to spread into the spectrum. However, the cross peaks are then in double dispersion, which further reduces their intensity and makes it more difficult to spot the characteristic anti-phase square arrays.

Both of these problems are neatly avoided (in large part) by a simple modification of COSY called *double-quantum filtered* COSY (DQF COSY). As we shall see in the next section, in the modified experiment both the diagonal- and cross-peak multiplets are in anti-phase and have the *same* lineshape.

8.4   DQF COSY

Fig. 8.15 The pulse sequence for double-quantum filtered COSY (DQF COSY). In this experiment it is arranged that the only signal observed comes from double-quantum coherence present between the second and third 90° pulses – hence the name 'double-quantum filtered'.

The pulse sequence for DQF COSY is shown in Fig. 8.15. The key point about this sequence is that we arrange things so that the signals observed during t_2 *all* derive from double-quantum coherence present between the second and third 90° pulses. In other words, all of the observed signals have been passed through, or been filtered through, a state of double-quantum coherence – hence the name of the experiment.

The way in which we ensure that the observed signals all derive from double-quantum coherence is to use a coherence selection method such as phase cycling or pulsed field gradients. These methods are described in detail in Chapter 11. For the present purposes we will simply assume that the selection can be made, and leave the details of how to later.

Starting with equilibrium magnetization on spin one, $\hat{I}_{1z}$, the evolution during t_1 and the effect of the second 90° pulse are exactly as for COSY: we thus obtain the four terms [1]–[4] listed on page 191. Of these, it is only term [2] which contains double-quantum coherence, so this is the only term of interest to us at present:

$$ - \sin\left(\pi J_{12} t_1\right) \cos\left(\Omega_1 t_1\right) 2\hat{I}_{1x}\hat{I}_{2y}. \qquad [2] $$

We saw in section 7.12.1 on page 174 that $2\hat{I}_{1x}\hat{I}_{2y}$ is a mixture of double- and zero-quantum coherence. In the table on page 175 the pure double-quantum operator $\hat{DQ}_y$ and the pure zero-quantum operator $\hat{ZQ}_y$ were defined as

$$ \hat{DQ}_y \equiv (2\hat{I}_{1x}\hat{I}_{2y} + 2\hat{I}_{1y}\hat{I}_{2x}) \qquad \hat{ZQ}_y \equiv (2\hat{I}_{1y}\hat{I}_{2x} - 2\hat{I}_{1x}\hat{I}_{2y}). $$

From these definitions we can see that

$$ \hat{DQ}_y - \hat{ZQ}_y = 2\left(2\hat{I}_{1x}\hat{I}_{2y}\right), $$

(a)   (b)

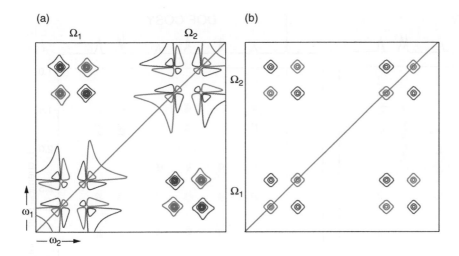

Fig. 8.16 Comparison of a conventional COSY, (a), and a DQF COSY, (b), for a two-spin system. The conventional COSY has been processed so that the cross peaks have the double absorption lineshape; as a result, the diagonal peaks have the double dispersion lineshape. The overall result is that the broad dispersive diagonal peaks rather dominate the spectrum. In contrast, both the diagonal- and cross-peak multiplets of the DQF COSY spectrum are anti-phase square arrays of absorption mode peaks. This results in the diagonal-peak multiplets being much less dominant.

or put the other way round

$$2\hat{I}_{1x}\hat{I}_{2y} = \tfrac{1}{2}\left(\hat{D}Q_y - \hat{Z}Q_y\right).$$

It follows that the pure double-quantum part of $2\hat{I}_{1x}\hat{I}_{2y}$ is $\tfrac{1}{2}\hat{D}Q_y$:

$$\begin{aligned}\text{double-quantum part of } 2\hat{I}_{1x}\hat{I}_{2y} &= \tfrac{1}{2}\hat{D}Q_y \\ &= \tfrac{1}{2}\left(2\hat{I}_{1x}\hat{I}_{2y} + 2\hat{I}_{1y}\hat{I}_{2x}\right).\end{aligned}$$

Using this result we see that the pure double-quantum part of term [2] is

$$-\tfrac{1}{2}\sin\left(\pi J_{12}t_1\right)\cos\left(\Omega_1 t_1\right)\left(2\hat{I}_{1x}\hat{I}_{2y} + 2\hat{I}_{1y}\hat{I}_{2x}\right).$$

The third 90° pulse rotates both of these terms into observable anti-phase magnetization:

$$\begin{aligned}-\tfrac{1}{2}\sin\left(\pi J_{12}t_1\right)\cos\left(\Omega_1 t_1\right)\left(2\hat{I}_{1x}\hat{I}_{2y} + 2\hat{I}_{1y}\hat{I}_{2x}\right) &\xrightarrow{(\pi/2)(\hat{I}_{1x}+\hat{I}_{2x})} \\ -\tfrac{1}{2}\sin\left(\pi J_{12}t_1\right)\cos\left(\Omega_1 t_1\right)&\left(2\hat{I}_{1x}\hat{I}_{2z} + 2\hat{I}_{1z}\hat{I}_{2x}\right).\end{aligned}$$

This brings us to the start of t_2.

The two terms $2\hat{I}_{1x}\hat{I}_{2z}$ and $2\hat{I}_{1z}\hat{I}_{2x}$ represent anti-phase magnetization on spins one and two, respectively; both terms are modulated in t_1 at Ω_1. Thus, the first term represents the diagonal-peak multiplet and the second the cross-peak multiplet. The really important thing is that both terms have the *same* modulation in t_1, and both appear along the x-axis; it will thus be possible to choose processing which results in all the lines in the spectrum

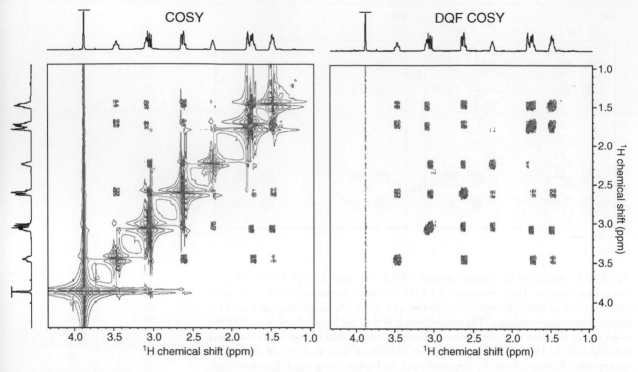

Fig. 8.17 Experimental COSY and DQF COSY spectra of quinine, recorded at 500 MHz; only a small part of the spectrum is shown and the conventional proton spectrum is plotted along the two axes. The COSY spectrum is dominated by the in-phase dispersive diagonal-peak multiplets. In contrast, in the DQF COSY spectrum both the diagonal- and cross-peak multiplets are in anti-phase and have absorption mode lineshapes. This results in a much better balance of intensity between the diagonal and cross peaks. The intense singlet at around 3.9 ppm does not appear in the DQF COSY spectrum since the methyl group responsible for this peak has no couplings to other spins.

having the double absorption mode lineshape. This is in contrast to the simple COSY experiment where the cross and diagonal peaks cannot be phased to have the same lineshape.

Expanding the t_1 modulation in the usual way gives:

$$-\tfrac{1}{2}\sin\left(\pi J_{12}t_1\right)\cos\left(\Omega_1 t_1\right) \equiv -\tfrac{1}{2}\times\tfrac{1}{2}\left[\sin\left(\Omega_1 t_1 + \pi J_{12}t_1\right) - \sin\left(\Omega_1 t_1 - \pi J_{12}t_1\right)\right].$$

This corresponds to an anti-phase doublet on spin one. So, we conclude that *both* the diagonal- and cross-peak multiplets show anti-phase structure in both dimensions. This is the second important property of the DQF COSY experiment. However, it should be noted that there is a price to pay, which is the loss of signal intensity by the factor of one half which arose from taking only the double-quantum part of the coherence present between the second and third pulses.

If we repeat the calculation starting with equilibrium magnetization on spin two, $\hat{I}_{2z}$, we find a further diagonal- and cross-peak multiplet, this time centred at Ω_2 in the ω_1 dimension. These multiplets have the same phase properties as those already described.

Figure 8.16 on page 201 compares a conventional COSY spectrum with a DQF COSY spectrum – the difference is dramatic. In the conventional COSY spectrum, the diagonal-peak multiplets are rather dominant and, due to their dispersive lineshapes, spread into the spectrum. In contrast, in the DQF COSY spectrum all of the multiplets are in anti-phase and all the peaks are in absorption mode. This results in a much nicer looking spectrum, with a better balance of intensity between the cross and diagonal peaks.

For systems involving several coupled spins the multiplets are of course more complex than those shown here for a two-spin system. However, the phase properties of the diagonal- and cross-peak multiplets remain substantially the same. We will return to this point in section 10.2 on page 325 which describes the detailed form of these multiplets for three and more coupled spins.

An additional benefit of the DQF COSY experiment is that singlets (i.e. peaks from uncoupled spins) do not appear in the spectrum. This is because the creation of double-quantum coherence requires the presence of a coupling in order to generate the anti-phase terms. The magnetization from uncoupled spins cannot therefore pass through the double-quantum filter, and so such spins do not contribute to the spectrum. Often, such singlets arise from solvents and the suppression of what can be rather intense peaks is a useful feature of the DQF COSY experiment.

Generally, the DQF COSY experiment is to be preferred to COSY as it gives a much nicer spectrum for hardly any complication of the experiment. The only case where one might not choose DQF COSY is when sensitivity is at a premium.

Figure 8.17 on the preceding page compares the experimental COSY and DQF COSY spectra of quinine. We immediately see how the COSY spectrum is dominated by the intense diagonal peak multiplets which, on account of their dispersive lineshapes, spread far out into the spectrum and thus obscure some of the cross peaks. In contrast, in the DQF COSY spectrum both the diagonal- and cross-peak multiplets are in absorption and have anti-phase multiplet structures. The result is that the diagonal-peak multiplets do not spread out so much, and there is a much more even balance in the intensity between the two types of peaks. In addition, the strong singlet at around 3.9 ppm is absent from the DQF COSY spectrum since there are no couplings to the methyl group which is responsible for this peak. However, due to experimental imperfections this strong peak does leave a trace of so-called 't_1 noise' parallel to ω_1 in the spectrum.

8.5 Double-quantum spectroscopy

It is important to remember that in two-dimensional NMR no observations are made during the evolution time t_1. It is therefore possible to follow the evolution of unobservable coherences, such as multiple-quantum coherence, using such experiments. Indeed, the advent of two-dimensional NMR opened up the possibility of studying such multiple-quantum coherences and led to a growing interest in their properties.

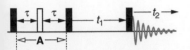

Fig. 8.18 Pulse sequence for recording two-dimensional spectra in which multiple-quantum coherence evolves during t_1. Anti-phase magnetization, generated during a spin echo (period **A**), is transferred into multiple-quantum coherences by the second 90° pulse. After evolution for t_1, the multiple-quantum coherence is transferred into observable magnetization by the final 90° pulse. As usual, filled in rectangles represent 90° pulses, whilst the open rectangle represents a 180° pulse.

In structure determination, the only two-dimensional experiment involving multiple quantum evolution which has widespread application is a double-quantum experiment. In this section we will describe how the experiment works, and the useful information which can be gleaned from the spectrum.

A simple pulse sequence for following the evolution of multiple-quantum coherence is shown in Fig. 8.18. Broadly speaking, what happens is the following:

(a) Transverse magnetization is generated by the first pulse.

(b) During the following spin echo, period **A**, some anti-phase magnetization is generated, the amount depending on the delay τ and the size of the coupling.

(c) The second 90° pulse converts this anti-phase magnetization to multiple-quantum coherence, which then evolves for time t_1.

(d) The final 90° pulse converts the multiple-quantum coherence back into observable magnetization, which is then recorded for t_2.

We thus expect to see a spectrum in which there are multiple-quantum frequencies in ω_1, and the frequencies of the normal spectrum in ω_2.

8.5.1   Detailed analysis of the pulse sequence

We will start with equilibrium magnetization on spin one, $\hat{I}_{1z}$, which is rotated by the first 90° pulse to $-\hat{I}_{1y}$. Next, we have a spin echo covering period **A**. As has been described in section 7.8.1 on page 159, the spin echo refocuses the offset, but the coupling evolves for time 2τ. The overall result of the spin echo is equivalent to evolution of the coupling for time 2τ, followed by a 180° pulse. So, at the end of period **A** we have

$$\cos\left(2\pi J_{12}\tau\right)\hat{I}_{1y} - \sin\left(2\pi J_{12}\tau\right)2\hat{I}_{1x}\hat{I}_{2z}.$$

The 90° pulse rotates these two terms to give

$$\cos\left(2\pi J_{12}\tau\right)\hat{I}_{1z} + \sin\left(2\pi J_{12}\tau\right)2\hat{I}_{1x}\hat{I}_{2y}.$$

We will suppose that we are able to select just the double-quantum part of the coherence at this point. This is the same thing that we did when analysing the DQF COSY (section 8.4 on page 200), where it was shown that:

$$\begin{aligned}\text{double-quantum part of } 2\hat{I}_{1x}\hat{I}_{2y} &= \tfrac{1}{2}\hat{\mathrm{DQ}}_y \\ &= \tfrac{1}{2}\left(2\hat{I}_{1x}\hat{I}_{2y} + 2\hat{I}_{1y}\hat{I}_{2x}\right).\end{aligned}$$

So, at the start of t_1 we have

$$\tfrac{1}{2}\sin\left(2\pi J_{12}\tau\right)\hat{\mathrm{DQ}}_y.$$

The evolution of this double-quantum term during t_1 can be worked out using the rules from section 7.12.3 on page 176:

$$\hat{\mathrm{DQ}}_y \xrightarrow{\text{evolution for } t_1} \cos\left([\Omega_1 + \Omega_2]t_1\right)\hat{\mathrm{DQ}}_y - \sin\left([\Omega_1 + \Omega_2]t_1\right)\hat{\mathrm{DQ}}_x,$$

where

$$\hat{DQ}_x \equiv (2\hat{I}_{1x}\hat{I}_{2x} - 2\hat{I}_{1y}\hat{I}_{2y}) \quad \text{and} \quad \hat{DQ}_y \equiv (2\hat{I}_{1x}\hat{I}_{2y} + 2\hat{I}_{1y}\hat{I}_{2x}).$$

$(\Omega_1 + \Omega_2)$, the sum of the offsets, is the double-quantum frequency.

Therefore, at the end of t_1 we have

$$\tfrac{1}{2}\sin(2\pi J_{12}\tau)\Big[\cos([\Omega_1 + \Omega_2]t_1)\,\hat{DQ}_y - \sin([\Omega_1 + \Omega_2]t_1)\,\hat{DQ}_x\Big].$$

The final thing to consider is the effect of the third 90° pulse. The terms $\hat{DQ}_x$ and $\hat{DQ}_y$ are affected differently by this pulse:

$$\hat{DQ}_x \equiv 2\hat{I}_{1x}\hat{I}_{2x} - 2\hat{I}_{1y}\hat{I}_{2y} \xrightarrow{(\pi/2)(\hat{I}_{1x}+\hat{I}_{2x})} 2\hat{I}_{1x}\hat{I}_{2x} - 2\hat{I}_{1z}\hat{I}_{2z},$$

and

$$\hat{DQ}_y \equiv 2\hat{I}_{1x}\hat{I}_{2y} + 2\hat{I}_{1y}\hat{I}_{2x} \xrightarrow{(\pi/2)(\hat{I}_{1x}+\hat{I}_{2x})} 2\hat{I}_{1x}\hat{I}_{2z} + 2\hat{I}_{1z}\hat{I}_{2x}.$$

From these we see that the term $\hat{DQ}_x$ does not lead to any observable magnetization, whereas the term $\hat{DQ}_y$ gives two anti-phase terms, one on each spin. So, at the start of t_2 the observable terms are

$$\tfrac{1}{2}\sin(2\pi J_{12}\tau)\cos([\Omega_1 + \Omega_2]t_1)\Big[2\hat{I}_{1x}\hat{I}_{2z} + 2\hat{I}_{1z}\hat{I}_{2x}\Big].$$

In the ω_2 dimension we have an anti-phase doublet on spin one, and a similar doublet on spin two. There is just one modulating frequency in t_1, which is the double-quantum frequency $(\Omega_1 + \Omega_2)$. The result is that all four peaks have this as the ω_1 frequency. A schematic spectrum is shown in Fig. 8.19. Since both the observable terms appear along the x-axis, and there is just one modulating frequency in t_1, it is possible to process the spectrum in such a way that all of the peaks are in absorption.

The overall intensity of the peaks in the spectrum depends on, amongst other things, the factor $\sin(2\pi J_{12}\tau)$ which arose in our calculation. This factor determines the amount of anti-phase magnetization present at the end of the spin echo (period **A**), and hence the overall amount of double-quantum coherence which is created. As we saw before, if $\tau = 1/(4J_{12})$ there is complete conversion to anti-phase, and in the present context this choice of τ will give the maximum amount of double-quantum coherence, and hence the strongest peaks in the spectrum. Since we have used a spin echo to create the anti-phase magnetization, the result is independent of the offsets of either spin.

8.5.2   Interpretation and application of double-quantum spectra

From the double-quantum spectrum we can determine the double-quantum evolution frequency, $(\Omega_1 + \Omega_2)$. However, as this is just the sum of the two offsets, which we already know, it is not a very significant piece of information.

What is more useful is to note that the peaks from coupled spins must share the *same* double-quantum frequency in the ω_1 dimension. This is because the double-quantum coherence present during t_1 is transferred back

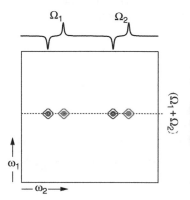

Fig. 8.19 Schematic double-quantum spectrum, recorded using the pulse sequence of Fig. 8.18 on the preceding page, of a two-spin system. In the ω_2 dimension we have anti-phase doublets on both spin one and spin two. All four peaks have the same ω_1 frequency, $(\Omega_1 + \Omega_2)$, which is the double-quantum frequency. From our detailed calculation, the overall intensity of these peaks is proportional to $\sin(2\pi J_{12}\tau)$.

to *both* of the spins involved. Furthermore, if the coupling is zero, no anti-phase magnetization is created, and so no double-quantum coherence is generated. As a result, there are no peaks in the spectrum.

The double-quantum spectrum therefore enables us to identify which pairs of spins are coupled. This is exactly the same information as we can find from a COSY spectrum, it is just that in the double-quantum spectrum the information is presented in a slightly different way.

If all we want to know is which spins are coupled to which, then a COSY spectrum is probably a better choice than a double-quantum spectrum, as the former is simpler to interpret and does not suffer from any complications. However, there are some special circumstances in which the double-quantum spectrum is useful – one of which we will look at in the next section.

8.5.3 INADEQUATE

INADEQUATE: Incredible Natural Abundance DoublE QUAntum Transfer Experiment

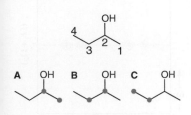

Fig. 8.20 Illustration of the occurrence of molecules containing two adjacent ^{13}C atoms in 2-butanol, whose structure is shown at the top. There are three possible isotopomers in which two ^{13}C atoms occupy *adjacent* positions: these are shown in structures **A**, **B** and **C** (the presence of a ^{13}C is indicated by the blue dot). In each isotopomer it is possible to generate double-quantum coherence between the two adjacent ^{13}C atoms.

The two-dimensional INADEQUATE experiment is a very elegant way of identifying the chemical shifts of directly bonded ^{13}C atoms. In favourable cases, it is possible to trace out the entire network of C–C bonds from such a two-dimensional spectrum. The experiment simply involves recording a two-dimensional double-quantum spectrum, just as was described in the previous section. However, instead of observing protons we observe ^{13}C.

The INADEQUATE experiment relies on two key ideas. The first is to note that the natural abundance of ^{13}C is rather low, about 1%. This means that the probability of any one carbon in a molecule being ^{13}C is 0.01, whereas the probability of there being two ^{13}C atoms in a molecule is $0.01 \times 0.01 = 0.0001 = 10^{-4}$. Experimentally, it is possible to detect spectra from molecules containing two ^{13}C atoms, but we can ignore molecules containing three such atoms as the probability of these occurring is simply too low.

The second point is that the one-bond carbon–carbon coupling constant is quite large, and covers a modest range (40–60 Hz); this coupling is also much larger than that for two- or three-bond couplings. It is therefore possible to generate double-quantum coherence between two *adjacent* ^{13}C atoms by setting the delay τ in the pulse sequence of Fig. 8.18 on page 204 to $1/(4\,^1J_{CC}) \approx 0.005$ s. The resulting double-quantum spectrum will contain only responses from *adjacent pairs* of ^{13}C atoms.

How these ideas enable us to trace out the carbon framework is best described using an example. Let us consider the simple molecule 2-butanol, shown in Fig. 8.20. There are three ways in which two ^{13}C atoms can appear in adjacent positions, and these three *isotopomers* (as they are called) are illustrated in the diagram.

The form of the INADEQUATE spectrum for 2-butanol is shown in Fig. 8.21 on the next page. In this diagram, the offsets of the ^{13}C atoms in the molecule are denoted Ω_1, Ω_2, Ω_3 and Ω_4, according to the numbering shown in Fig. 8.20. This spectrum can be understood by realizing that each of the three isotopomers **A**, **B** and **C** gives rise to a pattern of peaks of the form shown in Fig. 8.19 on the previous page.

The spectrum from isotopomer **A** thus shows, in the ω_2 dimension, two anti-phase doublets centred at Ω_1 and Ω_2, and in the ω_1 dimension

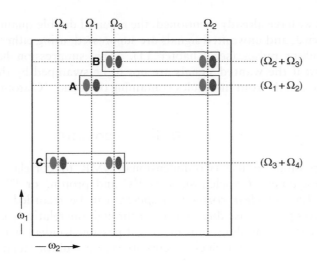

Fig. 8.21 Schematic double-quantum (INADEQUATE) spectrum of 2-butanol. The offsets of the carbons are denoted Ω_1, Ω_2 etc. according to the numbering shown in Fig. 8.20 on the facing page. Each of the three isotopomers shown in that figure gives rise to a pattern of peaks of the form shown in Fig. 8.19 on page 205. The peaks which belong to the same isotopomer can be identified as they all have the same ω_1 frequency; in addition, the ω_1 frequency must be the sum of the offsets of the two doublets in the ω_2 dimension. Using this approach, the peaks from each of the three isotopomers can be identified, and are indicated by the boxes. In the ω_2 dimension the peaks appear in anti-phase, which is indicated schematically by the blue and dark grey ovals.

a single peak at the double-quantum frequency $(\Omega_1 + \Omega_2)$. As was explained above, the fact that these peaks share the same double-quantum frequency indicates that spins one and two are coupled, which in the INADEQUATE experiment implies that they are directly bonded.

Isotopomer **B** gives anti-phase doublets at Ω_2 and Ω_3 in ω_2, and $(\Omega_2 + \Omega_3)$ in ω_1. This pattern of peaks implies that spins two and three are coupled, and hence that carbons two and three are directly bonded. Finally, isotopomer **C** gives doublets at Ω_3 and Ω_4 in ω_2, and $(\Omega_3 + \Omega_4)$ in ω_1, which implies that carbons 3 and 4 are directly bonded.

Thus, by looking for peaks which share a common ω_1 frequency, we can identify pairs of adjacent ^{13}C atoms, and hence trace out the C–C framework of the molecule. Note that we can be sure that this process identifies only directly bonded ^{13}C atoms as we have set the delay in the pulse sequence so that double-quantum coherence is generated only as a result of the evolution of the large one-bond coupling.

The technical problem with this experiment is that the signals due to molecules containing two ^{13}C atoms are around 100 times weaker than those from molecules containing one such atom. However, double-quantum coherence cannot be generated in molecules containing only one ^{13}C, so signals from such molecules do not contribute to the spectrum. Thus, by recording a double-quantum spectrum we are able to focus on the signals from the molecules containing two ^{13}C atoms, and reject all others.

As we have already mentioned, the required double-quantum coherence is selected, and unwanted signals are suppressed, using either phase cycling or gradient pulses (see Chapter 11). This suppression has to be very effective if the wanted signals are not to be swamped by the much more intense signals from molecules containing only one ^{13}C atom.

8.6   Heteronuclear correlation spectra

It is possible to use two-dimensional NMR to correlate the shifts of different types of nuclei, such as ^{13}C and proton, or ^{15}N and proton. These *heteronuclear correlation spectra* can be recorded in such a way that cross peaks arise due to transfer through the relatively large one-bond heteronuclear coupling, thus making it possible to identify the shifts of directly attached nuclei. Such spectra are very useful for tackling assignment problems.

It is also possible to arrange for the correlations to occur via smaller long-range couplings (typically over two or three bonds). The resulting spectra are more complicated than those from one-bond correlation experiments as there are likely to be many more couplings, and hence more cross peaks. Nevertheless, long-range correlation spectra have proved to be invaluable in tackling more difficult assignment problems.

In principle, any pair of heteronuclei can be correlated in a two-dimensional experiment, but by far the most popular combinations include protons as one of the nuclei. In the description which follows it is helpful to keep in mind that the I spin is likely to be a proton and the S spin a heteronucleus.

Before looking at particular experiments, we first need to consider which of the two nuclei we are going to observe, which is the topic of the next section.

8.6.1   Normal or inverse correlation

If we are going to use two-dimensional NMR to correlate protons and ^{13}C, then we have the choice of devising an experiment in which we observe *either* ^{13}C *or* proton. The question therefore arises as to which is the best choice.

Usually, achieving the highest sensitivity is the most important thing. In this regard, it turns out that observing the nucleus with the highest Larmor frequency gives the best sensitivity. So in the case of proton and ^{13}C, it is best to observe the protons.

However, there is a big problem when it comes to observing the protons, which derives from the fact that the natural abundance of ^{13}C is only 1%. So, only 1% of the molecules in the sample contain any ^{13}C atoms, and it is this small fraction which contributes signals to the two-dimensional correlation experiment. These wanted signals are all too easily overwhelmed by the much more intense contribution from the 99% of molecules containing no ^{13}C.

When we look at particular experiments we will see that there are ways of suppressing these unwanted signals. However, they must be suppressed very well if they are not to swamp the much weaker signals we are interested in. In the early days of two-dimensional NMR, achieving the required degree of suppression was, for technical reasons, beyond the capabilities of most spectrometers. However, the situation has now changed and it is possible to achieve excellent suppression on a routine basis.

If ^{13}C observation is used, then no such problems arise as molecules not containing ^{13}C simply do not contribute to the signal. So, the original heteronuclear correlation experiments all used ^{13}C detection, despite the fact that this gives lower sensitivity.

When they were first introduced, experiments using proton observation were termed *inverse* in order to distinguish them from the then 'normal' experiments which involved ^{13}C observation. The current state of play is that proton-observe experiments have become routine and practically the norm, so to describe them as 'inverse' is not entirely logical. However, this historic term is still widely used to describe heteronuclear experiments with proton observation.

8.7 HSQC

The HSQC experiment is widely used for recording one-bond correlation spectra between ^{13}C and proton, with the proton being the observed nucleus (i.e. it is an inverse experiment). It is also much used for correlating ^{15}N and proton in molecules of biological interest (peptides, proteins and nucleic acids) which can quite easily be enriched in ^{15}N.

HSQC: Heteronuclear Single-Quantum Correlation

The HMQC experiment, described in the next section, gives an essentially identical spectrum to HSQC. However, the way in which relaxation affects the two experiments is somewhat different. It is generally held that HSQC is the superior experiment for larger molecules, whereas for small to medium-sized molecules the two experiments give comparable results.

Two HSQC pulse sequences are shown in Fig. 8.22 on the following page; they are identical up until the end of period **D** and only differ in the details of how the I-spin signal is observed. Broadly speaking the sequence works by first transferring magnetization from the I spin to the S spin using the same method as in INEPT (section 7.10 on page 167). The S-spin magnetization then evolves for t_1, during which time it acquires a frequency label according to the offset of S. Finally this magnetization is transferred back to I, where it is observed. The resulting spectrum thus has peaks centred at the offset of the S spin in the ω_1 dimension, and at the offset of the I spin in the ω_2 dimension.

Let us start our analysis with equilibrium magnetization on the I spin, $\hat{I}_z$. This is made transverse by the first pulse, and there then follows a spin echo, period **A**, during which the coupling evolves, but the offset is refocused. Thus, at the end of this period we have:

$$\cos\left(2\pi J_{IS}\tau_1\right)\hat{I}_y - \sin\left(2\pi J_{IS}\tau_1\right)2\hat{I}_x\hat{S}_z.$$

The subsequent two 90° pulses, period **B**, transfer the anti-phase term to the S spin, and leave the in-phase term unaffected. We are only interested

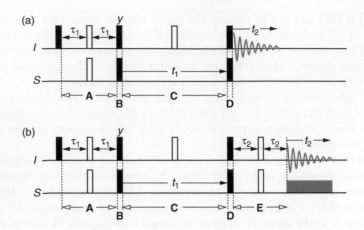

Fig. 8.22 Pulse sequences for heteronuclear correlation using the HSQC experiment. Both sequences start with equilibrium magnetization on the I spin, which is first transferred to the S spin using an INEPT-like sequence formed by periods **A** and **B** (see section 7.10 on page 167). The S-spin magnetization then evolves for t_1, with the centrally placed I spin 180° pulse refocusing the evolution of the coupling. Finally, this magnetization is transferred back to the I spin, where it is observed. In sequence (a), the signal is observed immediately. In sequence (b), a spin echo, period **E**, allows the anti-phase signals to become in-phase, so that they can then be observed using broadband S spin decoupling (indicated by the blue rectangle). The optimum value for both τ_1 and τ_2 is $1/(4J_{IS})$.

in the term which is transferred to S:

$$- \sin\left(2\pi J_{IS}\tau_1\right) 2\hat{I}_x\hat{S}_z \xrightarrow{(\pi/2)\hat{I}_y} \sin\left(2\pi J_{IS}\tau_1\right) 2\hat{I}_z\hat{S}_z$$

$$\xrightarrow{(\pi/2)\hat{S}_x} -\sin\left(2\pi J_{IS}\tau_1\right) 2\hat{I}_z\hat{S}_y.$$

Note that the pulse to the I spin must be about the y-axis for there to be any transfer. Periods **A** and **B** of the HSQC sequence are *identical* to those in the INEPT sequence, Fig. 7.15 on page 168, and so this initial part of the HSQC pulse sequence is often called an INEPT transfer.

Period **C** is the t_1 evolution, but note that the centrally placed 180° pulse to the I spin forms a spin echo, so that the evolution of the coupling during period **C** is refocused (see section 7.8.4 on page 164). Thus, the evolution during this period is the same as the offset evolving for t_1, followed by a 180° pulse to I.

$$- \sin\left(2\pi J_{IS}\tau_1\right) 2\hat{I}_z\hat{S}_y \xrightarrow{\Omega_S t_1 \hat{S}_z} -\cos\left(\Omega_S t_1\right) \sin\left(2\pi J_{IS}\tau_1\right) 2\hat{I}_z\hat{S}_y$$

$$+ \sin\left(\Omega_S t_1\right) \sin\left(2\pi J_{IS}\tau_1\right) 2\hat{I}_z\hat{S}_x.$$

$$\xrightarrow{\pi\hat{I}_x} \cos\left(\Omega_S t_1\right) \sin\left(2\pi J_{IS}\tau_1\right) 2\hat{I}_z\hat{S}_y - \sin\left(\Omega_S t_1\right) \sin\left(2\pi J_{IS}\tau_1\right) 2\hat{I}_z\hat{S}_x.$$

Next follows 90° pulses to both spins; these transfer the term $2\hat{I}_z\hat{S}_y$ to $-2\hat{I}_y\hat{S}_z$, which is anti-phase magnetization on the I spin. The term $2\hat{I}_z\hat{S}_x$ becomes $2\hat{I}_y\hat{S}_x$, which is unobservable multiple-quantum coherence. So, at the start of period **E** we have

$$- \cos\left(\Omega_S t_1\right) \sin\left(2\pi J_{IS}\tau_1\right) 2\hat{I}_y\hat{S}_z. \tag{8.9}$$

8.7.1 Coupled or decoupled acquisition

There are two alternatives at this point. The first is to observe the signal straight away, using pulse sequence (a) from Fig. 8.22 on the facing page. Looking at Eq. 8.9 on the preceding page, we can see that in the ω_2 dimension there is an anti-phase doublet centred at the shift of the I spin. There is a single modulating frequency in t_1, so that both components of the doublet appear at the offset of the S spin in the ω_1 dimension; a schematic spectrum is shown in Fig. 8.23 (a).

The peaks in this spectrum show the correlation between the offsets (shifts) of the two spins. Note, too, that the intensity of these peaks depends on the factor $\sin(2\pi J_{IS}\tau_1)$, which will be a maximum when $\tau_1 = 1/(4J_{IS})$. This is not surprising, as it is this value of the delay which gives complete conversion to anti-phase magnetization during period **A**.

Using sequence (a) results in a spectrum in which each correlation gives rise to two peaks, separated by J_{IS} in the ω_2 dimension. The spectrum can be simplified if we apply broadband decoupling of the S spin during acquisition, but recall from the discussion in section 7.10.3 on page 169 that we cannot simply apply decoupling after the final two 90° pulses as the anti-phase multiplet would collapse to zero. Rather we need to interpose another spin echo, period **E** of sequence (b), in order to allow the anti-phase terms to become in-phase.

As before, during this spin echo the coupling evolves, but the offset does not, so at the end of period **E** we have

$$-\cos(2\pi J_{IS}\tau_2)\cos(\Omega_S t_1)\sin(2\pi J_{IS}\tau_1)\,2\hat{I}_y\hat{S}_z$$
$$+\sin(2\pi J_{IS}\tau_2)\cos(\Omega_S t_1)\sin(2\pi J_{IS}\tau_1)\,\hat{I}_x.$$

If the signal is now observed while broadband decoupling is applied to the S spin, then just the term in $\hat{I}_x$ contributes. This results in a single peak at Ω_I in the ω_2 dimension, and Ω_S in ω_1, as shown in Fig. 8.23 (b). As before, the optimum value for τ_2 is $1/(4J_{IS})$. The overall result of sequence (b) is a very simple spectrum containing just one peak, whose coordinates allow us to read off the offsets (shifts) of the two coupled spins.

8.7.2 Suppressing unwanted signals in HSQC

We remarked at the beginning of this section that HSQC is usually an inverse experiment, with the proton being the observed nucleus (the I spin). If the S spin has low natural abundance, then we have to address the issue of how to suppress the intense signals which arise from protons which are *not* coupled to the heteronucleus. This can be achieved by using a difference experiment, an idea we have encountered before in the context of the INEPT experiment (section 7.10.4 on page 170).

Looking at the pulse sequences of Fig. 8.22 on the facing page, and the analysis we made of them, it can be seen that the first 90° pulse applied to the S spin only affects the I-spin magnetization which has become anti-phase with respect to the coupling. This pulse has *no effect* on in-phase I-spin magnetization, which will include all of the magnetization from nuclei which are *not* coupled to S.

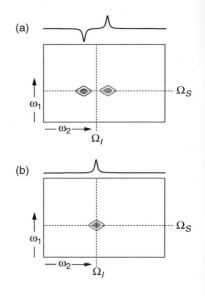

Fig. 8.23 Schematic HSQC spectra arising from pulse sequences (a) and (b) of Fig. 8.22 on the preceding page. In sequence (a), there is no spin echo just prior to acquisition, so that an anti-phase doublet is seen in the ω_2 dimension, as shown in spectrum (a). Sequence (b) uses broadband decoupling of the S spin during acquisition, so the I-spin doublet collapses to a single line, as shown in spectrum (b).

Thus, if we change the phase of this first S spin 90° pulse from x to $-x$, the sign of the wanted terms will be inverted, whereas the unwanted signals will be unaffected. So, all we need to do is repeat the experiment twice, once with the phase of the first S spin 90° pulse set to $+x$, and one with the phase set to $-x$. Subtracting the signals recorded from the two experiments will suppress the unwanted signals, and preserve the wanted ones.

The same effect can also be achieved by shifting the phase of the second I spin 90° pulse from $+y$ to $-y$, or of the second S spin 90° pulse from $+x$ to $-x$. Which of these is chosen in practice is largely a technical matter.

8.7.3   Sensitivity

If we step back and take a broad-brush look at the HSQC experiment we see that it involves three steps:

(a) Transfer of the equilibrium magnetization of the I spin to the S spin.

(b) Evolution of the S-spin magnetization for t_1.

(c) Transfer of the S-spin magnetization back to the I spin for observation.

It has already been explained that, if the I spin is proton, it is advantageous from the sensitivity point of view to make it the observed nucleus. However, the question arises as to why we need the first step, the transfer from I to S. After all, there is equilibrium magnetization on the S spin which could simply be excited and allowed to evolve for t_1.

In our discussion of the INEPT sequence (section 7.10 on page 167), it was explained that the equilibrium magnetization is larger for spins with higher Larmor frequencies. It is for this reason that we want to start with the equilibrium magnetization on proton, rather than on the heteronucleus. Ultimately, we will observe stronger signals, and therefore have higher sensitivity, by starting with the larger equilibrium magnetization.

There is one further advantage to starting with proton equilibrium magnetization. This is that, generally speaking, the proton magnetization returns to its equilibrium value somewhat more quickly than does the magnetization from heteronuclei, such as ^{13}C or ^{15}N. Between experiments, we need to allow sufficient time for the spins to come back to equilibrium; by starting with proton magnetization this time is minimized. So, we are able to repeat the experiment more quickly, and thus achieve greater signal-to-noise per unit time, by starting with proton magnetization.

8.8   HMQC

HMQC: Heteronuclear Multiple-Quantum Correlation

The pulse sequence for the HMQC experiment is shown in Fig. 8.24 (a) on the facing page. A detailed analysis of this sequence will show that it gives a spectrum identical to that for the decoupled HSQC experiment, Fig. 8.22 (b) on page 210. However the HMQC sequence works in rather a different way to HSQC.

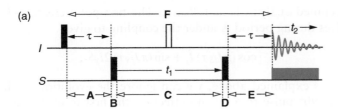

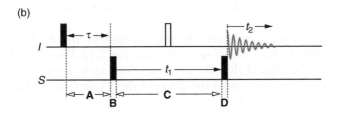

Fig. 8.24 Pulse sequence for: (a) the HMQC experiment; and (b) the HMBC experiment. In both experiments, equilibrium magnetization of the I spin is excited and allowed to become anti-phase during period **A**. It is then transferred to multiple-quantum coherence by the first S-spin pulse, period **B**. After evolution for t_1, the coherence is transferred back into anti-phase magnetization on the I spin by the second S-spin pulse, period **D**. In HMQC, the anti-phase magnetization evolves back into in-phase magnetization during period **E**. It is then observed under conditions of broadband S-spin decoupling. In HMBC, the signals are observed immediately after the coherence transfer step, and no broadband decoupling is used. The optimum value for τ is $1/(2J_{IS})$.

Broadly speaking the HMQC sequence consists of three steps:

(a) I-spin equilibrium magnetization is excited and then allowed to become anti-phase (period **A**).

(b) In period **B** this anti-phase magnetization is converted into heteronuclear multiple-quantum coherence, which then evolves for t_1 (period **C**).

(c) The multiple-quantum coherence is converted back into observable magnetization on the I spin, the coupling is allowed to rephase (period **E**), and then the signal is acquired under conditions of broadband S-spin decoupling.

A step-by-step analysis of this pulse sequence is rather involved, as both the offset and the coupling evolve during periods **A** and **E**. In addition, we need to cope with the multiple-quantum evolution during period **C**. However, things are simplified greatly by realizing that the 180° pulse which is placed in the centre of t_1 in fact creates a spin echo over the *whole* of period **F**, from the first pulse to the start of acquisition. The offset of the I spin is therefore refocused over the whole of this period, and so can be ignored in our calculations. You might be concerned that the pulses to the S spin will interfere with this refocusing, but it turns out that, because these pulses are disposed symmetrically about the 180° pulse, they do not cause a problem.

The detailed analysis is as follows. The first pulse creates $-\hat{I}_y$, which then evolves during period **A** under the coupling to give

$$- \cos\left(\pi J_{IS}\,\tau\right)\hat{I}_y + \sin\left(\pi J_{IS}\,\tau\right)2\hat{I}_x\hat{S}_z;$$

note that, as explained above, we can ignore the evolution of the I-spin offset. The 90° pulse to S has no effect on the first term, but rotates the second term into multiple quantum, to give $-\sin\left(\pi J_{IS}\,\tau\right)2\hat{I}_x\hat{S}_y$. We will ignore the first term as it does not give rise to any useful peaks in the spectrum.

The multiple-quantum term is not affected by the evolution of the coupling between the I and S spins (see section 7.12.3 on page 176), and we have already argued that the offset of the I spin is refocused, so that just leaves the evolution under the offset of the S spin to consider. This just affects the S spin operator, $\hat{S}_y$:

$$- \sin\left(\pi J_{IS}\,\tau\right)2\hat{I}_x\hat{S}_y \xrightarrow{\Omega_S t_1 \hat{S}_z}$$
$$- \cos\left(\Omega_S t_1\right)\sin\left(\pi J_{IS}\,\tau\right)2\hat{I}_x\hat{S}_y + \sin\left(\Omega_S t_1\right)\sin\left(\pi J_{IS}\,\tau\right)2\hat{I}_x\hat{S}_x.$$

Next comes a 90° pulse to S, period **D**. As this is about the x-axis, this pulse has no effect on the term $2\hat{I}_x\hat{S}_x$, but rotates $-2\hat{I}_x\hat{S}_y$ into $-2\hat{I}_x\hat{S}_z$, which is anti-phase magnetization on the I spin. So, the only observable term at the start of period **E** is

$$- \cos\left(\Omega_S t_1\right)\sin\left(\pi J_{IS}\,\tau\right)2\hat{I}_x\hat{S}_z.$$

During period **E** the coupling evolves, converting the anti-phase term to in-phase:

$$- \cos\left(\Omega_S t_1\right)\sin\left(\pi J_{IS}\,\tau\right)2\hat{I}_x\hat{S}_z \xrightarrow{2\pi J_{IS}\,\tau\hat{I}_z\hat{S}_z}$$
$$- \cos\left(\pi J_{IS}\,\tau\right)\cos\left(\Omega_S t_1\right)\sin\left(\pi J_{IS}\,\tau\right)2\hat{I}_x\hat{S}_z$$
$$- \sin\left(\pi J_{IS}\,\tau\right)\cos\left(\Omega_S t_1\right)\sin\left(\pi J_{IS}\,\tau\right)\hat{I}_y.$$

As we are using broadband decoupling of the S spin during acquisition, only the in-phase term is observable. Hence the sole observable term is

$$-\left[\sin\left(\pi J_{IS}\,\tau\right)\right]^2 \cos\left(\Omega_S t_1\right)\hat{I}_y.$$

Apart from a trivial phase shift from x to y, the result is identical to that for the decoupled HSQC experiment, and so the spectrum will be just as shown in Fig. 8.23 (b) on page 211.

The intensity of the peaks depends on the delay τ, for which the optimum value is $1/(2J_{IS})$, as this makes $\sin\left(\pi J_{IS}\,\tau\right) = 1$. As with HSQC, the optimum value for these fixed delays is that which leads to a complete interconversion of in- and anti-phase magnetization. Note that in HSQC the coupling evolves for $2\tau_1$ or $2\tau_2$, whereas in HMQC it evolves for τ; this is why the optimum value of τ_1 or τ_2 is $1/(4J_{IS})$, whereas that for τ is $1/(2J_{IS})$.

As with HSQC, we need to think about how we are going to suppress signals from protons (the I spin) which are not coupled to the heteronucleus

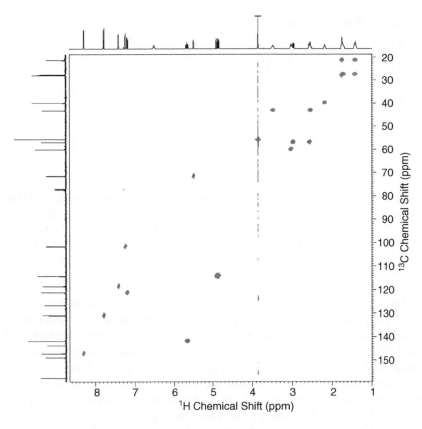

Fig. 8.25 ^{1}H–^{13}C HMQC spectrum of quinine recorded at 500 MHz for proton and with broadband ^{13}C decoupling during t_2; the conventional proton and ^{13}C spectra have been plotted along the relevant axes. Some carbons, such as the one at 57 ppm, have two inequivalent attached protons, and so show correlations to two different proton shifts.

(the S spin). Once again, we can use a difference experiment and, based on the earlier discussion, it is easy to spot that changing the phase of either of the S-spin pulses from $+x$ to $-x$ will change the sign of the wanted signals, but leave those from uncoupled spins unaffected. A simple difference experiment will therefore suppress these unwanted signals.

The HMQC experiment starts with equilibrium magnetization on the I spin so, if this is proton, all the sensitivity advantages that we described for HSQC will also apply to HMQC.

Figure 8.25 shows a ^{1}H-^{13}C HMQC spectrum of quinine, recorded using broadband ^{13}C decoupling during acquisition. The decoupled ^{13}C spectrum is plotted along the ω_1 axis, and tracing across at each carbon shift, we can find the shift of the attached proton; note that some of the carbons are quaternaries, and have no cross peak associated with them. The spectrum shows, in a very clean way, the connection between the ^{13}C and ^{1}H assignments.

8.9 Long-range correlation: HMBC

In both HMQC and HSQC, there are fixed delays whose durations need to be set according to the value of the coupling constant between the two nuclei which are being correlated. The values of one-bond ^{13}C–^{1}H or ^{15}N–^{1}H couplings cover quite a small range, so it is possible to find a value for these fixed delays which is a reasonable compromise for all pairs of nuclei.

HMBC: Heteronuclear Multiple-Bond Correlation

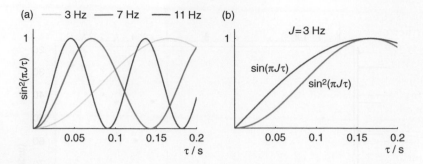

Fig. 8.26 Plot (a) shows the theoretical intensity of the correlations in an HMBC experiment as a function of the delay τ for three different values of the coupling constant. It is clear that there is no value of τ which will ensure good intensity for all three couplings. Plot (b) compares the functions $\sin^2(\pi J_{IS}\tau)$ and $\sin(\pi J_{IS}\tau)$ for $J_{IS} = 3$ Hz. It is clear that when the delay τ is much less than its optimum value, the $\sin^2$ function is significantly less than the sin function.

However, long-range coupling constants are much smaller than one-bond couplings, and also cover a much wider range. In order to see correlations through these smaller couplings it is necessary to lengthen the fixed delays considerably, but the presence of a wide range of couplings causes difficulties in choosing a suitable value for these delays.

It was shown above that, for the HMQC sequence, the intensity of the correlations goes as $\sin^2(\pi J_{IS}\tau)$. In Fig. 8.26 (a) this function is plotted against τ for J_{IS} of 3, 7 and 11 Hz. From these plots, we see that $\tau \approx 0.055$ s will give near to maximum intensity for the correlations through couplings of 7 and 11 Hz, but that the correlation through the 3 Hz couplings will be rather weak. Lengthening the delay to around 0.17 s increases the intensity of the latter correlation to near its maximum, but such a value of τ results in low intensity for the other two couplings. It is clear that there is no single value of τ which will give good intensity over such a wide range of couplings.

The only sure way around this problem is to record several spectra with different values of τ. If time does not permit this, then another approach is to set the value of τ according to the *largest* expected long-range coupling, and accept that much smaller couplings will lead to low intensity correlations.

There are two further problems with the HMQC experiment. The first arises from the fact that the intensity of the correlations goes as $\sin^2(\pi J_{IS}\tau)$. If τ is considerably less than its optimum value, then $\sin(\pi J_{IS}\tau)$ will be much less than one, and so its square will be very much less than one. This point is illustrated in Fig. 8.26 (b). The second problem is that relaxation during the long τ delays causes a loss of magnetization, and hence a reduction in the intensity of the correlations.

One way of minimizing these two problems is to use the modified HMQC sequence, usually known as HMBC, shown in Fig. 8.24 (b) on page 213. In HMBC the second τ delay is omitted, acquisition is started immediately after the final pulse, and broadband decoupling is not used. As a result of omitting the second delay τ, the intensity of the correlations goes as $\sin{(\pi J_{IS}\tau)}$, rather than the $[\sin{(\pi J_{IS}\tau)}]^2$ function found for HMQC. Furthermore, the losses due to relaxation are reduced as the total fixed delay is halved from 2τ to τ.

However, there is a price to pay for these two advantages. First, we cannot use decoupling during acquisition as the wanted operators are in anti-phase and so would collapse to zero. Secondly, although the centrally placed 180° pulse refocuses the evolution of the offset of the I spin during t_1, the evolution of the offset during τ is not refocused. The correlation peaks thus acquire a phase in ω_2 which depends on their offset and the value of τ. These phase distortions and the presence of the anti-phase terms simply have to be tolerated as a by-product of the improved sensitivity of HMBC. We return to this point in section 10.6 on page 347, where the effect of proton–proton couplings on the appearance of the HMBC multiplets is also considered.

The final point we need to make about the HMBC experiment is how relaxation affects the choice of the delay τ. In the absence of relaxation, the intensity of the correlations goes as $\sin{(\pi J_{IS}\tau)}$, but when relaxation is taken into account there will be a reduction in intensity which can be modelled by adding an exponential damping term:

$$\sin{(\pi J_{IS}\tau)}\exp{(-R\tau)}.$$

In this expression, faster relaxation corresponds to an increase in R.

Figure 8.27 illustrates the effect of this relaxation term. The plot shows the above function, with $J_{IS} = 3$ Hz, plotted against τ for different amounts of relaxation i.e. values of R.

If there is no relaxation (black line) the optimum value for τ is simply $1/(2J_{IS})$, or 0.17 s in this case. However, when relaxation is included, the maximum intensity is reached for a shorter value of τ. The faster the relaxation, the shorter the value of τ at which the maximum occurs and the lower the height of the maximum. What this implies is that to observe correlations through small couplings with the greatest intensity, one needs to use a value of τ which is significantly shorter than $1/(2J_{IS})$.

In principle, we can imagine modifying the HSQC experiment in a similar way we did for HMQC in order to optimize the observation of correlations through long-range couplings. However, in practice it is found that the resulting experiment has no particular advantages over HMBC.

Figure 8.28 on the following page shows part of the $^1\text{H}-^{13}\text{C}$ HMBC spectrum of quinine, recorded using the pulse sequence of Fig. 8.24 (b) on page 213 with a value of X ms for the delay τ. Some of the carbon atoms show correlations to more than one proton as a result of the large number of long-range C–H couplings present. The correlations to quaternary carbons are particularly useful, as such carbons do not appear in the HMQC spectrum.

Also present in this spectrum are some correlations through one-bond couplings; these features have been picked out in grey boxes. These correlations are easy to spot as the large one-bond coupling is present in the ω_2 dimension, resulting in two symmetrically placed features, of opposite sign, either side of the proton shift. Sometimes the presence of these peaks causes problems in that they overlap the wanted long-range

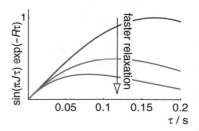

Fig. 8.27 In the presence of relaxation, the intensity of the correlations in an HMBC goes as $\sin{(\pi J_{IS}\tau)}\exp{(-R\tau)}$, where R is a relaxation time constant. This function is plotted against τ for $J_{IS} = 3$ Hz. The black line is for the case of no relaxation, and as expected the maximum occurs at $\tau = 1/(2J_{IS})$ or 0.17 s. The dark grey and light grey lines are for increasing rates of relaxation. As the rate of relaxation increases, the maximum moves to shorter values of τ and its height is reduced.

Fig. 8.28 Part of the ^1H–^{13}C HMBC spectrum of quinine recorded at 500 MHz for proton. The sequence of Fig. 8.24 (b) on page 213 has been used, with a delay τ of 33 ms (the optimum value for a 15 Hz long-range coupling). In this spectrum a number of the carbons, such as those at 143.5 ppm and 149 ppm, show several correlations due to long-range couplings. The grey boxes highlight correlations due to transfer through one-bond couplings. These features are distinctive as they are split by the large one-bond coupling in the proton dimension.

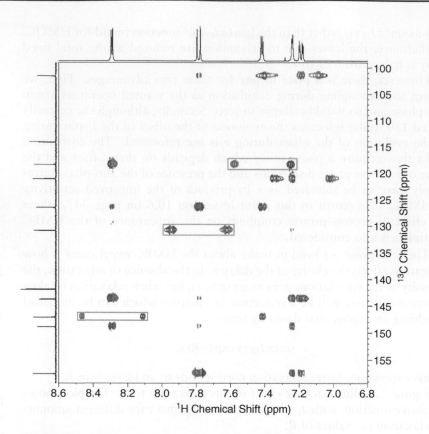

correlations. Fortunately, it is easy to suppress these one-bond correlations, as is described in the next section.

8.9.1   Suppressing one-bond peaks in HMBC spectra

In HMBC we make τ long enough that long-range I–S couplings will have gone anti-phase, and so give rise to cross peaks in the spectrum. However, in our sample, there are still I–S spin pairs which have a *one-bond* coupling between them, and these will also give rise to cross peaks whose intensity depends on $\sin(\pi J_{IS}\tau)$ in the usual way. This intensity will be a maximum when $\tau = 1/(2J_{IS})$ or at any *odd* multiple of this time i.e. $\tau = n/(2J_{IS})$, $n = 1, 3, 5 \ldots$. Depending on the exact values of τ and the one-bond coupling, it is quite possible that one-bond cross peaks will have significant intensity in an HMBC spectrum, as can be seen in Fig. 8.28.

If we know the value of the one-bond coupling $^1J_{IS}$, we can simply choose τ to be an *even* multiple of $1/(2\,^1J_{IS})$, as this makes $\sin(\pi\,^1J_{IS}\tau)$ go to zero. For example, if $^1J_{IS} = 160$ Hz, $1/(2\,^1J_{IS}) = 3.125$ ms, and so choosing $\tau = 20 \times 3.125 = 62.5$ ms will ensure that the one-bond cross peak has zero intensity.

The problem is that there is a range of values for the one-bond coupling. The function $\sin(\pi\,^1J_{IS}\tau)$ will have gone through so many oscillations by the time τ is large enough to generate long-range correlations, that even a small difference in the values of the one-bond couplings will result in the zero crossings of the corresponding sine curves getting out of step with

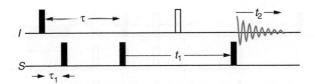

Fig. 8.29 Modified HMBC pulse sequence in which one-bond correlations are suppressed. The delay τ_1 is set to $1/(2 \, ^1J_{IS})$, and τ is the usual long delay needed for generating long-range correlations. One-bond coupled I–S pairs will give rise to anti-phase magnetization at the end of time τ_1, and this will be turned into multiple-quantum coherence by the first S-spin pulse. The sequence is repeated twice, with the phase of this pulse set to $+x$ and then to $-x$; as explained in the text, adding together the results of these two experiments cancels the one-bond correlations, but leaves the long-range correlations unaffected.

one another. There is therefore no way of choosing a single value of τ at which $\sin(\pi \, ^1J_{IS}\tau)$ will be close to zero for a range of values of the one-bond coupling.

Luckily, there is a simple modification to the HMBC pulse sequence which suppresses these one-bond correlations rather effectively; the modified pulse sequence is shown in Fig. 8.29. The idea is quite simple: after the initial I spin pulse, we leave a delay τ_1 which is set to $1/(2 \, ^1J_{IS})$, just as we would in an HMQC experiment. This means that at the end of this delay, the magnetization from an I spin which is one-bond coupled to an S spin will be anti-phase, and so the first S spin 90° pulse will transform this magnetization into multiple-quantum coherence.

In contrast, any magnetization from an I spin which is long-range coupled to an S spin will be unaffected by the first S-spin pulse, as the delay τ_1 will be insufficient for any anti-phase magnetization to develop. So, this I-spin magnetization continues to evolve through the rest of the pulse sequence, which, from this point on, is identical to HMBC.

We need to make sure that the heteronuclear multiple-quantum coherence generated by this extra S-spin pulse does not, as a result of the effect of some later pulse, contribute to the spectrum. This aim is achieved by repeating the experiment twice, once with the phase of this S-spin pulse set to $+x$, and once with it set to $-x$; the results from the two experiments are then added together. From the point of view of long-range coupled I–S pairs, this pulse has no effect, and so changing its phase is unimportant. However, for one-bond coupled pairs, this pulse creates multiple-quantum coherence, whose sign will be altered by altering the phase of the pulse. As a result, *adding* the signals recorded with this phase set to $+x$ and $-x$ will cancel any signals arising from this multiple-quantum coherence.

The extra delay τ_1 and associated S-spin pulse is often called a 'low-pass J filter' as only magnetization from spin pairs with low (small) values of the coupling passes through to the rest of the sequence.

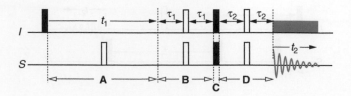

Fig. 8.30 Pulse sequence for the HETCOR experiment; note that, in contrast to HSQC and HMQC, the signal is observed on the S spin – the heteronucleus. Equilibrium magnetization of the I spin is excited by the first pulse, and then evolves under the influence of the offset of I for t_1. Period **B** is a spin echo, during which anti-phase magnetization develops. This magnetization is transferred to the S spin by the two 90° pulses which form period **C**. A further spin echo, period **D**, allows the anti-phase terms to become in-phase. They are then observed under conditions of broadband decoupling of the I spin. The optimum values for the delays τ_1 and τ_2 are both $1/(4J_{IS})$.

8.10   HETCOR

The last heteronuclear correlation experiment we will consider is the HET-COR experiment. In contrast to HSQC and HMQC, it is the heteronucleus (e.g. ^{13}C) which is observed in a HETCOR experiment. Generally speaking, this results in lower sensitivity than for the inverse experiments, which accounts for the popularity of the latter. Nevertheless, HETCOR is historically important in the development of two-dimensional NMR and remains in use to a significant extent.

The HETCOR pulse sequence is shown in Fig. 8.30. Broadly speaking the way the sequence works can be summarized as follows:

(a) Transverse magnetization of the I spin, excited by the first pulse, evolves for time t_1; the centrally placed 180° pulse refocuses the coupling.

(b) Anti-phase magnetization develops during the spin echo which constitutes period **B**, and then this magnetization is transferred to the S spin by the two 90° pulses of period **C**.

(c) The anti-phase terms evolve back into in-phase terms during the spin echo which forms period **D**. Finally, these in-phase terms are observed while broadband decoupling is applied to the I spin.

The detailed analysis proceeds as follows. We start with equilibrium magnetization of the I spin, which is rotated to $-\hat{I}_y$ by the first pulse. Period **A** is a spin echo in which the coupling is refocused, but the offset continues to evolve for the whole time t_1. So, at the end of t_1 we have

$$-\cos\left(\Omega_I t_1\right)\hat{I}_y + \sin\left(\Omega_I t_1\right)\hat{I}_x.$$

These in-phase terms cannot be transferred to the S spin. They need to be made anti-phase, which is the purpose of the spin echo which forms period **B**. During this echo the coupling evolves, but the offset is refocused. Following the usual procedure of replacing the echo by evolution of the coupling for $2\tau_1$ followed by $180°$ pulses to both spins, we find the following state at the end of period **B**:

$$\cos\left(2\pi J_{IS}\tau_1\right)\cos\left(\Omega_I t_1\right)\hat{I}_y - \sin\left(2\pi J_{IS}\tau_1\right)\cos\left(\Omega_I t_1\right)2\hat{I}_x\hat{S}_z$$
$$\cos\left(2\pi J_{IS}\tau_1\right)\sin\left(\Omega_I t_1\right)\hat{I}_x + \sin\left(2\pi J_{IS}2\tau_1\right)\sin\left(\Omega_I t_1\right)2\hat{I}_y\hat{S}_z.$$

Next comes $90°$ pulses to both spins. Of the four terms above, only the last leads to observable magnetization on the S spin, becoming

$$-\sin\left(2\pi J_{IS}\tau_1\right)\sin\left(\Omega_I t_1\right)2\hat{I}_z\hat{S}_y.$$

Finally we have the spin echo of period **D**, during which this anti-phase term goes in-phase. As we are going to observe the signal in the presence of broadband decoupling of the I spin, it is only the in-phase term present at the end of this period that is relevant. This term is

$$\sin\left(2\pi J_{IS}\tau_2\right)\sin\left(2\pi J_{IS}\tau_1\right)\sin\left(\Omega_I t_1\right)\hat{S}_x.$$

The resulting spectrum shows a single peak at Ω_I in the ω_1 dimension and at Ω_S in ω_2, as shown schematically in Fig. 8.31. This is the same as HSQC or HMQC, with the exception that the two dimensions are swapped around.

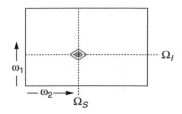

Fig. 8.31 Schematic HETCOR spectrum. As for HSQC, Fig. 8.23 (b) on page 211, there is a single peak whose coordinates give the offsets of the I and S spins. However, compared with HSQC, the two dimensions are swapped round.

8.11 TOCSY

TOCSY (**TO**tal **C**orrelation **S**pectroscop**Y**) is a homonuclear experiment, generally used for protons, which gives a spectrum in which a coupling between two spins is indicated by the presence of a cross-peak multiplet. To this extent, TOCSY is similar to COSY. However, in TOCSY we also see cross peaks between spins which are connected by an *unbroken chain of couplings*. So, for example, if spin A is coupled to spin B, and B is coupled to spin C, then in a TOCSY spectrum we will see a cross peak between A and C, even though there is no coupling between these two spins. This idea is illustrated in Fig. 8.32, which shows the TOCSY spectrum of the same spin system whose COSY spectrum is shown in Fig. 8.7 on page 191.

TOCSY is very useful for identifying the spins which belong to an extended network of couplings. In principle, such information is available by tracing out the sequence of cross peaks in a COSY, but in complex overlapping spectra it is not always possible to identify unambiguously such a series of related cross peaks. In TOCSY, a single cross-section taken at the shift of one spin should, in principle, show the multiplets of *all* of the spins which are part of the network of couplings to which this spin belongs. TOCSY also differs from COSY in one further important respect. As we shall show, both the diagonal- and cross-peak multiplets are in-phase and can be phased to pure absorption.

The pulse sequence for TOCSY is shown in Fig. 8.33 on the following page. The key part of the experiment is the period of *isotropic mixing*, for time τ_{mix}, which forms the mixing period in this sequence. In a two-spin system, such a period of isotropic mixing causes the following evolution of z-magnetization:

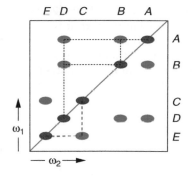

Fig. 8.32 Schematic TOCSY spectrum for the same spin system as for the COSY spectrum shown in Fig. 8.7 on page 191. The couplings present are A–B, B–D and C–E. Despite there being no coupling between A and D, the TOCSY spectrum contains a cross peak between these two spins as they are connected by an unbroken chain of couplings. A cross-section taken parallel to ω_1 at the shift of spin A in ω_2 shows multiplets from all of the spins which are in the same network of couplings as A.

$$\hat{I}_{1z} \longrightarrow \tfrac{1}{2}\left[1 + \cos\left(2\pi J_{12}\tau_{\text{mix}}\right)\right]\hat{I}_{1z} + \tfrac{1}{2}\left[1 - \cos\left(2\pi J_{12}\tau_{\text{mix}}\right)\right]\hat{I}_{2z}$$
$$-\sin\left(2\pi J_{12}\tau_{\text{mix}}\right)\tfrac{1}{2}\underbrace{\left(2\hat{I}_{1y}\hat{I}_{2x} - 2\hat{I}_{1x}\hat{I}_{2y}\right)}_{\text{ZQ}_y}. \tag{8.10}$$

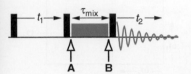

Fig. 8.33 Pulse sequence for the TOCSY experiment. The heart of the sequence is the period of isotropic mixing, indicated by the blue rectangle, which transfers magnetization between spins which are connected via an unbroken network of couplings. In practice, isotropic mixing is achieved by the use of a specially designed multiple-pulse sequence, such as DIPSI-2. As explained in the text, it is arranged that only z-magnetization present at points **A** and **B** contributes to the spectrum.

The important thing here is that z-magnetization on spin one is transferred to z-magnetization on spin two at a rate which depends on the coupling and the mixing time, τ_{mix}. It is this transfer which gives rise to cross peaks in the spectrum.

In addition to transfer to $\hat{I}_{2z}$, isotropic mixing generates zero-quantum coherence, specifically the term ZQ$_y$. This is not of any particular interest and, as we will see later, its presence causes phase distortions in the spectrum.

In practice, isotropic mixing is achieved by using specially designed pulse sequences, such as DIPSI-2. Such sequences involve applying a carefully crafted set of pulses of various phases and flip angles in a repetitive sequence. The way in which these sequences are designed is outside the scope of this book.

The reason why this kind of mixing is called 'isotropic' is that it transfers not only z-, but also x- and y-magnetization between spins in an essentially identical way. So, for example, the evolution of x-magnetization under isotropic mixing can be found from Eq. 8.10 by making the cyclic permutation of z to x, x to y, and y to z:

$$\hat{I}_{1x} \longrightarrow \tfrac{1}{2}\left[1 + \cos\left(2\pi J_{12}\tau_{\text{mix}}\right)\right]\hat{I}_{1x} + \tfrac{1}{2}\left[1 - \cos\left(2\pi J_{12}\tau_{\text{mix}}\right)\right]\hat{I}_{2x}$$
$$-\sin\left(2\pi J_{12}\tau_{\text{mix}}\right)\tfrac{1}{2}\left(2\hat{I}_{1z}\hat{I}_{2y} - 2\hat{I}_{1y}\hat{I}_{2z}\right).$$

We will see in due course that it is generally only desirable to transfer one component of the magnetization between spins, and here we choose this to be the z-component.

8.11.1   TOCSY for two spins

We will now analyse the pulse sequence of Fig. 8.33 in detail so that we can determine the expected form of the spectrum. In the pulse sequence, we have to arrange things so that *only* z-magnetization present at points **A** and **B**, just before and just after the period of isotropic mixing, contributes to the spectrum. Why this is necessary is easier to understand once the analysis has been completed.

The required z-magnetization can be selected at these points by using one of the coherence selection methods which will be described in Chapter 11. However, it turns out that separating the z-magnetization from the zero-quantum coherence is quite a difficult, but not impossible, task. Again, we defer discussion of this to section 11.15 on page 426.

In the sequence, the state of the system at point **A** is just the same as after the COSY sequence ($90° - t_1 - 90°$), so we can reuse the results of our earlier calculation. Of the four terms given on page 191, only term [1] contains the required z-magnetization:

$$-\cos\left(\pi J_{12}t_1\right)\cos\left(\Omega_1 t_1\right)\hat{I}_{1z}. \qquad [1]$$

As expected, the size of this magnetization reflects the evolution of the offset and coupling during t_1 i.e. the magnetization is modulated by these parameters.

The effect of isotropic mixing on $\hat{I}_{1z}$ is given by Eq. 8.10 on the facing page. However, we are only interested in the terms in $\hat{I}_{1z}$ and $\hat{I}_{2z}$ produced at point **B**, which are

$$-A_{1\to1}\cos(\pi J_{12}t_1)\cos(\Omega_1 t_1)\,\hat{I}_{1z} - A_{1\to2}\cos(\pi J_{12}t_1)\cos(\Omega_1 t_1)\,\hat{I}_{2z}.$$

For brevity we have introduced the transfer functions $A_{1\to1}$ and $A_{1\to2}$, defined as

$$A_{1\to1} = \tfrac{1}{2}\left[1 + \cos(2\pi J_{12}\tau_{\mathrm{mix}})\right] \qquad A_{1\to2} = \tfrac{1}{2}\left[1 - \cos(2\pi J_{12}\tau_{\mathrm{mix}})\right].$$

The final pulse rotates both $\hat{I}_{1z}$ and $\hat{I}_{2z}$ into in-phase observable magnetization:

$$A_{1\to1}\cos(\pi J_{12}t_1)\cos(\Omega_1 t_1)\,\hat{I}_{1y} + A_{1\to2}\cos(\pi J_{12}t_1)\cos(\Omega_1 t_1)\,\hat{I}_{2y}. \qquad (8.11)$$

The first term is modulated at the offset of spin one in t_1 and appears on spin one during t_2: it therefore gives rise to the diagonal-peak multiplet. The second term has the same modulation in t_1, but appears on spin two during t_2: this gives the cross-peak multiplet.

The overall intensity of the cross-peak multiplet depends on $A_{1\to2}$. From its definition, we can see that this term is at a maximum when

$$\cos(2\pi J_{12}\tau_{\mathrm{mix}}) = -1,$$

which, since $\cos\pi = -1$, occurs when

$$2\pi J_{12}\tau_{\mathrm{mix}} = \pi \qquad \text{i.e. } \tau_{\mathrm{mix}} = 1/(2J_{12}).$$

In fact, with this optimum value of τ_{mix}, the diagonal peak intensity $(A_{1\to1})$ goes to zero.

We can determine the detailed form of the cross- and diagonal-peak multiplets by analysing the modulation using the same approach as in section 8.3.2 on page 192. What we will find is that the diagonal- and cross-peak multiplets are in-phase in both dimensions, and can be processed in such a way that *all* the peaks have the double absorption lineshapes.

The fact that the cross-peak multiplets are in-phase is a substantial difference to COSY, where the multiplets are anti-phase. As was commented on in section 8.3.4 on page 198, the cancellation caused by this anti-phase structure reduces the overall intensity of the cross peak, and so places a lower limit on the size of the coupling which can be detected by COSY.

On the face of it, the absence of such cancellation in TOCSY cross peaks means that we might expect to be able to detect smaller couplings using TOCSY than we can using COSY. However, two extra factors must be taken into account. First, the overall intensity of the TOCSY cross peaks depends on the transfer function $A_{1\to2}$; secondly, relaxation will take place during the period of isotropic mixing, leading to loss of signal intensity.

If we are trying to detect a correlation through a small coupling, we will need to use a long period of mixing. The losses due to relaxation will thus be more severe than when looking for correlations through larger couplings. Whether or not the overall intensity of the COSY or TOCSY cross peaks is greater depends intimately on the choice of the experimental parameters and on the molecule being studied. However, the general experience is that for small to medium-sized molecules, COSY is the preferred experiment for detecting small couplings.

Before moving on to consider the TOCSY spectra of more extended spin systems, it is worthwhile looking at what would happen to the zero-quantum terms which are created by the isotropic mixing. At point **B** in the sequence, these terms will be

$$\sin\left(2\pi J_{12}\tau_{\mathrm{mix}}\right)\cos\left(\pi J_{12}t_1\right)\cos\left(\Omega_1 t_1\right)\tfrac{1}{2}\left(2\hat{I}_{1y}\hat{I}_{2x} - 2\hat{I}_{1x}\hat{I}_{2y}\right).$$

The final 90° pulse results in two anti-phase terms:

$$\sin\left(2\pi J_{12}\tau_{\mathrm{mix}}\right)\cos\left(\pi J_{12}t_1\right)\cos\left(\Omega_1 t_1\right)\tfrac{1}{2}\left(2\hat{I}_{1z}\hat{I}_{2x} - 2\hat{I}_{1x}\hat{I}_{2z}\right).$$

What we have, therefore, is an anti-phase contribution to both the cross- and diagonal-peak multiplets. Furthermore, this contribution is along x, whereas the in-phase term is along y, so the anti-phase contributions will be in dispersion (assuming the in-phase contribution is in absorption), resulting in a phase distorted spectrum with a mixed lineshape. It is to avoid this undesirable outcome that it is necessary to suppress the contributions from zero-quantum coherence (see section 11.15 on page 426).

It was mentioned above that the isotropic mixing sequence affects x-, y- and z-magnetization in essentially the same way. In the pulse sequence, we chose to allow only z-magnetization at point **A** to contribute to the spectrum, and it has been shown that this results in a spectrum with in-phase absorption multiplets. It is interesting to consider what would happen to x-magnetization present at this point.

From page 191, the x-magnetization is given by term [3]

$$\cos\left(\pi J_{12}t_1\right)\sin\left(\Omega_1 t_1\right)\hat{I}_{1x}. \qquad [3]$$

After the period of isotropic mixing this term goes to

$$A_{1\to 1}\cos\left(\pi J_{12}t_1\right)\sin\left(\Omega_1 t_1\right)\hat{I}_{1x} + A_{1\to 2}\cos\left(\pi J_{12}t_1\right)\sin\left(\Omega_1 t_1\right)\hat{I}_{2x}. \qquad (8.12)$$

Both of these terms are unaffected by the final pulse (which is about x), and so contribute directly to the spectrum.

If we determine the detailed form of the multiplets arising from these terms (using the approach of section 8.3.2 on page 192), we find that as before both the cross- and diagonal-peak multiplets are in phase. However, the problem is that the lineshapes of the peaks from the x-magnetization are 90° out of phase with those from the z-magnetization.

We can see this by comparing Eqs 8.11 and 8.12. The terms arising from z-magnetization appear along the y-axis in t_2, and are modulated in t_1 as $\cos\left(\Omega_1 t_1\right)$. In contrast, the terms arising from x-magnetization appear along the x-axis and have t_1 modulation which goes as $\sin\left(\Omega_1 t_1\right)$. There is thus a 90° shift in each dimension. So, if the peaks arising from z-magnetization

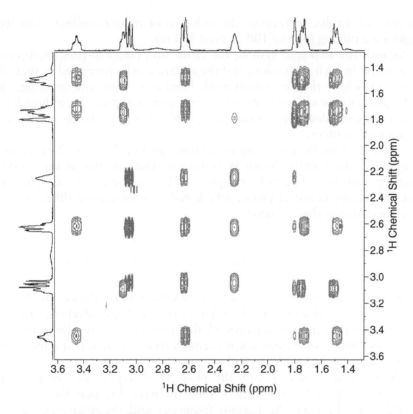

Fig. 8.34 Part of the TOCSY spectrum of quinine, recorded at 500 MHz and using a mixing time of 20 ms. Zero-quantum contributions present before and after the mixing time have been suppressed using the method described in section 11.15 on page 426.

are phased to double absorption, those arising from x-magnetization will give peaks in double dispersion.

Therefore, if we are to retain absorption mode lineshapes, we must restrict the transfer to either the z-magnetization or the x-magnetization. It has been shown that the relaxation losses during the mixing time are somewhat less for the mixing of z-magnetization, which is why we chose this in the first place.

8.11.2   TOCSY for more extended spin systems

Our analysis of TOCSY for two spins does not reveal what is probably the most interesting feature of the experiment which is the appearance of cross peaks between spins which are not directly coupled. Unfortunately, for more than two coupled spins, the evolution under isotropic mixing is rather complicated and cannot be expressed in such a simple form as Eq. 8.10 on page 222. It is, however, possible to make numerical calculations (e.g. with a computer) of the behaviour of particular spin systems.

Generally speaking it is found that cross peaks due to direct couplings build up to their maximum intensity in a time which is of the order of $1/(2J)$. Peaks which arise from transfer through two successive couplings take longer to build up, and if three successive couplings are involved, even longer mixing times are needed.

If we are only interested in seeing cross peaks between directly coupled spins, then a relatively short mixing time is used – say of the order of $1/(2J)$ for the largest expected coupling. However, if we are interested in seeing the cross peaks due to transfer through two or more couplings, then we might use a mixing time of 100 or even 200 ms.

As with the two-spin system, the cross- and diagonal-peak multiplets are in phase in each dimension, and the spectra can be processed so that all of the peaks have the absorption mode lineshape. Generally speaking, all of the peaks in the spectrum have the same sign (e.g. positive), although in some more extended spin systems it is possible for peaks to be negative at certain mixing times.

Figure 8.34 on the preceding page shows part of the TOCSY spectrum of quinine; the region shown is similar to that plotted in the COSY spectrum shown in Fig. 8.17 on page 202. In contrast to COSY, the TOCSY multiplets are in phase, which makes a substantial difference to the appearance of the spectrum.

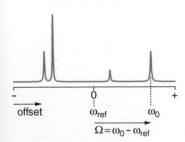

Optional section ⇒

Fig. 8.35 In a one-dimensional spectrum the receiver reference frequency ω_{ref} is usually placed somewhere in the middle of the spectrum. As was explained in section 4.6 on page 60, a peak gives rise to a detected signal oscillating not at the Larmor frequency, ω_0, but at the offset frequency Ω, which is the difference between the Larmor frequency and the receiver reference frequency: $\Omega = \omega_0 - \omega_{\text{ref}}$. As a result, there are peaks with both positive and negative offsets present in the spectrum.

8.12   Frequency discrimination and lineshapes

We now need to tackle the somewhat awkward and technical matter of frequency discrimination and its relation to lineshape selection. Up to now, we have been skirting around this problem in our discussion of two-dimensional NMR, but when it comes to any practical spectroscopy it is an issue which must be considered.

The problem arises because the offset Ω of a peak can be positive or negative. Recall from section 4.4.2 on page 55, that the offset is the difference between the Larmor frequency and the receiver reference frequency. In a multi-line spectrum, it is usual to set the receiver reference frequency to be somewhere in the middle of the spectrum, so there will be peaks with both positive and negative offsets, as shown in Fig. 8.35.

In one-dimensional NMR this does not represent a problem as we detect both the x- and y-components of the magnetization, and use them to construct a complex time-domain function of the form

$$\exp\left(i\Omega t\right)\exp\left(-Rt\right).$$

Fourier transformation of this signal gives, in the real part of the spectrum, an absorption mode peak at frequency Ω.

In the spectrum arising from Fourier transformation of a complex time-domain function, positive and negative frequencies are clearly separated. The spectrum is usually plotted with the scale running from negative frequencies, through zero, to positive frequencies. A peak at frequency $+\Omega$ will appear in a different place to a peak at frequency $-\Omega$. Such a spectrum is said to be *frequency discriminated*. It is clear that this frequency discrimination arises from the fact that $\exp\left(+i\Omega t\right)$ and $\exp\left(-i\Omega t\right)$ are different functions, i.e. the time-domain data are sensitive to the sign of the offset.

As we have seen, in two-dimensional NMR the modulation in the t_1 dimension is typically either of the form $\cos\left(\Omega t_1\right)$ or $\sin\left(\Omega t_1\right)$. Let us start out by considering the cosine modulated signal. Due to the property of a cosine that $\cos\left(+\theta\right) \equiv \cos\left(-\theta\right)$, it follows that

$$\cos\left(-\Omega t_1\right) \equiv \cos\left(+\Omega t_1\right).$$

As a result, a modulation at $+\Omega$ is indistinguishable from one at $-\Omega$.

In a spectrum containing peaks with both positive and negative offsets, the inability to discriminate the sign of the offset will lead to confusion. Each peak will appear twice, and we have no way of knowing whether a peak at 100 Hz is from a spin with an offset of +100 Hz or from one with offset of −100 Hz.

If the data are sine modulated, the situation is a little different. Recalling that $\sin(-\theta) \equiv -\sin(\theta)$, it follows that

$$\sin(-\Omega t_1) \equiv -\sin(\Omega t_1).$$

What this means is that a peak at $-\Omega$ is indistinguishable from a peak at $+\Omega$ but with negative intensity. In a spectrum with peaks at both positive and negative offsets the result will be very confusing. Not only could a negative peak overlap with, and so cancel out, a positive peak, but additional confusion could arise in spectra which genuinely have both positive and negative peaks.

Overall, we can see that two-dimensional experiments which have cosine or sine modulation with respect to t_1 are going to lead to problems in spectra containing peaks with positive and negative offsets. Modulation as cosine or sine in t_1 lead to spectra which are said to lack *frequency discrimination* in the ω_1 dimension.

However, all is not lost as there are straightforward ways of remedying this problem. These methods all rely on the ability to generate *both* sine and cosine modulated data; how this is done is the topic of the next section.

8.12.1   Obtaining cosine and sine modulated data

We saw in section 8.7 on page 209 that in the HSQC experiment the modulation in t_1 was of the form $\cos(\Omega t_1)$. If we repeat the analysis of the pulse sequence with the phase of the first 90° pulse applied to the S spin changed from x to y we will find that the modulation changes from $\cos(\Omega t_1)$ to $\sin(\Omega t_1)$.

In fact, in any of the experiments we have described so far, we can always change the t_1 modulation from cosine to sine (or vice versa) by shifting the phase of a suitably chosen pulse or pulses. Generally a 90° phase shift of *all* of the pulses which precede t_1 usually has the desired effect. In heteronuclear experiments, it is usually only necessary to shift the phases of the pulses on one of the spins. For example, in HMQC all we need to do is change the phase of the first 90° pulse to the S spin.

The overall result is that we will be able to record two data sets, one with cosine and one with sine modulation, simply by changing the phase of one or more pulses. The simplest form of these two data sets is:

$$S_c(t_1, t_2) = \cos(\Omega_A t_1) \exp(-R^{(1)} t_1) \exp(i\Omega_B t_2) \exp(-R^{(2)} t_2)$$
$$S_s(t_1, t_2) = \sin(\Omega_A t_1) \exp(-R^{(1)} t_1) \exp(i\Omega_B t_2) \exp(-R^{(2)} t_2).$$
$$(8.13)$$

From now on, we will work with these prototype data sets. It is the availability of these two data sets which makes it possible to achieve frequency discrimination.

8.12.2   P- and N-type selection: phase-twist lineshapes

The simplest way to achieve frequency discrimination is to make the t_1 modulation of the form $\exp(i\Omega t_1)$ i.e. of the same form as it is in t_2. As was explained above, this kind of modulation is sensitive to the sign of Ω, and so frequency discrimination is achieved.

Noting that $\exp(i\theta) \equiv \cos\theta + i\sin\theta$, we form the new time-domain function $S_P(t_1, t_2)$ from the combination $S_c(t_1, t_2) + i\,S_s(t_1, t_2)$:

$$
\begin{aligned}
S_P(t_1, t_2) &= S_c(t_1, t_2) + i\,S_s(t_1, t_2) \\
&= [\cos(\Omega_A t_1) + i\sin(\Omega_A t_1)]\exp(-R^{(1)}t_1)\exp(i\Omega_B t_2)\exp(-R^{(2)}t_2) \\
&= \exp(i\Omega_A t_1)\exp(-R^{(1)}t_1)\exp(i\Omega_B t_2)\exp(-R^{(2)}t_2).
\end{aligned}
$$

Thus, simply by combining the cosine and sine modulated data in this way we can generate modulation with respect to t_1 of the form $\exp(i\Omega t_1)$, which is sensitive to the sign of the offset. The resulting spectrum will be frequency discriminated.

We now need to work out the detailed form of the two-dimensional spectrum which arises from this time-domain signal. The first step is a Fourier transform with respect to t_2, to give the function $S_P(t_1, \omega_2)$. As we saw in section 8.2 on page 187, this gives an absorption mode line in the real part and the corresponding dispersion mode line in the imaginary part:

$$
S_P(t_1, \omega_2) = \exp(i\Omega_A t_1)\exp(-R^{(1)}t_1)[A_2(\Omega_B) + iD_2(\Omega_B)].
$$

As before, $A_2(\Omega)$ is an absorption mode peak at frequency Ω in the ω_2 dimension, and $D_2(\Omega)$ is the corresponding dispersion mode peak.

Next we compute the Fourier transform with respect to t_1. In contrast to the approach we took earlier in this chapter, we need to use the regular Fourier transform, as opposed to the cosine or sine transform, as the data are of the form $\exp(i\Omega t_1)$. As in the ω_2 dimension, we obtain an absorption mode line in the real part and a dispersion mode line in the imaginary part. The resulting spectrum, $S_P(\omega_1, \omega_2)$, is

$$
S_P(\omega_1, \omega_2) = [A_1(\Omega_A) + iD_1(\Omega_A)][A_2(\Omega_B) + iD_2(\Omega_B)],
$$

where $A_1(\Omega)$ is an absorption mode peak at frequency Ω in the ω_1 dimension, and $D_1(\Omega)$ is the corresponding dispersion mode peak.

Multiplying out the bracket and separating the result into real and imaginary parts gives

$$
S_P(\omega_1, \omega_2) = \underbrace{[A_1(\Omega_A)A_2(\Omega_B) - D_1(\Omega_A)D_2(\Omega_B)]}_{\text{real}}
$$
$$
+ i\underbrace{[A_1(\Omega_A)D_2(\Omega_B) + D_1(\Omega_A)A_2(\Omega_B)]}_{\text{imaginary}}.
$$

The real part of the spectrum, contained in the first square braces, consists of a double-absorption line centred at $\{\omega_1, \omega_2\} = \{\Omega_A, \Omega_B\}$ *and* a double dispersion line at the same frequency. This combination is called the *phase-twist* lineshape, and is illustrated in Fig. 8.36 on the facing page.

As can readily be appreciated from the diagram, this lineshape is not really suitable for high-resolution work on account of it having both positive

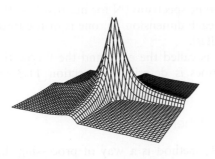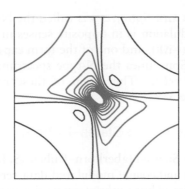

Fig. 8.36 Two views of the phase-twist lineshape: on the left is shown a perspective view, and on the right is shown a contour plot (positive contours are blue, negative contours are dark grey). This lineshape is the sum of a double absorption line (Fig. 8.6 on page 189) and a double dispersion line (Fig. 8.12 on page 197). The central part of the lineshape is dominated by the double absorption mode line, but as we move further away from the centre the absorption mode tails away leaving the broader double dispersion lineshape dominant. A cross-section taken parallel to either axis and through the centre of the peak shows the absorption mode lineshape, but as we move away from the centre of the peak, the lineshape becomes a mixture of absorption and dispersion. This phase-twist lineshape is very unsuitable for high-resolution work.

and negative parts, and also as a result of the broadness of the dispersion mode contribution. The phase-twist lineshape is an *inevitable* consequence of Fourier transforming a data set such as $S_P(t_1, t_2)$ in which the modulation is of the form $\exp(i\Omega t)$ in each dimension. Thus, although we have achieved the desired frequency discrimination, an unwanted by-product has been the appearance of the phase-twist lineshape.

A data set which is modulated as $\exp(i\Omega t_1)$ in the t_1 dimension is said to be *phase modulated* on account of the fact that as t_1 increases it is the phase of the observed signal which changes. Double Fourier transformation of such a phase modulated data set gives a frequency discriminated spectrum, but inevitably yields the phase-twist lineshape.

When the cosine and sine modulated data sets were combined, we chose the combination $S_c(t_1, t_2) + i S_s(t_1, t_2)$. However, we could just as well have made the combination $S_c(t_1, t_2) - i S_s(t_1, t_2)$:

$$
\begin{aligned}
S_N(t_1, t_2) &= S_c(t_1, t_2) - iS_s(t_1, t_2) \\
&= \exp(-i\Omega_A t_1)\exp(-R^{(1)} t_1)\exp(i\Omega_B t_2)\exp(-R^{(2)} t_2).
\end{aligned}
$$

Following through the same argument as before, the resulting spectrum is

$$
\begin{aligned}
S_N(\omega_1, \omega_2) = &[A_1(-\Omega_A)A_2(\Omega_B) - D_1(-\Omega_A)D_2(\Omega_B)] \\
&+ i[A_1(-\Omega_A)D_2(\Omega_B) + D_1(-\Omega_A)A_2(\Omega_B)].
\end{aligned}
$$

This is just the same as before, except that the sign of the frequency of the peak in the ω_1 dimension has changed. The difference is essentially a trivial one.

$S_P(\omega_1, \omega_2)$ is called the P-type spectrum (P for positive), as the modulation is in the same sense in each dimension i.e. both are of the form

$\exp(i\Omega t)$. $S_N(\omega_1, \omega_2)$ is called the N-type spectrum (N for negative), as the modulation is in opposite senses in each dimension i.e. one is of the form $\exp(-i\Omega t)$, and one of the form $\exp(i\Omega t)$.

Sometimes the N-type spectrum is called the *echo* and the P-type the *anti-echo*. The origin of these names is described in section 11.3 on page 389.

8.12.3   The States–Haberkorn–Ruben method

The States–Haberkorn–Ruben (SHR) method is a way of processing the sine and cosine modulated data sets such that frequency discrimination is achieved but *without* giving rise to the unfavourable phase-twist lineshape.

We start by Fourier transforming the cosine modulated data set with respect to t_2, resulting in the usual absorption and dispersion lineshapes:

$$S_c(t_1, \omega_2) = \cos(\Omega_A t_1) \exp(-R^{(1)} t_1) [A_2(\Omega_B) + i D_2(\Omega_B)].$$

We then take the real part of the signal:

$$
\begin{aligned}
S_{c,R}(t_1, \omega_2) &= \mathrm{Re}[S_c(t_1, \omega_2)] \\
&= \cos(\Omega_A t_1) \exp(-R^{(1)} t_1) A_2(\Omega_B).
\end{aligned}
$$

The same process is repeated for the sine modulated data set:

$$S_s(t_1, \omega_2) = \sin(\Omega_A t_1) \exp(-R^{(1)} t_1) [A_2(\Omega_B) + i D_2(\Omega_B)].$$

$$S_{s,R}(t_1, \omega_2) = \sin(\Omega_A t_1) \exp(-R^{(1)} t_1) A_2(\Omega_B).$$

Now we use $S_{c,R}(t_1, \omega_2)$ to form the real part, and $S_{s,R}(t_1, \omega_2)$ to form the imaginary part, of a new data set $S_{SHR}(t_1, \omega_2)$:

$$
\begin{aligned}
S_{SHR}(t_1, \omega_2) &= S_{c,R}(t_1, \omega_2) + i S_{s,R}(t_1, \omega_2) \\
&= [\cos(\Omega_A t_1) + i \sin(\Omega_A t_1)] \exp(-R^{(1)} t_1) A_2(\Omega_B) \\
&= \exp(i\Omega_A t_1) \exp(-R^{(1)} t_1) A_2(\Omega_B).
\end{aligned}
$$

The modulation in t_1 is of the form $\exp(i\Omega t_1)$, so frequency discrimination in the ω_1 dimension has been achieved.

In the last step we compute the usual Fourier transform with respect to t_1 to give the final spectrum:

$$
\begin{aligned}
S_{SHR}(\omega_1, \omega_2) &= [A_1(\Omega_A) + i D_1(\Omega_A)] A_2(\Omega_B) \\
&= A_1(\Omega_A) A_2(\Omega_B) + i D_1(\Omega_A) A_2(\Omega_B).
\end{aligned}
$$

The real part of $S_{SHR}(\omega_1, \omega_2)$ contains the required double absorption lineshape, $A_1(\Omega_A) A_2(\Omega_B)$. Therefore overall the SHR method achieves frequency discrimination *without* introducing the unwanted phase-twist lineshape.

The data processing for the SHR method is slightly more complex than for generating the N- or P-type spectrum, but the software on modern spectrometers offers such processing as a standard option. Note, however, that we need to record and store away separately a cosine and a sine modulated data set for each t_1 increment.

It is probably worth noting at this point that the SHR method does *not* get round the problem in COSY that the cross- and diagonal-peaks have different lineshapes. This is a fundamental property of the experiment, and not a function of the way the data are processed.

8.12.4   The TPPI or Redfield method

Suppose we *knew* that the offsets of all the peaks are positive, then we would not need to worry about frequency discrimination as there would be no ambiguity about the sign of the offset of a particular peak. Since we are at liberty to place the receiver reference frequency where we like, we can ensure that all of the offsets are positive by placing the receiver reference frequency just to the side of the peaks, as shown in Fig. 8.37.

Appealing though this simple method is, there are two good reasons *not* to use it. The first is that half the spectrum – the part where peaks with negative offsets normally appear – will be empty; this is a waste of data space. The second point is that it is usual to make the transmitter frequency the same as the receiver reference frequency, so the transmitter would end up being placed to the side of all of the resonances. This is unfavourable as it increases the maximum offset present in the spectrum, making off-resonance effects more likely.

The TPPI or Redfield method is a neat trick which enables us to leave the receiver reference frequency (and hence the transmitter frequency) in the middle of the spectrum, but make it *look* as if all of the offsets in the ω_1 dimension are positive. The method involves incrementing the phase of one of the pulses in the sequence in concert with the incrementation of t_1 – hence the name, *time proportional phase incrementation*, TPPI. We will show that such an approach adds a constant frequency to the offsets of all of the lines in the spectrum. By choosing this frequency appropriately, we can make it appear that all of the offsets are positive.

Like the SHR method, TPPI relies on the ability to change the form of the modulation in t_1 by changing the phase of an appropriate pulse (or pulses) in the sequence. Recall from section 8.12.1 on page 227 that we have access to cosine and sine modulated data sets of the form

$$S_c(t_1, t_2) = \cos(\Omega_A t_1) \exp(-R^{(1)} t_1) \exp(i\Omega_B t_2) \exp(-R^{(2)} t_2)$$
$$S_s(t_1, t_2) = \sin(\Omega_A t_1) \exp(-R^{(1)} t_1) \exp(i\Omega_B t_2) \exp(-R^{(2)} t_2).$$

Although cosine and sine are different functions, they are related by a simple shift in time (or phase). This is illustrated in Fig. 8.38 on the following page, where we see that shifting a cosine to the right by one quarter of a period gives us a sine wave. Recall that in a whole period the phase changes through $360°$, so a shift by one quarter of a period is the same thing as a phase shift of $90°$ or $\pi/2$ radians.

We saw in section 2.5.4 on page 17 that a cosine wave phase shifted by ϕ can be written

$$\cos(\Omega t + \phi).$$

Using the trigonometric identity $\cos(A + B) \equiv \cos A \cos B - \sin A \sin B$,

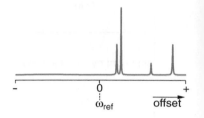

Fig. 8.37 If we deliberately place the receiver reference frequency to one side of all of the peaks, then all of the offsets will be positive. There is then no ambiguity over the sign of the offsets, and so frequency discrimination is not an issue. However, half of the spectrum is now empty – a waste of data space. Furthermore, as the transmitter is usually at the same frequency as the receiver reference, placing the latter to the side of the spectrum increases the likelihood of off-resonance effects being a problem with the RF pulses.

$\cos(\Omega t + \phi)$ can be written

$$\cos(\Omega t)\cos\phi - \sin(\Omega t)\sin\phi.$$

From this it is clear that a phase shift of $\phi = -\pi/2$ results in sine modulation:

$$\cos(\Omega t)\cos(-\pi/2) - \sin(\Omega t)\sin(-\pi/2) = \sin(\Omega t),$$

where we have used $\sin(-\pi/2) = -1$ and $\cos(-\pi/2) = 0$.

If we use this idea to write a sine wave as a phase shifted cosine wave, then the cosine and sine modulated data sets can be written in a single expression

$$S(\phi, t_1, t_2) = \cos(\Omega_A t_1 + \phi)\exp(-R^{(1)}t_1)\exp(i\Omega_B t_2)\exp(-R^{(2)}t_2), \quad (8.14)$$

where $\phi = 0$ for the cosine modulated data set, and $\phi = -\pi/2$ for the sine modulated data set.

We saw that shifting the phase of an appropriate pulse (or pulses) in the pulse sequence enables us to change the t_1 modulation from cosine to sine. Another way of describing this is to say that the phase ϕ in Eq. 8.14 can be changed by altering the phase of an appropriate pulse in the sequence. We can make ϕ any value we like by choosing an appropriate phase for the pulse. For example in HSQC, $\phi = \pi$ can be achieved by shifting the phase of the first S-spin pulse by π or $180°$.

The final, and key, step is to make ϕ proportional to t_1: $\phi \propto t_1$ or $\phi = \omega_{add}t_1$, where ω_{add} is the constant of proportion between the phase and t_1. It is clear that ω_{add} must be a frequency since phase = (frequency × time). Substituting this expression for ϕ into Eq. 8.14 we have

$$\begin{aligned} S(\phi, t_1, t_2) &= \cos(\Omega_A t_1 + \omega_{add}t_1)\exp(-R^{(1)}t_1)\exp(i\Omega_B t_2)\exp(-R^{(2)}t_2) \\ &= \cos([\Omega_A + \omega_{add}]t_1)\exp(-R^{(1)}t_1)\exp(i\Omega_B t_2)\exp(-R^{(2)}t_2). \end{aligned}$$

What we have achieved is the addition of a frequency ω_{add} to the modulation frequency in t_1.

This is the essence of how the TPPI method works. By making the phase ϕ proportional to t_1 we can add a constant frequency to the offsets of all of the peaks, and if this frequency is chosen appropriately we can make it appear that all of the offsets are positive. Our final task is to work out exactly how the phase has to be varied.

The data in t_1 are recorded at regularly spaced intervals of time, so in successive experiments t_1 take the values $0, \Delta_1, 2\Delta_1 \ldots$, where Δ_1 is the t_1 sampling interval. For reasons which are explained in section 13.5.2 on page 490, the maximum frequency which can be represented correctly by such a sampled signal is $1/(2\Delta_1)$ Hz; we will call this frequency f_{max}.

If we have frequency discrimination in t_1, then the range of offsets which can be represented correctly is from $-f_{max}$ to $+f_{max}$. In such a situation we would choose the receiver reference frequency and Δ_1 so that all of the offsets present fall within this range.

However, the problem is that we do not have frequency discrimination, and so need to make sure that all of the offsets are positive. We can do this by adding f_{max} to all of the offsets, as shown in Fig. 8.39 on the next page.

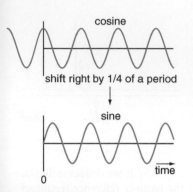

cosine

shift right by 1/4 of a period

sine

0   time

Fig. 8.38 Illustration of the relationship between a cosine and sine wave. At the top is shown a cosine wave; if this is shifted to the right by one quarter of a period, the result is a sine wave, shown underneath. Shifting the cosine wave to the right brings in data from negative times, here indicated in pale blue. A shift by one quarter of a period is equivalent to a phase shift of $90°$ or $\pi/2$ radians.

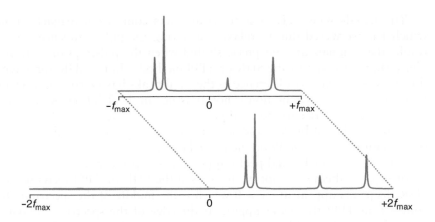

Fig. 8.39 If we have frequency discrimination then all we need to do is choose the receiver reference frequency and the t_1 sampling interval, Δ_1, such that all of the peaks have offsets which fall in the range $-f_{max}$ to $+f_{max}$, as shown in the top spectrum. Adding f_{max} to all of the offsets makes sure that they are all positive, but increases the maximum offset to $2f_{max}$, as shown in the lower spectrum.

Now the maximum frequency is $2f_{max}$, so to represent this range properly the value of Δ_1 will need to be halved, giving a new value of $\Delta_1' = 1/(4f_{max})$.

We saw above that the frequency of all the peaks can be shifted by ω_{add} by making the phase proportional to t_1: $\phi = \omega_{add}t_1$. The additional frequency we require is f_{max} Hz, which is $2\pi f_{max}$ rad s^{-1}. So, the phase is given by

$$\phi = 2\pi f_{max}t_1.$$

Recall that t_1 is incremented through the series of values 0, Δ_1', $2\Delta_1'$ ..., which can be written $(i-1)\Delta_1'$ where $i = 1, 2, 3 \ldots$. So the phase for the ith increment of t_1 is

$$\phi_i = 2\pi f_{max}(i-1)\Delta_1'.$$

Now we recall that $\Delta_1' = 1/(4f_{max})$, so

$$
\begin{aligned}
\phi_i &= 2\pi f_{max}(i-1)\frac{1}{4f_{max}} \\
&= (i-1)\frac{\pi}{2}.
\end{aligned}
$$

What this says is that as t_1 is incremented in steps of Δ_1', the phase must be incremented in steps of $\pi/2$. So, as t_1 goes through the series of values 0, Δ_1', $2\Delta_1'$, $3\Delta_1'$ ..., the phase moves in synchrony through the values 0, $\pi/2$, π, $3\pi/2$ ....

For the case of the COSY experiment, incrementing the phase of the first pulse by $\pi/2$ results in the phase ϕ changing by $\pi/2$. So, in the TPPI method the phase of first pulse is incremented by $\pi/2$ each time t_1 is incremented. This is illustrated in Fig. 8.40.

Once the TPPI method has been applied, all of the offsets in ω_1 will appear to be positive, and so we can process the data in the t_1 dimension using a cosine (or sine) Fourier transform. Such a transform only computes that part of the spectrum with positive frequencies, and so no data space is wasted on the empty part of the spectrum covering negative frequencies.

8.12.5 The States–TPPI method

Two-dimensional spectra often contain what are called *axial peaks*, which are a series of peaks at $\omega_1 = 0$. The origin of these peaks is described in more detail in section 11.7 on page 403, as is a way of suppressing them.

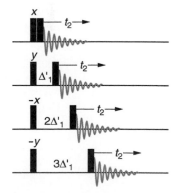

Fig. 8.40 Illustration of how the TPPI method is applied to the COSY pulse sequence. Each time that t_1 is incremented, the phase of the first pulse is incremented by $\pi/2$ radians or 90°.

The signals which give rise to axial peaks come from magnetization which has recovered due to relaxation during the pulse sequence. As a result, these signals *are not* phase shifted when the pulses prior to t_1 are phase shifted as part of the SHR or TPPI methods. In the SHR approach, this does not make any difference as the axial peaks have zero frequency in t_1, so frequency discrimination is not an issue. In the TPPI method, as the signals which give rise to the axial peaks do not experience the phase shift, they are not shifted by f_{max} and so remain at $\omega_1 = 0$, which is the *edge* of the spectrum (see Fig. 8.39 on the preceding page).

The way in which axial peaks appear in the spectrum is thus different according to whether or not we have used the SHR or TPPI methods. In SHR the peaks appear at $\omega_1 = 0$ which is the middle of the spectrum. In contrast, in TPPI the peaks appear at the edge of the spectrum. Having axial peaks in the middle of the spectrum is not really acceptable, as they might well fall on top of peaks we are interested in. However, having axial peaks at the edge of the spectrum, where there are probably no real peaks of interest, is acceptable.

The realization that TPPI is advantageous because of where the axial peaks appear led to the development of the States–TPPI procedure. In this, frequency discrimination is achieved by the usual SHR method i.e. we record separate sine and cosine modulated data sets and process them in the way already described. The new feature is that each time t_1 is incremented, we *invert* the phase of the pulses which precede t_1, and at the same time change the sign of the data we have recorded. Of course, these two things cancel one another out, and so make no difference, except for the signals which give rise to the axial peaks which are *not* inverted when the phase of the pulses prior to t_1 are inverted.

As a result, the sign of the signals which give rise to the axial peaks alternate each time t_1 is advanced. The effect of this phase alternation is similar to that used in TPPI, except that as the phase changes by 180° each time t_1 advances by Δ_1, the frequency shift is by f_{max}. As a result, the axial peaks are shifted to the edge of the spectrum and hence out of the way.

8.12.6   Phase in two-dimensional spectra

The prototype data sets given in Eq. 8.13 on page 227 are idealized in that there are no phase errors in either dimension. As we saw in section 5.3.2 on page 85, for instrumental reasons there is often a phase error in the acquisition dimension (here t_2), and the same applies to the t_1 dimension. So, a more realistic representation of the two data sets includes a phase error of ϕ_1 in t_1, and ϕ_2 in t_2:

$$S_c(t_1, t_2) = \cos(\Omega_A t_1 + \phi_1)\exp(-R^{(1)}t_1)\exp(i\Omega_B t_2)\exp(i\phi_2)\exp(-R^{(2)}t_2)$$

$$S_s(t_1, t_2) = \sin(\Omega_A t_1 + \phi_1)\exp(-R^{(1)}t_1)\exp(i\Omega_B t_2)\exp(i\phi_2)\exp(-R^{(2)}t_2).$$

To obtain absorption mode lineshapes, we will need to apply phase corrections to the final two-dimensional spectrum in just the same way as we do to one-dimensional spectra. The only difference is that there are two phases to adjust, one for each dimension.

The software provided with modern spectrometers makes it possible to apply such phase corrections. However, phasing a two-dimensional spectrum is not quite so straightforward as phasing in one dimension, as it is not feasible to recompute the *whole* spectrum after each trial phase correction is applied – to do so would simply be too time consuming. So, we cannot 'drag the mouse' to alter the phase and see right away how the two-dimensional spectrum changes in the way we are accustomed to for one-dimensional spectra.

The way two-dimensional spectra are phased in practice is to take out some selected cross-sections parallel to one of the dimensions, determine the correct phase using these, and then apply the correction to the whole spectrum. The process is then repeated for the other dimension.

Sometimes, particularly in routine work, it is not considered worthwhile going to all the trouble of using the SHR or TPPI methods to obtain absorption mode lineshapes, and then in addition having to phase correct the two-dimensional spectrum. Rather, the simpler approach, involving calculating the absolute value spectrum, is used. How this works is considered in the next section.

8.12.7 Absolute value spectrum

The simplest way to achieve frequency discrimination is to use P- or N-type data selection. The resulting time-domain signal, including phase errors, is of the form

$$S_P(t_1, t_2) = \exp(i\Omega_A t_1) \exp(i\phi_1) \exp(-R^{(1)}t_1)$$
$$\times \exp(i\Omega_B t_2) \exp(i\phi_2) \exp(-R^{(2)}t_2).$$

Noting that $\exp(i\phi_1) \exp(i\phi_2) = \exp(i[\phi_1 + \phi_2])$, enables us to simplify this expression to

$$S_P(t_1, t_2) = \exp(i\Omega_A t_1) \exp(-R^{(1)}t_1) \exp(i\Omega_B t_2) \exp(-R^{(2)}t_2) \exp(i\phi_{\text{tot}}),$$

where $\phi_{\text{tot}} = \phi_1 + \phi_2$.

As was shown in section 8.12.2 on page 228, double Fourier transformation of this time-domain function gives the spectrum

$$S_P(\omega_1, \omega_2) = \exp(i\phi_{\text{tot}})\Big\{ [A_1(\Omega_A)A_2(\Omega_B) - D_1(\Omega_A)D_2(\Omega_B)]$$
$$+ i[A_1(\Omega_A)D_2(\Omega_B) + D_1(\Omega_A)A_2(\Omega_B)] \Big\}.$$

Note that the phase factor $\exp(i\phi_{\text{tot}})$ simply multiplies the whole spectrum.

Finally, we compute the magnitude of this spectrum which is defined as

$$|S_P(\omega_1, \omega_2)| = \sqrt{S_P(\omega_1, \omega_2)\, S_P^{\star}(\omega_1, \omega_2)},$$

where $S_P^{\star}$ is the complex conjugate of the spectrum, found by changing the sign of the imaginary part:

$$S_P^{\star}(\omega_1, \omega_2) = \exp(-i\phi_{\text{tot}})\Big\{ [A_1(\Omega_A)A_2(\Omega_B) - D_1(\Omega_A)D_2(\Omega_B)]$$
$$- i[A_1(\Omega_A)D_2(\Omega_B) + D_1(\Omega_A)A_2(\Omega_B)] \Big\}.$$

Fig. 8.41 Two views of the absolute value lineshape: on the left is shown a perspective view, and on the right is shown a contour plot. The lineshape is strictly positive, but has broad features which derive from the dispersion contributions to the lineshape.

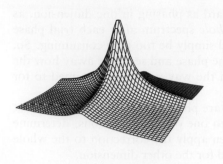
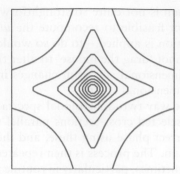

A few lines of algebra show us that

$$|S_P(\omega_1, \omega_2)| = \left(\begin{array}{c} [A_1(\Omega_A)A_2(\Omega_B) - D_1(\Omega_A)D_2(\Omega_B)]^2 \\ + [A_1(\Omega_A)D_2(\Omega_B) + D_1(\Omega_A)A_2(\Omega_B)]^2 \end{array} \right)^{\frac{1}{2}}.$$

The lineshape is rather an involved mixture of absorption and dispersion in each dimension. However, the nice thing about it is that it is *not* affected by the phase errors: these cancel out when computing the magnitude.

Figure 8.41 shows two views of this lineshape, known as the *absolute value* lineshape. As can be seen, it has undesirable broad features which continue well away from the centre of the peak – these are due to the dispersion mode contributions. In addition, because of the calculation of the square root, all of the peaks in the spectrum will turn out positive, which results in a loss of useful information in spectra such as COSY.

The combination of computing a P-type spectrum and then plotting the absolute value gets round some of the complexities of processing and phasing two-dimensional spectra. For routine work on unchallenging molecules, this approach is certainly convenient and probably sufficient. However, for more difficult problems, or where the highest resolution is required, there is really no substitute for computing a proper absorption mode spectrum.

8.13   Further reading

Two-dimensional NMR:

Chapter 5 from P. J. Hore, J. A. Jones and S. Wimperis, *NMR: The Toolkit* (Oxford University Press, 2000).

Chapter 8 from R. Freeman, *Spin Choreography* (Spektrum, 1997).

Chapters 5 and 16 from M. H. Levitt, *Spin Dynamics* (2nd edition, John Wiley & Sons, Ltd, 2008).

Chapter 4 from F. J. M. van de Ven, *Multidimensional NMR in Liquids* (VCH, 1995).

Chapters 6 and 8 from R. R. Ernst, G. Bodenhausen and A. Wokaun, *Principles of Nuclear Magnetic Resonance in One and Two Dimensions* (Oxford University Press, 1987).

Low-pass J-filters:

T. Schulte-Herbrüggen, A. Meissner, A. Papanikos, M. Meldal and
O. W. Sørensen, *J. Magn. Reson.*, **156**, 282–294 (2002).

TOCSY:

Chapter 6 (and references therein) from J. Cavanagh, W. J. Fairbrother,
A. G. Palmer III, M. Rance and N. J. Skelton, *Protein NMR
Spectroscopy* (2nd edition, Academic Press, 2007).

The SHR and TPPI methods:

J. Keeler and D. Neuhaus, *J. Magn. Reson.*, **63**, 454–472 (1985).

8.14 Exercises

8.1   Identify the preparation and mixing periods for each two-dimensional pulse sequence described in this chapter.

8.2   The diagram below is the same as Fig. 8.5 (b) on page 187 i.e. the result of Fourier transforming the time-domain data along t_2. Sketch the form of the cross-sections indicated by the arrows at positions 1, 2 and 3, and 4, 5, and 6. Explain carefully any differences and similarities between these cross-sections.

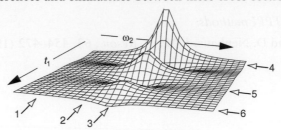

8.3   Starting with equilibrium magnetization on spin two, $\hat{I}_{2z}$, determine the form of the observable operators present at $t_2 = 0$ in the COSY sequence. Describe the kinds of peaks (cross or diagonal) which each observable term gives rise to, and work out the detailed form of the two-dimensional multiplets. In each case, choose phase corrections and appropriate Fourier transforms so as to give absorption mode peaks.

8.4   Repeat the previous exercise, but this time for the DQF COSY pulse sequence.

8.5   A two-dimensional zero-quantum spectrum can be recorded using the same pulse sequence as for double-quantum spectroscopy, Fig. 8.18 on page 204. The only difference is that zero-quantum coherence is selected during t_1. Given that

$$\text{zero-quantum part of } 2\hat{I}_{1x}\hat{I}_{2y} = -\tfrac{1}{2}\hat{Z}Q_y$$
$$= -\tfrac{1}{2}\left(2\hat{I}_{1y}\hat{I}_{2x} - 2\hat{I}_{1x}\hat{I}_{2y}\right),$$

determine the form of the two-dimensional spectrum. The evolution of this zero-quantum term is given in section 7.12.3 on page 176.

Compare the form of the zero-quantum and double-quantum spectra. What information is available from the zero-quantum spectrum?

8.6   Why must the second I spin 90° pulse in the HSQC sequence be applied about the y-axis? What would the effect be of applying this pulse about the $-y$-axis?

8.7   For the HSQC sequence of Fig. 8.22 (a) on page 210, show that changing the phase of the first 90° pulse to the S spin from x to y results in the modulation of the observed signal changing from $\cos(\Omega_S t_1)$ to $\sin(\Omega_S t_1)$.

8.8   For the HMQC sequence, as applied to proton and ^{13}C, explain how difference spectroscopy can be used to suppress the contributions from protons which are not coupled to ^{13}C. Make an explicit calculation to demonstrate that your proposal works.

8.9   Explain why, in an HMQC, the peaks will have their maximum intensity when $\tau = n/(2J_{IS})$ for $n = 1,\ 3,\ 5\ \ldots$, and zero intensity for $n = 2,\ 4,\ 6\ \ldots$.

8.10 Discuss how you would modify the HSQC pulse sequence to make it suitable for detecting correlations through long-range ^{13}C–^{1}H couplings.

8.11 It was shown that for the TOCSY sequence (applied to a two-spin system) the observable signals at the start of t_2 are given by Eq. 8.11 on page 223:

$$A_{1\to1}\cos(\pi J_{12}t_1)\cos(\Omega_1 t_1)\,\hat{I}_{1y} + A_{1\to2}\cos(\pi J_{12}t_1)\cos(\Omega_1 t_1)\,\hat{I}_{2y}.$$

Using the same approach as in section 8.3.2 on page 192, show that both the cross- and diagonal-peak multiplets in a TOCSY are in phase in each dimension and can be processed in such a way that the peaks in both multiplets have the absorption lineshape.

8.12 A peak in the P-type spectrum can be represented as

$$S_{\text{P}}(\omega_1,\omega_2) = \underbrace{[A_1(\Omega_A)A_2(\Omega_B) - D_1(\Omega_A)D_2(\Omega_B)]}_{\text{real}}$$
$$+\, \mathrm{i}\,\underbrace{[A_1(\Omega_A)D_2(\Omega_B) + D_1(\Omega_A)A_2(\Omega_B)]}_{\text{imaginary}}.$$

The real part of this phase-twist lineshape is plotted in Fig. 8.36 on page 229. What does the imaginary part look like? Sketch a contour plot.

8.13 Describe how you would apply (a) the SHR, and (b) the TPPI method to the HSQC sequence.

8.14 For the double-quantum experiment whose pulse sequence is shown in Fig. 8.18 on page 204, it turns out that shifting the phase of *all* the pulses which precede t_1 by ϕ results in the t_1 modulation being modified from $\cos([\Omega_1 + \Omega_2]t_1)$ to $\cos([\Omega_1 + \Omega_2]t_1 + 2\phi)$.

Describe how you would implement the TPPI method in this sequence so as to generate a frequency discriminated spectrum.

9

Relaxation and the NOE

In this chapter we are going to look at the phenomenon of *relaxation*, which is how the bulk magnetization from the spins reaches its equilibrium value. As we have already seen, at equilibrium we have magnetization along z, but none in the transverse plane: it is relaxation which drives the spins to this equilibrium state.

This chapter starts out by describing how relaxation comes about, and how its rate is determined by molecular properties, such as shape and motion. We then move on to consider some important applications of relaxation phenomena, such as the nuclear Overhauser effect (NOE) and cross correlation.

Relaxation in NMR is unusually slow when compared with that of other molecular energy levels. For example, the lifetime of an excited electronic state is typically a few microseconds, and the vibrational and rotational energies of molecules are likely to change at every collision, giving them lifetimes of a few nanoseconds. In contrast, it takes between milliseconds and seconds for the equilibrium magnetization to be established; in extreme cases this process can take minutes.

The slow relaxation is at once both an advantage and a disadvantage. The advantage of slow relaxation is that it means that any transverse magnetization (or, more generally, coherences) we generate will survive for long enough for it to be manipulated and observed. Multiple-pulse NMR, with its sequences of pulses and delays lasting many milliseconds, would simply not be possible if relaxation were not so relatively slow. Furthermore, slow relaxation means that the FID persists for long enough for us to obtain high-resolution spectra – that is, the linewidths are narrow.

The disadvantage of slow relaxation is that it sets a lower limit on the rate at which an experiment can be repeated. Remember that, in even the simplest pulse–acquire experiment, we usually repeat the experiment several times in order to improve the signal-to-noise ratio. The problem is that, before we can repeat the experiment, we have to allow sufficient time for the equilibrium magnetization to be re-established. If we do not do this, the z-magnetization will be reduced in size, and so will the observed signal.

Understanding NMR Spectroscopy, Second Edition James Keeler
© 2010 John Wiley & Sons, Ltd

We will see in this chapter that in NMR the rate of relaxation is sensitive to the physical environment the nuclei find themselves in and the nature of the motion which the molecule is undergoing. Thus relaxation can be used as a probe for studying both of these things. It turns out that, when compared with other techniques, NMR is sensitive to rather slow molecular motions, and so NMR can be a unique source of such information.

When it comes to structural studies, the one really important manifestation of relaxation is the NOE. This allows us to identify which nuclei are close in space (for protons, say within 5 to 6 Å), thus providing complementary information to scalar coupling, which relates to the bonding network. Although the NOE is only semi-quantitative when it comes to estimating distances, it nevertheless forms a key element in the NMR spectroscopist's arsenal of techniques. Much of this chapter will therefore be devoted to understanding the origin of the NOE, and to describing the experiments which are used to measure it.

We will start with a description of what relaxation is, focusing on how what is happening to the individual spins leads to a particular behaviour of the bulk magnetization of the sample. Our discussion will reveal the importance of the timescale of molecular motions, and introduce the key idea of the *correlation time*, which is used to characterize these motions.

Having established the underlying physical phenomena which are responsible for relaxation, we will go on to describe the origin of the NOE. To do this we will need to consider in detail the relaxation properties of a pair of spins, something which is conveniently described using the *Solomon equations*. These equations are also a convenient way of understanding how the experiments used to detect the NOE work.

Having discussed the NOE, we then go on to look at the relaxation of transverse magnetization; this kind of relaxation has some similarities and some important differences to the relaxation of z-magnetization. The chapter closes with a discussion of the phenomena of *cross correlation* which results from the interference between different sources of relaxation.

Describing the effects of relaxation is relatively straightforward, but understanding the origin of relaxation and its dependence on motion is rather more difficult. It is probably the part of NMR theory that most people, including the author, find most difficult to grasp. Do not be surprised, therefore, if you find this chapter more challenging than those which have preceded it.

9.1 The origin of relaxation

Relaxation is the process by which, over time, the bulk magnetization returns to its equilibrium position. We have already established that at equilibrium there is *no* transverse magnetization, but there is z-magnetization along the direction of the applied magnetic field. The size of this equilibrium z-magnetization depends on the number of spins, their gyromagnetic ratio and the size of the applied field.

Thus, relaxation drives the transverse magnetization to zero and the longitudinal (that is, along z) magnetization to a particular steady value, as illustrated in Fig. 9.1. The description of the equilibrium magnetization

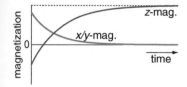

Fig. 9.1 Relaxation drives the z-magnetization to its equilibrium value, indicated here by the dashed line, and the transverse (x or y) magnetization to its equilibrium value of zero.

as being 'steady' is important, as by definition at equilibrium all quantities cease to be time dependent i.e. nothing is changing.

However, this description tells us nothing about what is happening to the individual spins which results in this behaviour of the bulk magnetization. How the two are connected is the topic of the next few sections.

9.1.1 Behaviour of individual magnetic moments

It is helpful at this point to remind ourselves how the bulk magnetization of the sample is related to the magnetic moments of individual spins. Recall the discussion in section 4.1 on page 47, where we described how each spin has associated with it a magnetic moment. This moment is a vector quantity, having a particular magnitude and direction. The magnitude is determined by the gyromagnetic ratio of the nucleus, but there are no restrictions on the direction in which a particular moment can point, so in general each moment has an x-, y- and z-component.

The bulk z-magnetization is found by adding together the z-components of the magnetic moment from each spin. Similarly, the x- and y-magnetizations are found by adding together the corresponding components from each spin. At equilibrium, the x- and y-components of the individual spins are distributed randomly, so adding them up results in no bulk transverse magnetization.

For the z-magnetization the story is rather different. As was described in section 4.1 on page 47, there is an energetic preference for the z-component of the magnetic moment to be aligned along the field direction. However, the energy preference is rather small and so the alignment is easily disrupted by the thermal motion. Nevertheless, when summed over the whole sample there is a net magnetization along the z-direction.

When an RF pulse is applied to equilibrium magnetization, the z-magnetization is rotated towards the transverse plane. As a result, the z-component is reduced in size. After the pulse, the resulting transverse magnetization precesses about the field direction, and it is this precession which we detect in the form of the FID.

The question is, what is happening to the magnetic moments of the individual spins during the pulse and period of free precession? It turns out that each magnetic moment behaves in *exactly* the same way as the bulk magnetization. That this is true can be seen from a quantum mechanical treatment of the motion of a single spin – the details of which are given in Chapter 6. However, even without looking into the details of the quantum mechanics, it seems reasonable that each magnetic moment behaves in the same way as the bulk magnetization, as the latter is composed of the former.

For example, if we apply a 90° pulse about the x-axis, the z-component of the magnetic moment is rotated to $-y$, the x-component is unaffected, and the y-component is rotated to $+z$. During free precession, the magnetic moment precesses about the z-axis at the Larmor frequency, sweeping out a cone of constant angle, as illustrated in Fig. 9.2. As a result we have a constant z-component, and oscillating x- and y-components.

At equilibrium, there is a net alignment of the individual moments along the z-axis. The effect of a $90°(x)$ pulse is to rotate the z-component of each magnetic moment onto the $-y$-axis. Therefore, after the pulse there is net

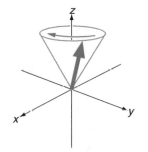

Fig. 9.2 Each spin has associated with it a magnetic moment, which is a vector quantity, having a magnitude and a direction. The moment can point in any direction, with its energy being determined by the angle between the moment and the applied magnetic field, which is along z. The moment precesses about the z-axis at the Larmor frequency, describing a cone of constant angle. As a result, the x- and y-components of the moment oscillate at the Larmor frequency. If a transverse magnetic field, oscillating at the Larmor frequency, is applied, the magnetic moment will be rotated in such a way that the angle it makes to the z-axis will be altered.

y-magnetization. The crucial thing is that the pulse affects each magnetic moment *in the same way*, so all the *z*-components are rotated onto −*y*. It is for this reason that the pulse is able to rotate the net *z*-magnetization into the transverse plane.

After the pulse, the magnetic moments of the individual spins precess about the *z*-axis at the Larmor frequency. Once again, the key thing is that each magnetic moment is precessing at the same frequency, so they do not get out of alignment with one another as a result of this precessional motion. The alignment between the moments is therefore maintained.

After the 90° pulse the spins are definitely not at equilibrium. The question is, how can equilibrium be restored? One simple option is to apply another 90° pulse which, if its phase is chosen correctly, will rotate the magnetization back to its equilibrium position along the *z*-axis. However, this is not relaxation – this is just us manipulating the spins. Relaxation is a natural process which takes place without any intervention from us, so we must identify another way in which the magnetization can be driven to its equilibrium position.

9.1.2 Local fields

From all we have seen so far, it is clear that to rotate the magnetization towards the *z*-axis we need a transverse magnetic field which is oscillating at or near to the Larmor frequency. During an RF pulse we apply such a field deliberately, but it turns out that such oscillating fields also occur *naturally* within the sample. These fields can interact with the individual magnetic moments, and thus rotate them to new positions, in just the same way as an RF pulse.

There are various mechanisms which lead to the generation of magnetic fields within the sample, and we will discuss these in more detail later on. At this stage it is helpful to describe just one source of these fields, which is from other spins in the sample. Each spin has a magnetic moment which, as we have noted before, behaves like a small bar magnet, generating its own magnetic field. A spin therefore experiences not only the static applied field but also magnetic fields from nearby spins; this is illustrated in Fig. 9.3

This field generated by a spin – called the *local field* – falls off rapidly with distance, and even at the closest approach, is many orders of magnitude weaker than the applied field. The local field varies in *magnitude* and *orientation* as the molecules move around due to thermal agitation. If this motion results in the transverse component of the field oscillating at close to the Larmor frequency, a magnetic moment which experiences the local field will be rotated to a new direction, just as it would be by a pulse.

The local field acts like a pulse but, rather than all of the spins being affected in the same way, the effect is highly localized. The local field is different in different parts of the sample, so each spin is affected in a *different* way. This is in complete contrast to a pulse, which affects all of the spins in the *same* way.

As these local fields can change the orientation of individual magnetic moments, both the magnitude and orientation of the bulk magnetization will be affected. What we now need to explore is how these local fields drive the bulk magnetization to its equilibrium position.

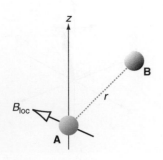

Fig. 9.3 Spin A experiences a local field, B_{loc}, due to the magnetic moment of a nearby spin B. The magnitude and direction of the local field depends on the distance r between the two spins, and the orientation of the vector joining the two spins (shown dashed) with respect to the applied field, which is along the *z*-axis.

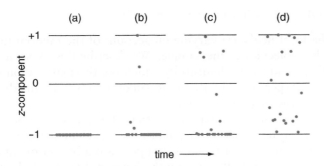

Fig. 9.4 Visualization of how the bulk z-magnetization is driven to zero by random changes in the z-component of the magnetic moments of individual spins. The z-component of each of 20 spins is represented by a dot; if the spin is aligned along $+z$, the component takes the value $+1$, whereas if the spin is aligned along $-z$, the component takes the value -1. The initial state, shown in (a), has all of the z-components along $-z$, such that the total z-magnetization, found by adding up the individual z-components, is -20. After a short time, some of the individual magnetic moments will be reoriented, thus changing their z-components. This is shown in (b), where the z-magnetization is -16. If we wait longer, more of the magnetic moments will have been reoriented, giving the arrangement shown in (c), for which the z-magnetization is -12. After sufficient time the z-magnetization will become zero, as is the case for arrangement (d). See text for discussion of why the magnetization is driven to zero, rather than to its proper equilibrium value.

If we assume, not unreasonably, that the local fields are varying randomly in their magnitude and orientation, then we expect that the bulk z-magnetization will eventually be driven to zero. The reason for this assertion is that if the individual magnetic moments are rotated by random amounts at random time intervals, after sufficient time the moments will be randomly oriented, resulting in no bulk magnetization.

This idea is illustrated in Fig. 9.4, in which we see how the z-components of just 20 spins are affected by random reorientations of the individual moments. So that we can see more clearly what is going on, at time zero the magnetic moments of all of the spins have been aligned along $-z$, as shown in (a). After some time, the magnetic moments of a few of the spins have been reoriented, leading to a change in their z-components; this is shown in (b). It is clear that the result of these reorientations is to reduce the size of the z-magnetization.

Leaving longer times, as shown in (c) and (d), results in more of the magnetic moments being reoriented, and so a further reduction in the total z-magnetization. It is clear that, after sufficient time, these random reorientations will drive the z-magnetization to zero.

There is clearly a problem with this description, as it predicts that at equilibrium there is *no* z-magnetization, which is certainly not the case. To resolve this problem with our argument we need to think about what it is that the spins are coming to equilibrium with, which is the topic of the next section.

9.1.3   Coming to equilibrium with the lattice

Recall that the local fields are varying on account of the random thermal agitation of the molecules in the sample. We describe this by saying that the local fields provide a mechanism by which the thermal motion of the molecules can be experienced by the spins – in other words it puts the spins into *contact* with the thermal motion.

We know from everyday experience that if we put two objects in contact, they will come to thermal equilibrium by exchanging heat energy. For example, if a hot lump of metal is dropped into a bucket of water the metal cools down as energy flows from the hot metal to the cooler water. As the metal cools, the water is heated, and eventually both come to the same temperature. They are then in thermal equilibrium with one another.

In our NMR sample, the nuclear spins have a certain amount of energy on account of the interaction of the magnetic moments with the applied magnetic field. Of course, the nuclei have other sorts of energy as well, but we are not concerned with these here, and so when we talk about the 'energy of the spins' we will just mean this energy of interaction between the spins and the field.

The interaction between the spins and the field can therefore be thought of as leading to a *reservoir* of (spin) energy. The thermal motion of the molecules is also a reservoir of energy, which is put in contact with the spin energy reservoir via the mediation of the local fields, as is illustrated in Fig. 9.5. As a result, the two reservoirs are able to come to thermal equilibrium with one another, just in the same way that two physical objects can come to thermal equilibrium.

When the bulk magnetization is rotated away from the z-axis, the energy of the spins is increased. This is because the number of spins whose magnetic moments are aligned with the z-axis, the low energy arrangement, is decreased. For the spins to come back to equilibrium they therefore need to lose energy.

The amount of energy that the spins need to lose to come to equilibrium is minuscule when compared with the energy of thermal motion. Bringing the spins back to equilibrium is analogous to dropping our hot metal into a swimming pool rather than a bucket.

The process by which the z-magnetization is returned to its equilibrium value is called *longitudinal relaxation*. You will also find it referred to as *spin–lattice relaxation*. This latter term arises from the understanding that this kind of relaxation involves the flow of energy between the spins and the molecular motion. The 'lattice' is a generic term used to describe a reservoir of energy, which in this case is the molecular motion.

The equilibrium z-magnetization

In the previous section we discussed a simple picture in which the reorientation of individual magnetic moments by random fields drove the bulk z-magnetization to zero – a result which is plainly wrong, as we know that relaxation results in the establishment of a finite z-magnetization at equilibrium. It is not that this picture of the magnetic moments being reoriented is wrong, it is just that there is an additional subtlety which

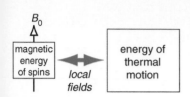

Fig. 9.5 The interaction of the spins with the applied magnetic field, and the thermal motion of the molecules, can both be thought of as leading to reservoirs of energy. The two reservoirs are able to come to equilibrium as a result of the mediation provided by the local fields i.e. the local fields create thermal contact between the two reservoirs.

we need to take account of in order to arrive at the correct equilibrium position.

If the local field at a particular spin results in its magnetic moment being rotated *towards* the z-axis, the energy of interaction between the spin and the applied field is reduced, and so there must be a flow of energy from the spin to the surroundings. In contrast, if the magnetic moment is turned *away* from the z-axis, the magnetic energy of the spin increases, and so there must be a flow of energy from the surroundings to the spin.

The key point to understand here is that there is a higher probability for a certain amount of energy to flow into the surroundings than there is for the same amount of energy to flow out of the surroundings. The reason for this is connected to the fact that the surroundings are at thermal equilibrium, a situation described by the Boltzmann distribution. This distribution tells us that as the energy of a particular state (i.e. energy level) of the surroundings goes up, the probability of that state being occupied goes down.

When energy is absorbed, the surroundings move from a lower energy state to a higher energy one; in contrast, when energy is given out, the surroundings move from a higher energy state to a lower energy one. As the lower energy state is the more occupied of the two, it follows that it is more probable that the surroundings will absorb, rather than give out, a certain amount of energy.

The overall result of this asymmetry between the probability of giving out and receiving energy is that events in which the magnetic moments are moved towards the z-axis are more probable than those in which the moment moves away from the z-axis. As a result, after many such events, the z-components of the moments are not distributed randomly, but in such a way as to lead to a net z-magnetization.

This discussion relies on the fact that the energy of interaction between the spins and the magnetic field is minuscule compared with the energy associated with the thermal motion of the molecules. We can therefore safely assume that once the surroundings are at equilibrium, the flow of energy to and from the spins will not perturb this equilibrium in any perceptible way. It also follows that the difference in the probability of these tiny amounts of energy flowing into or out of the surroundings is very small, so that the distribution of the z-components of the magnetic moments is only slightly perturbed from a random arrangement.

9.1.4 Transverse relaxation

The process by which transverse magnetization decays away to its equilibrium value of zero is called *transverse relaxation*. It is also called *spin–spin relaxation*, a confusing and unhelpful term which we will avoid.

When we talk about the individual magnetic moments changing orientation as a result of interacting with an oscillating transverse local field, it is not only the z-, but also the transverse components of the moments which can be changed. For example, if the local field has a component along x, then the y-component of the magnetic moment can be changed. Similarly, a local field along y can change the x-component. Overall therefore, these local fields not only drive the z-magnetization to its equilibrium value,

The origin of the terms *secular* and *non-secular* will be described on page 293.

but also do the same for the transverse magnetization. This is called the *non-secular* contribution to transverse relaxation.

There is, however, a second way in which the transverse magnetization is driven to equilibrium. Remember that the magnetic moments from individual spins precess about the applied magnetic field B_0 at the Larmor frequency (Fig. 9.2 on page 243). However, the field experienced by a particular spin is not just B_0, but is this field *plus* the z-component of the local field. As a result the precession frequency is changed, albeit by rather a small amount as the local field is much smaller than B_0.

The local field varies from spin to spin, so the precession frequency is slightly different for each spin. Consequently, over time the precession of the individual moments will get out of step with one another, and thus the net transverse magnetization will shrink. This is the *secular* contribution to transverse relaxation.

If the local fields did not change over time, then this decay of the transverse magnetization could be reversed using a spin echo. This comes about because the frequency at which each spin is evolving is constant, so a spin echo will refocus this, in just the same way that an offset term is refocused. However, it is *not* the case that the local fields are time independent – on the contrary, they are changing rapidly due to molecular motion. As a result, a spin echo cannot refocus the evolution at this ever changing frequency, and so the decay of the transverse magnetization cannot be reversed. A more detail discussion of transverse relaxation is deferred to section 9.8 on page 286.

9.1.5 Summary

Rather a lot of new ideas have been introduced in this section, so it is helpful to summarize the key steps in the discussion.

• The magnetic moment from each individual spin behaves in the same way as the bulk magnetization i.e. it can be rotated by a transverse field oscillating at the Larmor frequency, and precesses about the field along the z-axis.

• In the sample there are sources of local magnetic fields which vary randomly in orientation and magnitude. These fields are highly localized.

• If the local field experienced by a spin has a transverse component which is oscillating at the Larmor frequency, then the orientation of the magnetic moment of the spin will change. In particular, the components of the magnetic moment which are perpendicular to the transverse local field will be changed.

• These changes in orientation of the magnetic moments of individual spins result in changes in the bulk magnetization of the sample.

• The local fields provide a thermal contact between the spins and the random thermal motion of the molecules. This drives the z-magnetization back to its equilibrium value.

- The return of the z-magnetization to its equilibrium value is called longitudinal relaxation. Such relaxation is brought about by the transverse components of local magnetic fields which are oscillating at the Larmor frequency.

- The decay of transverse magnetization to its equilibrium value of zero is called transverse relaxation.

- There are two contributions to transverse relaxation: the non-secular contribution, like longitudinal relaxation, is brought about by the transverse components of local fields which are oscillating at the Larmor frequency; the secular contribution is caused by there being a distribution of the z-components of the local fields.

Our task now is to describe in more detail the various sources of these local fields, and then to go on to investigate how we can characterize the random motions which give rise to the required time dependence.

9.2 Relaxation mechanisms

A particular source of a local magnetic field is called a *relaxation mechanism*. While there are quite a lot of these, two tend to be dominant for spin-half nuclei: the *dipolar* mechanism and the *chemical shift anisotropy* mechanism. We will therefore largely confine our attention to these mechanisms.

9.2.1 The dipolar mechanism

In this mechanism the local field is due to the magnetic moment (or magnetic dipole, as it is sometimes called) of another spin, as was outlined in section 9.1.2 on page 244 and illustrated in Fig. 9.3 on the same page. There are thus two spins, or dipoles, involved: one generating the field, and one experiencing it. This is why the mechanism is also called the *dipole–dipole* mechanism.

The local field due to the neighbouring spin depends on a number of parameters:

- The distance r between the two spins: in fact, the interaction falls off rather quickly as $(1/r^3)$.

- The gyromagnetic ratio of the spin: the larger the gyromagnetic ratio, the larger the magnetic moment and the larger the local field. So, for example, protons give rise to larger local fields than ^{13}C nuclei.

- The orientation of the vector joining the two spins relative to the applied magnetic field (the z-axis).

With this strong distance dependence, the dipolar mechanism is only effective over rather short distances, say less than 5 Å for two protons. However, note that local fields can be contributed by more than one spin, so equally effective relaxation can be brought about by either a single nearby spin or a larger number of more remote spins.

The final point to note is that the effect of a given local field B_{loc} on the nucleus depends on the gyromagnetic ratio *of that nucleus*. This is because the rate at which the local field rotates the magnetic moment goes as γB_{loc}, in just the same way that an RF field B_1 causes the bulk magnetization to precess at a frequency γB_1. Therefore overall the strength of the dipolar interaction depends on the gyromagnetic ratio of the spin which is generating the local field, and the gyromagnetic ratio of the spin which is experiencing that local field.

9.2.2   Chemical shift anisotropy

The usual description given for the chemical shift is to say that, in the presence of a strong applied field, the electrons in the molecule give rise to a small induced (local) field at the nucleus. The nucleus therefore experiences the sum of the applied field and this induced field, thus shifting the Larmor frequency by an amount which depends on the size of the induced field.

For most molecules, the size of the induced field, and hence the size of the chemical shift, depends on the orientation of the molecule with respect to the applied magnetic field. This is described by saying that the chemical shift is *anisotropic*. In liquid samples the molecules are tumbling so rapidly that the nuclei see an average local field, and hence have an average chemical shift, called the isotropic shift. Nevertheless, at any instant, the local field is different for molecules at different orientations.

What is not usually commented on is that the local field due to the chemical shift is not necessarily parallel to the applied field – in fact in general this local field can point in any direction, as illustrated in Fig. 9.6. We see that this local field is therefore a relaxation mechanism.

The extent to which the local field varies as the molecule tumbles depends on the anisotropy of the chemical shift i.e. the extent to which the shift varies with orientation. With the exception of nuclei at sites of high symmetry, such as isolated atoms or nuclei in octahedral or tetrahedral sites, all chemical shifts are anisotropic. The extent of the anisotropy does, however, vary greatly between different isotopes.

As a rough guide, the shift anisotropy is of the order of the chemical shift range for that nucleus. So, for protons the shift anisotropy is rarely more than a few ppm, whereas for ^{13}C the anisotropy can easily be 100 ppm, or more. Nuclei such as ^{31}P, which have wide chemical shift ranges, can also have large anisotropies.

The local field due to the chemical shift is proportional to the applied field. In turn the interaction of this field with the nucleus depends on the gyromagnetic ratio of the nucleus, so overall the interaction goes as γB_0.

9.2.3   Relaxation by paramagnetic species

In the dipolar mechanism, the source of the local field is the magnetic moments of other nuclear spins in the sample. Unpaired electrons also generate local fields in an analogous way, and so are a potential source of relaxation. The magnetic moment of the electron is very much greater than that of the proton, so the local field generated by an electron is correspondingly much greater. Unpaired electrons can, therefore, be a

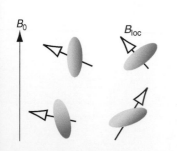

Fig. 9.6 When placed in a strong applied field B_0, a local field is generated as a result of the interaction between the electrons and B_0. It is this field which is responsible for the chemical shift. However, on account of the anisotropy of the electron distribution, the local field varies in direction and size as the molecule tumbles in solution. The resulting variation in the local field can be a source of relaxation.

particularly efficient source of relaxation, causing a significant effect even when present at low concentrations.

In preparing an NMR sample it is inevitable that small amounts of O_2 gas are dissolved in the solvent, and since O_2 has unpaired electron spins (i.e. it is paramagnetic), the dissolved oxygen is a source of relaxation. Sometimes, the sample is 'degassed', for example by bubbling pure nitrogen gas through the sample, in order to remove the oxygen and hence reduce the rate of relaxation.

9.3   Describing random motion – the correlation time

The next task is to work out how to describe and quantify the random thermal motion which provides the all-important time dependence to the local fields. Recall from our previous discussion that to cause longitudinal relaxation the transverse component of the local field must be oscillating at or near the Larmor frequency.

A molecule is likely to be executing two distinct kinds of motion: vibrations and overall rotation. During a vibration, the individual bonds and bond angles are oscillating back and forth about their equilibrium positions, and in principle this will modulate the dipolar interaction as the distance, and possibly the angle between the internuclear vector and the field direction, will change. However, such oscillations typically take place at frequencies of 10^{11} to 10^{13} Hz, which is much higher than even the highest achievable Larmor frequency of around 10^9 Hz. We can thus discount such vibrations as being effective sources of relaxation.

The overall rotation of a molecule will result in the modulation of the local fields due to both the dipolar interaction and the chemical shift anisotropy (CSA), therefore such motion can be a source of relaxation. The question is, are these motions at the right frequencies to cause relaxation?

In a gas, small molecules rotate freely at frequencies of around 10^8 to 10^9 Hz, which is certainly in the right ball-park for relaxation. However, in liquids the picture is quite different, and it is certainly not the case that even small molecules are free to rotate. The density of a liquid means that the frequency of collisions between molecules is rather high, but the high density, combined with the interactions between the molecules, means that the ability of a molecule to rotate is rather constrained. So, each collision is only capable of changing the orientation of a molecule by a small amount.

One way of visualizing this situation is to imagine that the solvent exerts a viscous drag on the molecule which is so large that even an energetic collision is only able to rotate the molecule by a small amount, if at all. An analogy would be trying to make a floating football rotate by throwing ping-pong balls at it.

Let us concentrate on just one molecule, and place an imaginary vector in it, such that the arrow starts out pointing along $+z$. As the molecule experiences collisions it will start to be rotated, in small steps, away from its starting position. However, since the collisions are random, the vector will not move steadily away from $+z$, but will execute a jerky motion in which the direction it moves, and the angle through which it moves, is different on each step. Such motion is called *rotational diffusion*.

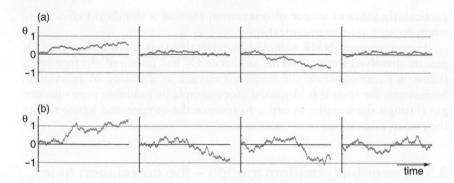

Fig. 9.7 Plots of the orientation, specified by the angle θ (rad), of a disc undergoing rotational diffusion. Each graph contains 300 equally spaced time steps, and the angle always starts at zero. In (a) the maximum allowed change in the angle on each step is ±0.04 radians, with the actual change being chosen at random in this interval. For the plots shown in (b) the maximum change is ±0.08 radians. Each plot (trajectory) is different on account of the random nature of the process. However, on average the trajectories shown in (b) deviate further from the starting position, as a result of the greater change in the angle allowed on each time step. The rotational correlation time for motion of which (a) is a representative sample of trajectories is therefore longer than for those shown in (b).

If we wait a 'long time', the vector will have wandered through more or less all possible orientations, whereas after a 'short time' the vector will barely have moved. What 'long' and 'short' mean can be quantified by defining a *rotational correlation time* τ_c, which is the *average* time it takes for a molecule to end up at an orientation about 1 radian from its starting position. Remember that the molecule does not jump to this new position in a few steps, but takes a tortuous and wandering path of many tiny steps before it finally finds itself 1 radian from where it started.

Figure 9.7 provides a visualization of the nature of rotational diffusion and the meaning of the correlation time. To make this diagram, we have simplified things by considering a two-dimensional case, for which only one angle is needed to specify the orientation i.e. we are thinking about the rotational diffusion of a disc rather than a sphere. Each graph shows the angle for 300 time steps, where at each step the angle is allowed to change by a random amount.

For the plots shown in (a), the maximum change in the angle allowed on any one step is ±0.04 radians, whereas in (b) the maximum change is ±0.08 radians. On any one time step, the size of the change is chosen at random within the permitted range.

The first thing to note is that, although the maximum change in the angle allowed on each step was the same for all the plots shown in (a), the actual sequence of angles is different in each case. This is expected as it is a random process, so each molecule heads off on its own unique trajectory. In none of the trajectories does θ reach one radian, so the correlation time must be somewhat longer than the time represented by these 300 steps.

For the plots shown in (b), the maximum change in the angle on each step is twice that for (a), so not surprisingly the excursions away from the starting position are greater. In fact, within the time frame the angle reaches 1 radian for two of the trajectories, and almost makes it for a third. It is clear that the correlation time for this set of trajectories must be shorter than that for those shown in (a).

It is important to realize that the correlation time is the *average* time needed to achieve an orientation 1 radian away from the starting position, not the time taken by one particular molecule to achieve this orientation. This average is taken over a large number of molecules in the sample.

For small molecules τ_c turns out to be around 10 ps, rising to 10 ns for small proteins. The reciprocal of the correlation time, $1/\tau_c$, is a rough estimate of the 'average frequency' of the motion (in rad s^{-1}), so $1/(2\pi \tau_c)$ gives this frequency in Hz. Correlation times in the range 10 ns to 10 ps give average frequencies in the range of 10^7 to 10^{10} Hz, which is comparable with typical Larmor frequencies. Rotational diffusion therefore appears to give motion in the right range to cause relaxation.

The problem we have is in trying to quantify the amount of this motion which is actually at the right frequency – that is, at the Larmor frequency. To do this we need to introduce the *correlation function* and the *spectral density*.

9.3.1   The correlation function

The correlation function is a way of characterizing the time dependence of the random motion in our sample, and hence finding out how much of the motion is present at the Larmor frequency. We will start out by describing what the correlation function is, and then, in the following section, explain how this function can be used to assess the amount of motion at particular frequencies.

Imagine that a particular spin i experiences a local field $B_{loc,i}(t)$ which is varying in time, so that at a time τ later the local field is $B_{loc,i}(t + \tau)$. The correlation function $G(t, \tau)$ is defined as the average over the sample of the product $B_{loc}(t)B_{loc}(t + \tau)$:

$$
\begin{aligned}
G(t, \tau) &= \frac{1}{N}[B_{loc,1}(t)B_{loc,1}(t + \tau) + B_{loc,2}(t)B_{loc,2}(t + \tau) + \ldots] \quad (9.1) \\
&= \frac{1}{N}\sum_{i=1}^{N} B_{loc,i}(t)B_{loc,i}(t + \tau) \\
&= \overline{B_{loc}(t)B_{loc}(t + \tau)},
\end{aligned}
$$

where N is the number of spins in the sample. On the last line we have used the overbar to indicate the average over the sample, known as the ensemble average.

Generally $B_{loc}(t)$ can take positive or negative values, which are equally distributed either side of zero, so that its average over the sample is zero. It is also usually the case that the magnitude of the local field from a particular source has a maximum value.

The local field varies due to the thermal motion in the sample, so $B_{loc}(t)$ is a random function of time. It turns out that the properties of this random function do not depend on the point from which time is measured; such functions are said to be *stationary* random functions. For such a random

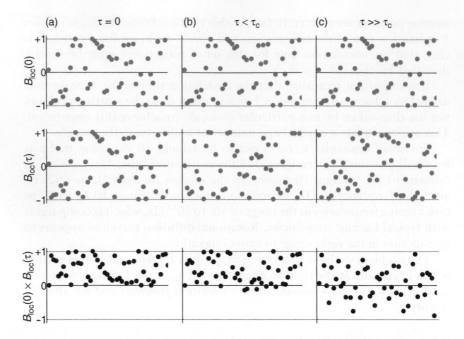

Fig. 9.8 Visualization of the behaviour of the correlation function $G(\tau)$ as a function of the time interval τ. In each of (a), (b) and (c), the data points represent: in the top row, $B_{\mathrm{loc}}(0)$; in the middle row, $B_{\mathrm{loc}}(\tau)$; and in the bottom row, the product $B_{\mathrm{loc}}(0)B_{\mathrm{loc}}(\tau)$. The values for each of 50 spins are represented by a dot, and it has arbitrarily been assumed that B_{loc} can only take values between -1 and $+1$. Note that the horizontal axis is *not* time, but is just used to spread out the values from the 50 spins. $G(\tau)$ is found by summing the points $B_{\mathrm{loc}}(0)B_{\mathrm{loc}}(\tau)$ (shown in the bottom row), and dividing by the number of points. In (a) $\tau = 0$, so $B_{\mathrm{loc}}(\tau) = B_{\mathrm{loc}}(0)$, and as a consequence all of the points in the lower plot are positive. $G(\tau)$ is therefore a maximum when $\tau = 0$. In (b) τ is shorter than correlation time τ_c, and so although for each spin $B_{\mathrm{loc}}(\tau)$ is different to $B_{\mathrm{loc}}(0)$, the change is not large. Most of the points in the lower plot are positive, but in the few cases where the local field has changed sign between time 0 and time τ, the point is negative, and as a result $G(\tau)$ is less than $G(0)$ (in this case $0.87 \times G(0)$). Finally, in (c) $\tau \gg \tau_c$, so that the local field of each spin has changed significantly between time zero and time τ. As a result, there are a significant number of negative points in the lower plot, and so $G(\tau)$ is significantly less than $G(0)$.

function, the value of the correlation function does not therefore depend on the time t, but only on the time *interval* τ. So, from now on we will write the correlation function as $G(\tau)$.

Figure 9.8 helps us to visualize how the correlation function varies with τ. The points in this figure represent the values of $B_{\mathrm{loc}}(0)$, $B_{\mathrm{loc}}(\tau)$, and the product $B_{\mathrm{loc}}(0)B_{\mathrm{loc}}(\tau)$ for different values of τ, and for 50 spins. For each value of τ, the top graph shows $B_{\mathrm{loc}}(0)$, the middle shows $B_{\mathrm{loc}}(\tau)$, and the lower shows $B_{\mathrm{loc}}(0)B_{\mathrm{loc}}(\tau)$. If follows from the definition of $G(\tau)$, Eq. 9.1 on the preceding page, that the value of $G(\tau)$ is found by summing the values of $B_{\mathrm{loc}}(0)B_{\mathrm{loc}}(\tau)$ in the lower plot and then dividing by the number of values.

The plots shown under (a) are for the case $\tau = 0$. Here $B_{\mathrm{loc}}(0)B_{\mathrm{loc}}(\tau) = B_{\mathrm{loc}}^2(0)$, so all of the points in the lower plot are positive, and hence $G(\tau)$ will be a maximum. In general, we can see that it will always be the case that the maximum value of $G(\tau)$ occurs at $\tau = 0$.

The plots shown under (b) are computed for a value of τ which is shorter that the correlation time. Careful comparison of the plots of $B_{\mathrm{loc}}(0)$ and $B_{\mathrm{loc}}(\tau)$ will show that, although the local fields have changed between the two times, the change is small. As a result $B_{\mathrm{loc}}(0)B_{\mathrm{loc}}(\tau)$ is still positive for most spins, but there are a few where the local field has changed sign, and for these $B_{\mathrm{loc}}(0)B_{\mathrm{loc}}(\tau)$ is negative. So, when $G(\tau)$ is found by computing the sum of the points in the lower plot the result will be somewhat less than $G(0)$.

Finally, the plots shown under (c) are for the case where the time τ is much longer than the correlation time. Now the local fields have changed significantly between time zero and time τ, so there are a significant number of negative points in the lower plot. Summing these points, that is computing $G(\tau)$, gives a value which is much less that $G(0)$.

Our overall picture for how the correlation function varies with τ is that it starts at a maximum at $\tau = 0$, and then falls away smoothly to zero at a rate determined by the value of the correlation time τ_c. The value of $G(0)$ can be computed from the definition of $G(\tau)$ given in Eq. 9.1 on page 253:

$$
\begin{aligned}
G(0) &= \overline{B_{\mathrm{loc}}(t)B_{\mathrm{loc}}(t)} \\
&= \overline{B_{\mathrm{loc}}^2}.
\end{aligned}
$$

This value is simply the average of the square of the local field. The time at which we compute this average is irrelevant for a stationary random function, so $G(0)$ just depends on the average size of the interaction.

The exact form of the correlation function depends on the details of the interaction between the molecule and the solvent. The simplest case to analyse is where the molecule is assumed to be spherical and the solvent simply provides a medium with a certain viscosity. Rotational diffusion in such a case, turns out to be described by a correlation function which is a simple exponential whose decay rate is set by τ_c:

$$
G(\tau) = \overline{B_{\mathrm{loc}}^2}\exp\left(-|\tau|/\tau_c\right).
$$

A plot of this function for three different correlation times is shown in Fig. 9.9. Note that the smaller the correlation time, the faster $G(\tau)$ decays. The function depends on the modulus of the time τ, as the same behaviour is expected for positive times τ as it is for negative times i.e. the behaviour for a particular time interval in the future is the same as for the same interval of time in the past.

The exponential part of the correlation function is independent of the source of the local fields, which only determines the overall magnitude of $G(\tau)$ via the term $\overline{B_{\mathrm{loc}}^2}$. It is therefore common to define a *reduced correlation function*, $g(\tau)$, which is independent of the size of the local fields:

$$
g(\tau) = \exp\left(-|\tau|/\tau_c\right). \tag{9.2}
$$

With this definition, $G(\tau) = \overline{B_{\mathrm{loc}}^2}g(\tau)$.

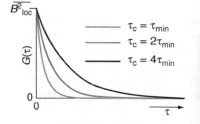

Fig. 9.9 Plot of the correlation function $G(\tau) = \overline{B_{\mathrm{loc}}^2}\exp\left(-|\tau|/\tau_c\right)$ for three different values of the correlation time, τ_c. The curve for shortest value of τ_c, $\tau_{\min}$, is shown in grey, the blue line is for a τ_c of twice this value, and the black line is for twice this value again. Note that for all three cases, $G(\tau)$ has the same maximum value of $\overline{B_{\mathrm{loc}}^2}$ at $\tau = 0$.

9.3.2   The spectral density

From the discussion in section 9.1.2 on page 244, you will recall that, in order to be effective at causing longitudinal relaxation, the local field must be oscillating close to the Larmor frequency. We have now seen that the time dependence of the local field results from molecular motion, specifically rotational diffusion, and that this motion can be characterized by a correlation function. The next task is to find out exactly how much of the motion is present at the required frequency. We will see in this section that the *spectral density* function, which is related to the correlation function, provides this information.

The correlation function is a function of *time*, analogous to the FID we observe in an NMR experiment. If we Fourier transform the FID we obtain a function of frequency, the spectrum, which tells us how much intensity there is at each frequency. In an analogous way, if we Fourier transform the correlation function we will obtain a function of frequency, and from this we will be able to find the amount of motion which is at the Larmor frequency.

The Fourier transform of the correlation function is called the *spectral density*, $J(\omega)$:

$$G(\tau) \xrightarrow{\text{FT}} J(\omega).$$

The amount of motion at the Larmor frequency is simply found by evaluating $J(\omega)$ at $\omega = \omega_0$.

In the case of the exponential correlation function, the Fourier transform is the familiar absorption mode Lorentzian, centred at $\omega = 0$:[1]

$$\overline{B_{\text{loc}}^2} \exp\left(-|\tau|/\tau_c\right) \xrightarrow{\text{FT}} \overline{B_{\text{loc}}^2} \frac{2\tau_c}{1 + \omega^2 \tau_c^2},$$

so

$$J(\omega) = \overline{B_{\text{loc}}^2} \frac{2\tau_c}{1 + \omega^2 \tau_c^2}. \tag{9.3}$$

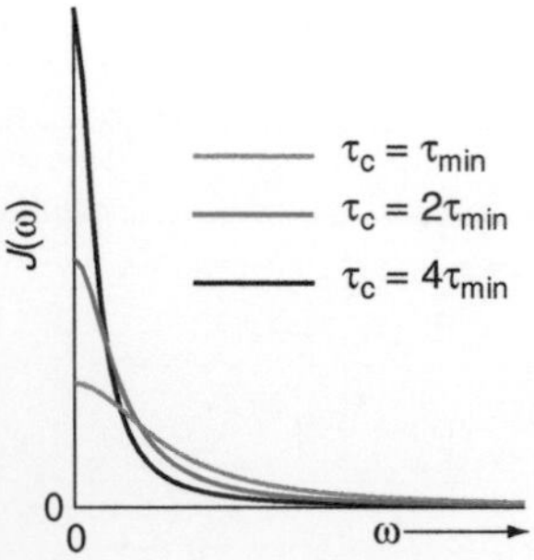

Figure 9.10 shows plots of this function for different values of the correlation time; only positive frequencies are shown. The spectral density has its maximum value of $2\overline{B_{\text{loc}}^2}\tau_c$ at $\omega = 0$, and then falls off steadily as ω increases, with the rate of the fall-off being determined by τ_c. As τ_c becomes shorter, the spectral density spreads out to higher frequencies. Note also that the value of $J(0)$ increases as τ_c increases.

A particularly important feature of the spectral density is that, if we plot it against ω, we find that the area under the curve is independent of τ_c. Expressed mathematically, the area under $J(\omega)$ is the integral

$$\begin{aligned} \text{area} &= \int_0^\infty \overline{B_{\text{loc}}^2} \frac{2\tau_c}{1 + \omega^2 \tau_c^2} \, \mathrm{d}\omega \\ &= \pi \overline{B_{\text{loc}}^2}. \end{aligned}$$

A consequence of the area being constant is that, although $J(\omega)$ is always a maximum at $\omega = 0$, as τ_c becomes smaller the size of the maximum

Fig. 9.10 Plots of the spectral density function, $J(\omega)$, for three different values of the correlation time, τ_c; the graph is plotted only for positive values of ω. The grey line is for the shortest value of τ_c, $\tau_{\min}$, the blue line for the case where $\tau_c = 2\tau_{\min}$, and the black line for the case where $\tau_c = 4\tau_{\min}$. Note that the shorter τ_c becomes, the higher the frequencies to which $J(\omega)$ spreads. As explained in the text, it turns out that the areas under these curves are all the same.

---

[1]The factor of two comes in as an FID only exists for time greater than zero, whereas the correlation function extends symmetrically either side of $\tau = 0$.

decreases and at the same time the spectral density spreads out to higher frequencies. This can be seen in Fig. 9.10 on the preceding page.

This behaviour has an important consequence if we think about the value of the spectral density at the Larmor frequency as a function of the correlation time. Remember that it is $J(\omega_0)$ which will determine the rate of longitudinal relaxation.

The spectral density at the Larmor frequency is given by

$$J(\omega_0) = \overline{B_{\mathrm{loc}}^2} \frac{2\tau_{\mathrm{c}}}{1 + \omega_0^2 \tau_{\mathrm{c}}^2}.$$

If we make a plot of this as a function of τ_{c}, as is shown in Fig. 9.11, we see that there is clearly a value of the correlation time which makes $J(\omega_0)$ a maximum. It is a relatively easy calculation to show that this maximum occurs when $\omega_0 \tau_{\mathrm{c}} = 1$ i.e. $\tau_{\mathrm{c}} = 1/\omega_0$.

It follows that the rate of longitudinal relaxation will also be a maximum when $\tau_{\mathrm{c}} = 1/\omega_0$, as this is the value of the correlation time which gives the maximum spectral density at the Larmor frequency. So, both correlation times which are shorter or longer than this optimum value will give slower relaxation. Just like Goldilocks and the porridge, for the most efficient longitudinal relaxation the correlation time must be neither too long nor too short, but 'just right'.

In the previous section we introduced (Eq. 9.2 on page 255) the idea of the reduced correlation function $g(\tau)$, which does not depend on the size of the local fields. The Fourier transform of this reduced correlation function is the *reduced spectral density*, $j(\omega)$:

$$g(\tau) \xrightarrow{\text{FT}} j(\omega)$$
$$\exp\left(-|\tau|/\tau_{\mathrm{c}}\right) \xrightarrow{\text{FT}} \frac{2\tau_{\mathrm{c}}}{1 + \omega^2 \tau_{\mathrm{c}}^2}. \tag{9.4}$$

With this definition, it follows that $J(\omega) = \overline{B_{\mathrm{loc}}^2} j(\omega)$.

9.3.3 Motional regimes

The comparison between the Larmor frequency and the correlation time turns out to be rather important in the theory of relaxation – indeed, we have already seen an example of this in Fig. 9.11 where the maximum in $J(\omega_0)$ occurs when $\tau_{\mathrm{c}} = 1/\omega_0$.

In making this comparison it is usual to distinguish two *motional regimes*. The first is called *fast motion* or *extreme narrowing*, and is defined mathematically as when $\omega_0 \tau_{\mathrm{c}} \ll 1$. Physically, this is the limit in which motion is very fast i.e. the correlation time is very short, such as would be the case for small molecules.

The reduced spectral density at the Larmor frequency is given by

$$j(\omega_0) = \frac{2\tau_{\mathrm{c}}}{1 + \omega_0^2 \tau_{\mathrm{c}}^2}.$$

If the fast motion limit applies, $\omega_0 \tau_{\mathrm{c}} \ll 1$, it follows that $(1 + \omega_0^2 \tau_{\mathrm{c}}^2) \approx 1$, and so $j(\omega_0)$ is given by

$$\text{fast motion:} \quad j(\omega_0) = 2\tau_{\mathrm{c}}.$$

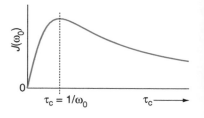

Fig. 9.11 A plot of the spectral density at the Larmor frequency, $J(\omega_0)$, as a function of the correlation time, τ_{c}. The plot shows a maximum at $\tau_{\mathrm{c}} = 1/\omega_0$. This implies that the rate of longitudinal relaxation will be a maximum for this value of the correlation time.

In words, what this fast motion limit means is that the spectral density is independent of the Larmor frequency. It should also be noted that as $j(0) = 2\tau_c$ (for all values of τ_c), in the fast motion limit it follows that $j(\omega_0) = j(0)$.

The other limit is called the *slow motion* or *spin diffusion* limit, and is when $\omega_0\tau_c \gg 1$. In this limit, $(1 + \omega_0^2\tau_c^2) \approx \omega_0^2\tau_c^2$, so the reduced spectral density becomes

$$\text{slow motion:} \quad j(\omega_0) = \frac{2}{\omega_0^2\tau_c}.$$

Recalling that $j(0) = 2\tau_c$, this expression can be written

$$j(\omega_0) = \frac{j(0)}{\omega_0^2\tau_c^2}.$$

As $\omega_0^2\tau_c^2 \gg 1$, it follows that $j(\omega_0)$ is very much smaller than $j(0)$.

For a Larmor frequency of 500 MHz ($\omega_0 = 3.1 \times 10^9$ rad s^{-1}), a small molecule with a correlation time of 10 ps has $\omega_0\tau_c = 0.03$, which is clearly in the fast motion limit. In contrast, a small protein with a correlation time of 10 ns has $\omega_0\tau_c = 31$, which is clearly in the slow motion limit.

9.3.4   Summary

To summarize what we have found so far:

- Rotational diffusion in a liquid causes the local fields to vary at rates which are suitable for causing relaxation.

- This random diffusive motion can be described using a correlation time τ_c, which is the average time it takes for a molecule to move to a position at an angle of about 1 radian from its starting position.

- The spectral density gives the frequency distribution of the motion. The simplest example of such a spectral density function is

$$J(\omega) = \overline{B_{\text{loc}}^2}\,\frac{2\tau_c}{1 + \omega^2\tau_c^2}.$$

  Note that $J(\omega)$ depends on the correlation time, τ_c.

- The rate of longitudinal relaxation depends on the spectral density at the Larmor frequency. The relaxation is most rapid when $\omega_0\tau_c = 1$.

We are now in a position to explore the details of how the z-magnetization behaves as a result of relaxation. To do this, we will first need to introduce the concept of the populations of energy levels.

9.4   Populations

In section 4.1 on page 47 we were at pains to point out that the magnetic moment from a single spin-half can point in any direction, and that its energy depends on the angle between the magnetic moment and the applied field. However, if we measure the energy of a particular spin, we will always find a value which corresponds to one of the two energy levels of a spin-half i.e. that corresponding to the α state or the β state (see section 3.1 on page 24 for a discussion of this point). It turns out that there is a certain probability p_α of finding the energy corresponding to the α state, and a certain probability p_β of finding the energy corresponding to the β state.

If we imagine measuring the energy of every spin in the sample, then for each there is a probability $p_{\alpha,i}$ of finding the energy corresponding to the state α. The sum of all these probabilities is the same as the number of spins in the sample which were found to have the energy of the α state. This number can be identified as the *population* of this state, n_α:

$$n_\alpha = p_{\alpha,1} + p_{\alpha,2} + p_{\alpha,3} \cdots$$

In a similar way, we can compute the population of the β state, n_β.

This language is rather dangerous, as the moment we start talking about the 'populations of the α and β states', it is easy to fall into the trap of imagining that each spin is in one of these states, which is certainly not true. Nevertheless, thinking about the sum of these individual probabilities as a population is a very useful concept when it comes to describing relaxation and so we will use it throughout the rest of this chapter.

9.4.1   The z-magnetization in terms of populations

If we accept this description of the spins in terms of the populations of the α and β states, it is easy to determine the bulk z-magnetization. A spin in the α state contributes $+\frac{1}{2}\hbar\gamma$ to magnetization, and one in the β state contributes $-\frac{1}{2}\hbar\gamma$, so the bulk z-magnetization, M_z, is

$$M_z = \tfrac{1}{2}\hbar\gamma(n_\alpha - n_\beta), \tag{9.5}$$

where n_α and n_β are the populations of the α and β states, respectively. What this equation says is that the magnetization is proportional to the population difference and to the gyromagnetic ratio. This latter factor comes about because the magnetic moment of each individual spin is proportional to its gyromagnetic ratio.

At equilibrium we know that the populations must be those given by the Boltzmann distribution, which in this case predicts the following:[2]

$$n_\alpha^0 = \tfrac{1}{2}N\exp\left(-E_\alpha/k_BT\right) \qquad n_\beta^0 = \tfrac{1}{2}N\exp\left(-E_\beta/k_BT\right), \tag{9.6}$$

where E_α and E_β are the energies of the α and β states, k_B is Boltzmann's constant, and n_α^0 and n_β^0 are the equilibrium populations.

The energies E_α and E_α are tiny compared with k_BT, so the fraction $(E_{\alpha/\beta}/k_BT)$ is very much less than one. We can therefore approximate the exponential by taking just the first two terms of the series expansion

$$\exp\left(-x\right) \approx 1 - x \qquad x << 1.$$

---

[2]The factor of $\frac{1}{2}$ in these expressions is $1/q$, where q is the partition function. In this case $q = 2$, as there are two accessible states.

Using this, the populations are

$$n_\alpha^0 = \tfrac{1}{2}N(1 - E_\alpha/k_B T) \qquad n_\beta^0 = \tfrac{1}{2}N(1 - E_\beta/k_B T),$$

and hence the equilibrium z-magnetization, M_z^0, is

$$\begin{aligned} M_z^0 &= \tfrac{1}{2}\hbar\gamma(n_\alpha^0 - n_\beta^0) \\ &= \tfrac{1}{4}\hbar\gamma N \frac{E_\beta - E_\alpha}{k_B T}. \end{aligned}$$

Not surprisingly, the population difference depends on the energy difference between the two levels. The energies of these two levels were given in section 3.2.7 on page 30 as

$$E_\alpha = -\tfrac{1}{2}\hbar\gamma B_0 \qquad E_\beta = +\tfrac{1}{2}\hbar\gamma B_0,$$

where we have included a factor of $\hbar$ so that the energies are in Joules, rather than in rad s^{-1}. Using these, the equilibrium magnetization can be written

$$M_z^0 = \frac{\gamma^2 \hbar^2 N B_0}{4 k_B T}. \tag{9.7}$$

This expression says that the bulk magnetization is proportional to the number of spins, the square of the gyromagnetic ratio, and the strength of the applied field. Consequently the largest equilibrium magnetization, and hence the strongest signals, come from nuclei with the greatest gyromagnetic ratios and from using the highest magnetic field strengths.

The expression for M_z^0 given in Eq. 9.7 goes as γ^2. This is because the energies of the states, which determine their populations, are proportional to γ, and the magnetic moment contributed by each spin is also proportional to γ.

Since it is not possible to measure that absolute size of an NMR signal, the absolute size of the z-magnetization is not important in our calculations. As a result it is usually acceptable to drop all the constants in Eq. 9.5 on the previous page and simply write

$$M_z = n_\alpha - n_\beta \quad \text{and} \quad M_z^0 = n_\alpha^0 - n_\beta^0.$$

Although these expressions are not correct in a formal sense, they capture the essence of the situation, which is that the magnetization is proportional to the population difference.

9.4.2   Relaxation in terms of populations

Longitudinal relaxation drives the z-magnetization towards its equilibrium value, and we have now seen that the z-magnetization is proportional to the population difference between the α and β energy levels. It therefore follows that the approach to equilibrium involves changing the populations of the two levels, so that longitudinal relaxation can be described as arising from *transitions* between the two levels. For example, a transition from α to β will decrease the population difference and hence reduce the z-magnetization, whereas a transition from β to α will increase the population difference, and hence increase the z-magnetization.

The simplest assumption we can make is that the rate of transitions from α to β is proportional to the population of the α state, n_α:

$$\text{rate from } \alpha \text{ to } \beta = W_{\alpha \to \beta} \, n_\alpha,$$

where we have written the constant of proportion as $W_{\alpha \to \beta}$. This constant is in fact a rate constant, entirely analogous to a first-order rate constant in kinetics; both have units of $(\text{time})^{-1}$ e.g. s^{-1}.

Similarly, the rate of transitions from β to α is

$$\text{rate from } \beta \text{ to } \alpha = W_{\beta \to \alpha} \, n_\beta.$$

Note that this time the rate is proportional to the population of the β state, as it is from this state that the transitions originate, and that we have also allowed the rate constant to be different. The two processes are illustrated in Fig. 9.12.

Transitions from α to β *decrease* the population of the α state, whereas transitions from β to α *increase* the population of the α state. Therefore the overall rate of change of the population of the α state is:

$$\text{rate of change of } n_\alpha = \underbrace{+W_{\beta \to \alpha} \, n_\beta}_{\text{increase in } n_\alpha} \; \underbrace{-W_{\alpha \to \beta} \, n_\alpha}_{\text{decrease in } n_\alpha} . \tag{9.8}$$

The first term is positive as it describes the rate of a process which increases the population of the α state, whereas the second term is negative as it describes a process which reduces the population of the state.

We can write an analogous equation for the rate of change of the population of the β state:

$$\text{rate of change of } n_\beta = \underbrace{+W_{\alpha \to \beta} \, n_\alpha}_{\text{increase in } n_\beta} \; \underbrace{-W_{\beta \to \alpha} \, n_\beta}_{\text{decrease in } n_\beta} . \tag{9.9}$$

At equilibrium the populations will be constant i.e. the rate of change will be zero. In addition, these populations will have their equilibrium values, as predicted by Eq. 9.6 on page 259. Applying these two conditions to the expressions of Eqs 9.8 and 9.9, we have

$$0 = W_{\beta \to \alpha} \, n_\beta^0 - W_{\alpha \to \beta} \, n_\alpha^0 \qquad 0 = +W_{\alpha \to \beta} \, n_\alpha^0 - W_{\beta \to \alpha} \, n_\beta^0.$$

From either of these it follows that

$$\frac{n_\alpha^0}{n_\beta^0} = \frac{W_{\beta \to \alpha}}{W_{\alpha \to \beta}}. \tag{9.10}$$

There is a difficulty with this expression, in that the theory most commonly used to predict values for the rate constants $W_{\beta \to \alpha}$ and $W_{\alpha \to \beta}$ comes up with the answer that these two are *equal*. As a result, it follows from Eq. 9.10 that $n_\alpha^0 = n_\beta^0$ i.e. the equilibrium populations are equal, which is certainly not correct.

The problem here turns out to be a deficiency in the theory, in which the spins are treated quantum mechanically, but the surroundings (the lattice) are treated classically. To obtain the correct result, consistent with the

Fig. 9.12 The population of the energy level associated with the spin being in the α state, n_α, will be decreased by transitions from the α to the β state, and increased by transitions in the reverse sense. The (first-order) rate constants for these two processes are $W_{\alpha \to \beta}$ and $W_{\beta \to \alpha}$, respectively.

expected equilibrium populations, a much more complete theory is needed which treats both the spins and the lattice quantum mechanically. However, such an approach is both complicated and laborious.

What we are going to do is avoid this problem entirely by rewriting the rate equations, Eq. 9.8 and Eq. 9.9 on the previous page, in the following way:

$$\text{rate of change of } n_\alpha = W_{\alpha\beta}(n_\beta - n_\beta^0) - W_{\alpha\beta}(n_\alpha - n_\alpha^0)$$
$$\text{rate of change of } n_\beta = -W_{\alpha\beta}(n_\beta - n_\beta^0) + W_{\alpha\beta}(n_\alpha - n_\alpha^0). \tag{9.11}$$

What we have done is made all the rate constants equal $(W_{\alpha\beta})$, and instead of these multiplying the populations they multiply the *deviation* of the populations from their equilibrium values i.e. $(n_\alpha - n_\alpha^0)$ and $(n_\beta - n_\beta^0)$. You can see immediately from Eq. 9.11 that, if $n_\alpha = n_\alpha^0$ and $n_\beta = n_\beta^0$, the rate of change of both n_α and n_β are zero, as is required at equilibrium.

As the z-magnetization is equal to the population difference $(n_\alpha - n_\beta)$, the rate of change of the z-magnetization is equal to the difference between the rate of change of n_α and that of n_β:

$$\text{rate of change of } M_z = \text{rate of change of } n_\alpha - \text{rate of change of } n_\beta$$

We can then use Eq. 9.11 to substitute in expressions for the rates of change of the populations:

$$
\begin{aligned}
\text{rate of change of } M_z &= 2W_{\alpha\beta}(n_\beta - n_\beta^0) - 2W_{\alpha\beta}(n_\alpha - n_\alpha^0) \\
&= -2W_{\alpha\beta}\left[(n_\alpha - n_\beta) - (n_\alpha^0 - n_\beta^0)\right] \\
&= -2W_{\alpha\beta}(M_z - M_z^0).
\end{aligned}
$$

This equation predicts what we expect: when $M_z = M_z^0$ the rate of change is zero.

The usual way to write this equation is using the language of calculus, in which we identify the rate of change of M_z as its derivative with respect to time, dM_z/dt:

$$\frac{dM_z(t)}{dt} = -R_z\left[M_z(t) - M_z^0\right]. \tag{9.12}$$

We have written the z-magnetization as $M_z(t)$ to remind ourselves that it is a function of time. R_z, the rate constant for longitudinal relaxation, is equal to $2W_{\alpha\beta}$.

This equation is often written

$$\frac{dM_z(t)}{dt} = -\frac{1}{T_1}\left[M_z(t) - M_z^0\right],$$

where T_1 is the *time constant* for longitudinal relaxation: $T_1 = 1/R_z$. The use of the symbol T_1 for this time constant is so ubiquitous that longitudinal relaxation is often called 'T_1 relaxation'.

The rate of relaxation, and hence the value of the relaxation rate constant R_z, will depend, as we have seen, on the average of the square of the local fields, and the spectral density at the Larmor frequency. For the moment we will not look in detail at the calculations which lead to a prediction of the rate constant, but rather look at the implications of Eq. 9.12 for the behaviour of the z-magnetization.

9.5   Longitudinal relaxation behaviour of isolated spins

In this section we are going to investigate the practical consequences of the fact that the z-magnetization relaxes according to

$$\frac{\mathrm{d}M_z(t)}{\mathrm{d}t} = -R_z \left[M_z(t) - M_z^0 \right].$$

This differential equation says that the rate of change of the z-magnetization is proportional to the deviation of the magnetization from its equilibrium value. The minus sign ensures that the z-magnetization changes in such a way that it approaches its equilibrium value as time increases. The rate constant R_z determines the rate at which equilibrium is approached. All this can best be seen by using an example.

Suppose that at time zero the z-magnetization is $M_z(0)$. What we will now work out, by integrating the above differential equation, is precisely how the z-magnetization moves back to its equilibrium position. We start by separating out the parts of the differential equation so that the variable $M_z(t)$ is on the left, and time is on the right:

$$\frac{\mathrm{d}M_z(t)}{\mathrm{d}t} = -R_z \left[M_z(t) - M_z^0 \right]$$

$$\text{hence} \quad \frac{1}{\left[M_z(t) - M_z^0 \right]} \, \mathrm{d}M_z(t) = -R_z \, \mathrm{d}t.$$

We can now integrate the left-hand side with respect to $M_z(t)$ and the right-hand side with respect to t:

$$\int \frac{1}{\left[M_z(t) - M_z^0 \right]} \, \mathrm{d}M_z(t) = \int -R_z \, \mathrm{d}t$$

$$\ln \left(M_z(t) - M_z^0 \right) = -R_z t + \text{const.} \tag{9.13}$$

To compute the integrals we have used the fact that M_z^0 and R_z are constants. As usual, a constant of integration has been included, but we can find out its value by noting that at $t = 0$ the magnetization is $M_z(0)$. So at $t = 0$ Eq. 9.13 becomes

$$\ln \left(M_z(0) - M_z^0 \right) = \text{const.}$$

Putting this value of the constant back into Eq. 9.13 we have

$$\ln \left(M_z(t) - M_z^0 \right) = -R_z t + \ln \left(M_z(0) - M_z^0 \right).$$

Tidying this up gives

$$\ln \left(\frac{M_z(t) - M_z^0}{M_z(0) - M_z^0} \right) = -R_z t. \tag{9.14}$$

If we take exponentials of both sides we have

$$\frac{M_z(t) - M_z^0}{M_z(0) - M_z^0} = \exp \left(-R_z t \right),$$

If you are not familiar with the method used to solve this kind of differential equation, then just skip forward to the result, Eq. 9.15 on the next page.

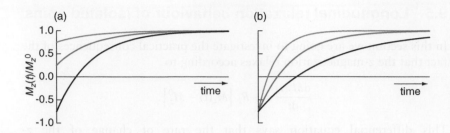

Fig. 9.13 Plots showing the way in which the z-magnetization approaches equilibrium, starting from a particular value, as predicted by Eq. 9.15. The quantity plotted along the vertical axis is $M_z(t)/M_z^0$, which takes the value one when the z-magnetization is at equilibrium. In (a), the three lines are computed for different values of the z-magnetization at time zero. Note that the further the system is from equilibrium, the more rapidly the magnetization changes. The plot in (b) shows the effect of increasing the rate constant R_z. The black line has the smallest value of R_z, the blue line has twice the value of the rate constant, and the grey line twice the value again. As expected, the greater the rate constant, the more rapidly the z-magnetization approaches its equilibrium value.

which can be rearranged to

$$M_z(t) = \left[M_z(0) - M_z^0 \right] \exp(-R_z t) + M_z^0. \qquad (9.15)$$

Figure 9.13 shows plots of $M_z(t)$, as predicted by this equation, for different initial z-magnetizations, $M_z(0)$, and for different values of the rate constant R_z. The common feature of all of these plots is that eventually the magnetization ends up at its equilibrium value. However, as shown in Fig. 9.13 (a), the further the magnetization is away from equilibrium, the faster the initial rate of change of the magnetization. Also, as shown in (b), the time that it takes the magnetization to reach its equilibrium value becomes shorter as the rate constant R_z becomes larger. This behaviour is in agreement with the qualitative predictions we made earlier.

9.5.1   Estimating the rate constant for longitudinal relaxation

It is often important to have an estimate for the rate constant for longitudinal relaxation, R_z, and we will see in this section that the analysis which led to Eq. 9.15 provides a convenient framework for doing this.

The method most commonly used for estimating R_z is the *inversion–recovery* experiment, whose pulse sequence is shown in Fig. 9.14. Initially, the magnetization is inverted by a 180° pulse, so that $M_z(0) = -M_z^0$. This inverted magnetization is then allowed to relax for a time τ. Using Eq. 9.15 with $t = \tau$ and $M_z(0) = -M_z^0$, we can see that the z-magnetization at time τ is

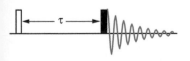

Fig. 9.14 Pulse sequence for the inversion–recovery experiment, used to estimate the value of the rate constant for longitudinal relaxation, R_z (or, alternatively, the time constant T_1).

$$
\begin{aligned}
M_z(\tau) &= -2M_z^0 \exp(-R_z \tau) + M_z^0 \\
&= M_z^0 \left[1 - 2\exp(-R_z \tau) \right].
\end{aligned}
$$

After time τ a 90° pulse is applied, the resulting FID observed, and then Fourier transformed to give the spectrum. A typical set of spectra

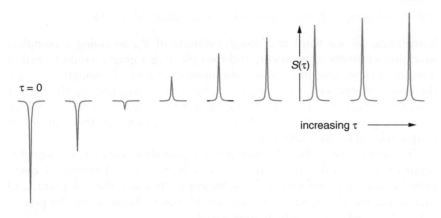

recorded for different values of τ is shown in Fig. 9.15. Note how the line starts negative, passes through zero and then becomes more positive as τ increases.

The height, $S(\tau)$, of the peak in the spectrum will be proportional to the size of the z-magnetization present just before the 90° pulse. Thus, $S(\tau)$ can be written

$$S(\tau) = c\left[1 - 2\exp\left(-R_z\tau\right)\right], \tag{9.16}$$

where c is some constant of proportion.

From Eq. 9.16, it follows that the peak height at time $\tau = 0$, $S(0)$, is $-c$. We can therefore replace the constant c in this equation by $-S(0)$ to give

$$S(\tau) = S(0)\left[2\exp\left(-R_z\tau\right) - 1\right],$$

which can be tidied up to

$$\frac{S(\tau)}{S(0)} = 2\exp\left(-R_z\tau\right) - 1.$$

This can be rearranged into a form which will give a straight-line plot in the following way:

$$
\begin{aligned}
\frac{S(\tau)}{S(0)} &= 2\exp\left(-R_z\tau\right) - 1 \\[2mm]
\frac{S(\tau)}{S(0)} + 1 &= 2\exp\left(-R_z\tau\right) \\[2mm]
\frac{S(\tau) + S(0)}{2S(0)} &= \exp\left(-R_z\tau\right) \\[2mm]
\ln\left(\frac{S(\tau) + S(0)}{2S(0)}\right) &= -R_z\tau,
\end{aligned}
$$

where, to go to the last line, we have taken natural logarithms of both sides.

This last line implies that if we plot $\ln\left(\left[S(\tau) + S(0)\right]/2S(0)\right)$ against τ we will obtain a straight line of slope $-R_z$. So, all we need to do is repeat the experiment over a suitable range of τ values, measure the peak heights, and make this plot.

9.5.2   Making a quick estimate of the relaxation rate constant

Sometimes, all we want is a rough estimate of R_z, so doing a complete inversion–recovery experiment, and then plotting a graph would be rather a waste of time. One method for obtaining such a rough estimate is to use the inversion–recovery pulse sequence, but rather than varying the delay τ systematically, we just try a few values until we locate the time at which the signal goes through a null. As we will see in a moment, this null time is simply related to the value of R_z.

It is quite easy to find the null point as, for short times τ, the signal is negative, whereas for longer times it will be positive. A couple of quick experiments are therefore usually sufficient to 'bracket' the null point, and then a few more trial values usually enables us to home in on the precise value of the delay, τ_{null}, which gives a null.

From Eq. 9.16 on the previous page, the peak height $S(\tau)$ will be zero when

$$0 = c\left[1 - 2\exp\left(-R_z\tau_{null}\right)\right].$$

This can be rearranged to

$$\exp\left(-R_z\tau_{null}\right) = \tfrac{1}{2}$$
$$\text{hence } R_z = -\ln\tfrac{1}{2}/\tau_{null} \qquad \text{or} \qquad R_z = \ln 2/\tau_{null}.$$

So, simply by finding the value of τ_{null}, we can obtain an estimate of R_z. If we want T_1 instead, then the relationship is simply $T_1 = \tau_{null}/\ln 2$.

9.5.3   How long do I have to leave between experiments?

As we noted at the start of this chapter, the rate of longitudinal relaxation determines the time we have to leave between experiments in order to allow the system to come to equilibrium. We can now make some estimates, in terms of R_z (or T_1), as to how long this time actually has to be.

The first thing to realize is that the time it takes to get back to equilibrium depends on where you start from. For example, if we start out with the magnetization being inverted (i.e. along $-z$), then it will take longer to get back to equilibrium than if we start out with no z-magnetization.

Just exactly where the magnetization ends up at the end of a pulse sequence depends on the details of that sequence, so we cannot come to any general conclusions. However, assuming that there is no z-magnetization at the end of the sequence is a reasonable choice, as most sequences finish with a 90° pulse followed by data acquisition. The chances are that this pulse will rotate all of the magnetization into the transverse plane.

Let us suppose, therefore, that at the end of data acquisition there is no z-magnetization i.e. $M_z(0) = 0$. What we are going to work out is the relaxation delay, t_r, needed for the z-magnetization to return to a fraction f of its equilibrium value, that is when $M_z(t_r) = f M_z^0$. For complete return to equilibrium $f = 1$, whereas for a return to 90% of the equilibrium magnetization, $f = 0.9$.

Equation 9.14 on page 263 is a useful place to start, with $t = t_r$:

$$\ln\left(\frac{M_z(t_r) - M_z^0}{M_z(0) - M_z^0}\right) = -R_z t_r.$$

We put $M_z(0) = 0$ and $M_z(t_r) = fM_z^0$ to give

$$\ln\left(\frac{fM_z^0 - M_z^0}{-M_z^0}\right) = -R_z t_r.$$

This simplifies to

$$t_r = -\frac{1}{R_z}\ln(1 - f) \quad \text{or} \quad t_r = -T_1 \ln(1 - f).$$

What this tells us is that it takes an infinite amount of time for the spins to return completely to equilibrium ($f = 1$), a result which comes about because the rate of change of the magnetization gets slower the closer the magnetization is to equilibrium.

If, however, we lower our sights and accept 99% of the equilibrium magnetization, then the above expression predicts for $f = 0.99$ that $t_r = 4.6 \times T_1$. If 95% is acceptable, then $t_r = 3.0 \times T_1$, and for 90% $t_r = 2.3 \times T_1$. These values are illustrated in Fig. 9.16.

Therefore leaving a delay of around five times T_1 will guarantee that, to all intents and purposes, the magnetization has recovered to equilibrium. If we are more impatient, then three times T_1 will give us 95% of the magnetization – a value which is often regarded as being an acceptable compromise. Note that the time during which the FID is recorded can actually be counted into the relaxation delay t_r, as longitudinal relaxation is taking place during the FID.

Difficulties arise when not all of the spins in the sample have the same relaxation time T_1. The most conservative approach is to set the relaxation delay according to the longest T_1, but if this results in an unacceptably long delay, we might choose to set the relaxation delay according to an average value of T_1.

Often it is the solvent which has the slowest relaxation, and hence the longest T_1. Since we are not usually interested in the solvent signal, it is common to set the relaxation delay according to the behaviour of the interesting solute signals, and simply ignore the behaviour of the solvent.

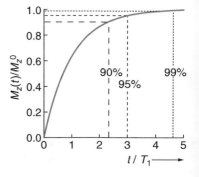

Fig. 9.16 Illustration of how long it takes to recover a certain proportion of the equilibrium z-magnetization, starting from $M_z = 0$. The horizontal time axis is expressed as a multiple of the longitudinal relaxation time, T_1. For the magnetization to recover to 99% of its equilibrium value (i.e. $M_z/M_z^0 = 0.99$) takes almost $5 \times T_1$, but recovery to 95% only takes about $3 \times T_1$. The difference in these values arises from the fact that the rate of change of the z-magnetization gets slower the closer we are to equilibrium.

9.6 Longitudinal dipolar relaxation of two spins

We now turn to a very important topic which is the relaxation behaviour of two spins which are interacting via the dipolar mechanism described in section 9.2.1 on page 249. The relaxation in this system has features which do not occur in a one-spin system, in particular it is only in a two-spin system that the phenomenon of *cross relaxation*, which gives rise to the very important nuclear Overhauser effect (NOE), is seen. We will therefore spend quite some time exploring dipolar relaxation and its consequences.

9.6.1 Energy levels and transition rates

We introduced and derived the energy levels for two coupled spins in section 3.5 on page 35. For the present discussion, it is not necessary for the spins to be coupled, but as we saw in section 3.5, the presence of the scalar

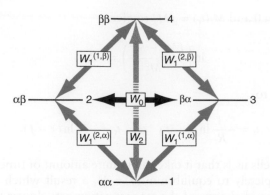

Fig. 9.17 Dipolar relaxation causes transitions between all four energy levels of a two-spin system. Four of the transitions involve a change in the total magnetic quantum number, M, by ± 1. The corresponding rate constants are thus given a subscript '1', and a superscript showing which spin is flipping and the spin state of the passive spin. So the transition from $\beta\alpha$ to $\beta\beta$ has rate constant $W_1^{(2,\beta)}$ as spin two is flipping, and spin one is in the β state. There is one double-quantum transition, with rate constant W_2, and one zero-quantum transition, with rate constant W_0; as before, the subscript gives the value of ΔM. The arrangement of the energy levels shown here is appropriate for a homonuclear spin system; in a heteronuclear system the $\alpha\beta$ and $\beta\alpha$ levels will not have the same energy. However, the same set of relaxation-induced transitions occur.

coupling does not change the wavefunctions associated with the levels, but only shifts their energies very slightly.

These four energy levels can be labelled with the spin states of each spin, so the level $\alpha\alpha$ has spin one and spin two in the α state, whereas $\beta\alpha$ has spin one in the β state and spin two in the α state.

The dipolar interaction between two spins can lead to relaxation-induced transitions between *any* of these four levels, as is illustrated in Fig. 9.17. To demonstrate that this is so we would need to look at the detailed form of the Hamiltonian which describes the dipolar interaction – something which is beyond the level of this text.

Each transition has associated with it a rate constant, $W_{\Delta M}$, where the subscript gives the change, ΔM, in the total magnetic quantum number associated with the transition. Four of the transitions can be characterized as single quantum, $\Delta M = 1$, and we can further distinguish them according to which of spin one or spin two is flipping, and the spin state of spin which is not flipping (the passive spin). So, the transition from $\alpha\alpha$ to $\beta\alpha$ has rate constant $W_1^{(1,\alpha)}$ as it is spin one which is flipping, and spin two is in the α state. Similarly, the transition from $\alpha\alpha$ to $\alpha\beta$ has rate constant $W_1^{(2,\alpha)}$ as spin two is flipping, and the passive spin is in the α state.

The single-quantum relaxation rates depend on the spectral density at the frequency corresponding to the transition, which are, of course, the Larmor frequencies of the spin which is flipping. So $W_1^{(1,\alpha)}$ depends on the spectral density at the Larmor frequency of spin one, $\omega_{0,1}$, whereas $W_1^{(2,\beta)}$ depends on the spectral density at the Larmor frequency of spin two, $\omega_{0,2}$.

There is one double-quantum transition, between states $\alpha\alpha$ and $\beta\beta$, with transition rate constant W_2. This depends on the spectral density at the sum of the Larmor frequencies of spins one and two: $(\omega_{0,1} + \omega_{0,2})$. The zero-quantum transition rate constant, between states $\alpha\beta$ and $\beta\alpha$, is W_0, and this depends on the spectral density at the difference of the two Larmor frequencies: $(\omega_{0,1} - \omega_{0,2})$.

9.6.2 Rate equations for the populations and z-magnetizations

Just as we did for a single spin, we can work out differential equations for the rate of change of the population of each level. Assuming that the rate is proportional to the deviation from the equilibrium population, we can write the rate of change of the population of level 1 as

$$\frac{dn_1}{dt} = \underbrace{-W_1^{(2,\alpha)}(n_1 - n_1^0) - W_1^{(1,\alpha)}(n_1 - n_1^0) - W_2(n_1 - n_1^0)}_{\text{loss from level 1}}$$

$$+ \underbrace{W_1^{(2,\alpha)}(n_2 - n_2^0)}_{\text{gain from level 2}} + \underbrace{W_1^{(1,\alpha)}(n_3 - n_3^0)}_{\text{gain from level 3}} + \underbrace{W_2(n_4 - n_4^0)}_{\text{gain from level 4}},$$

where n_i is the population of the ith level, and n_i^0 is the equilibrium population of that level. The first three terms are all negative as they represent processes in which population is lost from level 1. In addition, the rates all depend on the deviation of the population of level 1 from its equilibrium value, as it is from this level that the transition is coming.

The first positive term represents a process by which the population of level 1 is increased as a result of transitions from level 2. The rate of the process therefore depends on the population of level 2. Similarly, the second and third positive terms all represent transitions in which the population of level 1 is increased.

We can write similar differential equations for each of the populations. The resulting four equations, although simple to construct, are certainly rather complex in form. However, things can be improved by rewriting the populations in terms of the z-magnetization of the two spins.

For example, following what we did before for a single spin, the z-magnetization of spin one depends on the population difference between levels 1 and 3, and between 2 and 4. Both of these population differences contribute to the spin one z-magnetization, as both transitions belong to spin one. Leaving out any constants of proportion, we can write the z-magnetization of spin one as

$$I_{1z} = (n_1 - n_3) + (n_2 - n_4).$$

We have written this as I_{1z} rather than $M_{1,z}$ in order to emphasize the connection between this z-magnetization and the operator $\hat{I}_{1z}$, which represents it in quantum mechanics.

Recognizing that transitions 1–2 and 3–4 belong to spin two, we can write

$$I_{2z} = (n_1 - n_2) + (n_3 - n_4),$$

for the z-magnetization from the second spin. It turns out that we need another magnetization term which depends on the *difference* between the population differences across the two spin-one transitions:

$$2I_{1z}I_{2z} = (n_1 - n_3) - (n_2 - n_4).$$

As with I_{1z} and I_{2z}, this term is denoted $2I_{1z}I_{2z}$ as it is related to the product operator $2\hat{I}_{1z}\hat{I}_{2z}$. In fact, simply by rearranging the populations on the right of this equation, we can see that $2I_{1z}I_{2z}$ is also the difference between the population differences across the two spin-two transitions.

$$2I_{1z}I_{2z} = (n_1 - n_2) - (n_3 - n_4).$$

$2I_{1z}I_{2z}$ is often called a 'zz term'.

The magnetizations also have equilibrium values defined in terms of the equilibrium populations:

$$I_{1z}^0 = n_1^0 - n_3^0 + n_2^0 - n_4^0,$$

and similarly for I_{2z}^0. It turns out that the equilibrium value of $2I_{1z}I_{2z}$ is zero.

After a lot of tedious algebra, the rate equations for the populations can be rewritten in terms of rate equations for the magnetizations:

$$\begin{aligned}
\frac{dI_{1z}}{dt} &= -R_z^{(1)}(I_{1z} - I_{1z}^0) - \sigma_{12}(I_{2z} - I_{2z}^0) - \Delta^{(1)}\,2I_{1z}I_{2z} \\
\frac{dI_{2z}}{dt} &= -\sigma_{12}(I_{1z} - I_{1z}^0) - R_z^{(2)}(I_{2z} - I_{2z}^0) - \Delta^{(2)}\,2I_{1z}I_{2z} \qquad (9.17) \\
\frac{d\,2I_{1z}I_{2z}}{dt} &= -\Delta^{(1)}(I_{1z} - I_{1z}^0) - \Delta^{(2)}(I_{2z} - I_{2z}^0) - R_z^{(1,2)}\,2I_{1z}I_{2z}.
\end{aligned}$$

The various rate constants are defined as follows in terms of the rate constants for the individual transitions:

$$\begin{aligned}
R_z^{(1)} &= W_1^{(1,\alpha)} + W_1^{(1,\beta)} + W_2 + W_0 \\
R_z^{(2)} &= W_1^{(2,\alpha)} + W_1^{(2,\beta)} + W_2 + W_0 \\
\sigma_{12} &= W_2 - W_0 \\
\Delta^{(1)} &= W_1^{(1,\alpha)} - W_1^{(1,\beta)} \\
\Delta^{(2)} &= W_1^{(2,\alpha)} - W_1^{(2,\beta)} \\
R_z^{(1,2)} &= W_1^{(1,\alpha)} + W_1^{(1,\beta)} + W_1^{(2,\alpha)} + W_1^{(2,\beta)}.
\end{aligned}$$

The rate constant $R_z^{(1)}$ describes the *self relaxation* of spin one, meaning that this rate constant simply determines the rate at which the magnetization I_{1z} approaches equilibrium without the involvement of I_{2z} and $2I_{1z}I_{2z}$. Similarly, $R_z^{(2)}$ is the self-relaxation rate constant for spin two.

The rate constant σ_{12} describes the rate at which magnetization from spin one is transferred, by relaxation processes, to spin two. We make this interpretation as, in the differential equation for I_{1z}, there is a term $\sigma_{12}(I_{2z} - I_{2z}^0)$ which says that the rate of change of I_{1z} is proportional to I_{2z}, with σ_{12} as the constant of proportion. There is a similar process which transfers magnetization from spin one to spin two; it turns out to have the same rate constant.

This relaxation-mediated transfer of z-magnetization from one spin to another is called *cross relaxation*. It is an (almost) unique feature of dipolar relaxation and is, as we shall see shortly, responsible for the NOE. It is interesting to note that in scalar coupled systems we saw that it was possible to transfer transverse anti-phase magnetization from one spin to another using RF pulses. Now we have a second mechanism for transfer, but this time it involves dipolar relaxation and z-magnetization; no scalar coupling is required.

Following the same discussion as above, $R_z^{(1,2)}$ is the rate constant for self relaxation of the $2I_{1z}I_{2z}$ term. Finally $\Delta^{(1)}$ and $\Delta^{(2)}$ are the rate constants for the interconversion of I_{1z} with $2I_{1z}I_{2z}$, and of I_{2z} with $2I_{1z}I_{2z}$. These various relaxation pathways are visualized in Fig. 9.18.

For dipolar relaxation between two spins it turns out that $W_1^{(1,\alpha)} = W_1^{(1,\beta)}$ and $W_1^{(2,\alpha)} = W_1^{(2,\beta)}$; however, in more complex situations, such as that discussed in section 9.11 on page 306, these rate constants can be different. Writing these rate constants as $W_1^{(1)}$ and $W_1^{(2)}$, respectively, the above differential equations and expressions for the rate constants simplify considerably as $\Delta^{(1)} = 0$ and $\Delta^{(2)} = 0$:

$$\frac{dI_{1z}}{dt} = -R_z^{(1)}(I_{1z} - I_{1z}^0) - \sigma_{12}(I_{2z} - I_{2z}^0)$$

$$\frac{dI_{2z}}{dt} = -\sigma_{12}(I_{1z} - I_{1z}^0) - R_z^{(2)}(I_{2z} - I_{2z}^0) \qquad (9.18)$$

$$\frac{d\,2I_{1z}I_{2z}}{dt} = -R_z^{(1,2)}\,2I_{1z}I_{2z}.$$

The rate constants simplify to:

$$R_z^{(1)} = 2W_1^{(1)} + W_2 + W_0$$

$$R_z^{(2)} = 2W_1^{(2)} + W_2 + W_0$$

$$\sigma_{12} = W_2 - W_0 \qquad (9.19)$$

$$R_z^{(1,2)} = 2W_1^{(1)} + 2W_1^{(2)}.$$

We now see that there is still cross relaxation between I_{1z} and I_{2z}, but there is no relaxation-induced transfer between $2I_{1z}I_{2z}$ and either of I_{1z} or I_{2z}. Equations 9.18 are often called the *Solomon equations*; we will use them extensively in the following discussion.

9.6.3   Relaxation rate constants

The theory of how the rate constants W for transitions between individual levels are computed is beyond the scope of this text. No new quantum mechanical concepts are needed to make these calculations – it is just that the details are rather involved because we are dealing with time-dependent random processes.

What we find is that the theory predicts that the transition rate constant between two levels i and j, W_{ij}, is always the product of three terms:

$$W_{ij} = A_{ij} \times Y^2 \times j(\omega_{ij}).$$

We will consider each of these terms in turn.

A_{ij} is a number which arises from the details of the Hamiltonian which represents the particular interaction which is causing relaxation. Y^2 is a term which is related to the magnitude of the local fields which are causing

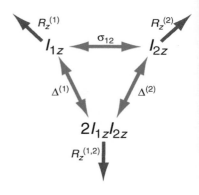

Fig. 9.18 Visualization of the relaxation pathways between different kinds of z-magnetization for two spins undergoing relaxation via the dipolar mechanism. The dark grey arrows pointing to the edges of the diagram show the self-relaxation processes i.e. those in which the magnetization simply returns to its equilibrium value. In addition, there are processes, indicated by the blue arrows, which connect the z-magnetization terms of the two spins. The most important of these is cross relaxation, with rate constant σ_{12}, which transfers magnetization between the two spins.

relaxation. The term is quite deliberately written as the square since it turns out that the rate of relaxation always depends on the average of the *square* of the local fields. Generally, Y^2 depends on the physical details of the interaction, such as the distance between two spins or the size of the chemical shift anisotropy.

Finally, $j(\omega_{ij})$ is the reduced spectral density at the frequency of the transition between the two energy levels. As we have already seen in section 9.3.2 on page 256, this is a measure of the amount of the random motion which is at the correct frequency, here ω_{ij}, needed to cause transitions between the two levels.

In the case of dipolar relaxation between two spins, the rate constants defined above are given by:

$$W_1^{(1)} = \tfrac{3}{40} b^2 j(\omega_{0,1}) \qquad W_1^{(2)} = \tfrac{3}{40} b^2 j(\omega_{0,2})$$
$$W_2 = \tfrac{3}{10} b^2 j(\omega_{0,1} + \omega_{0,2}) \qquad W_0 = \tfrac{1}{20} b^2 j(\omega_{0,1} - \omega_{0,2}),$$

where b is

$$b = \frac{\mu_0 \gamma_1 \gamma_2 \hbar}{4\pi r^3}.$$

In this expression μ_0 is a physical constant called the permeability of vacuum and which has the value $4\pi \times 10^{-7}$ H m^{-1}; 'H' stands for Henries, the unit of inductance. γ_1 and γ_2 are the gyromagnetic ratios of the two spins, and r is the distance between them. As described above, each rate constant consists of the product of a number, the square of a size factor and a spectral density.

Note that W_2 depends on the spectral density at the sum of the Larmor frequencies of the two spins, as this is the frequency of the transition between the $\alpha\alpha$ and $\beta\beta$ states, which are connected by W_2. Similarly, W_0 depends on the spectral density at the difference of the two Larmor frequencies, as this is the frequency of the $\alpha\beta \leftrightarrow \beta\alpha$ transition.

Using these expressions for the various rate constants between levels, the rate constants given in Eq. 9.19 on the preceding page can be written as

$$R_z^{(1)} = b^2 \left[\tfrac{3}{20} j(\omega_{0,1}) + \tfrac{3}{10} j(\omega_{0,1} + \omega_{0,2}) + \tfrac{1}{20} j(\omega_{0,1} - \omega_{0,2}) \right]$$
$$R_z^{(2)} = b^2 \left[\tfrac{3}{20} j(\omega_{0,2}) + \tfrac{3}{10} j(\omega_{0,1} + \omega_{0,2}) + \tfrac{1}{20} j(\omega_{0,1} - \omega_{0,2}) \right]$$
$$\sigma_{12} = b^2 \left[\tfrac{3}{10} j(\omega_{0,1} + \omega_{0,2}) - \tfrac{1}{20} j(\omega_{0,1} - \omega_{0,2}) \right] \tag{9.20}$$
$$R_z^{(1,2)} = b^2 \left[\tfrac{3}{20} j(\omega_{0,1}) + \tfrac{3}{20} j(\omega_{0,2}) \right].$$

Note how each rate constant depends on the spectral density at more than one frequency.

9.6.4   Cross relaxation in the two motional regimes

The self-relaxation rate constants $R_z^{(1)}$ and $R_z^{(2)}$ are always positive, as they are the sum of positive terms. However, the cross-relaxation rate constant σ_{12} is the difference of two terms, and so it may be positive or negative. As we will see in this section, it turns out that the sign of σ_{12} depends on an interplay between the sizes of the Larmor frequencies and the correlation time. We will also discover in a subsequent section that the sign of the cross-relaxation rate constant is of considerable importance in the theory of the NOE.

The situation which is of most interest is when the two spins are of the same type e.g. both protons, in which case they both have the same Larmor frequency (the tiny difference due to chemical shifts is of no significance for relaxation). In this case $j(\omega_{0,1} + \omega_{0,2})$ becomes $j(2\omega_0)$ and $j(\omega_{0,1} - \omega_{0,2})$ becomes $j(0)$. So, from Eq. 9.20 on the preceding page the cross-relaxation rate constant is

$$\sigma_{12} = \underbrace{b^2 \tfrac{3}{10} j(2\omega_0)}_{W_2} - \underbrace{b^2 \tfrac{1}{20} j(0)}_{W_0}.$$

The underbraces remind us of the origin of the two terms. It is interesting to examine the value of σ_{12} in the two motional regimes, described in section 9.3.3 on page 257.

In the fast motion limit, the reduced spectral density is simply $2\tau_c$ at all frequencies, so:

$$\begin{aligned}
\text{fast motion:}\quad \sigma_{12} &= \underbrace{b^2 \tfrac{3}{10} j(2\omega_0)}_{W_2} - \underbrace{b^2 \tfrac{1}{20} j(0)}_{W_0} \\
&= b^2 \tfrac{3}{10} 2\tau_c - b^2 \tfrac{1}{20} 2\tau_c \\
&= \tfrac{1}{2} b^2 \tau_c.
\end{aligned}$$

From the final expression we see that in the fast motion limit σ_{12} is clearly positive. Looking back through the calculation we can see that this comes about because $W_2 > W_0$.

In the slow motion limit $j(0)$ is still $2\tau_c$, but $j(2\omega_0)$ is negligible compared with $j(0)$, so:

$$\begin{aligned}
\text{slow motion:}\quad \sigma_{12} &= \underbrace{b^2 \tfrac{3}{10} j(2\omega_0)}_{W_2} - \underbrace{b^2 \tfrac{1}{20} j(0)}_{W_0} \\
&= 0 - b^2 \tfrac{1}{20} 2\tau_c \\
&= -\tfrac{1}{10} b^2 \tau_c.
\end{aligned}$$

Now, σ_{12} is negative, which we can see comes about because $W_0 > W_2$.

The cross-over point, where $\sigma_{12} = 0$, occurs when $W_2 = W_0$:

$$\text{cross-over:}\quad \underbrace{b^2 \tfrac{3}{10} j(2\omega_0)}_{W_2} = \underbrace{b^2 \tfrac{1}{20} j(0)}_{W_0}.$$

Simply by substituting in the expression for $j(\omega)$ (Eq. 9.4 on page 257) we can see that this cross-over occurs when

$$\omega_0 \tau_c = \sqrt{\frac{5}{4}}.$$

For protons at a Larmor frequency of 500 MHz, the correlation time at this cross-over point is 360 ps, as is illustrated in Fig. 9.19. This value of the correlation time is typical of a medium-sized molecule in a more viscous solvent such as water. Small to medium-sized molecules in less

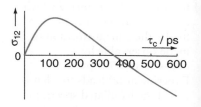

Fig. 9.19 Illustration of how the cross-relaxation rate constant, σ_{12}, changes sign from positive to negative as the correlation time is increased. The graph is computed for two protons with a Larmor frequency of 500 MHz, so the cross-over point is at $\tau_c \approx 360$ ps.

viscous solvents, such as $CDCl_3$, will have smaller correlation times than this, whereas large molecules, such as proteins and polysaccharides, will have much longer correlation times than this.

In the next section we will discover that the sign of the cross-relaxation rate constant has important consequences when it comes to the NOE.

9.7 The NOE

The differential equation of the rate of change of the z-magnetization:

$$\frac{dI_{1z}}{dt} = -R_z^{(1)}(I_{1z} - I_{1z}^0) - \sigma_{12}(I_{2z} - I_{2z}^0), \qquad (9.21)$$

tells us that if spin two is not at equilibrium i.e. $(I_{2z} - I_{2z}^0) \neq 0$, the rate of change of the z-magnetization on spin one will have a contribution which is proportional to the cross-relaxation rate constant, σ_{12}. This rate constant will only be non-zero if there is dipolar relaxation between the two spins, so if we find that the behaviour of the z-magnetization of spin one is affected by the amount of z-magnetization on spin two, we can deduce that cross relaxation must be taking place.

In section 9.6.3 on page 271, we saw that $\sigma_{12} \propto b^2$, so that the cross-relaxation rate goes as $1/r^6$. The rate thus falls off rather rapidly with distance, so in practice we are only likely to be able to observe the effects of cross relaxation between spins which are reasonably close to one another. In practice, for two protons, this means a distance of less than about 5 Å. Thus, if we see evidence of cross relaxation between two spins, we can be sure that they are reasonably close in space.

Cross relaxation leads to what is called the nuclear Overhauser effect (NOE), which is an exceptionally important tool when it comes to structural studies by NMR. In the following sections we will look at different ways in which cross relaxation, and hence the NOE, can be detected. All of these experiments have it in common that one spin is perturbed away from equilibrium, and then the effect of this on the z-magnetization of the other spin is determined.

9.7.1 The transient NOE experiment

The pulse sequence for the simplest transient NOE experiment is shown in Fig. 9.20. This is a difference experiment, in which the spectrum arising from sequence (b) is subtracted from that arising from sequence (a). As we shall see, taking this difference reveals the presence of the NOE in a particularly convenient way.

In sequence (a) the first thing which happens is that the z-magnetization from spin two is inverted by a selective 180° pulse (see section 4.11 on page 67); the z-magnetization from spin one is not affected by the pulse. Therefore immediately after this pulse (time $\tau = 0$) the z-magnetization of the two spins can be written as

$$\text{at } \tau = 0: \qquad I_{1z}(0) = I_{1z}^0 \quad \text{and} \quad I_{2z}(0) = -I_{2z}^0. \qquad (9.22)$$

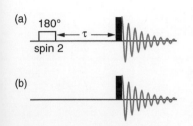

Fig. 9.20 Pulse sequence for the simple transient NOE experiment. There are two parts of the experiment. In the first part, (a), the z-magnetization of spin two is inverted by a selective 180° pulse. There then follows a delay τ during which cross relaxation takes place, and finally the z-magnetization is made observable by a non-selective 90° pulse. Experiment (a) leads to what is called the irradiated spectrum. The second experiment, (b), is simply a pulse–acquire sequence; this gives us the reference spectrum. By subtracting the reference spectrum from the irradiated spectrum, we obtain an NOE difference spectrum in which the presence of any cross relaxation to spin two is revealed.

These are the initial conditions which we would need to take into account in solving the differential equation, Eq. 9.21 on the facing page, which describes how I_{1z} varies with time.

Solving this differential equation in general terms is not fundamentally very difficult, but is a more complex task than we want to get involved in here. We are going to take a simpler approach, by solving the equation in what is called the *initial rate limit*, which only applies at short times. In this approach we assume that on the right-hand side of Eq. 9.21 I_{1z} and I_{2z} have their initial values i.e. the values at time zero given in Eq. 9.22 on the preceding page. Applying this idea gives

$$
\begin{aligned}
\left(\frac{dI_{1z}}{dt}\right)_{\text{init}} &= -R_z^{(1)}\left[I_{1z}(0) - I_{1z}^0\right] - \sigma_{12}\left[I_{2z}(0) - I_{2z}^0\right] \\
&= -R_z^{(1)}(I_{1z}^0 - I_{1z}^0) - \sigma_{12}(-I_{2z}^0 - I_{2z}^0) \\
&= 2\sigma_{12}I_{2z}^0.
\end{aligned}
\tag{9.23}
$$

We have had to put the subscript 'init' on the differential to remind us that this expression only applies in the initial rate limit. This approximation is valid for times short enough that the z-magnetizations have not changed by very much from their initial values.

It is now easy to separate the variables (as we did in section 9.5 on page 263) in Eq. 9.23, and then integrate both sides:

$$
\begin{aligned}
\frac{dI_{1z}(t)}{dt} &= 2\sigma_{12}I_{2z}^0 \\
dI_{1z}(t) &= 2\sigma_{12}I_{2z}^0\,dt \\
\int dI_{1z}(t) &= \int 2\sigma_{12}I_{2z}^0\,dt \\
I_{1z}(t) &= 2\sigma_{12}tI_{2z}^0 + \text{const.}
\end{aligned}
$$

In these manipulations we have written $I_{1z}(t)$ to remind ourselves that I_{1z} depends on time. To go to the second line, we have simply taken the term dt over to the right: this separates the variables into $I_{1z}(t)$ on the left and t on the right. The two integrals are easy to compute as $\sigma_{12}I_{2z}^0$ is just a constant.

We can find the constant of integration since we know that at time zero $I_{1z}(0) = I_{1z}^0$; the constant is therefore I_{1z}^0. Thus overall the z-magnetization of spin one at time $t = \tau$ is

$$
I_{1z}(\tau) = 2\sigma_{12}\tau I_{2z}^0 + I_{1z}^0.
$$

What this says is that there will be a contribution to the z-magnetization of spin one which is proportional to the time τ and the cross-relaxation rate constant σ_{12}: this is what leads to the NOE.

The next step is to work out what has happened to the z-magnetization from spin two. Again, we start with the differential equation

$$
\frac{dI_{2z}}{dt} = -R_z^{(2)}(I_{2z} - I_{2z}^0) - \sigma_{12}(I_{1z} - I_{1z}^0),
$$

and impose the initial conditions as before $I_{1z}(0) = I_{1z}^0$ and $I_{2z}(0) = -I_{2z}^0$ to give

$$
\left(\frac{dI_{2z}}{dt}\right)_{\text{init}} = 2R_z^{(2)}I_{2z}^0.
$$

This has the same form as Eq. 9.23 on the preceding page, but with a different rate constant. Integrating as before gives

$$I_{2z}(t) = 2R_z^{(2)}tI_{2z}^0 + \text{const.}$$

The initial condition is $I_{2z}(0) = -I_{2z}^0$, so the constant of integration is $-I_{2z}^0$. Thus, the z-magnetization from spin two as a function of the time τ is given by

$$I_{2z}(\tau) = 2R_z^{(2)}\tau I_{2z}^0 - I_{2z}^0.$$

Since $R_z^{(2)}\tau$ is positive, this equation says that as τ increases the initially inverted z-magnetization becomes less negative, which is the expected result as the magnetization is moving towards equilibrium.

So far we have computed the z-magnetization for each spin as a function of the time τ in pulse sequence (a) of Fig. 9.20 on page 274. For pulse sequence (b), the situation is very simple as both spins are at equilibrium just prior to the 90° pulse. The results for both experiments are summarized in the following table:

experiment	spin-one magnetization	spin-two magnetization
(a)	$I_{1z}(\tau) = 2\sigma_{12}\tau I_{2z}^0 + I_{1z}^0$	$I_{2z}(\tau) = 2R_z^{(2)}\tau I_{2z}^0 - I_{2z}^0$
(b)	I_{1z}^0	I_{2z}^0

The 90° pulse in both experiments rotates any z-magnetization into the transverse plane, where it is observed as an FID. Fourier transformation of the FID will give us a spectrum in which there is one peak at the offset (shift) of spin one, and similarly a peak at the offset of spin two.

The height of the peak for spin one, $S_1(\tau)$, will be proportional to $I_{1z}(\tau)$, and similarly the height of the spin-two peak, $S_2(\tau)$, will be proportional to $I_{1z}(\tau)$. Furthermore, if both spins are of the same type (e.g. proton), their equilibrium z-magnetizations will be equal: $I_{1z}^0 = I_{2z}^0$. Thus, the data in the table above can be converted into peak heights (in the table, c is the constant of proportion):

spectrum	$S_1(\tau)$	$S_2(\tau)$
irradiated: (a)	$c(1 + 2\sigma_{12}\tau)$	$c\left(2R_z^{(2)}\tau - 1\right)$
reference: (b)	c	c
NOE difference: (a) – (b)	$c\,2\sigma_{12}\tau$	$c\left(2R_z^{(2)}\tau - 2\right)$

Spectrum (a) is called the *irradiated* spectrum, as it is from the experiment in which one of the spins was inverted. Spectrum (b) is called the *reference* spectrum; it is simply the normal spectrum. The table also shows the peak heights in what is known as the *NOE difference* spectrum, found by taking (a) – (b).

To interpret these results we will assume, for the sake of argument, that σ_{12} is positive. We must also remember that these results are only valid in the initial rate limit, which in practice means that $\sigma_{12}\tau \ll 1$ and $R_z^{(2)}\tau \ll 1$.

In spectrum (a), the peak for spin one is, as a result of cross relaxation, a little higher than the corresponding peak in the reference experiment, (b). We say that the peak has received an *NOE enhancement* or, more loosely, that the peak 'has an NOE'. The presence of this enhancement is revealed by subtracting the reference spectrum (b) from the irradiated spectrum (a), as doing so just leaves the intensity which arose due to cross relaxation i.e. the NOE enhancement. The difference (a) − (b) is called the *NOE difference spectrum*.

The peak for spin two in spectrum (a) is negative, on account of the fact that the spin was inverted, with a peak height which is very close to minus the peak height in the reference experiment (b). Thus, in the NOE difference spectrum, the spin-two peak is negative, with a peak height of more or less twice that in the reference spectrum. The process of computing the NOE difference spectrum is shown in Fig. 9.21.

If there is no cross relaxation i.e. $\sigma_{12} = 0$, then the intensity of the spin-one peak in the irradiated spectrum (a) will be the same as in the reference spectrum (b), and so the spin-one peak will not appear in the NOE difference spectrum. This difference spectrum is therefore a very direct way of seeing which spins are cross relaxing with the spin which is inverted (in this case, spin two). In this spectrum, the only peaks we will see are those which receive an NOE enhancement and the one which was initially irradiated.

In practice, it is rather difficult to record high-quality NOE difference spectra using the simple pulse sequence described in this section. However, by replacing the selective 180° pulse with an alternative inversion sequence which uses pulsed field gradients, it is possible to obtain excellent spectra on a routine basis. The details of this modified experiment are given in section 11.16 on page 432.

The NOE enhancement

The size of the NOE enhancement, η, is expressed as a fraction in the following way:

$$\eta = \frac{\text{peak height in irradiated spectrum} - \text{peak height in reference spectrum}}{\text{peak height in reference spectrum}}.$$

For the above example the enhancement is computed as

$$\eta = \frac{c(1 + 2\sigma_{12}\tau) - c}{c}$$
$$= 2\sigma_{12}\tau.$$

For molecules in the fast motion limit, σ_{12} is positive and so is the enhancement. In the slow motion limit, σ_{12}, and therefore the enhancement, is negative.

The larger the cross-relaxation rate constant σ_{12}, the greater the enhancement. Recall from section 9.6.3 on page 271 that $\sigma_{12} \propto r^{-6}$, so spins which are closer will have faster cross relaxation, and hence show larger enhancements. It is possible, therefore, to use the size of the enhancement as an indicator of the internuclear distance.

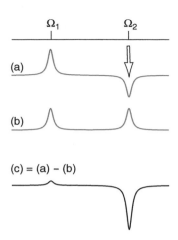

Fig. 9.21 Illustration of how NOE difference spectra are constructed. Spectrum (a) is the irradiated spectrum, recorded using sequence (a) of Fig. 9.20 on page 274. Spin two has been inverted, as indicated by the arrow and the negative intensity of the corresponding peak. In fact, in this spectrum the peak from spin one is slightly higher than in the reference spectrum, (b), on account of there being cross relaxation from spin two. This increase in peak height is most simply visualized by computing the NOE difference spectrum, (c), as (a) − (b). Now we can clearly see that spin one has received an NOE enhancement. In the difference spectrum, the irradiated peak appears with negative intensity. It has been assumed that the cross-relaxation rate constant is positive, so the NOE enhancement is positive.

To use the NOE enhancement as a *quantitative* measure of the distance turns out to be rather difficult due to a number of theoretical and practical limitations. To find out more about this, you should refer to the excellent text by Neuhaus and Williamson (see Further reading).

Longer mixing times

Remember that all of our calculations so far are in the initial rate limit. To work out what happens at longer times, we need to integrate the Solomon equations without making the restrictive assumptions we used above. We will not go into the details here but just describe the outcome.

To start with, the NOE enhancement increases linearly – this is what we predicted using the initial rate approximation. However, at longer times, the rate of increase of the enhancement starts to slow down, and eventually it reaches a maximum. After this, the enhancement steadily falls to zero.

The maximum value of the enhancement, and the time at which this occurs, is a function of the cross-relaxation and self-relaxation rate constants. Not surprisingly, the greater σ_{12}, the greater the maximum and the earlier time at which it occurs. The effect of increasing the self-relaxation rate constants is to decrease the maximum enhancement.

9.7.2 The steady-state NOE experiment

In this experiment, rather than inverting one of the spins, and then letting cross relaxation take place, the target spin is irradiated continuously with a weak RF field. The field is chosen to be weak enough that only the spin with which it is on-resonance is affected.

The result of this irradiation is to *saturate* the target spin, which means that its z-magnetization is forced to zero. The term saturation comes about from thinking about the populations of the two energy levels. Continuous irradiation eventually equalizes these populations, a situation which in spectroscopy is described as saturation. The population difference, and hence the z-magnetization, thus goes to zero.

The pulse sequence used to measure a steady-state NOE is shown in Fig. 9.22. As for the transient experiment, two spectra are recorded: one in which the target spin (here spin two) is saturated, and a reference spectrum in which all spins are at equilibrium.

As before, we can analyse the experiment using the Solomon equations. However, the approach is a little different to that used for the transient experiment. There are two key ideas needed here. The first is that as spin two is irradiated continuously it is kept saturated, and so we can assume that there is no spin-two magnetization i.e. $I_{2z} = 0$ at all times. The second is that if we saturate spin two for long enough we will reach a new steady state in which the spin-one magnetization is not changing with time, i.e.

$$\left(\frac{\mathrm{d}I_{1z}}{\mathrm{d}t}\right)_{SS} = 0;$$

the subscript 'SS' indicates the new steady state. As before we start with the Solomon equation for the spin-one magnetization (Eq. 9.21 on page 274):

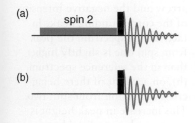

(a)   spin 2

(b)

Fig. 9.22 Pulse sequence used to record a steady-state NOE difference spectrum. Two experiments are needed: in (a) the target spin (here spin two) is irradiated with a weak field so as to saturate it. The irradiation is applied for long enough for a new steady state to be reached. Experiment (b) is just a normal pulse–acquire sequence, which leads to the reference spectrum. The NOE difference spectrum is found by taking (a) – (b).

$$\frac{dI_{1z}}{dt} = -R_z^{(1)}(I_{1z} - I_{1z}^0) - \sigma_{12}(I_{2z} - I_{2z}^0).$$

Applying the two conditions we have described gives

$$0 = -R_z^{(1)}(I_{1z,\text{SS}} - I_{1z}^0) - \sigma_{12}(-I_{2z}^0),$$

where $I_{1z,\text{SS}}$ is the steady-state value of the z-magnetization on spin one. Rearranging this last expression gives

$$I_{1z,\text{SS}} = \frac{\sigma_{12}}{R_z^{(1)}} I_{2z}^0 + I_{1z}^0.$$

As with the transient experiment, the z-magnetization on spin one is altered as a result of cross relaxation.

If we assume, as we did before, that $I_{1z}^0 = I_{2z}^0$, then the peak height of the spin-one resonance in the steady-state spectrum is $c(1 + \sigma_{12}/R_z^{(1)})$. Since spin two is saturated, no spin-two peak appears in the irradiated experiment. In the reference spectrum, both peaks have height c. Therefore in the NOE difference spectrum the spin-one peak will have height $c\sigma_{12}/R_z^{(1)}$, and the spin-two peak will have height $-c$; this is illustrated in Fig. 9.23. As before, the difference spectrum reveals the presence of cross relaxation in a very convenient way.

The NOE enhancement is given by

$$\begin{aligned}
\eta_{\text{SS}} &= \frac{c(1 + \sigma_{12}/R_z^{(1)}) - c}{c} \\
&= \frac{\sigma_{12}}{R_z^{(1)}}.
\end{aligned}$$

As in the transient experiment, an NOE enhancement will only be seen if the cross-relaxation rate constant is non-zero. However, in contrast to the transient experiment, the steady-state NOE depends not just on σ_{12}, but the ratio of this cross-relaxation rate constant to the self-relaxation rate constant of spin one (i.e. the spin which is enhanced).

Thus the size of the steady-state NOE does not simply reflect the amount of cross relaxation, but rather the balance between cross and self relaxation. Cross relaxation must be from the dipolar mechanism, but self relaxation can be from this and other sources (such as dissolved oxygen). We therefore have no way of relating the self-relaxation rate constant, and hence the NOE enhancement, to the internuclear distance. Steady-state NOE measurements can therefore only be used qualitatively.

The final point is to estimate how long it takes the spins to come to the new steady state once we start irradiating one of them. As we have seen, the steady-state enhancement depends on a balance between the self and cross relaxation of spin one, so the rate at which the system comes to a steady state clearly has something to do with these rate constants. It turns out that the smaller of these is always σ_{12}, so the value of this rate constant is the limiting factor in determining the rate at which the steady state is achieved.

For a first-order process, the reciprocal of the rate constant gives a related parameter known as the *time constant*. It can be shown that

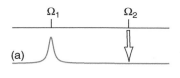

(a)

(b)

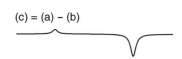

(c) = (a) − (b)

Fig. 9.23 Illustration of how a steady-state NOE difference spectrum is constructed. The spectra are similar to those of Fig. 9.21 on page 277 except that spin two is saturated, so does not appear in the irradiated spectrum (a). Subtracting the reference spectrum (b) from the irradiated spectrum (a) gives the NOE difference spectrum (c). As before, the peak from the irradiated spin is negative, and the enhancement is visible on spin one. It has been assumed that the cross-relaxation rate constant is positive.

equilibrium is only reached after waiting several times this time constant. So, in the case of the steady-state NOE, we need to wait several times $1/\sigma_{12}$. This can easily be several seconds. However, since there is little quantitative information that we can derive from the size of the steady-state NOE enhancements, it is not necessary to wait until the steady state has been reached.

9.7.3   Heteronuclear steady-state NOE

An important application of the steady-state NOE is to enhance the intensity of signals recorded from heteronuclei such as ^{13}C. The idea is that by irradiating the protons, the z-magnetization of the ^{13}C nuclei will be enhanced as a result of cross relaxation. Thus, when the z-magnetization is rotated into the transverse plane, stronger ^{13}C signals will be observed as a result of the NOE enhancement.

Since this is a heteronuclear experiment, we will switch to the IS notation, where the I spins are typically protons. Adapting the result of the previous section, the enhanced z-magnetization on the S spin (^{13}C) as a result of saturating the I spin (proton) is:

$$S_{z,SS} = \frac{\sigma_{IS}}{R_z^{(S)}} I_z^0 + S_z^0. \tag{9.24}$$

In section 9.6.2 on page 269 we defined I_z in terms of the population difference across the I spin transitions, so I_z^0 depends on the equilibrium population difference. It was shown in section 9.4.1 on page 259 that this difference is proportional to the gyromagnetic ratio of the spin, and so it follows that

$$\frac{S_z^0}{I_z^0} = \frac{\gamma_S}{\gamma_I}. \tag{9.25}$$

Therefore as γ for proton is about four times γ for ^{13}C, the equilibrium magnetization of proton is four times that of ^{13}C.

If we use Eq. 9.25 to write I_z^0 in terms of S_z^0 in Eq. 9.24, we find that the steady-state magnetization on the S spin is

$$S_{z,SS} = \frac{\gamma_I \, \sigma_{IS}}{\gamma_S \, R_z^{(S)}} S_z^0 + S_z^0.$$

Using this we can determine that the NOE enhancement of the S spin is

$$\eta_{SS} = \frac{\gamma_I \, \sigma_{IS}}{\gamma_S \, R_z^{(S)}}.$$

An estimate for this enhancement can be made if it is assumed that, for a ^{13}C with a directly attached proton, the self relaxation of the ^{13}C is dominated by its dipolar interaction with the proton. We will also assume that we are in the fast motion limit, so that all spectral densities are simply $2\tau_c$.

With these assumptions, the relevant expressions for the rate constants given in Eq. 9.20 on page 272 simplify to:

$$R_z^{(S)} = b^2 \tau_c \qquad \sigma_{IS} = \tfrac{1}{2} b^2 \tau_c.$$

Substituting these into the expression for η_{SS} we find

$$\text{fast motion:} \quad \eta_{SS} = \frac{\gamma_I}{2\gamma_S},$$

which, in the case of spin I being proton and spin S being ^{13}C, gives $\eta_{SS} \approx 2$. Clearly, there is a substantial enhancement of the ^{13}C signal to be obtained in this way.

In practice, the enhancement we obtain will be less than this estimate. There are two main reasons for this. First, there are probably other sources of relaxation of the S spin (the ^{13}C) than the attached proton. As a result, $R_z^{(S)}$ is increased and the enhancement is reduced. Secondly, we rarely have the patience to wait the rather long time which is needed for the steady state to be reached. Nevertheless, useful enhancements can be achieved with somewhat shorter times.

9.7.4   Two-dimensional NOESY

A two-dimensional NOESY spectrum looks very similar to a COSY spectrum, with the important exception that the cross peaks are generated, not by coherence transfer through couplings, but by cross relaxation. Thus, the appearance of a NOESY cross peak at $\{\Omega_i, \Omega_j\}$ tells us that there is cross relaxation between spins i and j – in other words, the two spins must be reasonably close in space. COSY and NOESY are therefore complementary experiments: the former tells us which spins are coupled, and so connected by the bonding network, whereas the latter tells us which spins are close in space.

The NOESY pulse sequence is shown in Fig. 9.24. We will analyse it for the case of two spins which are undergoing dipolar relaxation and, for simplicity, we will assume that there is no scalar coupling between the two spins.

The first part of the sequence, $90°-t_1-90°$, has already been analysed in detail when discussing the COSY experiment. We are only interested in the z-magnetization present after the second pulse, as it is such magnetization which can undergo transfer due to cross relaxation. Therefore we need term [1] from page 191:

$$-\cos(\pi J_{12} t_1)\cos(\Omega_1 t_1)\,\hat{I}_{1z}. \quad [1]$$

As we are assuming that there is no coupling, $J_{12} = 0$ and hence $\cos(\pi J_{12} t_1) = 1$, giving

$$-\cos(\Omega_1 t_1)\,\hat{I}_{1z}.$$

This term arises from the equilibrium magnetization on spin one; there is an analogous terms arising from the spin-two equilibrium magnetization:

$$-\cos(\Omega_2 t_1)\,\hat{I}_{2z}.$$

Therefore, just after the second pulse ($\tau = 0$) the z-magnetizations on the two spins are:

$$I_{1z} = -\cos(\Omega_1 t_1) I_{1z}^0 \quad \text{and} \quad I_{2z} = -\cos(\Omega_2 t_1) I_{2z}^0.$$

NOESY: Nuclear Overhauser Effect SpectroscopY

Fig. 9.24 The NOESY pulse sequence. During t_1, transverse magnetization acquires a phase label according to the offset; this transverse magnetization is rotated onto the z-axis by the second pulse. During the mixing time τ cross relaxation may transfer this labelled z-magnetization to other spins. The final pulse rotates the z-magnetization into the transverse plane, allowing a signal to be detected.

Essentially what we have here are z-magnetizations which carry a label with them, in the form of the modulation $\cos(\Omega_i t_1)$, identifying them as being from spin one or spin two. During the mixing time, these magnetizations may be transferred to another spin, carrying with them the label which identifies their source. We will see that such transfers are the origin of the cross peaks in the spectrum.

We are going to assume that the two spins are of the same type, so that their equilibrium z-magnetizations are equal; we will write these as I_z^0. Also, to save space we will use the notation

$$c_1 = -\cos(\Omega_1 t_1) \quad \text{and} \quad c_2 = -\cos(\Omega_2 t_1).$$

Further, as there is only one cross-relaxation rate constant, we will write it as σ, and we will also assume that the two self-relaxation rate constants are equal, with value R_z.

With all of these simplifications, the Solomon equations (Eq. 9.18 on page 271) become

$$\frac{dI_{1z}(t)}{dt} = -R_z\left[I_{1z}(t) - I_z^0\right] - \sigma\left[I_{2z}(t) - I_z^0\right]$$
$$\frac{dI_{2z}(t)}{dt} = -\sigma\left[I_{1z}(t) - I_z^0\right] - R_z\left[I_{2z}(t) - I_z^0\right]. \tag{9.26}$$

We are going to solve these using the initial rate approximation, just as we did for the transient NOE experiment in section 9.7.1 on page 274. The initial conditions are

$$I_{1z}(0) = c_1 I_z^0 \quad \text{and} \quad I_{2z}(0) = c_2 I_z^0. \tag{9.27}$$

Inserting these on the right-hand sides of Eq. 9.26 gives, in the initial rate limit,

$$\left(\frac{dI_{1z}(t)}{dt}\right)_{init} = -R_z(c_1 - 1)I_z^0 - \sigma(c_2 - 1)I_z^0$$
$$\left(\frac{dI_{2z}(t)}{dt}\right)_{init} = -\sigma(c_1 - 1)I_z^0 - R_z(c_2 - 1)I_z^0.$$

These equations can be integrated, just as we did before, to give

$$I_{1z}(t) = -R_z t(c_1 - 1)I_z^0 - \sigma t(c_2 - 1)I_z^0 + \text{const. 1}$$
$$I_{2z}(t) = -\sigma t(c_1 - 1)I_z^0 - R_z t(c_2 - 1)I_z^0 + \text{const. 2}$$

The initial conditions given in Eq. 9.27 enable us to find the constants as:

$$\text{const. 1} = c_1 I_z^0 \quad \text{and} \quad \text{const. 2} = c_2 I_z^0.$$

Putting these constants back into the integrated equations, tidying up and finally setting the time to τ (the mixing time) gives the following expressions for the z-magnetizations at the end of the mixing time:

$$\frac{I_{1z}(\tau)}{I_z^0} = c_1(1 - R_z\tau) - c_2\sigma\tau + (R_z + \sigma)\tau \tag{9.28}$$

$$\frac{I_{2z}(\tau)}{I_z^0} = c_2(1 - R_z\tau) - c_1\sigma\tau + (R_z + \sigma)\tau. \tag{9.29}$$

The final pulse in the sequence rotates this z-magnetization into the transverse plane, where it is then observed during t_2. Without any detailed calculations we can see that the spin-one z-magnetization in Eq. 9.28 on the facing page will give rise to peaks at the offset of spin one, Ω_1, in the ω_2 dimension. From the right-hand side of Eq. 9.28 on the preceding page, we can see that there are three terms modulating this peak in t_1. Writing out c_1 and c_2 in full, these terms are:

$$\underbrace{-\cos(\Omega_1 t_1)(1 - R_z\tau)}_{\text{diagonal}} + \underbrace{\cos(\Omega_2 t_1)\sigma\tau}_{\text{cross}} + \underbrace{(R_z + \sigma)\tau}_{\text{axial}}.$$

The first term is modulated at Ω_1 in t_1 and is therefore a diagonal peak, which will appear at $\{\omega_1, \omega_2\} = \{\Omega_1, \Omega_1\}$; its intensity is $(R_z\tau - 1)$. Recall that in this initial rate limit $R_z\tau \ll 1$, so this peak is negative. Our interpretation of the diagonal peak is that it arises from z-magnetization which started out on spin one and remained on that spin during the mixing time.

The second term is modulated at Ω_2 in t_1, and so gives rise to a cross peak at $\{\Omega_2, \Omega_1\}$. The intensity of this peak is $(\sigma\tau)$, which means that if σ is positive, it will be small and positive i.e. of *opposite* sign to the diagonal. On the other hand, if we are in the slow motion limit, which gives $\sigma < 0$, the cross peak will have the *same* sign as the diagonal. This term arises from magnetization which started on spin two at the beginning of the mixing time and was then transferred to spin one as a result of cross relaxation. If there is no cross relaxation ($\sigma = 0$), then there are no cross peaks.

Finally, the third term has no modulation in t_1, and so will appear at $\omega_1 = 0$; such peaks are called *axial peaks*. The magnetization responsible for the axial peak has lost its frequency label as this is the magnetization which has recovered as a result of relaxation during the mixing time. The intensity of the axial peaks goes as $(R_z + \sigma)\tau$, and as it turns out that $R_z > |\sigma|$, these peaks will be positive, i.e. opposite in sign to the diagonal. Figure 9.25 illustrates the way in which all three types of peaks appear in the spectrum.

From Eq. 9.29 on the preceding page we can see that there is a complementary set of peaks arising from the z-magnetization on spin two present at the end of the mixing time: a diagonal peak at $\{\Omega_2, \Omega_2\}$, a cross peak at $\{\Omega_1, \Omega_2\}$, and an axial peak at $\{0, \Omega_2\}$. The intensities of these peaks are the same as their counterparts appearing at Ω_1 in ω_2.

In many ways, NOESY is the two-dimensional counterpart of the transient NOE experiment described in section 9.7.1 on page 274. As in that experiment, at first increasing τ increases the intensity of the cross peaks. Then, the rate of increase decreases until we reach a maximum, after which the intensity falls off.

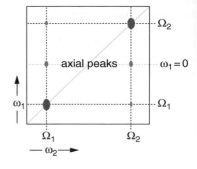

Fig. 9.25 Schematic NOESY spectrum for two spins undergoing cross relaxation. Positive peaks are shown in blue and negative in dark grey; the spectrum is shown for a molecule in the fast motion limit i.e. $\sigma > 0$. The spectrum is closely analogous to a COSY, except that in NOESY the cross peaks arise due to cross relaxation during the mixing time. In addition to the diagonal peaks and cross peaks, the spectrum shows axial peaks which appear at $\omega_1 = 0$.

Suppression of axial peaks

The axial peaks convey no useful information and can be somewhat troublesome if they obscure the wanted cross peaks. It would be a good idea, therefore, to have a method of suppressing them.

Looking back over the calculation, you can see that the distinguishing feature of the terms which give rise to the axial peaks is that they have no modulation as a function of t_1. The reason why this is so is that the magnetization which gives rise to the axial peaks is created as a result of relaxation during τ.

Fig. 9.26 Part of the NOESY
spectrum of quinine, recorded
at 500 MHz and with a
mixing time of 1 s. As
expected for this small
molecule, the cross and
diagonal peaks have opposite
signs. The region shown is
the same as for the DQF
COSY spectrum shown in
Fig. 8.17 on page 202. The
contribution from
zero-quantum coherence
present during the mixing
time has been suppressed
using the method described in
section 11.15 on page 426.

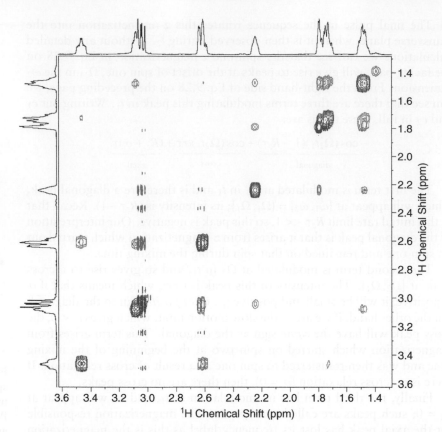

Fig. 9.26 Part of the NOESY spectrum of quinine, recorded at 500 MHz and with a mixing time of 1 s. As expected for this small molecule, the cross and diagonal peaks have opposite signs. The region shown is the same as for the DQF COSY spectrum shown in Fig. 8.17 on page 202. The contribution from zero-quantum coherence present during the mixing time has been suppressed using the method described in section 11.15 on page 426.

If the phase of the first pulse in the sequence in changed from x to $-x$, then the sign of the z-magnetization at the start of the mixing time is also changed. This sign change propagates through the subsequent calculations, and results in both the diagonal and cross peaks changing sign. However, the axial peaks do not change sign as they arise from recovered magnetization.

So, we can use a simple difference method to suppress the axial peaks. The experiment is repeated twice, once with the phase of the first pulse set to x and one with the phase set to $-x$. Subtracting the two experiments results in the cancellation of the axial peaks, while the cross and diagonal peaks add up.

Figure 9.26 shows part of the NOESY spectrum of quinine. For this small molecule the cross-relaxation rate constant is positive, and so the diagonal and cross peaks have opposite sign. Some of the cross peaks are quite strong, whereas others are much weaker, implying that they are between more distant spins.

9.7.5   The NOE in more extended spin systems

The dipolar interaction involves a pair of spins, so even in an extended spin system we can still think of the cross relaxation between two particular spins, and hence the NOE which will be seen between them. The presence of other spins does affect the NOE, however, as they provide additional sources of relaxation. An NOE is always a competition between the transfer due to cross relaxation and the general loss of magnetization caused by self relaxation of the two spins involved. If the self relaxation is fast enough, then it will outcompete the cross relaxation, and no NOE will be seen. It is essentially this competition which sets a limit on the maximum distance over which we are able to observe the effects of cross relaxation.

In an extended spin system it is possible that NOE enhancements between spins which are *not* cross relaxing one another will be observed. This comes about in the following way. Suppose we have three spins, **A**, **B** and **C**. **A** is close to **B**, so there is cross relaxation between them. Similarly, **B** and **C** are close to one another, and so cross relax. However, there is no cross relaxation between **A** and **C**; the situation is depicted in Fig. 9.27.

Let us assume that we are in the fast motion limit, so σ is positive. If we selectively invert **A** in a transient NOE experiment, then we will see a positive NOE enhancement on **B**. This enhancement means that the z-magnetization on spin **B** has been made *greater* than its equilibrium value.

Consequently, as spin **B** is no longer at equilibrium, cross relaxation with spin **C** will cause magnetization from **B** to be transferred to **C**. As a result, an NOE enhancement will be seen on spin **C**. Since the magnetization on **B** is *greater* than the equilibrium value, cross relaxation to **C** results in a negative enhancement of that spin, even though σ is positive. Figure 9.27 shows the resulting NOE difference experiment.

The overall result is that inversion of **A** results in a positive NOE enhancement of **B**, and a *negative* NOE enhancement on the remote spin **C**, even though this latter spin is not cross relaxing with **A**. Spin **C** is said to receive a *relayed* NOE from **A**. It turns out that in the slow motion limit, where σ is negative, both the direct and relayed NOE enhancements are negative.

Such relayed NOEs are potentially confusing as they break the simple interpretation that observation of an NOE between two spins means that they must be close in space. However, in the fast motion limit, the sign of the NOE enhancement does give us a way of distinguishing the direct from the relayed NOEs. In the slow motion limit, all the NOEs are negative, and therefore cannot be distinguished by their signs.

The final point is that these relayed NOEs generally build up much more slowly than the direct NOEs. This is because the relayed NOE requires first that a normal direct NOE be generated, and then that this enhancement causes a second NOE to the remote spin. By restricting ourselves to modest mixing times, it is rather unlikely that such transferred NOEs will have had time to build up.

In the slow motion limit, the cross-relaxation rate constant depends only on W_0, which in turn depends on $j(0)$, where $j(0) = 2\tau_c$. As the correlation time gets longer and longer, the cross-relaxation rate therefore also increases. So, for large molecules with long correlation times, such as proteins, cross relaxation can be quite efficient. In such cases, relayed NOE enhancements can build up quickly, and indeed multiple relays along a chain of spins are also possible. This phenomenon, in which magnetization is spread amongst the spins by efficient cross relaxation, is called *spin diffusion*. The presence of this effect leads to ambiguities in the interpretation of the observed NOE enhancements.

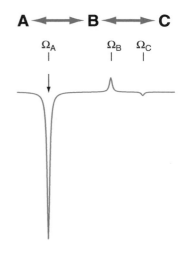

Fig. 9.27 Schematic NOE difference spectrum for three spins, **A**, **B** and **C**, in a line. Cross relaxation between the spins is indicated by the double-headed arrows; note that there is no cross relaxation between **A** and **C**. The lower part of the figure shows the NOE difference spectrum expected for the case where **A** is irradiated. As it is assumed that the cross-relaxation rate constant is positive, we see a positive NOE enhancement on **B**. Even though **A** and **C** are not cross relaxing one another, we see a negative enhancement on **C**. This is due to a two-stage transfer, first from **A** to **B**, and then from **B** to **C**. The NOE on **C** is described as a relayed NOE. Note that for the purposes of the diagram, the size of the relayed NOE has been greatly exaggerated.

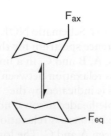

Fig. 9.28 In fluorocylohexane the fluorine can be either axial or equatorial, and these two conformers are in dynamic equilibrium with one another. F_{ax} and F_{eq} will have different chemical shifts, but if the exchange rate between the two conformers is fast compared with the frequency separation between the two fluorine resonances a single line is seen in the spectrum: this is called *fast exchange*. If the exchange rate is slow compared with the frequency separation, then separate lines are seen for the two fluorine environments: this is called *slow exchange*.

9.8 Transverse relaxation

We now turn our attention to a more detailed discussion of transverse relaxation. In section 9.1.4 on page 247 it was noted that there are two contributions to transverse relaxation: the non-secular part, which arises from the same fluctuations in the local magnetic fields which give rise to longitudinal relaxation, and the secular part, which arises from the z-components of these fields. Unsurprisingly, the rates of relaxation due to both of these contributions can be described in a similar way to that we have already used for longitudinal relaxation, with the spectral density and the correlation time playing a central role. However, there are some subtle differences between the secular and non-secular contributions, and so we first turn to a discussion of these. A useful way of understanding the origin of the secular contribution is to draw an analogy between it and the process of chemical exchange, which is therefore the topic of the next section.

9.8.1   Chemical exchange

Chemical exchange is exemplified by the simple equilibrium illustrated in Fig. 9.28. The fluorine in the axial position is expected to have a different chemical shift from that in the equatorial position. However, what we see in the spectrum of this molecule depends on the comparison between the rate constant for the exchange process and the frequency separation between the resonances corresponding to the two fluorine environments.

If the rate constant for exchange is much less that the frequency separation we expect in the (proton-decoupled) ^{19}F spectrum to see two lines, one from F_{ax} and one from F_{eq}. However, if the rate constant is much greater than the frequency separation we expect to see just one line at a position somewhere between the shifts of F_{ax} and F_{eq}. The former situation is called *slow exchange*, whereas the latter is called *fast exchange*.

Figure 9.29 on the facing page is a simulation showing how the spectra for a simple $A \rightleftharpoons B$ equilibrium change as the rate constant for the exchange process is increased. In the simulation it is assumed that the equilibrium populations of A and B are the same, so the rate constants for the forward and back processes are the same (k_{ex}). When there is no exchange ($k_{ex} = 0$ s^{-1}, shown on the bottom left) we see two lines, one at the shift of A and one at the shift of B. The width of these lines is determined by the rate constant for transverse relaxation, which has simply been set to some convenient value.

The spectra on the left-hand side of the figure show that as the exchange rate constant increases the two lines first begin to broaden, then overlap, and eventually merge to give a single line. The theory indicates that this coalescence of the two lines occurs when the exchange rate constant is 2.22 times the frequency separation of A and B. In this case the separation is 160 Hz and so coalescence occurs when $k_{ex} = 350$ s^{-1}.

Further increases in the exchange rate constant cause this single merged line to become narrower, as is illustrated in the spectra on the right-hand side of Fig. 9.29. Eventually, when the exchange rate constant becomes very much larger than the frequency separation, the line has the same width as it did in the absence of exchange. The position of the merged line depends on

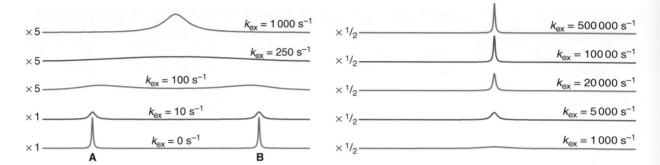

Fig. 9.29 Simulated spectra showing the effect of exchange between two species **A** and **B**. In the absence of exchange ($k_{ex} = 0$ s^{-1}, bottom left), separate lines are seen for the two species and the widths of these lines are determined by the transverse relaxation rate constant; in this simulation their frequency separation is 160 Hz. It is assumed that the populations of **A** and **B** are equal, so the two lines have the same intensity. As the exchange rate constant increases the lines first broaden and then merge into one, as shown in the left-hand series of spectra. A further increase in the exchange rate constant causes the merged line to narrow, as shown in the right-hand series of spectra. In the limit that the exchange rate constant is very much larger than the frequency separation the exchange process has no effect on the linewidth. Note that scale expansions have been used for some of the spectra on the left.

the populations of **A** and **B**, and since we have assumed these to be equal, the merged line lies halfway between the two separate resonances.

An increase in the linewidth as a result of an exchange process is called *exchange broadening*. If we start out from the slow exchange limit, increasing the rate constant for the exchange process increases the amount of exchange broadening. However, after coalescence further increases in the rate constant cause a reduction in the linewidth. This phenomena is called *exchange narrowing*.

A full discussion of the theory used to simulate the spectra in Fig. 9.29 is given in J. Cavanagh, W. J. Fairbrother, A. G. Palmer III, M. Rance and N. J. Skelton, *Protein NMR Spectroscopy* (2nd edition, Academic Press, 2007), pages 391 to 400.

Exchange processes from the point of view of single spins

So far we have described the overall spectra which arise from species undergoing chemical exchange. Necessarily, these spectra are observed on macroscopic samples containing vast numbers of spins, so what we see is the average behaviour of all of these spins. What we are now going to explore is the microscopic behaviour, at the level of an individual spin, which explains the spectra we observe from the whole sample.

As before, we will consider the simple equilibrium between two species **A** and **B**, but this time focus in on the behaviour of a particular nuclear spin. Let us assume that this nucleus finds itself in an **A** molecule. As a result, the magnetic moment of this spin will precess at the Larmor frequency $\omega_{0,A}$ appropriate for this environment, and this tiny contribution will be added to that made by all of the other spins so as to give the FID. At some point, the **A** molecule may undergo a chemical transformation to **B**. All this involves is a rearrangement of the nuclei, so our spin now finds itself in a **B** molecule and the magnetic moment now precesses at the Larmor frequency

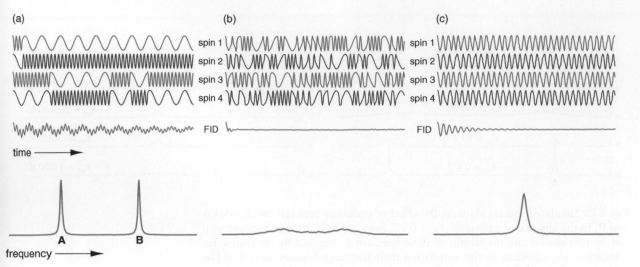

Fig. 9.30 Simulations showing how the behaviour of individual spins can account for the form of the spectra seen for the $A \rightleftharpoons B$ exchange system. At the top of (a), (b) and (c) are shown the contributions from four typical spins. Underneath, labelled FID, is the sum of the contributions from 1000 spins: this is a reasonable approximation to the FID which would be observed from a macroscopic sample. Beneath each FID is shown the corresponding spectrum. In (a) the number of transformations between A and B are rather few so distinct periods of oscillation at the two Larmor frequencies can be seen. The result is two clear lines in the spectrum, corresponding to the slow exchange limit. In (b) the transformations are more frequent, and as a result the contributions of the individual spins soon cancel one another out, leading to a quickly decaying FID and broad lines in the spectrum. In (c) the transformations are so frequent that each spin appears to be evolving at the average frequency of A and B. The individual contributions do not cancel one another as quickly as in (b), leading to a sharper line.

$\omega_{0,B}$, thus making a contribution to the FID at this frequency. At some time later, the molecule may transform back into A once again carrying the spin with it; now the spin makes a contribution at $\omega_{0,A}$.

Therefore, at the level of a single spin all that happens is that, as the molecule transforms back and forth between A and B, the Larmor frequency of the spin hops back and forth between $\omega_{0,A}$ and $\omega_{0,B}$. For a given spin, the time between transformations is essentially random, but the faster the overall rate of the chemical process the shorter the average time between transformations.

Figure 9.30 visualizes this process by showing the contribution from individual spins, as well as the sum of the contributions from many spins. Consider first the left-hand side of the diagram, (a). Shown at the top are the contributions to the transverse magnetization made by four individual spins. The Larmor frequencies in the two environments have been chosen to be sufficiently different that in the diagram they can be discerned by eye.

The abrupt change in the frequency of the oscillation comes about when the molecule transforms from A to B, or vice versa. Over the timescale of the diagram, only a few transformations take place, and as a result we can see extended periods during which an individual spin is precessing at one of the two Larmor frequencies. Note that the changes take place at random times and are not correlated between the different spins.

In order to work out what the spectrum looks like from a macroscopic sample we need to add together the contributions from a very large number of spins. In the figure, this has been done for just 1000 spins (in order to make the simulation feasible), giving the time-domain function marked FID. The Fourier transformation of this gives the spectrum shown.

Not surprisingly, for (a) the time-domain signal clearly shows the presence of two frequencies, as is confirmed by the two peaks in the spectrum. What is going on here is that each individual spin spends long enough precessing at one or other of the two Larmor frequencies that distinct contributions to the overall time-domain signal are seen at these two frequencies.

Now consider Fig. 9.30 (b), in which the transformations between **A** and **B** are more frequent than they were in (a). However, we can still just about discern periods of oscillation at the two different frequencies. Adding up 1000 of these contributions from individual spins gives the time-domain signal labelled FID. What has happened here is that the frequent jumps from one frequency to another have led the contributions from the individual spins to cancel one another out, leading to a rapidly decaying FID and hence broad lines in the spectrum.

Finally consider part (c). Here the $\mathbf{A} \rightleftharpoons \mathbf{B}$ transformations occur so frequently that we cannot see periods of oscillation at either of the two frequencies. Rather, what we have for each spin is an oscillation at a single frequency, intermediate between the two seen in (a), arising from the fact that the frequency at which the spin is evolving is alternating rapidly between $\omega_{0,A}$ and $\omega_{0,B}$. Admittedly the oscillation is not a pure cosine wave, but its period is nevertheless clear. Adding up contributions from 1000 such spins gives some cancellation at later times, but nothing like the extent seen in (b); in addition, it is clear that there is a single frequency present in the FID. The resulting spectrum has a line broader than in (a), but much narrower than in (b).

The interpretation of these results is that Fig. 9.30 (a) corresponds to slow exchange, (b) corresponds to a situation approaching coalescence in which there is considerable line broadening, and (c) corresponds to a regime in which, although the two lines have collapsed to give one, there is still a significant exchange contribution to the linewidth. The spectra are analogous to those shown in Fig. 9.29 on page 287, although they were generated by a completely different method. We can therefore explain the spectra arising from chemically exchanging systems in terms of the behaviour of individual spins.

The conditions for slow and fast exchange

It has already been mentioned that whether or not we are in the fast or slow exchange region depends on the comparison between the rate constant for exchange and the *difference* in the Larmor frequencies between the two sites. However, Fig. 9.30 perhaps does not quite convey the impression that it is this frequency difference which is the important parameter.

To explore this further consider a 'thought' experiment in which we are required to tell the difference between a wave oscillating at 10.0 Hz and

In the simulation used to create Fig. 9.30 an extra decay has been imposed on the FID in order to make it more realistic. This rate of this decay gives the linewidth in the absence of chemical exchange effects, and the same decay has been used for all parts of the figure.

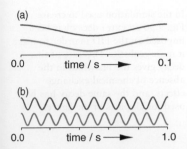

Fig. 9.31 Illustration of the time needed to differentiate between two cosine waves of different frequencies. In both (a) and (b) two cosine waves with frequencies 10.0 Hz and 10.5 Hz are plotted in black and blue, respectively. If the cosine waves are observed for 0.1 s, as in (a), it is very hard to see that they have different frequencies. However, if the observation is made for 1.0 s, as in (b), it is immediately obvious that the frequencies of the two waves are different.

one oscillating at 10.5 Hz. The question is, for how long would we need to observe these oscillations before we could be certain which was which?

The answer to this question is illustrated in Fig. 9.31. In (a) we see a plot of the two waves over a time interval of 0.1 s. Each oscillation barely completes one cycle, and we would have to look very hard to spot which is which. If there were any appreciable noise present, we surely would not be able to discern any difference between the two curves.

In contrast, if the oscillations are observed for a longer time, such as the 1.0 s shown in (b), it is quite easy to spot that the two oscillations have different frequencies since we can see that they are getting out of step with one another. By the time one second has elapsed the blue trace has completed almost half a cycle more than the black trace, so the former must be the higher frequency oscillation. Put more formally, by the end of the time period in (b) there is a significant phase difference between the two oscillations, whereas in (a) there is not. It is this difference in phase which makes it possible to tell the two waves apart.

The phase difference which develops over a time τ is $2\pi\Delta \times \tau$, where Δ is the difference in frequency (in Hz) between the two sites. For this phase difference to be significant τ must be comparable with or greater than $1/\Delta$. In other words, to be able to differentiate two frequencies we need to observe for a time which is greater than the reciprocal of the difference between the two frequencies.

In the case of chemical exchange characterized by a first-order rate constant k_{ex}, the average time before **A** is converted to **B** (or vice versa) is of the order of $1/k_{ex}$: this is often called the lifetime of the reaction, τ_{ex}. We are only able to observe the oscillations at a particular frequency for this time, because on average this is how long there is before the frequency changes.

If the rate of the process is such that $\tau_{ex} \gg 1/\Delta_{AB}$, where Δ_{AB} is the difference in frequency between the two sites, then it will be possible to distinguish between these two frequencies and so we expect to see two lines in the spectrum: this is the slow exchange limit. Replacing τ_{ex} with $1/k_{ex}$ and rearranging gives the condition in its familiar form:

$$\text{slow exchange} \quad k_{ex} \ll \Delta_{AB}.$$

Conversely, if $\tau_{ex} \ll 1/\Delta_{AB}$ the time is insufficient to distinguish between the two frequencies, and so we expect there to be one line in the spectrum: this is the fast exchange limit. As before, substitution for τ_{ex} and rearrangement gives the familiar result

$$\text{fast exchange} \quad k_{ex} \gg \Delta_{AB}.$$

9.8.2 The secular contribution to transverse relaxation

We can now draw on this discussion of the effects of chemical exchange to understand the origin of the secular contribution to transverse relaxation. Imagine a sample consisting of many spins, each of which experiences a local field due, for example, to interactions with nearby spins. The z-component of the local field at a particular spin will determine the Larmor frequency of that spin. Since we expect there to be a distribution of local fields, there will be a distribution of Larmor frequencies across the sample.

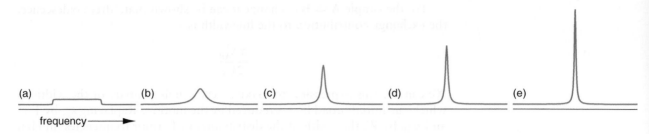

(a)   (b)   (c)   (d)   (e)

frequency⟶

Fig. 9.32 Simulations showing the effect on the lineshape of exchange between spins with a continuous range of Larmor frequencies. In (a) is depicted the situation in which there is no exchange between the spins. As a result the line is broad, reflecting the range of Larmor frequencies present. Spectra (b)–(e) show the effect of increasing the exchange rate between the spins. The faster the exchange, the narrower the line becomes in a way analogous to two-site exchange after coalescence. In (e), although the line in much narrower than in (a), there is still a contribution arising from the original spread of Larmor frequencies shown in (a): this is the origin of the secular part of transverse relaxation.

First consider the situation in which the local fields do not change with time. Each spin will make a contribution to the FID at its own Larmor frequency, and so the distribution of Larmor frequencies across the sample will result in the spectrum consisting of a very broad line. This is shown schematically in Fig. 9.32 (a). The width of the line depends on the spread of values of the local field. For example, in the case where the fields are caused by dipolar interaction between two protons, the line could easily be 100 kHz wide.

Of course, in a liquid the local fields are not constant but, as described in section 9.3 on page 251, change rather rapidly as a result of the thermal motion (rotational diffusion) in the sample. An individual spin therefore experiences a local field which, on average, changes after a time of the order of the correlation time τ_c. The change in the component of the local field along the z-axis results in a change in the Larmor frequency. This is entirely analogous to chemical exchange in which the transformation of the molecule results in a change in the Larmor frequency experienced by a spin. The only difference is that rather than the spin experiencing two Larmor frequencies it can experience a whole range of Larmor frequencies.

In the previous section it was noted that the rate of an exchange process could be characterized by a lifetime τ_{ex} which is simply $1/k_{ex}$. In the case of rotational diffusion we can identify τ_{ex} with the correlation time τ_c, since both characterize the timescale of the process. For a non-viscous liquid τ_c might be around 100 ps, which corresponds to a 'rate constant' of $1/(100 \times 10^{-12}) = 10^{10}$ s^{-1} – a very large value. The rate constant is very much larger than the spread of Larmor frequencies due to the local fields, we are clearly in the exchange narrowing region. Therefore we expect the rapid motion to result in significant narrowing of the broad line. The faster the motion, that is the shorter the correlation time, the greater the narrowing effect, as shown in Fig. 9.32 (b)–(e).

For the simple $A \rightleftharpoons B$ exchange it can be shown that, after coalescence, the exchange contribution to the linewidth is

$$\frac{\pi \Delta_{AB}^2}{2 k_{ex}}.$$

We can use this expression to make a very rough estimate of the width to which our initial broad line is reduced by the motion. Replacing k_{ex} by $1/\tau_c$ and Δ_{AB} by W, the width of the distribution of Larmor frequencies (in Hz), we obtain

$$\text{width of narrowed line} \approx W^2 \times \tau_c. \qquad (9.30)$$

This estimate is so rough that we have simply dropped the factor of $\pi/2$ as being irrelevant.

Putting $W = 100\,\text{kHz}$ and $\tau_c = 100\,\text{ps}$, we obtain a value of 1 Hz for the width of the line. Clearly the motion has had a dramatic narrowing effect, taking the linewidth from 100 kHz in the absence of motion to just 1 Hz when motion is taken into account. However, for high-resolution NMR of liquids, 1 Hz is still a significant contribution to the linewidth. Note that the linewidth scales directly with the correlation time. Therefore if $\tau_c = 10\,\text{ns}$, a value typical for a medium-size protein, the linewidth becomes 100 Hz, which is very significant.

In section 9.3.2 on page 256 we introduced the reduced spectral density $j(\omega)$ which gives a measure of the amount of motion at frequency ω. For the simplest kind of rotational diffusion this spectral density is given by:

$$j(\omega) = \frac{2\tau_c}{1 + \omega^2 \tau_c^2}.$$

The spectral density at zero frequency is therefore $2\tau_c$. The rough estimate given in Eq. 9.30 can therefore be written as

$$\text{width of narrowed line} \approx \tfrac{1}{2} W^2 \times j(0),$$

where $j(0)$ is the reduced spectral density at zero frequency.

We can now see that the origin of the secular part of transverse relaxation lies in the distribution of local fields along the z-axis, and hence in the resulting distribution of Larmor frequencies. However, rapid motion in the liquid means that the observed line is very much narrower than the original distribution of Larmor frequencies. By drawing an analogy with chemical exchange we have been able to predict that the width of the narrowed line will be proportional to τ_c i.e. the value of the reduced spectral density at zero frequency, $j(0)$.

This brings us back to the discussion at the start of section 9.1.4 on page 247 in which the two contributions to transverse relaxation were described. The non-secular contribution involves reorientation of the magnetic moments of individual spins and so is identical to the process which leads to longitudinal relaxation. It therefore depends on the spectral density at the Larmor frequency. In contrast, the secular contribution arises from the motionally narrowed distribution of Larmor frequencies arising from the z-components of the local fields. The motionally narrowed linewidth, and hence the contribution to transverse relaxation, depends on the spectral density at zero frequency.

Terminology

The terms secular and non-secular come from quantum mechanics. In this theory, a secular perturbation is one which changes the energy, but not the wavefunction, whereas a non-secular perturbation changes both. An oscillating transverse magnetic field causes an individual magnetic moment to change orientation, and so changes the wavefunction – it is thus a non-secular effect. A field along the z-axis will alter the rate of precession of the magnetic moment, which is the same as altering the energy, but does not alter the orientation with respect to the z-axis – this is a secular effect.

9.8.3 Effect of transverse relaxation

In section 9.4.2 on page 260 we showed that, for a sample consisting of single spins, the relaxation behaviour of the z-magnetization is given by (Eq. 9.12 on page 262)

$$\frac{\mathrm{d}M_z(t)}{\mathrm{d}t} = -R_z\left[M_z(t) - M_z^0\right].$$

This equation predicts that relaxation will drive the z-magnetization to its equilibrium value of M_z^0.

The transverse magnetization, for example the x-magnetization, obeys the following differential equation:

$$\frac{\mathrm{d}M_x(t)}{\mathrm{d}t} = -R_{xy}\,M_x(t), \tag{9.31}$$

where R_{xy} is the rate constant for transverse relaxation and, for simplicity, it has been assumed that the offset is zero. There is a similar equation for the y-magnetization, with the same rate constant. We will now show that this equation predicts that the x-magnetization decays exponentially to its equilibrium value of zero.

All that we have to do is to integrate Eq. 9.31 using the same approach as in section 9.5 on page 263:

$$\begin{aligned}
\frac{\mathrm{d}M_x(t)}{\mathrm{d}t} &= -R_{xy}\,M_x(t) \\
\frac{\mathrm{d}M_x(t)}{M_x(t)} &= -R_{xy}\,\mathrm{d}t \\
\int \frac{\mathrm{d}M_x(t)}{M_x(t)} &= \int -R_{xy}\,\mathrm{d}t \\
\ln M_x(t) &= -R_{xy}\,t + \text{const.}
\end{aligned}$$

We can find a value for the constant by taking the value of the x-magnetization at time $t = 0$ to be $M_x(0)$. Substituting this into the last line shows that the constant of integration is $\ln M_x(0)$, so the solution to the differential equation is

$$\ln M_x(t) = -R_{xy}\,t + \ln M_x(0).$$

This can be rearranged as follows:

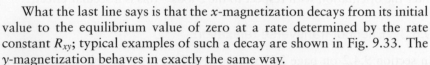

$$\ln M_x(t) - \ln M_x(0) = -R_{xy}\,t$$

$$\ln\left(\frac{M_x(t)}{M_x(0)}\right) = -R_{xy}\,t$$

$$\frac{M_x(t)}{M_x(0)} = \exp(-R_{xy}\,t)$$

$$M_x(t) = M_x(0)\exp(-R_{xy}\,t).$$

What the last line says is that the x-magnetization decays from its initial value to the equilibrium value of zero at a rate determined by the rate constant R_{xy}; typical examples of such a decay are shown in Fig. 9.33. The y-magnetization behaves in exactly the same way.

The reciprocal of the rate constant R_{xy} is the time constant for the decay of transverse magnetization. Usually this time constant is denoted T_2, where $T_2 = 1/R_{xy}$. The use of this symbol is so widespread that transverse relaxation is often called 'T_2 relaxation'.

If the offset is not zero then, in addition to decaying due to transverse relaxation, the x- and y-components of the magnetization are interconverted due to their precession in the transverse plane. If we are only interested in the relaxation part of this process, we can eliminate the effect of precession by defining M_{xy} as

$$M_{xy} = \sqrt{M_x^2 + M_y^2}.$$

The value of M_{xy} does not change as a result of precession, so its time dependence is just determined by relaxation:

$$\frac{dM_{xy}(t)}{dt} = -R_{xy}\,M_{xy}(t).$$

9.8.4 Relaxation by random fields

One of the simplest problems we can deal with in the theory of relaxation is to imagine that our spin experiences a random local field. We do not ask where the field comes from, but just accept that it is there.

To make things as simple as possible we will also assume that the mean square of the local field is the same in the x-, y- and z-directions:

$$\overline{B_{\text{loc},x}^2} = \overline{B_{\text{loc}}^2}, \qquad \overline{B_{\text{loc},y}^2} = \overline{B_{\text{loc}}^2}, \qquad \overline{B_{\text{loc},z}^2} = \overline{B_{\text{loc}}^2}.$$

With these assumptions, it can be shown that the rate constant for longitudinal relaxation is given by:

$$R_z = \gamma^2 \overline{B_{\text{loc}}^2}\, j(\omega_0), \tag{9.32}$$

where γ is the gyromagnetic ratio and ω_0 is the Larmor frequency (we are not able to go into the details of the theory used to compute these rate constants). As expected, the rate of longitudinal relaxation depends on the spectral density at the Larmor frequency.

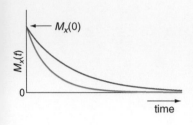

Fig. 9.33 Transverse magnetization decays exponentially at a rate determined by the transverse relaxation rate constant, R_{xy}. Curves are shown for two values of the rate constant, with the blue curve having twice the rate constant of the dark grey curve.

The rate constant for transverse relaxation, R_{xy}, can be shown to be given by

$$R_{xy} = \underbrace{\tfrac{1}{2}\gamma^2 \overline{B_{\text{loc}}^2}\, j(0)}_{\text{secular}} + \underbrace{\tfrac{1}{2}\gamma^2 \overline{B_{\text{loc}}^2}\, j(\omega_0)}_{\text{non-secular}}.$$

As expected, the rate constant has a secular part, depending on $j(0)$, and a non-secular part, depending on $j(\omega_0)$. For the secular part the term $\gamma^2 \overline{B_{\text{loc}}^2}$ is analogous to the square of the width of the distribution of Larmor frequencies caused by the local fields, as described in section 9.8.2 on page 290.

The new feature which these calculations reveal is that the non-secular part is precisely *half* the value of the longitudinal rate constant given in Eq. 9.32 on the facing page:

$$R_{xy} = \underbrace{\tfrac{1}{2}\gamma^2 \overline{B_{\text{loc}}^2}\, j(0)}_{\text{secular}} + \underbrace{\tfrac{1}{2}R_z}_{\text{non-secular}}.$$

It turns out that for *any* relaxation mechanism it is always true that there is a contribution to the transverse relaxation rate constant which is equal to one-half of the longitudinal rate constant. This factor of one-half cannot be a coincidence, so where does it come from? One argument goes along the following lines.

Imagine that a particular spin experiences a local field, the transverse component of which is aligned along the x-axis and oscillating at the Larmor frequency. The z-component of the magnetic moment of the spin will thus be rotated, as will its y-component. However, the x-component is unaffected, as it is parallel with the local field. Thus, the z-component of the moment is altered, leading to longitudinal relaxation, but only *one* of the transverse components is altered. We interpret this by saying that the field is half as effective at causing transverse relaxation as it is at causing longitudinal relaxation, as only one of the two transverse components of the moment is affected.

Relaxation rates in the two motional limits

It is interesting to see how the longitudinal and transverse relaxation rate constants compare in the fast and slow motion limits. Recall that in the fast motion limit $j(\omega_0) = 2\tau_c$ and, independent of the motional regime, $j(0) = 2\tau_c$. The two rate constants are thus

$$\text{fast motion:} \quad R_z = \gamma^2 \overline{B_{\text{loc}}^2} j(\omega_0)$$
$$= 2\gamma^2 \overline{B_{\text{loc}}^2} \tau_c,$$

and

$$\text{fast motion:} \quad R_{xy} = \tfrac{1}{2}\gamma^2 \overline{B_{\text{loc}}^2}\, j(0) + \tfrac{1}{2}\gamma^2 \overline{B_{\text{loc}}^2}\, j(\omega_0)$$
$$= \gamma^2 \overline{B_{\text{loc}}^2}\, \tau_c + \gamma^2 \overline{B_{\text{loc}}^2}\, \tau_c$$
$$= 2\gamma^2 \overline{B_{\text{loc}}^2}\, \tau_c.$$

Therefore, in the fast motion limit the two rate constants are equal, which is a consequence of the spectral density being independent of frequency.

In the slow motion limit $j(0)$ is still $2\tau_c$, and as described in section 9.3.3 on page 257, $j(\omega_0)$ is given by

$$j(\omega_0) = \frac{2}{\omega_0^2 \tau_c}.$$

So, the rate constant for longitudinal relaxation becomes

$$\text{slow motion:} \quad R_z \; = \; \gamma^2 \overline{B_{loc}^2}\, j(\omega_0)$$

$$= \; \frac{2\gamma^2 \overline{B_{loc}^2}}{\omega_0^2 \tau_c}.$$

Note that this rate constant *decreases* as the correlation time increases.

In contrast, the rate constant for transverse relaxation goes on increasing as τ_c increases on account of the secular term:

$$\text{slow motion:} \quad R_{xy} \; = \; \tfrac{1}{2}\gamma^2 \overline{B_{loc}^2}\, j(0) + \tfrac{1}{2}\gamma^2 \overline{B_{loc}^2}\, j(\omega_0)$$

$$= \; \gamma^2 \overline{B_{loc}^2}\, \tau_c.$$

On the second line we have simply set $j(\omega_0)$ to zero, as it will be negligible compared with $j(0)$.

Figure 9.34 shows a plot of the two rate constants as a function of the correlation time. The fast motion limit, in which the two rate constants are equal, is seen when τ_c is small. As the correlation time increases, the longitudinal relaxation rate constant eventually reaches a maximum. It was shown in section 9.3.2 on page 256 that $j(\omega_0)$ is a maximum when $\omega_0\tau_c = 1$, so this is the condition for R_z being a maximum. As the correlation time increases further, R_z falls off steadily. In contrast, the transverse relaxation rate constant goes on increasing with τ_c.

The practical consequence of these observations is that for large molecules, which tumble slowly and so have long correlation times, transverse magnetization decays away to zero much more quickly than the z-magnetization recovers to equilibrium.

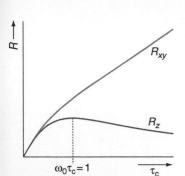

Fig. 9.34 Plot of the rate constants for longitudinal (R_z) and transverse (R_{xy}) relaxation, caused by random fields, as a function of the correlation time, τ_c. In the fast motion limit ($\omega_0\tau_c \ll 1$), the two rate constants are equal, and increase with τ_c. However, as the correlation time increases further the rate constant for longitudinal relaxation reaches a maximum and then falls off. In contrast, the rate constant for transverse relaxation goes on increasing indefinitely, eventually becoming a linear function of τ_c.

9.8.5 Transverse dipolar relaxation of two spins

For two spins relaxing by the dipolar interaction the following relaxation behaviour is found:

$$\frac{dI_{1x}}{dt} = -R_{xy}^{(1)} I_{1x} \qquad \frac{dI_{2x}}{dt} = -R_{xy}^{(2)} I_{2x},$$

where

$$R_{xy}^{(1)} = b^2 [\underbrace{\tfrac{1}{10} j(0)}_{\text{secular}} + \underbrace{\tfrac{3}{20} j(\omega_{0,2}) + \tfrac{3}{40} j(\omega_{0,1}) + \tfrac{3}{20} j(\omega_{0,1} + \omega_{0,2}) + \tfrac{1}{40} j(\omega_{0,1} - \omega_{0,2})}_{\text{non-secular} \, = R_z^{(1)}/2}],$$

and

$$R_{xy}^{(2)} = b^2 [\underbrace{\tfrac{1}{10} j(0)}_{\text{secular}} + \underbrace{\tfrac{3}{20} j(\omega_{0,1}) + \tfrac{3}{40} j(\omega_{0,2}) + \tfrac{3}{20} j(\omega_{0,1} + \omega_{0,2}) + \tfrac{1}{40} j(\omega_{0,1} - \omega_{0,2})}_{\text{non-secular} \, = R_z^{(2)}/2}].$$

As before, the non-secular contribution is equal to one-half of the corresponding longitudinal rate constant, expressions for which were given in Eq. 9.20 on page 272.

Of the two terms which have been labelled secular, the first needs no comment since, as in the case of random fields, it depends on the spectral density at zero frequency. In the expression for $R_{xy}^{(1)}$, that is the transverse relaxation rate constant for *spin one*, the second secular term depends on the spectral density at the Larmor frequency of *spin two*, $\omega_{0,2}$. Our interpretation of this term is as follows.

Spin one sees a magnetic field due to spin two and, as we have seen, this field depends on the distance between the two spins and the orientation of the vector joining them. However, the field at spin one also depends on the spin state of spin two i.e. whether it is up or down. Flipping spin two causes the local field experienced by spin one to change, thus making a contribution to the spread of local fields. The rate of flipping of spin two depends on the spectral density at the Larmor frequency of this spin, as it is motion at this frequency which is needed to flip the spin.

As with relaxation by random fields, in the fast motion limit $R_{xy} = R_z$, and in addition R_{xy} is the same for both spins involved in the dipolar interaction.

9.8.6 Transverse cross relaxation: ROESY

For two spins undergoing dipolar relaxation, we saw that the behaviour of the z-magnetization of spin one depends on the deviation of spin two from equilibrium, and vice versa. This effect appears in the Solomon equations via the cross relaxation term depending on σ_{12}:

$$\frac{\mathrm{d}I_{1z}}{\mathrm{d}t} = -R_z^{(1)}(I_{1z} - I_{1z}^0) - \sigma_{12}(I_{2z} - I_{2z}^0)$$

$$\frac{\mathrm{d}I_{2z}}{\mathrm{d}t} = -\sigma_{12}(I_{1z} - I_{1z}^0) - R_z^{(2)}(I_{2z} - I_{2z}^0).$$

However, for transverse relaxation, no such cross term is seen between the two transverse terms:

$$\frac{\mathrm{d}I_{1x}}{\mathrm{d}t} = -R_{xy}^{(1)}I_{1x} \qquad \frac{\mathrm{d}I_{2x}}{\mathrm{d}t} = -R_{xy}^{(2)}I_{2x}.$$

The question is, why not?

The answer to this question is that, in principle, there *is* such a cross-relaxation term between I_{1x} and I_{2x} – it is just that we cannot observe the effect of this term. To see why this is, imagine that spin one is on resonance and that its transverse magnetization is stationary along the x-axis. Spin two, however, does have an offset and so its transverse magnetization precesses around in the transverse plane.

Cross relaxation can cause some of the spin-two transverse magnetization to become spin-one transverse magnetization. This new contribution to spin one will appear in the direction in which the spin-two magnetization is pointing. A little time later, further cross relaxation generates some more spin-one transverse magnetization. As the spin-two magnetization has precessed around, the new contribution to spin one will be in a different direction to the first. As this process goes on, the overall result is that the contributions to the spin-one magnetization which come from cross relaxation of spin two are just spread around the transverse plane, and so simply cancel one another out. This is why, although there is cross relaxation, we cannot see its effect.

We can see from this description that if the two spins had the *same offset*, their transverse magnetizations would precess at the same frequency and so be aligned at all times. In such a situation, we would expect there to be net transfer of transverse magnetization due to cross relaxation. The problem is that as the two spins have the same offset, they will not be separated in the spectrum.

However, there is a way – called *spin locking* – of making it appear for a period of time that two spins have the same offset. How this works is the topic we will turn to next.

Spin locking

Fig. 9.35 A simple pulse sequence illustrating the idea of spin locking. The initial pulse rotates the equilibrium magnetization onto the x-axis. Immediately after the pulse, a strong RF field is applied along the x-axis. Provided the field is strong enough, the magnetization remains locked along x, even if the spin has an offset.

The idea of spin locking is quite simple. Imagine starting at equilibrium and first applying a 90° pulse about the y-axis; this brings the magnetization down along the x-axis. Immediately after the pulse, we apply a strong RF field, just as we would use for a pulse, along the x-axis; the sequence is shown in Fig. 9.35. If this field is sufficiently strong it is found that it 'locks' the magnetization and keeps it aligned with the RF field. Effectively, this strong RF field suppresses the offset term, thus keeping the magnetization stationary along the x-axis.

For the spin-locking field to be 'sufficiently strong' its field strength, as determined by the frequency ω_1, must be much larger than the greatest offset present. Clearly, if there is only one line in the spectrum, we can satisfy this condition with any field simply by placing the transmitter on resonance with the line. However, in general there will be more than one line on the spectrum, and hence a range of offsets.

As an example, consider a proton spectrum recorded at 500 MHz. If we place the transmitter in the middle of the spectrum, the largest offset is about 5 ppm, which corresponds to 2500 Hz. To be effective across the spectrum, the spin-locking field strength, $\omega_1/(2\pi)$, would need to be two or three times this maximum offset; three times the maximum offset gives $\omega_1/(2\pi) = 7500$ Hz, which is quite feasible. Of course, we can only hope to spin lock spins of the same isotope (e.g. both protons), as the frequency separation between spins of different isotopes would be too large.

If we assume that the transverse magnetization from both spins is spin locked, then the differential equations which describe the relaxation behaviour will include a cross-relaxation term:

$$\frac{dI_{1x}}{dt} = -R_{xy}^{(1)} I_{1x} - \sigma_{xy} I_{2x} \qquad \frac{dI_{2x}}{dt} = -\sigma_{xy} I_{2x} - R_{xy}^{(2)} I_{1x}.$$

The transverse cross-relaxation rate constant is given by

$$\sigma_{xy} = b^2 \left[\tfrac{1}{10} j(0) + \tfrac{3}{20} j(\omega_0) \right].$$

In this expression, ω_0 is the Larmor frequency of both spins – the difference caused by chemical shifts is too small to be important.

The really important thing about σ_{xy} is that, as it is the sum of two terms, it is positive for *all* values of the correlation time. This is in contrast to the corresponding cross-relaxation rate constant for z-magnetization which, as was described in section 9.6.4 on page 272, changes sign as τ_c changes, and therefore has a zero-crossing. This key observation is exploited in the two-dimensional ROESY experiment.

ROESY

ROESY is analogous to the NOESY experiment, except that instead of generating cross peaks by cross relaxation between the z-magnetization of different spins, the cross peaks in ROESY arise from cross relaxation between spin-locked transverse magnetization. The experiment is a useful alternative to NOESY as the cross peaks in a ROESY spectrum always have the same sign, regardless of the value of the correlation time. ROESY is therefore used to look for NOE enhancements in molecules whose correlation times make the conventional NOE zero or close to zero (see section 9.6.4 on page 272).

ROESY: Rotating frame Nuclear Overhauser Effect SpectroscopY

The pulse sequence for the two-dimensional ROESY experiment is shown in Fig. 9.36. It is quite similar to the NOESY pulse sequence (Fig. 9.24 on page 281) in that frequency-labelled magnetization is prepared during t_1. The difference comes in the mixing time. In NOESY, the frequency-labelled magnetization is rotated onto the z-axis, where cross relaxation takes place. In ROESY, it is the x-magnetization present at the end of t_1 which is spin-locked so that transverse cross relaxation can take place.

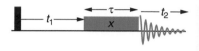

At the start of the mixing time (i.e. the spin locking), $\tau = 0$, the x-magnetizations of the two spins are

$$I_{1x}(0) = \sin(\Omega_1 t_1)\, I_z^0 \quad \text{and} \quad I_{2x}(0) = \sin(\Omega_2 t_1)\, I_z^0,$$

Fig. 9.36 Pulse sequence for the two-dimensional ROESY experiment. The mixing time consists of a period of spin locking during which cross relaxation between transverse magnetization takes place.

where we have assumed that the equilibrium z-magnetizations of the two spins are the same. Following the same approach as we used to analyse the NOESY experiment in the initial rate limit, we can show that at the end of the mixing time the magnetizations are:

$$I_{1x}(\tau) = \underbrace{(1 - R_{xy}^{(1)}\tau)\sin(\Omega_1 t_1)\, I_z^0}_{\text{diagonal peak}} \underbrace{-\sigma_{xy}\tau \sin(\Omega_2 t_1)\, I_z^0}_{\text{cross peak}}$$

$$I_{2x}(\tau) = \underbrace{(1 - R_{xy}^{(2)}\tau)\sin(\Omega_2 t_1)\, I_z^0}_{\text{diagonal peak}} \underbrace{-\sigma_{xy}\tau \sin(\Omega_1 t_1)\, I_z^0}_{\text{cross peak}}.$$

As before, we have two cross peaks, both with intensity $(-\sigma_{xy}\tau)$, and two diagonal peaks, with intensity $(1 - R_{xy}^{(1)}\tau)$ and $(1 - R_{xy}^{(2)}\tau)$. In the initial rate limit $R_{xy}\tau \ll 1$, so the diagonal peaks are positive, and the cross peaks are negative. The transverse cross-relaxation rate constant σ_{xy} is positive for all values of the correlation time, so it is *always* the case that the cross and diagonal peaks will have the opposite signs.

9.9   Homogeneous and inhomogeneous broadening

The picture we have developed so far is that transverse magnetization decays at a rate determined by the transverse relaxation rate constant, R_{xy}, so that a typical FID would be described by the function

$$S(t) = A \exp(i\Omega t) \exp(-R_{xy}t).$$

Fourier transformation of this function will give a peak of width (at half-height) $2R_{xy}$ rad s^{-1} or R_{xy}/π Hz. The faster the relaxation, the greater R_{xy}, and hence the broader the line.

A line whose width is determined by the transverse relaxation rate is said to be *homogeneously broadened*. Such broadening is a fundamental property of the molecule, its environment and motion, as it is these attributes which determine the rate of relaxation.

However, rapid relaxation is not the only way in which a line can become broad. For example, if the applied magnetic field is not uniform across the sample (i.e. the field is not the same everywhere), then spins in different parts of the sample will have different Larmor frequencies. As a result, there will be a spread of frequencies across the sample which will result in a broad line. A familiar example of this is when the shims are poorly adjusted, resulting in a lineshape which reflects the inhomogeneity of the B_0 field.

The way we usually think about the effect of a non-uniform field is to imagine dividing up the sample into volumes which are small enough that over each separate volume the magnetic field is uniform. Each of these little volumes thus contributes a line to the spectrum whose width is determined by the transverse relaxation rate constant R_{xy}, but whose Larmor frequency is determined by the precise value of the magnetic field in that volume. Therefore when we think about the spectrum from the whole sample, the lineshape we see is the sum of the lines from each of the separate volumes. As a consequence, the width of the overall line will be greater than the width of the line from each individual volume; Fig. 9.37 on the facing page illustrates this idea. The width and shape of the line from the whole sample will thus depend on the details of the non-uniformity of the applied field.

This kind of line broadening is called *inhomogeneous*. The term conveys the idea that different parts of the line are from different parts of the sample, rather than in the case of homogeneous broadening where the entire line comes from the whole sample. As the diagram illustrates, we can think of an inhomogeneous lineshape as resulting from the addition of many homogeneously broadened lines which are centred at slightly different frequencies.

Another way of thinking about this difference between homogeneous and inhomogeneous line broadening is to consider how the behaviour of the transverse magnetization is affected by the presence of these two types of line broadening.

In the case of inhomogeneous line broadening it is convenient to think, as we did before, about the sample being divided into small volumes, each of which has a different Larmor frequency. The transverse magnetization from each small volume therefore precesses at a different frequency. The total magnetization is found by adding up these individual contributions,

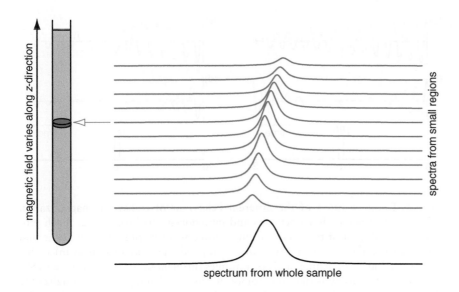

Fig. 9.37 Illustration of the idea of an inhomogeneously broadened lineshape. The NMR sample is shown on the left, and we imagine that the magnetic field is not uniform along the z-axis. Thus, at each position along z the magnetic field, and hence the Larmor frequency, is different. We can imagine a small volume, shown here as a disc, in which the magnetic field is uniform. Each disc gives rise to a homogeneously broadened line, as shown on the right. However, discs at different positions give lines at different frequencies. A range of such lines is shown on the right; their intensities are not equal, as it is typically the case that the extremities of the sample do not contribute as much signal. What we actually observe is the spectrum from the *whole* sample. This spectrum, shown at the bottom, is the sum of the lines contributed by each of the discs; note that the line from the whole sample is broader than the lines from each of the small regions. The line we observe is said to be inhomogeneously broadened as it is the sum of (homogeneously broadened) lines with different frequencies. Note that parts of the inhomogeneous line which appear at different frequencies correspond to different parts of the sample.

and it is clear that, on account of the spread of Larmor frequencies, the contributions from each volume will get out of step with one another. As a result, cancellation will occur, leading to a decay of the total magnetization.

The key thing is that this decay can be reversed simply by using a spin echo. Recall that the special property of the spin echo sequence $-\tau-180°-\tau-$ is that, at the end of the second τ delay, all the magnetization ends up in the same position *regardless* of the frequency at which it evolves during τ. So, although the magnetization from different parts of the sample is evolving at different frequencies, and so getting out of step with one another, at the end of the spin echo the magnetization from each part of the sample will be aligned in the same direction. This is usually described by saying that the decay due to the inhomogeneity has been refocused.

However, the decay of the transverse magnetization due to homogeneous broadening is quite different: it *cannot* be reversed by a spin echo. The reason for this is that the decay is due to transverse relaxation, which is a natural process, rooted in the random molecular motion and the approach to equilibrium. There is no way we can reverse its effects.

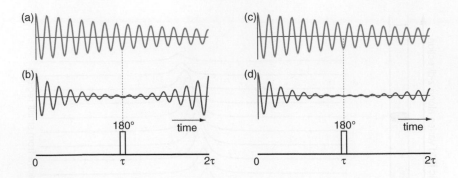

Fig. 9.38 Demonstration of the different behaviour of transverse magnetization under the influence of homogeneous and inhomogeneous broadening. Each plot shows the evolution of transverse magnetization as a function of time; at the mid point, that is after time τ, a 180° pulse is applied. The decay seen in (a) is due to homogeneous broadening, and so continues throughout, unaffected by the 180° pulse. In (b) the decay is due to inhomogeneous broadening; to help visualize the effect, the decay rate has been made faster than in (a). By the time the 180° pulse is applied, the magnetization has decayed to almost zero; this decay is attributed to the magnetization from different parts of the sample getting out of step with one another. However, after the 180° pulse, the overall magnetization grows back, and by time 2τ it has returned to its initial value. This is because the evolution due to the offset is refocused at time 2τ, so *at this point* there is no dephasing due to there being a spread of offsets. Decay (c) is the same as (a), i.e. homogeneous. In (d) we see the effect of having both inhomogeneous and homogeneous broadening. As in (b), the magnetization decays at first, but after the 180° pulse the magnetization grows back. However, at time 2τ the magnetization does not reach its initial value on account of the decay due to the homogeneous broadening. Note that at time 2τ, the magnetization in (c) and (d) are the same i.e. its size at this point is determined solely by the homogeneous term.

Figure 9.38 illustrates the different way in which transverse magnetization behaves during a spin echo, depending on whether the broadening is homogeneous or inhomogeneous. The magnetization in (a) decays due to homogeneous broadening i.e. relaxation; as a result, the 180° pulse has no effect on the decay. In contrast, the magnetization shown in (b) decays due to inhomogeneous line broadening. The 180° pulse reverses the decay, so that at time 2τ, when the refocusing is complete, the magnetization has the same size as it did at time zero.

Plot (d) shows what happens when homogeneous and inhomogeneous broadening are both present. During the time up to the 180° pulse we see decay due to both kinds of broadening. After the pulse, the decay due to the inhomogeneous broadening is reversed, and completely removed at time 2τ. However, the decay due to homogeneous broadening is not refocused, and continues throughout the echo, as shown in (c). At time 2τ, therefore, the size of the magnetization is determined only by the homogeneous broadening. At this time, the magnetization in (c) and (d) are therefore the same.

Seen in the time domain the crucial difference between homogeneous and inhomogeneous broadening is that the effects of the latter can be undone by a spin echo, whereas nothing can be done about the former.

9.9.1 Describing inhomogeneous broadening: $T_2^\star$

As we noted above, a typical time-domain function which decays due to relaxation can be written as

$$S(t) = A \exp(i\Omega t) \exp(-R_{xy}t).$$

Such a decay due to relaxation leads to homogeneous broadening. It is commonly *assumed* that inhomogeneous broadening can also be described by an exponential decay, with rate constant R_{inhom}, giving a time-domain signal of the form

$$S(t) = A \exp(i\Omega t) \underbrace{\exp(-R_{xy}t)}_{\text{homogeneous}} \underbrace{\exp(-R_{inhom}t)}_{\text{inhomogeneous}} \qquad (9.33)$$

$$= A \exp(i\Omega t) \exp(-[R_{xy} + R_{inhom}]t).$$

This assumption is convenient as the overall decay of the signal is determined by the sum of the two rate constants, $(R_{xy} + R_{inhom})$, so the corresponding linewidth is simply $(R_{xy} + R_{inhom})/\pi$ Hz. In other words, the overall linewidth is the sum of a homogeneous contribution, R_{xy}/π, and an inhomogeneous contribution, R_{inhom}/π.

   This assumption that the inhomogeneous broadening can be described by an exponential decay, as in Eq. 9.33, is simply not valid. The reason is that the inhomogeneous lineshape, and hence the corresponding decay of the time-domain function, depends in detail on how the applied magnetic field varies across the sample. There is absolutely no reason to assume that the result of this will be an exponential decay. We only have to think about the odd-looking lineshapes that we obtain when the shims are poorly adjusted to realize that the inhomogeneous part of the decay is certainly not exponential.

   Notwithstanding this, NMR spectroscopists are fond of measuring the linewidth in the spectrum, and then assuming that this is equal to $(R_{xy} + R_{inhom})/\pi$ Hz. Although not correct in any real sense, provided the lineshape is something like a Lorentzian, such an approach is probably not a bad way of estimating the inhomogeneous contribution to the decay.

   As we noted above, it is common to specify the relaxation rate constant R_{xy} in terms of its associated time constant $T_2 = 1/R_{xy}$. Similarly, the time constant $1/R_{inhom}$ is often denoted $T_2^\dagger$. Using these, Eq. 9.33 becomes

$$S(t) = A \exp(i\Omega t) \underbrace{\exp(-t/T_2)}_{\text{homogeneous}} \underbrace{\exp(-t/T_2^\dagger)}_{\text{inhomogeneous}}.$$

$$= A \exp(i\Omega t) \exp(-t/T_2^\star),$$

where

$$\frac{1}{T_2^\star} = \frac{1}{T_2} + \frac{1}{T_2^\dagger}.$$

$T_2^\star$ gives the overall decay rate, due to both homogeneous and inhomogeneous contributions. It is also not uncommon for people to talk of '$T_2^\star$ relaxation', which is rather a loose term as the inhomogeneous part of the decay is not due to relaxation.

Fig. 9.39 The simple spin echo pulse sequence used for measuring the rate constant for transverse relaxation. The spin echo refocuses any decay due to inhomogeneous contributions to the lineshape, so that at the start of acquisition the size of the magnetization only depends on the transverse relaxation which has taken place during 2τ. Therefore, the peak height also reflects just this relaxation-induced decay.

9.9.2   Measuring the transverse relaxation rate constant

In the absence of inhomogeneous broadening, a good estimate of the rate constant for transverse relaxation can be obtained by measuring the width of the corresponding line at half height. R_{xy} is simply π times the width in Hz. However, if there is a significant inhomogeneous contribution to the linewidth, we cannot use this method.

To get round this problem of the inhomogeneous contribution, we use the simple spin echo pulse sequence of Fig. 9.39 to measure R_{xy}. As was explained above, any effects of inhomogeneous broadening are refocused at the end of the echo, so the size of the transverse magnetization present at the start of data acquisition (after the second τ delay) *only* depends on the transverse relaxation rate constant and the time 2τ.

The experiment simply involves executing the pulse sequence, and recording the FID, for a series of times τ. The peak height in the spectrum, $S(\tau)$, will simply follow the decay of the transverse magnetization i.e.

$$S(\tau) = S(0) \exp(-2R_{xy}\tau),$$

where $S(0)$ is the peak height at time $\tau = 0$. Taking logarithms of both sides of this equation gives

$$\ln S(\tau) = \ln S(0) - 2R_{xy}\tau,$$

so a plot of $\ln S(\tau)$ against 2τ will be a straight line of slope $-R_{xy}$.

Optional section ⇒

9.10   Relaxation due to chemical shift anisotropy

The way in which chemical shift anisotropy (CSA) can become a relaxation mechanism was outlined in section 9.2.2 on page 250. In the present section, we will look in more detail first at how the size of the CSA is specified, and secondly at the resulting relaxation rate constants. Part of the reason for doing this is that CSA relaxation is important for heteronuclei such as ^{13}C, ^{15}N and ^{31}P, and also because the cross correlation between CSA and dipolar relaxation, which we will consider in the following section, is an important phenomenon in NMR of large (biological) molecules.

9.10.1   Specifying the CSA

When a molecule is placed in a static magnetic field, in order to work out the size and direction of the local field at a particular nucleus we need to know first the *chemical shielding tensor* of that nucleus, and secondly the orientation of the tensor with respect to the applied field. A full description of what a tensor is, how the shielding tensor can be measured, and the details of how the local field is calculated is beyond the level of this text.

However, for present purposes the important thing we need to know is that the 'size' of the tensor is specified by its three principal components, usually denoted σ_{xx}, σ_{yy} and σ_{zz}. Like the chemical shift, these components are quoted in ppm. The orientation of the tensor is fixed within the molecule, so that as the molecule tumbles, the tensor moves with it. To complete our description, we also need to know the orientation of the tensor with respect to the molecular framework.

Things become somewhat simpler if we have what is called an *axially symmetric* tensor, which is one in which two of the principal components are equal, but the third is different. Such a tensor can be represented by a three-dimensional ellipsoid, as shown in Fig. 9.40. The distance along the symmetry axis from the centre of the ellipsoid to the edge gives the *parallel component* of the shielding tensor, denoted $\sigma_\parallel$. The distance from the centre to the edge, measured perpendicular to the symmetry axis, gives the *perpendicular component* of the tensor, denoted $\sigma_\perp$.

When the tensor is oriented so that the symmetry axis is along the direction of the applied field, the chemical shift is $\sigma_\parallel$, whereas if the symmetry axis is perpendicular to the field, the chemical shift is $\sigma_\perp$. At other orientations, the shift is between these two values. Similarly, the direction and size of the local field depends on the orientation of the tensor with respect to the applied field.

It is often found that the chemical shielding tensor of the ^{13}C in a C–H group, or of the ^{15}N in an N–H group is, to a good approximation, axially symmetric, with the symmetry axis pointing along the direction of the C–H or N–H bond. We will have more to say about the significance of this in the subsequent section concerned with cross correlation.

9.10.2 Relaxation rate constants due to CSA

CSA relaxation only involves one spin, so there is no possibility of cross relaxation and hence transfer of magnetization between spins. The magnetizations thus decay with simple exponentials, just as for the case of relaxation by random fields (section 9.8.4 on page 294):

$$\frac{\mathrm{d}M_{xy}}{\mathrm{d}t} = -R_{xy}\,M_{xy} \qquad \frac{\mathrm{d}M_z}{\mathrm{d}t} = -R_z\,(M_z - M_z^0).$$

For an axially symmetric shielding tensor, the rate constants are as follows:

$$R_{xy} = c^2\left[\tfrac{2}{45}j(0) + \tfrac{1}{30}j(\omega_0)\right] \qquad R_z = c^2\tfrac{1}{15}j(\omega_0),$$

where

$$c = \gamma B_0(\sigma_\parallel - \sigma_\perp).$$

As we have come to expect, R_z only depends on $j(\omega_0)$, whereas R_{xy} has a secular contribution depending on $j(0)$ and a non-secular contribution, equal to $\tfrac{1}{2}R_z$, depending on $j(\omega_0)$.

The rate constants go as $(\gamma B_0)^2$, and so this kind of relaxation is likely to be more significant at higher field, and for higher gyromagnetic ratio nuclei. Finally, note that the rate constants depend on the difference between $\sigma_\parallel$ and $\sigma_\perp$, often rather loosely called the 'anisotropy', and is sometimes given the symbol Δ.

In contrast to the other mechanisms we have considered, the two rate constants are *not* equal in the fast motion limit. In fact, in this limit, $R_{xy}/R_z = 7/6$.

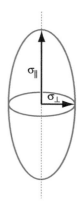

Fig. 9.40 Representation of an axially symmetric shielding tensor as a three-dimensional ellipsoid i.e. a sphere which has been 'stretched' along one axis; the symmetry axis is shown dotted. The distance between the centre and the edge of the ellipsoid, measured along the symmetry axis, is $\sigma_\parallel$. The distance measured in a perpendicular direction is $\sigma_\perp$. On account of the axial symmetry, this latter distance is the same in any direction which is perpendicular to the symmetry axis. In this diagram, $\sigma_\parallel > \sigma_\perp$, but it can just as well be the other way round.

Optional section ⇒

9.11   Cross correlation

A particular nucleus in a molecule is likely to experience random fields from more than one source, for example by dipolar interaction with several other nuclei, from its CSA or from paramagnetic species. The simplest assumption we can make is that these different sources of relaxation are *independent*, and so the total relaxation rate constant is the sum of the rate constants due to each source of local fields:

$$R_{tot} = R_1 + R_2 + R_3 + \ldots,$$

where R_i is the rate constant for relaxation caused by local fields from source i.

For two sources of relaxation to be independent of one another, the time dependence of the associated random fields must be completely different. In other words, there must be no correlation between these random functions. Whether or not this is the case will depend on the kind of molecular motion which is causing the fields to vary.

If it turns out that the random fields from two separate sources are not independent, then we say that there is *cross correlation* (or relaxation interference) between the relaxation mechanisms giving rise to the fields. As we shall see, the presence of such an effect alters the relaxation rate constants in rather a subtle way.

For the remainder of this section we are going to discuss cross correlation between dipolar and CSA relaxation in the ^{15}N–^{1}H fragment, such as would be found in the amide bond of a polypeptide (protein). We choose this example as it is a case where cross correlation has led to some very important applications in the area of biological NMR.

9.11.1   Cross correlation in longitudinal relaxation

The ^{15}N nucleus in an N–H group is relaxed by two mechanisms: the dipolar interaction with the proton and the CSA of the ^{15}N itself. In both cases, the time dependence of the local field experienced by the ^{15}N derives from the rotational reorientation of the molecule due to thermal motion. Such motion alters the orientation of both the N–H vector and the ^{15}N shift anisotropy tensor with respect to the applied field. It is therefore clear that the variations of the random fields arising from these two sources *must* be correlated to some extent as both are modulated by the same motion.

Once more, the details of how the rate constants are calculated for such a case is beyond the level of this text, so we will simply have to quote some results, and then discuss their interpretation. To start with, we will consider the longitudinal relaxation of our N–H pair, the dipolar part of which was discussed in section 9.6 on page 267. In our discussion, spin one will be the ^{15}N and spin two will be the proton; we will also assume, for simplicity, that only the ^{15}N has a significant CSA.

It turns out that, in the presence of cross correlation, the relaxation rate constants $W_1^{(1,\alpha)}$ and $W_1^{(1,\beta)}$, defined in Fig. 9.17 on page 268, are different. This is in contrast to the case of pure dipolar relaxation (section 9.6.3 on page 271), where these two rate constants are equal. The theory shows us that the rate constants are given by

$$W_1^{(1,\alpha)} = W_{\text{dipolar}} + W_{\text{CSA}} \underbrace{- \tfrac{1}{10} c_1 b P_2(\cos\theta) j(\omega_{0,1})}_{\text{cross-correlation term}}$$

$$W_1^{(1,\beta)} = W_{\text{dipolar}} + W_{\text{CSA}} \underbrace{+ \tfrac{1}{10} c_1 b P_2(\cos\theta) j(\omega_{0,1})}_{\text{cross-correlation term}}.$$

$$(9.34)$$

In these expressions, W_{dipolar} is the contribution from pure dipolar relaxation, W_{CSA} is the contribution from pure CSA relaxation, and $P_2(\cos\theta)$ is the second-order Legendre polynomial, given by

$$P_2(\cos\theta) = \tfrac{1}{2}(3\cos^2\theta - 1).$$

θ is the angle between the symmetry axis of the chemical shielding tensor (which is assumed to be axially symmetric) and the N–H vector. The constants b and c_1 are those we have defined before when discussing dipolar relaxation and CSA relaxation:

$$b = \frac{\mu_0 \gamma_1 \gamma_2 \hbar}{4\pi r^3} \qquad c_1 = \gamma_1 B_0(\sigma_{1,\|} - \sigma_{1,\perp}); \qquad (9.35)$$

we have added the subscript 1 to c_1 in order to indicate that it describes the CSA of spin one.

The crucial thing to note is that the cross-correlation term has the *opposite sign* in $W_1^{(1,\alpha)}$ to that in $W_1^{(1,\beta)}$. What this means is that the presence of cross correlation *decreases* the rate of relaxation for one of the transitions, and *increases* it for the other. Remember that the only thing that is different between these two spin-one transitions is the spin state of spin two (the passive spin), so what we have here is a spin-one relaxation rate constant whose value depends on the spin state of spin two.

This behaviour can be interpreted in the following way. Spin one experiences two local fields, one deriving from its CSA and one from the dipolar interaction with spin two; due to cross correlation the fluctuations in these random fields are partly correlated. Depending on the way in which the fields are correlated, they might reinforce one another to some extent, thus increasing the relaxation rate constant, or they might cancel one another to some extent, thus reducing the relaxation rate constant.

We commented before (section 9.8.5 on page 296) that the direction of the local field at spin one depends on, amongst other things, the spin state of spin two. Therefore if when spin two is in the one state (i.e α or β), the dipolar and CSA derived fields reinforce to some extent, we expect that when spin two is in the opposite state (i.e. β or α), the two local fields will cancel to some extent, simply as the direction of the dipolar-derived field has changed. This is the origin of the difference in the sign of the cross-correlation terms in Eq. 9.34.

The degree of cross correlation between the two random fields depends, via the $P_2(\cos\theta)$ term in Eq. 9.34, on the orientation of the shift anisotropy tensor with respect to the N–H vector. This term can be positive or negative, and has its maximum when $\theta = 0°$ or $180°$ i.e. when the tensor is aligned with, or against, the direction of the N–H bond. In practice the tensor and the bond vector are usually quite closely aligned, making the cross-correlation effect a maximum.

The question arises as to which of $W_1^{(1,\alpha)}$ and $W_1^{(1,\beta)}$ is the greater: referring to Eq. 9.34 on the preceding page we see that this depends on the *signs* of $P_2(\cos\theta)$, b and c_1. The sign of b depends on the signs of the two gyromagnetic ratios, γ_1 and γ_2. The sign of c_1 depends on the sign of γ_1 and of $(\sigma_{1,\parallel} - \sigma_{1,\perp})$. So, although it is clear that in the presence of cross correlation $W_1^{(1,\alpha)}$ and $W_1^{(1,\beta)}$ will be different, which is the greater depends on the type of nuclei and the details of the CSA. Note that pure dipolar relaxation depends on b^2, and pure CSA relaxation of spin one depends on c_1^2, so in these cases the signs of b and c_1 are not important.

The effect of $W_1^{(1,\alpha)}$ and $W_1^{(1,\beta)}$ not being the same can best be seen by looking at the Solomon equations in the form given in Eq. 9.17 on page 270:

$$\frac{dI_{1z}}{dt} = -R_z^{(1)}(I_{1z} - I_{1z}^0) - \sigma_{12}(I_{2z} - I_{2z}^0) - \Delta^{(1)}\,2I_{1z}I_{2z}$$

$$\frac{dI_{2z}}{dt} = -\sigma_{12}(I_{1z} - I_{1z}^0) - R_z^{(2)}(I_{2z} - I_{2z}^0) - \Delta^{(2)}\,2I_{1z}I_{2z}$$

$$\frac{d\,2I_{1z}I_{2z}}{dt} = -\Delta^{(1)}(I_{1z} - I_{1z}^0) - \Delta^{(2)}(I_{2z} - I_{2z}^0) - R_z^{(1,2)}\,2I_{1z}I_{2z}.$$

Recall that $\Delta^{(1)} = W_1^{(1,\alpha)} - W_1^{(1,\beta)}$ and $\Delta^{(2)} = W_1^{(2,\alpha)} - W_2^{(1,\beta)}$. For pure dipolar relaxation, $W_1^{(1,\alpha)} = W_1^{(1,\beta)}$ and so $\Delta^{(1)} = 0$; the same is true for $\Delta^{(2)}$. As a result there is no transfer between I_{1z} and $2I_{1z}I_{2z}$. However, in the presence of cross correlation $\Delta^{(1)}$ is not zero but, from Eq. 9.34 on the preceding page, is given by

$$\Delta^{(1)} = -\tfrac{1}{5}\,c_1 b P_2(\cos\theta)\,j(\omega_{0,1}).$$

Note that this rate constant can be positive or negative. $\Delta^{(2)}$ is still zero as we are assuming that there is no CSA on this spin. So the Solomon equations become

$$\frac{dI_{1z}}{dt} = -R_z^{(1)}(I_{1z} - I_{1z}^0) - \sigma_{12}(I_{2z} - I_{2z}^0) - \Delta^{(1)}\,2I_{1z}I_{2z}$$

$$\frac{dI_{2z}}{dt} = -\sigma_{12}(I_{1z} - I_{1z}^0) - R_z^{(2)}(I_{2z} - I_{2z}^0)$$

$$\frac{d\,2I_{1z}I_{2z}}{dt} = -\Delta^{(1)}(I_{1z} - I_{1z}^0) - R_z^{(1,2)}\,2I_{1z}I_{2z}.$$

In these equations the rate constants $R_z^{(1)}$ and $R_z^{(1,2)}$ have contributions from both the dipolar and CSA relaxation mechanisms.

What these particular Solomon equations tell us is that, in the presence of cross correlation, there will be a relaxation-induced transfer between I_{1z} and $2I_{1z}I_{2z}$; there is no such transfer in the case of pure dipolar relaxation. The observation of this transfer is thus a convenient way of detecting the presence of cross correlation.

9.11.2   Cross correlation in transverse relaxation

We now turn to the effect of cross correlation on transverse relaxation, and again restrict our attention to the $^{15}\text{N}-^1\text{H}$ spin pair. In this case, the theory predicts that the effect of cross correlation is to make the transverse

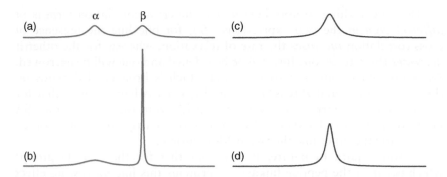

Fig. 9.41 Illustration of the effect of cross correlation on the spin-one doublet. In the absence of cross correlation, the two lines of the doublet have the same width, and hence the same intensity, as shown in (a). Cross correlation results in one of the lines becoming narrower, and the other becoming broader, as shown in (b); arbitrarily, we have assumed that the line associated with the α state of spin two is the broader of the two. Since the peak height goes inversely with the linewidth, the narrow line is rather dominant in this doublet. If the spin-one doublet is observed under conditions of broadband decoupling of spin two we will see a single line, whose width is the average of the widths of the two lines of the doublet. In the case of the symmetrical doublet (a), decoupling gives a line of the same width and twice the height, shown in (c). However, for the asymmetric doublet (b), decoupling gives a line of intermediate width, shown in (d). The advantage of having a narrow line in (c) is completely lost if decoupling is used.

relaxation rate constants *different* for the magnetization associated with different lines of the ^{15}N doublet. As with longitudinal relaxation, what we have here is relaxation of spin one which depends on the spin state of spin two (recall that the two lines of the doublet are associated with different spin states of spin two).

As the magnetization associated with the two lines relaxes at different rates, in the spectrum the two lines will have different widths. So, rather than having the usual doublet in which both lines have the same width and height, as shown in Fig. 9.41 (a), the presence of cross correlation results in an asymmetric doublet where one line has become broader, and the other has become sharper, as shown in Fig. 9.41 (b). Narrowing a line increases its peak height, whereas broadening the line decreases the height, so in the doublet shown in (b) the sharp line is very much taller than the broad line.

It turns out that the asymmetry increases as the correlation time becomes greater. In the slow motion limit, in which we can set all of the spectral densities to zero except that at zero frequency, the relaxation rate constants for the two lines are

$$\text{line 1}: \underbrace{\tfrac{1}{10}\,b^2 j(0)}_{\text{dipolar}} + \underbrace{\tfrac{2}{45}\,c_1^2 j(0)}_{\text{CSA}} + \underbrace{\tfrac{2}{15}\,c_1 b\, P_2(\cos\theta)\, j(0)}_{\text{cross correlation}}$$

$$\text{line 2}: \underbrace{\tfrac{1}{10}\,b^2 j(0)}_{\text{dipolar}} + \underbrace{\tfrac{2}{45}\,c_1^2 j(0)}_{\text{CSA}} - \underbrace{\tfrac{2}{15}\,c_1 b\, P_2(\cos\theta)\, j(0)}_{\text{cross correlation}}.$$

As before, we are assuming that spin one is the ^{15}N and spin two is the proton, and we are also assuming that only the ^{15}N has a significant CSA.

The crucial thing to note here is that the cross-correlation term is of *different sign* in the two expressions. So, for one line the presence of cross correlation *increases* the rate of relaxation, whereas for the other it *decreases* the rate i.e. one line will be broadened and one will be narrowed. As before, which line is narrowed and which is broadened depends on the signs of b, c_1 and $P_2(\cos\theta)$. Physically, what is happening is that for the line which is narrowed, the random fields from the dipolar and CSA interactions are correlated in such a way that they partly cancel one another, whereas for the other line the two fields reinforce.

By some quirk of nature, it turns out that for the $^{15}N-^1H$ groups which occur in the peptide linkages in proteins this line narrowing effect is particularly pronounced at high magnetic fields. Indeed, for typical values of the N–H bond length and ^{15}N CSA, it turns out that for one line the relaxation rate constant *goes to zero* at field strengths of around 25 T (assuming the slow motion limit). At the currently highest available field strengths of around 20 T the amount of line narrowing is still very significant, with the narrow line typically being one-twentieth of the width of the broad line. The presence of such unexpectedly narrow lines for large molecules greatly improves the sensitivity, and therefore makes it much easier to obtain structural information; the effect has been given the name TROSY.

TROSY: Transverse Relaxation Optimized SpectroscopY

In the $^{15}N-^1H$ group the proton also has a CSA, albeit a factor of ten or more smaller than that of the ^{15}N. However, as the gyromagnetic ratio of proton is some ten times larger than than of ^{15}N, the effect, in relaxation terms, of the CSA of the proton is comparable with that of the ^{15}N. As a result, the two lines of the proton doublet (due to the coupling to ^{15}N) also have different widths due to the presence of cross correlation. Once again, by some quirk of nature, the conditions which maximize the effect for the ^{15}N doublet also result in significant narrowing of one line of the proton doublet. The fact that both the proton and ^{15}N doublets contain a sharp line has been exploited extensively in devising high-sensitivity N–H correlation experiments, such as those discussed in section 10.9 on page 358.

Note that we can only see this difference in the linewidth between the two lines of the multiplet if the splitting is resolved. If there is no splitting, for example as a result of applying broadband decoupling to spin two, what we see is one line with a width determined by the *average* of the widths of the two lines of the doublet. This average linewidth will be much larger than the width of the narrow line, so there will be a large reduction in peak height, and we therefore lose any advantage in signal-to-noise ratio gained from having a tall sharp line. This idea is illustrated in Fig. 9.41 on the previous page. It is therefore important that experiments which are designed to exploit this TROSY effect retain the splitting throughout; we return to a discussion of suitable pulse sequences in section 10.9 on page 358.

9.12   Summary

- Local fields, generated by mechanisms such as the dipolar interaction or CSA, are responsible for relaxation. The time dependence of these local fields is characterized by the correlation function, which depends on the correlation time. The spectral density $j(\omega)$, which is the Fourier transform of the correlation function, gives the amount of motion present at frequency ω.

- The rate constant for longitudinal relaxation depends on the spectral density at the Larmor frequency. The rate constant for transverse relaxation depends on the spectral density at both the Larmor frequency and at zero frequency.

- The rate constant for longitudinal relaxation reaches a maximum when $\omega_0\tau_c \approx 1$. The rate constant for transverse relaxation increases indefinitely as the correlation time increases.

- Two motional limits can be distinguished: fast motion, when $\omega_0\tau_c \ll 1$, and slow motion, when $\omega_0\tau_c \gg 1$. In the fast motion limit the spectral density is independent of frequency and is given by $j(\omega) = 2\tau_c$ i.e. proportional to the correlation time.

- Dipolar relaxation is unique in giving rise to cross relaxation, a process which leads to the transfer of magnetization from one spin to another and hence the NOE. The cross-relaxation rate constant is positive in the fast motion limit and negative in the slow motion limit.

- The size of a steady-state NOE enhancement is determined by the competition between cross relaxation and self relaxation.

- We distinguish between homogeneous and inhomogeneously broadened lines. Homogeneous broadening is due to relaxation, and the corresponding decay of the magnetization cannot be reversed. In contrast, the dephasing due to inhomogeneous broadening can be reversed by a spin echo.

9.13   Further reading

Molecular motion and relaxation:

Chapters 19 and 20 from M. H. Levitt, *Spin Dynamics* (2nd edition, John Wiley & Sons, Ltd, 2008).

Theory of relaxation:

Chapter 6 from F. J. M. van de Ven, *Multidimensional NMR in Liquids* (VCH, 1995).

Chapter 5 from J. Cavanagh, W. J. Fairbrother, A. G. Palmer III, M. Rance and N. J. Skelton, *Protein NMR Spectroscopy* (2nd edition, Academic Press, 2007).

All aspects of the NOE, both theoretical and experimental:

D. Neuhaus and M. P. Williamson, *The Nuclear Overhauser Effect in Structural and Conformational Analysis* (2nd edition, John Wiley & Sons, Ltd, 2000).

Relaxation and the NOE, in the context of larger molecules:

Chapter 5 from J. Cavanagh, W. J. Fairbrother, A. G. Palmer III, M. Rance and N. J. Skelton, *Protein NMR Spectroscopy* (2nd edition, Academic Press, 2007).

TROSY:

C. Fernández and G. Wider, *Current Opinion in Structural Biology*, **13**, 570–580 (2003).

9.14 Exercises

Data:

gyromagnetic ratio for protons $\quad +2.675 \times 10^8$ rad s^{-1} T^{-1}

gyromagnetic ratio for ^{13}C $\quad +6.728 \times 10^7$ rad s^{-1} T^{-1}

gyromagnetic ratio for ^{15}N $\quad -2.713 \times 10^7$ rad s^{-1} T^{-1}

Boltzmann constant, k_B $\quad 1.381 \times 10^{-23}$ J K^{-1}

Planck constant, h $\quad 6.626 \times 10^{-34}$ J s

Planck constant / 2π, $\hbar$ $\quad 1.055 \times 10^{-34}$ J s

Permeability of vacuum, μ_0 $\quad 4\pi \times 10^{-7}$ H m^{-1}

$1 \text{ Å} = 10^{-10}$ m $\qquad 1 \text{ ns} = 10^{-9}$ s $\qquad 1 \text{ ps} = 10^{-12}$ s

9.1 Suppose that an NMR sample containing 10^{13} protons is placed in a magnetic field of 9.4 T. Using the Boltzmann distribution, calculate the equilibrium populations of the α and β levels, assuming that the temperature is 298 K.

When the sample is first placed in the magnetic field, half the protons will be in the α level and half will be in the β level. Calculate the total energy change of the spins when they go from this initial state to equilibrium. Compare your answer with the typical thermal energy possessed by N molecules, which is of the order of Nk_BT.

9.2 Show that, for a fixed frequency ω, the maximum in the reduced spectral density function $j(\omega)$ occurs at $\tau_c = 1/\omega$. What is the significance of this result?

9.3 For a sample consisting of isolated spins, explain *in words* why it is necessary for the rate constant for the relaxation-induced transitions from the α state to the β state to be less than the rate constant for the transitions in the opposite direction.

9.4 In an inversion–recovery experiment the following peak heights $S(\tau)$ (arbitrary units) were measured as a function of the delay τ:

τ / s	0.0	0.1	0.5	0.9	1.3	1.7	2.1	2.9
$S(\tau)$	−129.7	−93.4	7.6	62.6	93.4	109.5	118.9	126.4

Use a graphical method to analyse these data and hence determine a value for the rate constant for longitudinal relaxation and the corresponding value of the relaxation time, T_1.

9.5 In an experiment to estimate T_1 using the inversion–recovery sequence, three peaks in the spectrum were observed to go through a null at 0.5, 0.6 and 0.8 s, respectively. Explain how an estimate for T_1 can be obtained from such measurements, and give the value of T_1 for each line. A solvent resonance was still inverted after a delay of 1.5 s; what does this tell you about the relaxation time of the solvent?

9.6   An alternative to the inversion–recovery method for estimating R_z is the saturation–recovery experiment. This starts with the spin being irradiated for a long time so that it becomes saturated, i.e. $M_z = 0$. There then follows a delay τ, followed by a $90°$ pulse and then observation of the FID.

Show that the z-magnetization at the end of τ is given by

$$M_z(\tau) = M_z^0 \left[1 - \exp(-R_z\tau)\right];$$

make a sketch of $M_z(\tau)$ as a function of τ.

Explain how the data from a series of experiments recorded with increasing values of τ can be used, in conjunction with a graphical method, to estimate R_z.

9.7   Using the approach of section 9.6.2 on page 269, write down expressions for dn_i/dt for each of the four levels in terms of the populations and the transition rate constants, W.

9.8   In this exercise we will use the expressions given in section 9.6.3 on page 271 and in section 9.8.5 on page 296 to calculate the relaxation rate constants which arise from the dipolar relaxation of two spins. In principle these calculations are just a question of substituting in the appropriate values into the formulae, but in doing so it is all too easy to make mistakes, particularly over the units of the various quantities. This exercise takes you through the calculations step-by-step so that you can check your results at each stage.

(a)   Consider the two protons in a CH_2 group, which are separated by 1.8 Å. Remembering to put r in m, show that $b^2 = 1.675 \times 10^{10}$ s^{-2}. [Working out the units of b^2 is rather difficult, so we will just accept that they are s^{-2}.]

(b)   Assuming that the molecule is in the fast motion limit, so that $j(\omega) = 2\tau_c$, and taking $\tau_c = 20$ ps, show that the transition rate constants have the following values (all in units of s^{-1}): $W_1^{(1)} = 0.0503$, $W_1^{(2)} = 0.0503$, $W_2 = 0.201$, $W_0 = 0.0335$. Then, using Eq. 9.19 on page 271, show that $R_z^{(1)} = 0.335$, $R_z^{(2)} = 0.335$ and $\sigma_{12} = 0.168$ (all in units of s^{-1}).

(c)   Use Eq. 9.20 on page 272 to calculate $R_z^{(1)}$, $R_z^{(2)}$ and σ_{12}; you should, of course, obtain the same values as you did in the previous part.

(d)   Use the expressions in section 9.8.5 on page 296 to determine $R_{xy}^{(1)}$ and $R_{xy}^{(2)}$; you should find that both have the value 0.335 s^{-1}.

(e)   Comment on the values you have calculated, and the comparison between them.

(f) The next task is to repeat all of these calculations for a correlation time of 500 ps, and for a magnetic field strength of 11.74 T (a proton Larmor frequency of 500 MHz). Such a correlation time places the motion well outside the fast motion limit, so you will need to compute the reduced spectral densities explicitly for each frequency. First, show that the proton Larmor frequency is 3.140×10^9 rad s^{-1}, and the use this to show that $j(\omega_0) = 2.88 \times 10^{-10}$ s, $j(2\omega_0) = 9.20 \times 10^{-11}$ s and $j(0) = 1.00 \times 10^{-9}$ s. Note that to compute $j(\omega)$, ω must be in rad s^{-1}.

Use these values for the reduced spectral density, along with the value of b^2 computed earlier, to show that $R_z^{(1)} = 2.025$, $R_z^{(2)} = 2.025$ and $\sigma_{12} = -0.375$, $R_{xy}^{(1)} = 3.41$ and $R_{xy}^{(2)} = 3.41$, all in units of s^{-1}.

(g) Comment on the values you have obtained for the longer correlation time, and compare them with those obtained in the extreme narrowing limit.

9.9 The typical separation of a directly bonded ^{13}C–^{1}H pair is 1.1 Å. Assuming a correlation time of 20 ps (i.e. the fast motion limit), calculate values for the rate constants $R_z^{(1)}$, $R_z^{(2)}$, σ_{12}, $R_{xy}^{(1)}$ and $R_{xy}^{(2)}$ (take spin one to be ^{13}C and spin two to be ^{1}H). [If you have completed the previous exercise, then all you have to do is recognize that, in the fast motion limit, the only difference between a ^{1}H–^{1}H pair and a ^{13}C–^{1}H pair is the value of the constant b.]

Compare your answers with those in parts (b)–(d) of the previous exercise.

9.10 For a bonded ^{13}C–^{1}H pair a typical value of the chemical shift anisotropy $(\sigma_\parallel - \sigma_\perp)$ is 100 ppm. Assuming a correlation time of 20 ps (i.e. the fast motion limit), compute the contribution which CSA relaxation makes to the ^{13}C at static magnetic field strengths of 4.7 T and 11.74 T. In computing the constant c, remember to put $(\sigma_\parallel - \sigma_\perp) = 100 \times 10^{-6}$ on account of the fact that the value quoted is in ppm.

Comment on the values you obtain at the two fields, and compare them with the dipolar contributions calculated in the previous question.

[The values you should find at 4.7 T are $c^2 = 1.00 \times 10^9$ s^{-2}, $R_z = 0.00267$ s^{-1} and $R_{xy} = 0.00311$ s^{-1}; the values at 11.74 T are all larger by the ratio $(11.74/4.7)^2$.]

9.11 Protons have much smaller chemical shift anisotropies than heteronuclei, but at the very high magnetic fields which are now becoming available CSA relaxation may be significant. Assuming that a proton has $(\sigma_\parallel - \sigma_\perp) = 10$ ppm, calculate the CSA contribution to its transverse relaxation at fields of 4.7 T, 11.74 T and 23.5 T (the latter field corresponds to a proton Larmor frequency of 1000 MHz); in your calculation, assume that the fast motion limit applies, and take $\tau_c = 20$ ps. Compare your values with the rate constants for dipolar relaxation calculated for the same correlation time in exercise 9.8.

9.12 An alternative to the transient NOE experiment described in section 9.7.1 on page 274, is one in which rather than spin two being inverted by a selective 180° pulse at the start of the experiment, it is saturated. After this, there is a delay τ and then a 90° pulse followed by observation, just as in the experiment described in section 9.7.1 on page 274.

In this modified experiment, the initial conditions are:

$$\text{at } \tau = 0: \quad I_{1z}(0) = I_{1z}^0 \quad \text{and} \quad I_{2z}(0) = 0.$$

Using these initial conditions, analyse the experiment using the same approach as in section 9.7.1 on page 274. You should find that the NOE enhancement is given by $\eta = \sigma_{12}\tau$. Sketch the expected form of the irradiated, reference and NOE difference spectra.

9.13 Why is an NOE difference spectrum a convenient way of visualizing which resonances are receiving an NOE enhancement?

9.14 Explain the following observations, *in words*, concerning the NOE in a two-spin system.

(a) In a transient NOE experiment, and in the initial rate, the NOE enhancement depends only on the cross-relaxation rate constant.

(b) At longer times, the NOE enhancement in this transient experiment depends on the self-relaxation rate constants of both spins as well as on the cross-relaxation rate constant.

(c) In a steady-state NOE experiment, the NOE enhancement of spin two, observed when spin one is saturated, depends on the ratio of the cross-relaxation rate constant to the self-relaxation rate constant of spin two; the self-relaxation rate constant of spin one does not affect the size of the enhancement.

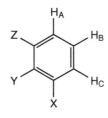

9.15 For the molecule shown opposite, a transient NOE experiment (recorded in the initial rate limit) in which H_B was inverted gave equal NOE enhancements (in the initial rate limit) on H_A and H_C. If H_A was inverted, the enhancement on H_B was the same as in the first experiment; no enhancement was seen on H_C.

In steady-state experiments, irradiation of H_B gave equal enhancements on H_A and H_C. However, irradiation of H_A gave a much smaller enhancement on H_B than for the case where H_B was the irradiated spin and the enhancement was observed on H_A. Explain these observations.

9.16 For a two-spin system, and in the initial rate, show that in the NOESY experiment, changing the phase of the first pulse from x

to $-x$, changes the sign of the diagonal and cross peaks, but leaves the axial peaks unaffected. Explain how the latter peaks can be suppressed.

9.17 In the case of relaxation caused by random fields, calculate the value of $\overline{B_{loc}^2}$ needed to give a proton T_1 of 1 s at a Larmor frequency of 500 MHz and for a correlation time of 10 ps. Comment on your answer.

9.18 In a spin echo experiment designed to measure the value of R_{xy}, the following peak heights $S(\tau)$ were measured as a function of the spin echo delay τ. Use a graphical method to estimate the value of R_{xy}.

$\tau\,/\,\mathrm{s}$	0	0.1	0.2	0.3	0.4	0.5	0.6	0.7
$S(\tau)$	65.0	39.4	23.9	14.5	8.80	5.34	3.24	1.96

Explain why it is not usually possible to estimate R_{xy} by simply measuring the linewidth in the spectrum.

10

Advanced topics in two-dimensional NMR

This chapter is something of a rag-bag of different experiments and techniques, with rather little connection between them. However, everything in this chapter is of importance to some particular application of two-dimensional NMR, and so the chances are that at some stage you will come across one or other of the techniques described here.

Some of the ideas we want to discuss in this chapter only show up in spin systems consisting of three or more coupled spins, so we will first have to extend the product operator approach of Chapter 7 from two to three spins. In addition, we will also introduce another operator basis, polarization operators, which provides a more convenient description of some experiments than do product operators.

The topics which are covered in this chapter are:

10.1 *Product operators for three spins*. This is a straightforward extension of the two-spin case discussed in Chapter 7.

10.2 *COSY for three spins*. In this section particular attention will be focused on the detailed form of the cross-peak multiplets. We will find that in these multiplets the splittings due to *active* and *passive* couplings appear in distinct, and potentially useful, ways.

10.3 *Reduced multiplets in COSY spectra*. It will be shown that, under some circumstances, the cross-peak multiplets in COSY can be simplified in a particularly useful way which makes it possible to determine the relative signs of coupling constants and, under some circumstances, measure the values of rather small coupling constants.

10.4 *Polarization operators*. This alternative operator expansion is particularly useful for understanding the detailed form of cross-peak multiplets, such as those seen in COSY and ZCOSY experiments.

10.5 *ZCOSY*. This is a modified COSY experiment which is the best practical method for recording reduced multiplets.

Understanding NMR Spectroscopy, Second Edition James Keeler
© 2010 John Wiley & Sons, Ltd

10.6 *HMBC.* The important HMBC experiment is revisited to see what effect the presence of proton–proton couplings has on the appearance of the spectrum. We will discover that such couplings result in rather complex cross-peak multiplets.

10.7 *Sensitivity-enhanced experiments.* In this section we will look at the way in which the sensitivity of some heteronuclear experiments can be improved by relatively simple modifications to the pulse sequence. Such an approach has proved to be very useful in biomolecular NMR.

10.8 *Constant time experiments.* The constant time pulse sequence element makes it possible to remove, from the ω_1 dimension, the splittings due to homonuclear couplings. Again, this is a modification which has proved to be particularly popular in biomolecular NMR, especially in three- and four-dimensional experiments.

10.9 *TROSY.* The TROSY technique exploits the fact that, due to cross correlation between CSA and dipolar relaxation, the two lines of a doublet can have very different linewidths – an effect which is very pronounced in the spectra of ^{15}N–^1H pairs in large biomolecules. Experiments such as HSQC can be modified so that only the correlation between the two narrow lines is seen, thus leading to a significant improvement in sensitivity and resolution.

10.10 *Double-quantum spectroscopy of a three-spin system.* Although we have already looked at double-quantum spectroscopy for two spins in Chapter 8, several new features arise for a three-spin system.

10.1   Product operators for three spins

In section 7.4 on page 149 we described how the product operators for two spins were constructed by taking any one of the four operators for spin one:

$$\hat{E}_1 \qquad \hat{I}_{1x} \qquad \hat{I}_{1y} \qquad \hat{I}_{1z},$$

and multiplying it by any one of the four operators for spin two

$$\hat{E}_2 \qquad \hat{I}_{2x} \qquad \hat{I}_{2y} \qquad \hat{I}_{2z}.$$

To extend this approach to three spins, all we need to do is to further multiply by any one of the four operators for spin three:

$$\hat{E}_3 \qquad \hat{I}_{3x} \qquad \hat{I}_{3y} \qquad \hat{I}_{3z}.$$

Recall that the $\hat{E}$ are unit operators which, for brevity, we do not bother to write out each time. For example, $\hat{I}_{1x}\hat{E}_2\hat{E}_3$ would usually be written as $\hat{I}_{1x}$.

With four operators for each spin, we can see that there will be a total of $4^3 = 64$ possible product operators for a three-spin system. This is rather a lot of operators to deal with, but we will find that in most calculations only a small sub-set of the operators are important, so that things are not as complex as they might appear at first.

Just as in the two-spin case, normalization factors are needed for some of the product operators. Those containing two operators which are not $\hat{E}$ have a factor of 2, whereas those with three such operators have a factor of 4. For example, $\hat{I}_{1x}\hat{I}_{2z}\hat{E}_3$ has a normalization factor of 2 and becomes $2\hat{I}_{1x}\hat{I}_{2z}$, whereas for $\hat{I}_{1x}\hat{I}_{2z}\hat{I}_{3z}$ the normalization factor is 4 to give $4\hat{I}_{1x}\hat{I}_{2z}\hat{I}_{3z}$.

10.1.1 Interpretation of the product operators for three spins

In thinking about the interpretation of the product operators for a three-spin system, it is useful to keep in mind what the spectrum of such a spin system looks like. As was discussed in section 3.7 on page 40, each spin gives rise to a doublet of doublets and, as is illustrated in Fig. 3.12 on page 42, each line in the multiplet can be labelled with the spin states of the two coupled spins. It is also important to remember that the appearance of the multiplet, and the labelling of the lines, depends on the relative sizes and signs of the coupling constants present.

As for two spins, $\hat{I}_{1z}$ represents z-magnetization on spin one and, setting aside any constant (see section 6.8.6 on page 131), the equilibrium z-magnetization on this spin can be written $\hat{I}_{1z}$. Similarly, $\hat{I}_{2z}$ and $\hat{I}_{3z}$ represent z-magnetization on spins two and three, respectively.

$\hat{I}_{1x}$ represents in-phase x-magnetization on spin one. If this operator is allowed to evolve, and the resulting FID Fourier transformed, the spectrum will consist of the spin-one multiplet (a doublet of doublets) in which all of the lines have the *same phase*. Just exactly what this multiplet looks like will depend on the size of the coupling constants to spins two and three, a point which is illustrated in Fig. 10.1 on the following page where the multiplets for three different combinations of couplings are shown. Note in particular that the multiplet shown in (c) is for the case $J_{12} = J_{13}$, and so appears as a 1:2:1 triplet.

$2\hat{I}_{1x}\hat{I}_{2z}$ gives rise to a multiplet in which those lines associated with spin two being in the α spin state are *negative*, while those associated with spin two being in the β state are *positive*. The result, also illustrated in Fig. 10.1 on the next page, is described as being anti-phase with respect to the coupling to spin two. Note once again that the detailed appearance of the multiplet depends on the relative size of the two coupling constants. If, as is shown in (a), $J_{12} > J_{13}$, the intensity pattern is $- - ++$, whereas if $J_{13} > J_{12}$, as shown in (b), the pattern is $- + -+$. If the two coupling constants are equal, shown in (c), the two centre lines cancel to give what appears to be an anti-phase doublet. However, it is important to realize that this is not a doublet, but a doublet of doublets in which two of the lines have cancelled one another.

In a similar way, $2\hat{I}_{1x}\hat{I}_{3z}$ represents x-magnetization on spin one which is anti-phase with respect to the coupling to spin three. The signs of the lines in the corresponding multiplet are affected by the spin state of spin *three*. Therefore, as is shown in Fig. 10.1 on the following page, we see a pattern of two positive and two negative lines, the exact form of which depends on the relative size of the two coupling constants. Once again, if these two coupling constants are equal, the two inner lines cancel.

Finally, the operator $4\hat{I}_{1x}\hat{I}_{1z}\hat{I}_{3z}$ gives rise to a multiplet in which lines associated with spins two and three being in the *same* spin state are positive,

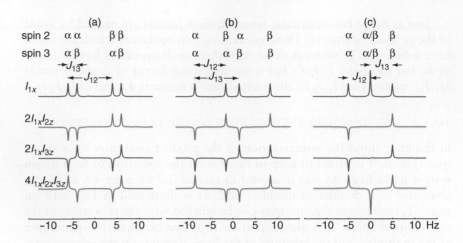

Fig. 10.1 Illustration of the form of the spin-one multiplets expected for four product operators which lead to observable signals on spin one; it has been assumed that x-magnetization will give rise to an absorption mode lineshape, and that the offset of spin one is 0 Hz. In (a) $J_{12} = 10$ Hz and $J_{13} = 2$ Hz, in (b) $J_{12} = 7$ Hz and $J_{13} = 10$ Hz, and in (c) $J_{12} = 5$ Hz and $J_{13} = 5$ Hz. Each line is labelled with the spin state of the two coupled spins, spins two and three. $\hat{I}_{1x}$ is described as an in-phase operator as it gives rise to multiplets in which all four lines have the same phase i.e. all positive. $2\hat{I}_{1x}\hat{I}_{2z}$ is described as being anti-phase with respect to the coupling to spin two; lines associated with spin two being in the α spin state are negative, while those associated with the β spin state are positive. Similarly, $2\hat{I}_{1x}\hat{I}_{3z}$ is anti-phase with respect to the coupling to spin three, and the signs of the lines are determined by the spin states of that spin. Finally, $4\hat{I}_{1x}\hat{I}_{2z}\hat{I}_{3z}$ is described as being doubly anti-phase; lines in which the spin states of the two coupled spins are the same are positive, whereas those in which the spin states are opposite are negative. Note that for the multiplets shown in (c), in which $J_{12} = J_{13}$, the in-phase multiplet becomes a 1:2:1 triplet, whereas the singly anti-phase multiplets *appear* to be anti-phase doublets on account of the cancellation of two of the lines. However, the doubly anti-phase term results in a $+1 : -2 : +1$ 'triplet', as the two centre lines reinforce.

whereas if the spins are in different states the lines are negative; such an arrangement is said to be *doubly anti-phase*. As can be seen in Fig. 10.1, the resulting pattern of intensities is $+--+$. In contrast to the singly anti-phase terms, if the two coupling constants are equal, two of the lines reinforce one another resulting in a $+1 : -2 : +1$ 'triplet'.

Similar interpretations can be made of the operators $\hat{I}_{2x}$, $2\hat{I}_{1z}\hat{I}_{2x}$, $2\hat{I}_{2x}\hat{I}_{3z}$ and $4\hat{I}_{1z}\hat{I}_{2x}\hat{I}_{3z}$: they all represent spin-two multiplets which are, respectively, in-phase, anti-phase with respect to the coupling to spin one, anti-phase with respect to the coupling to spin three, and doubly anti-phase with respect to the couplings to spins one and three. The corresponding operators $\hat{I}_{3x}$, $2\hat{I}_{1z}\hat{I}_{3x}$, $2\hat{I}_{2z}\hat{I}_{3x}$ and $4\hat{I}_{1z}\hat{I}_{2z}\hat{I}_{3x}$ give rise to multiplets on spin three. There are a further set of operators along y, such as $4\hat{I}_{1y}\hat{I}_{2z}\hat{I}_{3z}$, which give rise to multiplets which are phase shifted by 90° compared with those along x. It is usual to call product operators such as $2\hat{I}_{1x}\hat{I}_{2z}$ *singly* anti-phase, and those such as $4\hat{I}_{1x}\hat{I}_{2z}\hat{I}_{3z}$ *doubly* anti-phase.

The only observable product operators are those already described i.e. those containing just *one* transverse operator $\hat{I}_x$ or $\hat{I}_y$. Products containing two such transverse operators correspond to double- and zero-quantum coherence, and products containing three such operators correspond to triple-quantum coherence and a kind of single-quantum coherence associated with combination lines; none of these operators are observable.

10.1.2 Evolution due to offsets and pulses

The evolution under the influence of offsets and pulses follows the same rules as were established for two spins, and are summarized in Fig. 7.4 on page 148. As before, although we need a separate term for the offset of each spin, these can simply be applied one after another, in any order. So, for example, the evolution during a delay due to the offset is given by

$$\rho(0) \xrightarrow{\Omega_1 t \hat{I}_{1z}} \xrightarrow{\Omega_2 t \hat{I}_{2z}} \xrightarrow{\Omega_3 t \hat{I}_{3z}} .$$

Often, calculating the effect of these three rotations will be simpler than it might seem at first sight, as the offset for spin one only affects operators of that spin and *not* the operators for the other two spins. The effect of a pulse can similarly be calculated by considering three successive rotations; for example, for a 90° pulse about x:

$$\rho(0) \xrightarrow{(\pi/2)\hat{I}_{1x}} \xrightarrow{(\pi/2)\hat{I}_{2x}} \xrightarrow{(\pi/2)\hat{I}_{3x}} .$$

10.1.3 Evolution of couplings

The effect of coupling is in principle the same as for two spins, and is summarized in Fig. 7.6 on page 152. However, the results are a little more complicated than for two spins, and how this all turns out is best illustrated by means of an example.

Let us start with in-phase magnetization on spin one $\hat{I}_{1x}$, and allow it to evolve first under the coupling to spin two, and secondly under the coupling to spin three. We do not need to consider the coupling between spins two and three as this cannot affect the evolution of a spin-one operator.

The evolution due to the 1–2 coupling is just as before, leading to an anti-phase term along y:

$$\hat{I}_{1x} \xrightarrow{2\pi J_{12} t \hat{I}_{1z}\hat{I}_{2z}} \cos(\pi J_{12}t)\,\hat{I}_{1x} + \sin(\pi J_{12}t)\,2\hat{I}_{1y}\hat{I}_{2z}.$$

We will now consider the effect of the 1–3 coupling separately on each of the terms on the right of the previous equation. For the term in $\hat{I}_{1x}$ it is just the same as before, except that the coupling is between spins one and three, thus generating the anti-phase term $2\hat{I}_{1y}\hat{I}_{3z}$ rather than $2\hat{I}_{1y}\hat{I}_{2z}$:

$$\cos(\pi J_{12}t)\,\hat{I}_{1x} \xrightarrow{2\pi J_{13} t \hat{I}_{1z}\hat{I}_{3z}}$$
$$\cos(\pi J_{13}t)\cos(\pi J_{12}t)\,\hat{I}_{1x} + \sin(\pi J_{13}t)\cos(\pi J_{12}t)\,2\hat{I}_{1y}\hat{I}_{3z}.$$

The evolution of the term $\sin(\pi J_{12}t)\,2\hat{I}_{1y}\hat{I}_{2z}$ is a little more complicated. The first thing to realize is that the spin-two operator $\hat{I}_{2z}$ is unaffected by the 1–3 coupling, so as far as this part of the calculation is concerned, the operator $\hat{I}_{2z}$ is just a constant, like the factor of 2 and the sine term. So, it is just the term $\hat{I}_{1y}$ which will evolve under the 1–3 coupling; as before, in-phase along y gives rise to anti-phase along x:

$$\hat{I}_{1y} \xrightarrow{\ 2\pi J_{13} t \hat{I}_{1z}\hat{I}_{3z}\ } \cos\left(\pi J_{13}t\right)\hat{I}_{1y} - \sin\left(\pi J_{13}t\right)2\hat{I}_{1x}\hat{I}_{3z}. \tag{10.1}$$

Note that the anti-phase term which is produced is $2\hat{I}_{1x}\hat{I}_{3z}$, as it is the 1–3 coupling which is evolving.

The evolution of $\sin\left(\pi J_{12}t\right)2\hat{I}_{1y}\hat{I}_{2z}$ due to the 1–3 coupling is therefore found by multiplying both sides of Eq. 10.1 by $2\sin\left(\pi J_{12}t\right)\hat{I}_{2z}$:

$$\sin\left(\pi J_{12}t\right)2\hat{I}_{1y}\hat{I}_{2z} \xrightarrow{\ 2\pi J_{13} t \hat{I}_{1z}\hat{I}_{3z}\ }$$
$$\cos\left(\pi J_{13}t\right)\sin\left(\pi J_{12}t\right)2\hat{I}_{1y}\hat{I}_{2z} - \sin\left(\pi J_{13}t\right)\sin\left(\pi J_{12}t\right)4\hat{I}_{1x}\hat{I}_{2z}\hat{I}_{3z}.$$

The factor of 4 in the last term arises from one factor of 2 in the starting operator $2\hat{I}_{1y}\hat{I}_{2z}$, and a second factor of 2 from Eq. 10.1.

The overall result of the evolution of $\hat{I}_{1x}$ under coupling is best summarized in a table:

term	dependence on J_{12}	dependence on J_{13}	axis	description
$\hat{I}_{1x}$	$\cos\left(\pi J_{12}t\right)$	$\cos\left(\pi J_{13}t\right)$	x	in-phase
$2\hat{I}_{1y}\hat{I}_{2z}$	$\sin\left(\pi J_{12}t\right)$	$\cos\left(\pi J_{13}t\right)$	y	anti-phase with respect to J_{12}
$2\hat{I}_{1y}\hat{I}_{3z}$	$\cos\left(\pi J_{12}t\right)$	$\sin\left(\pi J_{13}t\right)$	y	anti-phase with respect to J_{13}
$-4\hat{I}_{1x}\hat{I}_{2z}\hat{I}_{3z}$	$\sin\left(\pi J_{12}t\right)$	$\sin\left(\pi J_{13}t\right)$	$-x$	doubly anti-phase with respect to J_{12} and J_{13}

The size of the singly anti-phase terms depends on a factor $\sin\left(\pi J_{\mathrm{active}}t\right)$, where J_{active} is the coupling with respect to which the term is anti-phase, and a factor $\cos\left(\pi J_{\mathrm{passive}}t\right)$, where J_{passive} is the coupling to the other spin. The in-phase term has two such cosine factors, whereas the doubly anti-phase term depends on two such sine factors. Note also that if the in-phase term starts along x, the singly anti-phase terms appear along y, and the doubly anti-phase term appears along $-x$.

There are lots of nice patterns here, which can also be expressed diagrammatically as shown in Fig. 10.2. At the top of the diagram we have the starting operator $\hat{I}_{1x}$. Under the influence of the 1–2 coupling, two operators are generated: the original operator, $\hat{I}_{1x}$, and the anti-phase operator $2\hat{I}_{1y}\hat{I}_{2z}$. Then, each of these operators splits into two as a result of the evolution of the 1–3 coupling. The cascade is arranged so that an operator which has split to the left has associated with it a cosine factor, whereas an operator which has split to the right has a sine factor. The signs of the operators are found simply by following around the rotations in Fig. 7.6 on page 152. Such a diagram is a convenient way of keeping track of the operators, their signs and the trigonometric factors.

We will do one more example, which is to start with the doubly anti-phase term $4\hat{I}_{1z}\hat{I}_{2z}\hat{I}_{3y}$ and allow it to evolve under the coupling between spins one and three, and between spins two and three; we need not concern

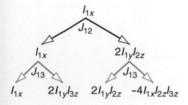

Fig. 10.2 Representation of the evolution of $\hat{I}_{1x}$ under the influence of coupling to spin two and spin three. An arrow to the left implies a factor of $\cos\left(\pi J_{ij}t\right)$, whereas an arrow to the right implies a factor of $\sin\left(\pi J_{ij}t\right)$. The top set of arrows is for the coupling between spins one and two, and the lower set for the coupling between spins one and three. The way in which each term splits up (including the sign) is found by following round the diagrams in Fig. 7.6 on page 152. From the diagram we can see that the factors associated with the term $2\hat{I}_{1y}\hat{I}_{2z}$ are $\sin\left(\pi J_{12}t\right)$ and $\cos\left(\pi J_{13}t\right)$, as to arrive at this term we first split to the right and then to the left.

ourselves with the coupling between spins one and two as this cannot affect a spin-three term.

The evolution of the 1–3 coupling is best determined by regarding $4\hat{I}_{1z}\hat{I}_{2z}\hat{I}_{3y}$ as $A \times 2\hat{I}_{1z}\hat{I}_{3y}$, where $A = 2\hat{I}_{2z}$; the term A has been separated out as it will not be affected by the evolution of the 1–3 coupling.

Allowing $2\hat{I}_{1z}\hat{I}_{3y}$ to evolve under the 1–3 coupling gives

$$2\hat{I}_{1z}\hat{I}_{3y} \xrightarrow{\;2\pi J_{13}t\hat{I}_{1z}\hat{I}_{3z}\;} \cos{(\pi J_{13}t)}\,2\hat{I}_{1z}\hat{I}_{3y} - \sin{(\pi J_{13}t)}\,\hat{I}_{3x}.$$

Putting back in the factor $A = 2\hat{I}_{2z}$ gives the result

$$\cos{(\pi J_{13}t)}\,4\hat{I}_{1z}\hat{I}_{2z}\hat{I}_{3y} - \sin{(\pi J_{13}t)}\,2\hat{I}_{2z}\hat{I}_{3x}. \qquad (10.2)$$

Now we need to consider the evolution of each term on the right-hand side of Eq. 10.2 under the 2–3 coupling. From the first term we can take out a factor $B = \cos{(\pi J_{13}t)}\,2\hat{I}_{1z}$ which will be unaffected by the evolution of this coupling; the remaining operator product evolves according to

$$2\hat{I}_{2z}\hat{I}_{3y} \xrightarrow{\;2\pi J_{23}t\hat{I}_{2z}\hat{I}_{3z}\;} \cos{(\pi J_{23}t)}\,2\hat{I}_{2z}\hat{I}_{3y} - \sin{(\pi J_{23}t)}\,\hat{I}_{3x}.$$

Putting back in the factor B gives

$$\cos{(\pi J_{23}t)}\cos{(\pi J_{13}t)}\,4\hat{I}_{1z}\hat{I}_{2z}\hat{I}_{3y} - \sin{(\pi J_{23}t)}\cos{(\pi J_{13}t)}\,2\hat{I}_{1z}\hat{I}_{3x}.$$

Finally, we need to consider the evolution of the term $-\sin{(\pi J_{13}t)}\,2\hat{I}_{2z}\hat{I}_{3x}$ from Eq. 10.2 under the 2–3 coupling; this is straightforward and gives

$$-\sin{(\pi J_{13}t)}\,2\hat{I}_{2z}\hat{I}_{3x} \xrightarrow{\;2\pi J_{23}t\hat{I}_{2z}\hat{I}_{3z}\;}$$
$$-\cos{(\pi J_{23}t)}\sin{(\pi J_{13}t)}\,2\hat{I}_{2z}\hat{I}_{3x} - \sin{(\pi J_{23}t)}\sin{(\pi J_{13}t)}\,\hat{I}_{3y}.$$

In summary, after the evolution of both couplings we have the four terms:

$$\cos{(\pi J_{23}t)}\cos{(\pi J_{13}t)}\,4\hat{I}_{1z}\hat{I}_{2z}\hat{I}_{3y}$$
$$-\cos{(\pi J_{23}t)}\sin{(\pi J_{13}t)}\,2\hat{I}_{2z}\hat{I}_{3x}$$
$$-\sin{(\pi J_{23}t)}\cos{(\pi J_{13}t)}\,2\hat{I}_{1z}\hat{I}_{3x}$$
$$-\sin{(\pi J_{23}t)}\sin{(\pi J_{13}t)}\,\hat{I}_{3y}.$$

As before, there is a nice pattern here, which is also illustrated in Fig. 10.3. The doubly anti-phase term has two cosine factors, the singly anti-phase terms have a cosine and a sine factor, and the in-phase term has two sine factors. Note also that the doubly anti-phase term is along y, the singly anti-phase terms are along $-x$, and the in-phase terms are along $-y$.

10.2 COSY for three spins

Based on our previous discussion of COSY (section 8.3 on page 190), we know what to expect for three mutually coupled spins. Assuming that all the couplings are non-zero, there will be cross peaks centred at $\{\omega_1, \omega_2\} = \{\Omega_1, \Omega_2\}$, $\{\Omega_1, \Omega_3\}$ and $\{\Omega_2, \Omega_3\}$, along with a symmetry-related set of cross peaks which have the ω_1 and ω_2 coordinates transposed. In addition, there will be diagonal peaks centred at $\{\Omega_1, \Omega_1\}$, $\{\Omega_2, \Omega_2\}$ and $\{\Omega_3, \Omega_3\}$. The overall form of the spectrum is shown in Fig. 10.4 on the following page.

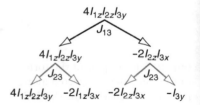

Fig. 10.3 Representation of the evolution of $4\hat{I}_{1z}\hat{I}_{2z}\hat{I}_{3y}$ under the influence of the coupling of spin three to spins one and two. The diagram is interpreted in the same way as Fig. 10.2 on the facing page.

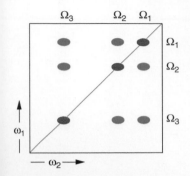

Fig. 10.4 Schematic COSY spectrum for three mutually coupled spins. The cross peaks are shown in blue, and the diagonal peaks in dark grey; multiplet structures are not shown.

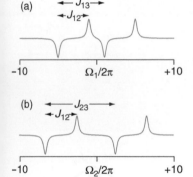

Fig. 10.5 Examples of the form of the multiplets expected along the two dimensions for the 1–2 cross peak. These multiplets have been computed for the particular case $J_{12} = 4$ Hz, $J_{13} = 6$ Hz and $J_{23} = 9$ Hz. Multiplet (a) is that along ω_1, and is from spin one; the multiplet is anti-phase with respect to the 1–2 coupling, but in-phase with respect to the 1–3 coupling. Multiplet (b) is that along ω_2, and is from spin two; the multiplet is also anti-phase with respect to the 1–2 coupling, but in-phase with respect to the 2–3 coupling.

What we are going to look at in this section is the detailed form of the cross-peak multiplets. We will see that they have rather an aesthetically pleasing arrangement of peaks, from which we can determine something about the relative size of the coupling constants which are responsible for the cross peak (the active coupling), and the other (passive) couplings.

Starting from equilibrium magnetization on spin one, $\hat{I}_{1z}$, and applying the COSY pulse sequence of Fig. 8.8 on page 192, gives us the following *observable* terms at time $t_2 = 0$:

$$\cos{(\pi J_{13}t_1)}\cos{(\pi J_{12}t_1)}\sin{(\Omega_1 t_1)}\,\hat{I}_{1x} \qquad [1]$$

$$-\cos{(\pi J_{13}t_1)}\sin{(\pi J_{12}t_1)}\sin{(\Omega_1 t_1)}\,2\hat{I}_{1z}\hat{I}_{2y} \qquad [2]$$

$$-\sin{(\pi J_{13}t_1)}\cos{(\pi J_{12}t_1)}\sin{(\Omega_1 t_1)}\,2\hat{I}_{1z}\hat{I}_{3y}. \qquad [3]$$

Looking at these, we can see that term [1] is the diagonal peak as it is modulated at the offset of spin one, Ω_1, in t_1, and appears as observable magnetization on spin one. Term [2] is also modulated at Ω_1 in t_1, but appears as observable magnetization on spin two: it therefore gives rise to the cross peak at $\{\Omega_1, \Omega_2\}$, which for short we will call the 1–2 cross peak. Similarly, term [3] gives rise to the cross peak at $\{\Omega_1, \Omega_3\}$, i.e. the 1–3 cross peak.

Note that if $J_{12} = 0$, term [2] goes to zero on account of the factor $\sin{(\pi J_{12}t_1)}$ being zero. Just as expected, there will be no 1–2 cross peak if the coupling between spins one and two is zero. On the other hand, having $J_{12} = 0$ does not force term [3] to be zero, so the 1–3 cross peak is still present.

10.2.1   Structure of the cross-peak multiplets

We will now focus on the 1–2 cross-peak multiplet, represented by term [2]. The ω_1 frequencies of the peaks in this multiplet are found by examining the modulation with respect to t_1. Just as we did for the two-spin case, we can use the usual trigonometric identities to transform the t_1 modulation, $-\cos{(\pi J_{13}t_1)}\sin{(\pi J_{12}t_1)}\sin{(\Omega_1 t_1)}$, into a sum of terms.

The first step is to combine the terms in the product $\sin{(\pi J_{12}t_1)}\sin{(\Omega_1 t_1)}$ using the identity

$$\sin A \sin B \equiv \tfrac{1}{2}\left[\cos{(A-B)} - \cos{(A+B)}\right].$$

Using this, as well as the identity $\cos{(-A)} \equiv \cos{(A)}$, gives

$$-\cos{(\pi J_{13}t_1)}\tfrac{1}{2}\left[\cos{(\Omega_1 t_1 - \pi J_{12}t_1)} - \cos{(\Omega_1 t_1 + \pi J_{12}t_1)}\right].$$

We then multiply out the square bracket and combine the product of two cosines using the identity

$$\cos A \cos B \equiv \tfrac{1}{2}\left[\cos{(A-B)} + \cos{(A+B)}\right].$$

The result of all these manipulations is four terms:

$$\tfrac{1}{4}\left[+ \cos{(\Omega_1 t_1 + \pi J_{12}t_1 + \pi J_{13}t_1)} + \cos{(\Omega_1 t_1 + \pi J_{12}t_1 - \pi J_{13}t_1)}\right.$$
$$\left. - \cos{(\Omega_1 t_1 - \pi J_{12}t_1 + \pi J_{13}t_1)} - \cos{(\Omega_1 t_1 - \pi J_{12}t_1 - \pi J_{13}t_1)}\right].$$

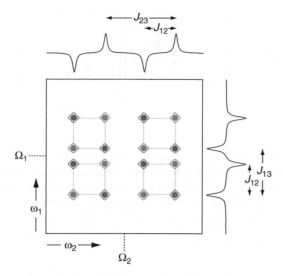

Fig. 10.6 Contour plot of the COSY cross-peak multiplet between spins one and two in the three-spin system; positive contours are shown in blue, and negative in dark grey. The couplings are $J_{12} = 4$ Hz, $J_{13} = 6$ Hz and $J_{23} = 9$ Hz, and the whole plot covers ±10 Hz from the centre of the multiplet. The multiplets plotted along the top and at the side are those shown in Fig. 10.5 on the preceding page. The cross-peak multiplet consists of four anti-phase square arrays, which are picked out by the grey boxes.

The four frequencies are the four lines of the spin-one multiplet (a doublet of doublets); however, two of the lines are positive and two are negative. The lines which are separated by J_{12} have opposite signs, whereas those which are separated by J_{13} have the same signs. Figure 10.5 (a) on the facing page shows an example of such a multiplet for a particular set of couplings; the multiplet is described as being anti-phase with respect to J_{12}, and in-phase with respect to J_{13}.

In the t_1 modulation, $-\cos(\pi J_{13} t_1) \sin(\pi J_{12} t_1) \sin(\Omega_1 t_1)$, it is the presence of the *sine* term, $\sin(\pi J_{12} t_1)$, which makes the multiplet anti-phase with respect to J_{12}. The *cosine* term, $\cos(\pi J_{13} t_1)$, results in an in-phase splitting with respect to J_{13}.

The operator in term [2] is $2\hat{I}_{1z}\hat{I}_{2y}$. Following the discussion in section 10.1.1 on page 321, this operator gives, in ω_2, a multiplet on spin two which is anti-phase with respect to the coupling to spin one, and in-phase with respect to the coupling to spin three. Figure 10.5 (b) shows an example of such a multiplet for a particular set of couplings. Comparing the two multiplets shown in (a) and (b), we see that *both* are anti-phase with respect to the 1–2 coupling, which is the coupling responsible for forming the cross peak, but are in-phase with respect to the coupling to the third spin.

Now that we have identified the form of the multiplets in each dimension, we can work out the detailed form of the cross peak by 'multiplying together' the ω_1 and ω_2 multiplets in the way which was introduced in Fig. 8.10 on page 194. Figure 10.6 shows the resulting two-dimensional multiplet.

Looking at the multiplet we can see immediately that it consists of four anti-phase square arrays of the type pictured in Fig. 8.10; in Fig. 10.6 these four arrays are picked out by the grey boxes. In each dimension, the peaks in the anti-phase square array are separated by J_{12}, which is the coupling responsible for the cross peak – termed the *active coupling*.

Relative to the centre of the cross-peak multiplet at $\{\Omega_1, \Omega_2\}$, the four anti-phase square arrays are centred at the following frequencies (for

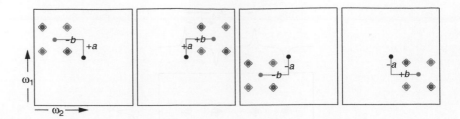

Fig. 10.7 Illustration of how the cross-peak multiplet shown in Fig. 10.6 on the preceding page is constructed from four anti-phase square arrays. The centre of the multiplet is indicated by the black dot and, relative to this point, the anti-phase square arrays are shifted by $\pm\frac{1}{2}J_{13}$ Hz in the ω_1 dimension, and $\pm\frac{1}{2}J_{23}$ Hz in the ω_2 dimension. These shifts are indicated by the blue lines which connect the centre (the black dot) to the centre of each anti-phase square array (the blue dot); in the diagram $a = \frac{1}{2}J_{13}$ and $b = \frac{1}{2}J_{23}$.

convenience, we have written the frequencies in Hz, rather than rad s^{-1}):

$$\{+\tfrac{1}{2}J_{13}, +\tfrac{1}{2}J_{23}\}, \quad \{-\tfrac{1}{2}J_{13}, +\tfrac{1}{2}J_{23}\}, \quad \{+\tfrac{1}{2}J_{13}, -\tfrac{1}{2}J_{23}\}, \quad \{-\tfrac{1}{2}J_{13}, -\tfrac{1}{2}J_{23}\}.$$

The location of these four anti-phase square arrays is illustrated in Fig. 10.7.

In the ω_1 dimension, the coupling J_{13} is described as *passive* as it it not responsible for the cross peak, but does involve one of the two spins the coupling between which is responsible for the cross peak. In the same way, in ω_2 the coupling J_{23} is passive.

The exact appearance of the multiplet depends on the relative sizes of the couplings involved. It is particularly easy to spot the four anti-phase square arrays in Fig. 10.6 on the preceding page, as the active coupling is smaller than both of the passive couplings. Other arrangements of couplings lead to cross-peak multiplets which are a little more difficult to disentangle.

Figure 10.8 on the next page shows a series of different 1–2 cross-peak multiplets all of which have the same active coupling, but in which the passive couplings are different; the values are given in the following table:

cross peak	J_{12} / Hz	J_{13} / Hz	J_{23} / Hz	comment
(a)	8	4	6	
(b)	8	4	7.5	two couplings similar
(c)	8	4	8	two couplings equal
(d)	8	2	6	
(e)	8	1	6	
(f)	8	0	6	one coupling zero

Multiplet (a) should be compared with that in Fig. 10.6 on the preceding page. The difference between these two multiplets is that in (a) the active coupling is *larger* than either of the two passive couplings, whereas in Fig. 10.6 on the previous page the active coupling is *smaller* than the passive couplings. The anti-phase square arrays in multiplet (a) are thus not

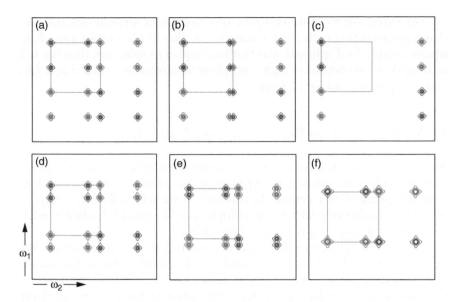

Fig. 10.8 Illustration of how the appearance of the 1–2 cross-peak multiplet depends on the relative sizes of the active and passive couplings. For each multiplet the active coupling, J_{12}, is 8 Hz; the passive couplings have the values given in the table in the text. In all cases, the area plotted is ±10 Hz from the centre of the cross-peak multiplet. In each multiplet, one of the four anti-phase square arrays is indicated by a grey box.

separate from one another, but overlap (to avoid confusion, in the diagram only one of the squares is indicated).

In multiplet (b) the passive coupling in the ω_2 dimension, J_{23}, is similar in size to the active coupling, J_{12}. As a result, the two columns of peaks in the centre of the multiplet come quite close together, and so the adjacent positive and negative peaks begin to cancel. In the limit that $J_{23} = J_{12}$, multiplet (c), this cancellation is complete and so the multiplet consists of only eight peaks, rather than the usual sixteen. This multiplet looks rather strange until you realize that, because of cancellation, some of the peaks which form the anti-phase square arrays are missing. For the case shown in (c), the spin-two multiplet in the conventional one dimensional spectrum would be a 1:2:1 triplet.

Multiplets (d), (e) and (f) illustrate what happens as the passive coupling J_{13} gets smaller and smaller. As expected, the individual multiplet components move closer together, but as the peaks which become adjacent have the *same* sign they reinforce one another. In the limit that $J_{13} = 0$, multiplet (f), two of the anti-phase square arrays lie on top of one another and so there are only eight individual peaks in the multiplet.

You can see from these examples that, although in principle each cross-peak multiplet is composed of four anti-phase square arrays, what the multiplet actually ends up looking like depends in detail on the relative sizes of the coupling constants. In addition, the extent to which individual peaks will cancel or reinforce one another depends on the linewidth, which may be different in the two dimensions. As a result, in practical spectroscopy one often needs a sharp eye and an inventive mind to disentangle the structure of a particular cross-peak multiplet.

If we are able to understand the form of the cross-peak multiplet, then it gives us extra information on the relative sizes of the couplings involved. However, in general we need to be cautious about actually trying to measure the values of couplings from these cross peaks since, as noted in section 8.3.4 on page 198, cancellation between nearby peaks results in the splittings between peaks of opposite signs not being equal to the coupling constants.

The extension to more complex spin systems is straightforward. All that happens is that each passive coupling further duplicates the anti-phase square arrays. So if spin one was further coupled to spin four, then the 1–2 cross peak would consist of eight anti-phase square arrays, shifted by $\pm \frac{1}{2} J_{13}$ and $\pm \frac{1}{2} J_{14}$ in ω_1, and by $\pm \frac{1}{2} J_{23}$ in ω_2.

10.3   Reduced multiplets in COSY spectra

As we have seen, for a system of three mutually coupled spins, each cross-peak multiplet contains sixteen separate peaks, which can be grouped into four anti-phase square arrays. In this section we will look at the ways in which the number of peaks in the multiplet can be reduced, leading to what are called, not surprisingly, *reduced multiplets*.

It turns out that from these reduced multiplets we can determine the *relative signs* of the two passive couplings. In addition, under favourable circumstances, we can measure the values of the coupling constants to the passive spins to high accuracy. The latter feature, often called the *ECOSY principle*, has been used very widely to measure values of coupling constants in labelled proteins and nucleic acids.

It is easiest to see how reduced multiplets arise by first thinking about a three-spin system in which one of the spins is of a different type to the others e.g. a heteronucleus. Once we have examined this case, we will go on to explain how reduced multiplets can be generated in homonuclear spin systems.

10.3.1   COSY for a three-spin system containing one heteronucleus

Imagine that, in our three-spin system, spin three is of a different type to the other two e.g. spins one and two are protons, whereas spin three is a heteronucleus, such as ^{13}C, ^{15}N, ^{31}P, or ^{19}F. Furthermore, in recording our COSY spectrum, we will apply pulses only to the first type of nucleus (spins one and two); spin three, being a heteronucleus, does not experience any pulses. For compatibility with the next section, we will denote the operators of the third spin $\hat{I}_{3x}$, $\hat{I}_{3y}$ and $\hat{I}_{3z}$, rather than $\hat{S}_x$, $\hat{S}_y$ and $\hat{S}_z$, as would be usual for a heteronucleus.

If we repeat the calculation at the start of section 10.2 on page 325, we will find that, starting from $\hat{I}_{1z}$, the following observable terms *on spin two* are present at the beginning of t_2:

$$- \cos\left(\pi J_{13} t_1\right) \sin\left(\pi J_{12} t_1\right) \sin\left(\Omega_1 t_1\right) 2\hat{I}_{1z}\hat{I}_{2y} \qquad [2]$$

$$- \sin\left(\pi J_{13} t_1\right) \sin\left(\pi J_{12} t_1\right) \cos\left(\Omega_1 t_1\right) 4\hat{I}_{1z}\hat{I}_{2y}\hat{I}_{3z}. \qquad [5]$$

Both of these terms are modulated at Ω_1 in the ω_1 dimension: they therefore both contribute to the 1–2 cross peak.

Term [2] is exactly the same as the one we found before on page 326. As is shown in Fig. 10.6 on page 327, this term gives rise to a cross-peak multiplet which is anti-phase with respect to J_{12} in each dimension, but is

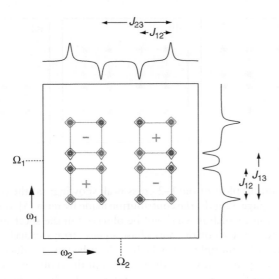

Fig. 10.9 Contour plot of the contribution made by term [5] to the cross-peak multiplet between spins one and two; the parameters of the spin system are the same as in Fig. 10.6 on page 327. The ω_1 multiplet, which is on spin one and doubly anti-phase with respect to the 1–2 and 1–3 couplings, is shown down the side. Similarly, along the top is shown the ω_2 multiplet, which is on spin two and doubly anti-phase with respect to the 1–2 and 2–3 couplings. Multiplying these two multiplets together gives us the form of the two-dimensional multiplet. As in Fig. 10.6 we can pick out four anti-phase square arrays, however the overall sign of two of these are opposite to those in Fig. 10.6. The overall sign of each anti-phase square array is shown by the symbol in the middle of the array.

in phase with respect to J_{13} in the ω_1 dimension, and with respect to J_{23} in the ω_2 dimension.

Term [5] did not appear in our original calculation because in that case the final 90° pulse was applied to all three spins, whereas in the present case this final pulse is not felt by spin three. You can see that if we did apply a $90°(x)$ pulse to spin three, this last term would be rotated from $4\hat{I}_{1z}\hat{I}_{2y}\hat{I}_{3z}$ to $-4\hat{I}_{1z}\hat{I}_{2y}\hat{I}_{3y}$, which is unobservable. Therefore the consequence of spin three being a heteronucleus is the presence of an additional contribution to the cross peak, as represented by term [5].

We now need to work out the form of the two-dimensional multiplet which arises from term [5]. As before, we have to expand the modulation in t_1, $-\sin(\pi J_{13}t_1)\sin(\pi J_{12}t_1)\cos(\Omega_1 t_1)$, using trigonometric identities; the result is

$$\tfrac{1}{4}[+\cos(\Omega_1 t_1 + \pi J_{12}t_1 + \pi J_{13}t_1) - \cos(\Omega_1 t_1 + \pi J_{12}t_1 - \pi J_{13}t_1)$$
$$-\cos(\Omega_1 t_1 - \pi J_{12}t_1 + \pi J_{13}t_1) + \cos(\Omega_1 t_1 - \pi J_{12}t_1 - \pi J_{13}t_1)].$$

The four lines are clearly those of the spin-one multiplet, however the pattern of intensities is that for a doubly anti-phase state (see Fig. 10.1 on page 322), this is in contrast to the multiplet from term [2], which is singly anti-phase with respect to the active coupling, J_{12}. The t_1 modulation of term [5] has sine factors depending on J_{12} and J_{13}: it is these which make the multiplet anti-phase with respect to both of these couplings.

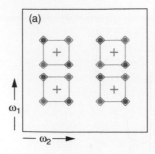
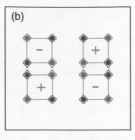
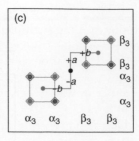

Fig. 10.10 Contour plots of the 1–2 cross peak showing: (a) the contribution made by term [2] on page 326; (b) the contribution made by term [5]; (c) the sum of these two contributions, which is what will be observed in the spectrum. As a result of adding (a) and (b), two of the anti-phase square arrays cancel, so that there are only two arrays in multiplet (c), which is described as a reduced multiplet. The two anti-phase square arrays are displaced from the centre by $\{+\frac{1}{2}J_{13}, +\frac{1}{2}J_{23}\}$, and $\{-\frac{1}{2}J_{13}, -\frac{1}{2}J_{23}\}$; in the diagram $a = \frac{1}{2}J_{13}$ and $b = \frac{1}{2}J_{23}$. In (c) the peaks are labelled according to the spin state of spin three, the passive spin to which both spins one and two are coupled; note that all the peaks present have the *same* spin state in both dimensions. The value of the coupling constants are the same as for Fig. 10.9 on the previous page.

In ω_2, the operator in term [5] is $4\hat{I}_{1z}\hat{I}_{2y}\hat{I}_{3z}$. This gives rise to a spin-two multiplet which is doubly anti-phase with respect to J_{12} and J_{23}. Now that we have worked out the form of the multiplets in the two dimensions we can 'multiply' them together in the way we did in Fig. 10.6 on page 327; the result is shown in Fig. 10.9 on the previous page.

Just as for term [2], the multiplet from which is shown in Fig. 10.6, there are four anti-phase square arrays, but this time two are of opposite overall sign to those in Fig. 10.6. Apart from these sign changes, the multiplets from terms [2] and [5] are identical.

The difference between the two multiplets, and the way in which they combine to give the overall form of the cross peak, is illustrated in Fig. 10.10. Here, multiplet (a) is from term [2] – note that the four anti-phase square arrays all have the same overall sign. Multiplet (b) is from term [5], and in this case two of the anti-phase square arrays are of opposite overall sign to the other two. In the COSY spectrum what we will see is the sum of the two contributions (a) and (b), which is what is shown in (c). Two of the anti-phase square arrays have cancelled one another, and two have reinforced. The result is called a *reduced multiplet*.

The form of the reduced multiplet can be described in the following way. There are two anti-phase square arrays, split by the active coupling, J_{12}, in each dimension. One array is shifted away from the nominal centre of the cross peak at $\{\Omega_1, \Omega_2\}$ by $+\frac{1}{2}J_{13}$ in the ω_1 dimension, and $+\frac{1}{2}J_{23}$ in the ω_2 dimension. The other array is shifted by $-\frac{1}{2}J_{13}$ and $-\frac{1}{2}J_{23}$ in the respective dimensions. Note that the anti-phase square array is split by the *active* coupling, whereas the shifts are determined by the sum of the *passive* couplings.

In Fig. 10.10 (c) the peaks are labelled according to the spin state of spin three, which is the passive spin to which both spins one and two are coupled. We see that the only peaks which are present in the two-dimensional multiplet are those in which the spin state of spin three is the *same* in both dimensions.

This observation gives us a way of thinking about what a reduced multiplet is. We imagine that one anti-phase square array comes from molecules in which spin three is in the α state, and as a result this array is centred at $\{v_1 - \frac{1}{2}J_{13}, v_2 - \frac{1}{2}J_{23}\}$.[1] The other array comes from molecules in which spin three is in the β state, and is centred at $\{v_1 + \frac{1}{2}J_{13}, v_2 + \frac{1}{2}J_{23}\}$.

This way of thinking about a reduced multiplet also indicates in what sort of experiments we can expect to find such multiplets. The key point is that to have a reduced multiplet, the spin state of the passive spin (here spin three) must remain the *same* throughout i.e. it must not change between the t_1 and t_2 periods. Therefore, no pulses can be applied to the passive spins, as these will scramble the spin states (an exception would be a $180°$ pulse which simply swaps the spin states, but does not scramble them). This is why a COSY of a homonuclear three-spin system does not show reduced multiplets, whereas if one of the spins is heteronuclear, and so is unaffected by the pulses, we do find reduced multiplets.

The final point to make is that for a reduced multiplet to appear between spins one and two, *both* must be coupled to the same third spin. If only spin one is coupled to spin three, then we will not see a reduced multiplet.

10.3.2   Determining the relative signs of the passive couplings

In section 3.6 on page 38 we explained that changing the *sign* of a coupling constant has no visible effect on the normal one-dimensional spectrum, but does affect the labelling of the lines in the multiplet according to the spin states of the coupled spins. This is illustrated in Fig. 10.11, where we see the effect on the labelling of the spin-one and spin-two multiplets of changing the sign of the coupling to the third spin.

In a reduced multiplet, the only peaks which appear are those in which the spin state of the third spin (the passive spin) is the same in each dimension. Where these peaks appear will therefore depend on the *sign* of the coupling constants to this passive spin since, as we have seen, this affects the labelling of the peaks. We therefore expect the appearance of the reduced multiplet to be affected by the signs of the coupling constants to the passive spins.

Figure 10.12 on the following page illustrates how the appearance of the reduced multiplet between spins one and two is affected by the signs of J_{13} and J_{23}. What we see here is that it is the *relative* signs of the couplings which is important. If both couplings have the same sign, then the two anti-phase square arrays are arranged such that the multiplet is 'tilted' to the right, whereas if they have opposite signs, the multiplet is tilted to the left. It is thus possible to determine the *relative* signs of the coupling constants simply by inspecting which way the reduced multiplet is tilted.

It is not possible to determine the absolute sign of the coupling constants by this method: only their relative signs can be determined. However, the sign of some couplings are known unambiguously from other consider-

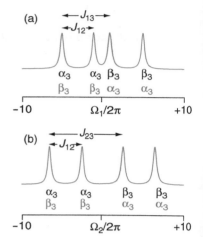

Fig. 10.11 Illustration of the effect on the spin-one and spin-two multiplets of changing the sign of the passive couplings to spin three. These multiplets have been computed for the particular case $|J_{12}| = 4$ Hz, $|J_{13}| = 6$ Hz and $|J_{23}| = 9$ Hz. Multiplet (a) is for spin one, and each line is labelled according to the spin state of spin three (denoted α_3 or β_3). The labels in black are for the case where $J_{13} = +6$ Hz, and those in blue are for $J_{13} = -6$ Hz; note that the multiplet does not change, but the labels do. Multiplet (b) is for spin two, and is similarly labelled for the case $J_{23} = +9$ Hz (in black), and $J_{23} = -9$ Hz (in blue).

---

[1] v_1 and v_2 are the offsets of spins one and two, respectively, in units of Hz.

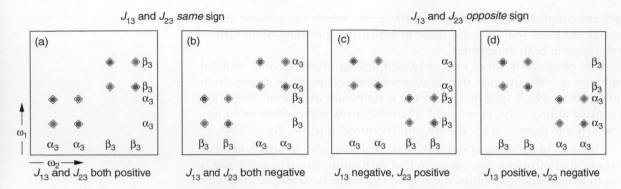

J_{13} and J_{23} *same* sign

(a) J_{13} and J_{23} both positive

(b) J_{13} and J_{23} both negative

J_{13} and J_{23} *opposite* sign

(c) J_{13} negative, J_{23} positive

(d) J_{13} positive, J_{23} negative

Fig. 10.12 Illustration of how the form of the reduced cross-peak multiplet between spins one and two is affected by the signs of the two passive couplings to spin three. The spin-one multiplet appears along ω_1, and the spin-two multiplet along ω_2; as in Fig. 10.11 on the preceding page, $|J_{12}| = 4$ Hz, $|J_{13}| = 6$ Hz and $|J_{23}| = 9$ Hz, and a range of ± 10 Hz is plotted from the centre of the cross peak. The spin state of the third (passive) spin is denoted by α_3/β_3. Note that the only components of the cross peak which will be present are those which have the *same* spin state of the passive spin in the two dimensions. If the two passive couplings J_{13} and J_{23} have the same sign, shown in (a) and (b), the reduced multiplet 'tilts' to the right. In contrast, if the couplings have opposite signs, as in (c) and (d), the multiplet tilts to the left.

ations, and if such a coupling is one of the passive couplings, then the absolute sign of the other passive coupling can be determined. For example, one-bond ^{13}C–^{1}H couplings are known to be positive, so from a reduced multiplet in which one passive coupling is a one-bond C–H coupling, and the other is a long-range C–H coupling, it will be possible to determine the sign of the long-range coupling.

10.3.3   Measuring the size of the passive coupling constants

We have already noted that, because of the way in which the positive and negative components of a cross-peak multiplet interfere with one another, it is not usually possible to measure values of coupling constants simply by measuring the splittings between the peaks in the multiplet (section 8.3.4 on page 198). However, in a reduced multiplet it is possible, under some circumstances, to measure the size of the coupling constants to the passive spin.

The idea is illustrated in Fig. 10.13 on the facing page. In (a) we have the 1–2 cross-peak multiplet as would be seen in a simple COSY spectrum of a homonuclear spin system. The 1–3 passive coupling is rather small (0.5 Hz), whereas the 2–3 passive coupling is large. As a result, two pairs of the anti-phase square arrays overlap extensively, and it is only just possible to see the splitting in the ω_1 dimension due to the 1–3 coupling.

Multiplet (b) is the corresponding reduced multiplet, as would be seen for the case where spin three is a heteronucleus. As has been described above, the two remaining anti-phase square arrays are shifted by $\{+\frac{1}{2}J_{13}, +\frac{1}{2}J_{23}\}$ and $\{-\frac{1}{2}J_{13}, -\frac{1}{2}J_{23}\}$ from the centre of the cross peak. Since J_{23} is so large, the two arrays are well separated in the ω_2 dimension, and as a result there is no overlap or interference *between* the two arrays. It

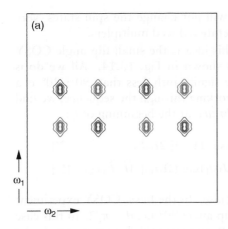
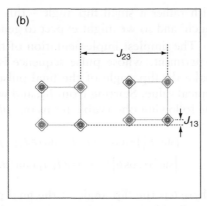

Fig. 10.13 Illustration of the use of a reduced multiplet to facilitate the measurement of a small coupling constant. Multiplet (a) is the normal 1–2 COSY cross peak, computed for the couplings $J_{12} = 4$ Hz, $J_{13} = 0.5$ Hz and $J_{23} = 9$ Hz; the spin-one multiplet appears along ω_1, and the spin-two multiplet along ω_2. The splitting due to the small passive coupling in the ω_1 dimension is barely visible. Shown in (b) is the corresponding reduced multiplet for the case where spin three is a heteronucleus, and so is unaffected by the final pulse of the COSY sequence. The remaining two anti-phase square arrays are clearly separated as a consequence of the large passive coupling J_{23}. As the two arrays no longer interfere with one another, it is now possible to measure J_{13} from the displacement, in the ω_1 dimension, between the two arrays.

is therefore possible to measure the small coupling J_{13} by measuring the displacement of the two arrays in ω_1, as is shown in the figure. Similarly, it is possible to measure the large coupling J_{23} in the other dimension.

Essentially what we are doing here is to use the large passive coupling to 'drag apart' the two anti-phase square arrays. Once they are well separated, there is no longer any interference between them, and so we can find the value of the small passive coupling by measuring the displacement between the two arrays.

This idea has been used to great effect in the NMR of proteins and nucleic acids. It is possible to label such samples with 100% ^{13}C and ^{15}N, thus making it relatively straightforward to see the effect of couplings to these nuclei on the proton spectra. Typically, the one-bond C–H or N–H coupling takes on the role of the 'large' passive coupling, enabling us to measure the much smaller long-range heteronuclear couplings from the reduced multiplets.

10.3.4   Reduced multiplets in homonuclear spin systems

We now turn to how reduced multiplets can be generated for purely homonuclear spin systems. A clue as to how this might be achieved comes from the heteronuclear case we have been discussing so far, in which we noted that the important thing was for the spin state of spin three to remain the *same* throughout the experiment. Any pulse (other than a 180° pulse) applied to spin three will violate this condition, but suppose we use a pulse

Fig. 10.14 Pulse sequence for small flip angle COSY. If the flip angle of the final pulse is significantly less than 90°, the resulting spectrum shows reduced multiplets.

with rather a small flip angle – this will not change the spin states 'very much' and so we might expect to generate reduced multiplets.

The simplest implementation of this idea is the small flip angle COSY experiment, whose pulse sequence is shown in Fig. 10.14. All we do is make the flip angle of the final pulse significantly less than 90°; 20° is a typical value. Starting from $\hat{I}_{1z}$, and working through the sequence, we find the following observable terms on *spin two* at the beginning of t_2:

$$\left[\sin^2\theta\right]\left[-\cos(\pi J_{13}t_1)\sin(\pi J_{12}t_1)\sin(\Omega_1 t_1)\right] 2\hat{I}_{1z}\hat{I}_{2y} \qquad [2']$$

$$\left[\sin^2\theta\cos\theta\right]\left[-\sin(\pi J_{13}t_1)\sin(\pi J_{12}t_1)\cos(\Omega_1 t_1)\right] 4\hat{I}_{1z}\hat{I}_{2y}\hat{I}_{3z}, \qquad [5']$$

where θ is the flip angle of the final pulse. In the basic COSY experiment described before the final pulse has flip angle 90° i.e. $\theta = \pi/2$. In this case term [5'] disappears since $\cos(\pi/2) = 0$, and term [2'] becomes exactly the same as term [2] on page 326 since $\sin(\pi/2) = 1$. We therefore regain the previous result.

In general the two terms [2'] and [5'] do not have the same overall size on account of their different dependence on the flip angle θ. They will not therefore combine to give a reduced multiplet in the same way as terms [2] and [5] on page 330. However, we will now show that if the flip angle θ is small, the two terms will have the same size, and so combine to give a reduced multiplet.

We first note that $\sin\theta$ and $\cos\theta$ can be expressed as power series in θ (here θ must be in radians):

$$\sin\theta = \theta - \tfrac{1}{3!}\theta^3 + \tfrac{1}{5!}\theta^5 \ldots \qquad \cos\theta = 1 - \tfrac{1}{2!}\theta^2 + \tfrac{1}{4!}\theta^4 \ldots$$

We have only written out the first three terms in each case. If θ is small (i.e. $\theta \ll 1$), then θ^2 is even smaller, and θ^3 even smaller still. Under these circumstances we can discard all of the terms in θ^2 and higher powers of θ. This gives the approximate results

$$\text{small } \theta: \qquad \sin\theta \approx \theta \qquad \cos\theta \approx 1.$$

If we now return to terms [2'] and [5'], and assume that θ is small enough that we can use $\sin\theta \approx \theta$ and $\cos\theta \approx 1$, we find

$$\left[\theta^2\right]\left[-\cos(\pi J_{13}t_1)\sin(\pi J_{12}t_1)\sin(\Omega_1 t_1)\right] 2\hat{I}_{1z}\hat{I}_{2y} \qquad [2']$$

$$\left[\theta^2\right]\left[-\sin(\pi J_{13}t_1)\sin(\pi J_{12}t_1)\cos(\Omega_1 t_1)\right] 4\hat{I}_{1z}\hat{I}_{2y}\hat{I}_{3z}. \qquad [5']$$

Both terms now have the same dependence on the flip angle θ and, apart from the factor θ^2, they are identical to terms [2] and [5] on page 330. Therefore the resulting cross peak will have a reduced multiplet structure of precisely the form described in section 10.3.1 on page 330.

There are some practical difficulties with using small flip angle COSY as a way of generating reduced multiplets. The overall intensity of the cross peak goes as θ^2 which, as it is necessary that $\theta \ll 1$, means that the cross peaks will be much weaker than in a COSY where the final pulse is 90°. Recalling the discussion in section 8.3.2 on page 192, one of the problems with COSY is that the diagonal peaks tend to be stronger than the cross peaks and, on account of their lineshape, the diagonal peaks tend to spread well away from the main diagonal. In a small flip angle COSY these problems are even worse as the cross peaks become weaker as the flip angle is reduced whereas, as we will show in section 10.4.4 on page 340, the components of the diagonal-peak multiplet which lie exactly on the diagonal get stronger. Thus for a small flip angle, the cross peaks can easily be swamped by intense tails from the diagonal peaks.

Generally speaking, small flip angle COSY is not a convenient method of generating reduced multiplets in homonuclear spin systems. Luckily, there are other experiments, such as ZCOSY and ECOSY which generate such multiplets, and largely side-step the problems which stem from the diagonal peaks in a conventional COSY.

ZCOSY is most simply described using a different set of operators to the ones we have been using so far. In the next section we will introduce these operators, and first illustrate how they can be used to describe small flip angle COSY. Then, we will use the same operators to show how ZCOSY works.

10.4 Polarization operators

Product operators are generally speaking an excellent way of analysing the outcome of multiple-pulse experiments, but they are perhaps not best-suited to analysing multiplet structures, or dealing with the effects of pulses whose flip angles are not 90° or 180°. Under some circumstances, a useful alternative is to construct our product operators from a different set of operators, called *polarization operators*. To keep the distinction clear, we will call the product operators we have been using so far *cartesian product operators*.

In this section we will introduce these operators, and then go on to see how they provide a convenient description of experiments such as small flip angle COSY and ZCOSY.

10.4.1 Construction and interpretation of polarization operators

The state of each spin is represented by one of four polarization operators; for spin one these are

$$\hat{I}_{1+} \quad \hat{I}_{1-} \quad \hat{I}_{1\alpha} \quad \hat{I}_{1\beta}. \tag{10.3}$$

We have come across the operators $\hat{I}_{1+}$ and $\hat{I}_{1-}$ before in section 7.12.1 on page 174. $\hat{I}_{1+}$ is called the raising operator, and $\hat{I}_{1-}$ the lowering operator; they are defined in terms of $\hat{I}_{1x}$ and $\hat{I}_{1y}$ as follows:

$$\hat{I}_{1+} \equiv \hat{I}_{1x} + i\hat{I}_{1y} \qquad \hat{I}_{1-} \equiv \hat{I}_{1x} - i\hat{I}_{1y}.$$

These identities can be expressed the other way round:

$$\hat{I}_{1x} \equiv \tfrac{1}{2}\left(\hat{I}_{1+} + \hat{I}_{1-}\right) \qquad \hat{I}_{1y} \equiv \tfrac{1}{2i}\left(\hat{I}_{1+} - \hat{I}_{1-}\right). \tag{10.4}$$

The operators $\hat{I}_{1\alpha}$ and $\hat{I}_{1\beta}$ are related to the unit operator $\hat{E}_1$ and $\hat{I}_{1z}$:

$$\hat{E}_1 \equiv \hat{I}_{1\alpha} + \hat{I}_{1\beta} \qquad \hat{I}_{1z} \equiv \tfrac{1}{2}\left(\hat{I}_{1\alpha} - \hat{I}_{1\beta}\right). \tag{10.5}$$

A similar set of four operators are needed for spins two, three and so on.

The basis operators are formed by making all possible products consisting of one of the operators of the type given in Eq. 10.3 on the preceding page for each spin. For example, in a three-spin system, we can have products such as:

$$\hat{I}_{1-}\hat{I}_{2\alpha}\hat{I}_{3\alpha} \qquad \hat{I}_{1+}\hat{I}_{2-}\hat{I}_{3\alpha} \qquad \hat{I}_{1\alpha}\hat{I}_{2\beta}\hat{I}_{3\alpha}. \tag{10.6}$$

Only products containing *one* operator of the type $\hat{I}_+$ or $\hat{I}_-$ are in principle observable. Therefore, of the three product operators in Eq. 10.6, only the first gives rise to observable magnetization and, since the operator is $\hat{I}_{1-}$, the magnetization appears on spin one. For reasons which will be described in section 11.1.4 on page 386, only operator products containing one operator of the type $\hat{I}_-$ are observable in practical experiments.

The second product in Eq. 10.6 is zero-quantum coherence between spins one and two as it contains the operators $\hat{I}_{1+}$ and $\hat{I}_{2-}$ (see section 7.12.1 on page 174). The final product represents the population of the $\alpha\beta\alpha$ energy level.

10.4.2   Free evolution

The really nice thing about these products of polarization operators is that free evolution simply results in a phase factor. In contrast to cartesian product operators, the number of operators does not increase.

For example, the evolution of the product $\hat{I}_{1-}\hat{I}_{2\alpha}\hat{I}_{3\alpha}$ for a time t gives:

$$\exp\left(+i[\Omega_1 - \pi J_{12} - \pi J_{13}]t\right)\hat{I}_{1-}\hat{I}_{2\alpha}\hat{I}_{3\alpha},$$

whereas the product $\hat{I}_{1-}\hat{I}_{2\beta}\hat{I}_{3\alpha}$ evolves according to

$$\exp\left(+i[\Omega_1 + \pi J_{12} - \pi J_{13}]t\right)\hat{I}_{1-}\hat{I}_{2\beta}\hat{I}_{3\alpha},$$

and the product $\hat{I}_{1-}\hat{I}_{2\beta}\hat{I}_{3\beta}$ goes to

$$\exp\left(+i[\Omega_1 + \pi J_{12} + \pi J_{13}]t\right)\hat{I}_{1-}\hat{I}_{2\beta}\hat{I}_{3\beta}.$$

In each case the operator simply acquires a phase factor which depends on the frequency in the square brace and the time. Since the operator is $\hat{I}_{1-}$, the frequencies all include the offset of spin one. If the operator for spin two is $\hat{I}_{2\alpha}$ then a term $-\pi J_{12}$ is included, whereas if the operator is $\hat{I}_{2\beta}$ a term $+\pi J_{12}$ is included. Likewise, $\hat{I}_{3\alpha}$ gives a term $-\pi J_{13}$ and $\hat{I}_{3\beta}$ gives a term $+\pi J_{13}$.

The corresponding operators containing $\hat{I}_{1+}$ simply evolve in the opposite sense, meaning that the argument of the exponential has a factor of $-i$ rather than $+i$; for example

$$\hat{I}_{1-}\hat{I}_{2\beta}\hat{I}_{3\alpha} \longrightarrow \exp\left(+i[\Omega_1 + \pi J_{12} - \pi J_{13}]t\right)\hat{I}_{1-}\hat{I}_{2\beta}\hat{I}_{3\alpha}$$
$$\hat{I}_{1+}\hat{I}_{2\beta}\hat{I}_{3\alpha} \longrightarrow \exp\left(-i[\Omega_1 + \pi J_{12} - \pi J_{13}]t\right)\hat{I}_{1+}\hat{I}_{2\beta}\hat{I}_{3\alpha}.$$

The same rules apply to operator products containing $\hat{I}_2$ and $\hat{I}_{2+}$; for example

$$\hat{I}_{1\alpha}\hat{I}_{2+}\hat{I}_{3\beta} \longrightarrow \exp\left(-\mathrm{i}[\Omega_2 - \pi J_{12} + \pi J_{23}]t\right)\hat{I}_{1\alpha}\hat{I}_{2+}\hat{I}_{3\beta}.$$

The offset term is Ω_2, as the raising operator present is for spin two. Since the spin-one operator is $\hat{I}_{1\alpha}$ we include a term $-\pi J_{12}$ in the frequency, and as the spin-three operator is $\hat{I}_{3\beta}$ we include a term $+\pi J_{23}$. Note that the latter term involves the coupling between spins two and three, as spin two is present as the raising operator.

You will have noticed by now that each of the four operators

$$\hat{I}_{1-}\hat{I}_{2\alpha}\hat{I}_{3\alpha} \qquad \hat{I}_{1-}\hat{I}_{2\beta}\hat{I}_{3\alpha} \qquad \hat{I}_{1-}\hat{I}_{2\alpha}\hat{I}_{3\beta} \qquad \hat{I}_{1-}\hat{I}_{2\beta}\hat{I}_{3\beta},$$

evolves at the frequency of one of the four lines of the spin-one multiplet. Each operator represents observable magnetization which corresponds to one of the four lines of the multiplet. In the same way the four operators

$$\hat{I}_{1\alpha}\hat{I}_{2-}\hat{I}_{3\alpha} \qquad \hat{I}_{1\beta}\hat{I}_{2-}\hat{I}_{3\alpha} \qquad \hat{I}_{1\alpha}\hat{I}_{2-}\hat{I}_{3\beta} \qquad \hat{I}_{1\beta}\hat{I}_{2-}\hat{I}_{3\beta},$$

represent the four lines of the spin-two multiplet.

10.4.3 Pulses

The effect of pulses on these operators is rather more complex than for cartesian product operators. For an x-pulse of flip angle θ to spin one we have:

$$\hat{I}_{1+} \xrightarrow{\theta\hat{I}_{1x}} \cos^2\left(\tfrac{1}{2}\theta\right)\hat{I}_{1+} + \sin^2\left(\tfrac{1}{2}\theta\right)\hat{I}_{1-} + \tfrac{1}{2}\mathrm{i}\sin\theta\left(\hat{I}_{1\alpha} - \hat{I}_{1\beta}\right)$$

$$\hat{I}_{1-} \xrightarrow{\theta\hat{I}_{1x}} \cos^2\left(\tfrac{1}{2}\theta\right)\hat{I}_{1-} + \sin^2\left(\tfrac{1}{2}\theta\right)\hat{I}_{1+} - \tfrac{1}{2}\mathrm{i}\sin\theta\left(\hat{I}_{1\alpha} - \hat{I}_{1\beta}\right)$$

$$\hat{I}_{1\alpha} \xrightarrow{\theta\hat{I}_{1x}} \cos^2\left(\tfrac{1}{2}\theta\right)\hat{I}_{1\alpha} + \sin^2\left(\tfrac{1}{2}\theta\right)\hat{I}_{1\beta} + \tfrac{1}{2}\mathrm{i}\sin\theta\left(\hat{I}_{1+} - \hat{I}_{1-}\right)$$

$$\hat{I}_{1\beta} \xrightarrow{\theta\hat{I}_{1x}} \cos^2\left(\tfrac{1}{2}\theta\right)\hat{I}_{1\beta} + \sin^2\left(\tfrac{1}{2}\theta\right)\hat{I}_{1\alpha} - \tfrac{1}{2}\mathrm{i}\sin\theta\left(\hat{I}_{1+} - \hat{I}_{1-}\right).$$

The effect of a pulse is to interconvert all the operators in a fairly complex way which depends on trigonometric functions of θ and $\tfrac{1}{2}\theta$. There are a similar set of relations for the operators of spins two, three and so on.

Two special cases are of interest. For a 180° pulse, $\theta = \pi$, the transformations become rather simple:

$$\hat{I}_{1+} \xrightarrow{\pi\hat{I}_{1x}} \hat{I}_{1-} \qquad \hat{I}_{1-} \xrightarrow{\pi\hat{I}_{1x}} \hat{I}_{1+} \qquad \hat{I}_{1\alpha} \xrightarrow{\pi\hat{I}_{1x}} \hat{I}_{1\beta} \qquad \hat{I}_{1\beta} \xrightarrow{\pi\hat{I}_{1x}} \hat{I}_{1\alpha}.$$

The second case we are going to be interested in is when the flip angle θ is small. As we discussed in section 10.3.4 on page 335, in such circumstances we can write $\sin\theta \approx \theta$ and $\cos\theta \approx 1$. Under these circumstances the trigonometric terms in the above equations become

$$\cos^2\left(\tfrac{1}{2}\theta\right) \approx 1 \qquad \sin^2\left(\tfrac{1}{2}\theta\right) \approx \tfrac{1}{4}\theta^2 \qquad \sin\theta \approx \theta.$$

With these approximations, the effect of a small flip angle pulse is

$$\hat{I}_{1+} \xrightarrow{\theta \hat{I}_{1x}} \hat{I}_{1+} + \tfrac{1}{4}\theta^2 \hat{I}_{1-} + \tfrac{1}{2}i\theta\left(\hat{I}_{1\alpha} - \hat{I}_{1\beta}\right)$$

$$\hat{I}_{1-} \xrightarrow{\theta \hat{I}_{1x}} \hat{I}_{1-} + \tfrac{1}{4}\theta^2 \hat{I}_{1+} - \tfrac{1}{2}i\theta\left(\hat{I}_{1\alpha} - \hat{I}_{1\beta}\right)$$

$$\hat{I}_{1\alpha} \xrightarrow{\theta \hat{I}_{1x}} \hat{I}_{1\alpha} + \tfrac{1}{4}\theta^2 \hat{I}_{1\beta} + \tfrac{1}{2}i\theta\left(\hat{I}_{1+} - \hat{I}_{1-}\right) \qquad (10.7)$$

$$\hat{I}_{1\beta} \xrightarrow{\theta \hat{I}_{1x}} \hat{I}_{1\beta} + \tfrac{1}{4}\theta^2 \hat{I}_{1\alpha} - \tfrac{1}{2}i\theta\left(\hat{I}_{1+} - \hat{I}_{1-}\right).$$

We are now in a position to use these polarization operators in some practical calculations.

10.4.4   Small flip angle COSY

As we did before, let us imagine a homonuclear three-spin system and start out with z-magnetization on spin one, $\hat{I}_{1z}$. We are going to assume that the phase of the first pulse is y, so that the operator generated by this pulse is $\hat{I}_{1x}$. This simplifies the calculation slightly when compared with starting with a $90°(x)$ pulse; the form of the spectrum is not changed, apart from an overall phase shift.

To express $\hat{I}_{1x}$ in terms of polarization operators we first have to realize that $\hat{I}_{1x}$ is really a shorthand for $\hat{I}_{1x}\hat{E}_2\hat{E}_3$ (see section 10.1 on page 320). These three operators can be expressed in terms of polarization operators by using Eq. 10.4 and Eq. 10.5 on page 338:

$$\hat{I}_{1x} \equiv \tfrac{1}{2}\left(\hat{I}_{1+} + \hat{I}_{1-}\right) \qquad \hat{E}_2 \equiv \hat{I}_{2\alpha} + \hat{I}_{2\beta} \qquad \hat{E}_3 \equiv \hat{I}_{3\alpha} + \hat{I}_{3\beta}.$$

Substituting in these expressions gives

$$\hat{I}_{1x}\hat{E}_2\hat{E}_3 \equiv \tfrac{1}{2}\left(\hat{I}_{1+} + \hat{I}_{1-}\right)\left(\hat{I}_{2\alpha} + \hat{I}_{2\beta}\right)\left(\hat{I}_{3\alpha} + \hat{I}_{3\beta}\right).$$

Multiplying this all out we find

$$\hat{I}_{1x} \equiv + \tfrac{1}{2}\left(\hat{I}_{1+}\hat{I}_{2\alpha}\hat{I}_{3\alpha} + \hat{I}_{1+}\hat{I}_{2\alpha}\hat{I}_{3\beta} + \hat{I}_{1+}\hat{I}_{2\beta}\hat{I}_{3\alpha} + \hat{I}_{1+}\hat{I}_{2\beta}\hat{I}_{3\beta}\right)$$

$$+ \tfrac{1}{2}\left(\hat{I}_{1-}\hat{I}_{2\alpha}\hat{I}_{3\alpha} + \hat{I}_{1-}\hat{I}_{2\alpha}\hat{I}_{3\beta} + \hat{I}_{1-}\hat{I}_{2\beta}\hat{I}_{3\alpha} + \hat{I}_{1-}\hat{I}_{2\beta}\hat{I}_{3\beta}\right). \qquad (10.8)$$

The four operators in the first bracket correspond to the four lines of the spin-one multiplet, which is hardly a surprise as we know from section 10.1.1 on page 321 that the cartesian operator $\hat{I}_{1x}$ represents an in-phase multiplet on spin one. Similarly, the four operators in the second bracket also correspond to the four lines of the multiplet: all that is different about this second group of four, as compared with the first, is a change in the sense of the evolution.

To predict the form of the spectrum we first have to allow each of these operators to evolve for time t_1, and then work out the effect of the final small flip angle pulse. This is not quite as complicated as it seems, since once we have done a couple of terms a pattern will develop, which means that we do not really have to work through them all.

To start with we will just consider the term $\hat{I}_{1+}\hat{I}_{2\alpha}\hat{I}_{3\alpha}$; during t_1 this term simply acquires a phase factor:

$$\exp\left(-i[\Omega_1 - \pi J_{12} - \pi J_{13}]t_1\right)\hat{I}_{1+}\hat{I}_{2\alpha}\hat{I}_{3\alpha}.$$

The final pulse will transfer this term, along with its modulation, to the other spins, thereby generating the cross peak. We will focus on the transfers to spin two, which lead to the 1–2 cross peak.

The 1–2 cross-peak multiplet

In order to contribute to the 1–2 cross peak, the final pulse must transfer $\hat{I}_{1+}\hat{I}_{2\alpha}\hat{I}_{3\alpha}$ into an operator product containing $\hat{I}_{2-}$ (recall that only $\hat{I}_{-}$ operators are observable). In addition, for the term to be observable the spin-one and spin-three operators must be $\hat{I}_{1\gamma}$ and $\hat{I}_{3\gamma'}$, where γ and γ' can be α or β:

$$\hat{I}_{1+}\hat{I}_{2\alpha}\hat{I}_{3\alpha} \longrightarrow \hat{I}_{1\gamma}\hat{I}_{2-}\hat{I}_{3\gamma'}.$$

The effect of the small flip angle pulse is given by Eq. 10.7 on the preceding page, from which we see that the transformation from $\hat{I}_{2\alpha}$ to $\hat{I}_{2-}$ carries with it the coefficient $-\frac{1}{2}i\theta$.

For spin one, if the operator goes from $\hat{I}_{1+}$ to $\hat{I}_{1\alpha}$ there is a coefficient of $+\frac{1}{2}i\theta$, whereas if the final operator is $\hat{I}_{1\beta}$, the coefficient is $-\frac{1}{2}i\theta$. For spin three, if the operator remains the same ($\hat{I}_{3\alpha}$), there is a coefficient of 1, whereas the transformation to $\hat{I}_{3\beta}$ gives a coefficient of $\frac{1}{4}\theta^2$. In summary, we have the following four possibilities

$$\hat{I}_{1+}\hat{I}_{2\alpha}\hat{I}_{3\alpha} \longrightarrow \left(+\tfrac{1}{2}i\theta\right)\left(-\tfrac{1}{2}i\theta\right)(1)\,\hat{I}_{1\alpha}\hat{I}_{2-}\hat{I}_{3\alpha}$$

$$\hat{I}_{1+}\hat{I}_{2\alpha}\hat{I}_{3\alpha} \longrightarrow \left(-\tfrac{1}{2}i\theta\right)\left(-\tfrac{1}{2}i\theta\right)(1)\,\hat{I}_{1\beta}\hat{I}_{2-}\hat{I}_{3\alpha}$$

$$\hat{I}_{1+}\hat{I}_{2\alpha}\hat{I}_{3\alpha} \longrightarrow \left(+\tfrac{1}{2}i\theta\right)\left(-\tfrac{1}{2}i\theta\right)\left(\tfrac{1}{4}\theta^2\right)\hat{I}_{1\alpha}\hat{I}_{2-}\hat{I}_{3\beta}$$

$$\hat{I}_{1+}\hat{I}_{2\alpha}\hat{I}_{3\alpha} \longrightarrow \left(-\tfrac{1}{2}i\theta\right)\left(-\tfrac{1}{2}i\theta\right)\left(\tfrac{1}{4}\theta^2\right)\hat{I}_{1\beta}\hat{I}_{2-}\hat{I}_{3\beta}.$$

If the flip angle is small, we can discount the last two terms as they go as θ^4, which will give them very low intensity compared with the first two terms, which go as θ^2; we will therefore ignore the terms in θ^4.

Note that the first two terms are the ones in which the spin-three operator is unchanged, whereas in the second two terms this operator changes from $\hat{I}_{3\alpha}$ to $\hat{I}_{3\beta}$. What is happening is that the small flip angle pulse is discriminating in favour of transfers in which spin three, the passive spin, *does not* change spin state.

Multiplying out the brackets for the first two terms gives us

$$\hat{I}_{1+}\hat{I}_{2\alpha}\hat{I}_{3\alpha} \longrightarrow +\tfrac{1}{4}\theta^2\,\hat{I}_{1\alpha}\hat{I}_{2-}\hat{I}_{3\alpha} \qquad \hat{I}_{1+}\hat{I}_{2\alpha}\hat{I}_{3\alpha} \longrightarrow -\tfrac{1}{4}\theta^2\hat{I}_{1\beta}\hat{I}_{2-}\hat{I}_{3\alpha}.$$

Notice that there is a sign change depending on whether spin one ends up as $\hat{I}_{1\alpha}$ or $\hat{I}_{1\beta}$.

As we saw above, during t_1 the product $\hat{I}_{1+}\hat{I}_{2\alpha}\hat{I}_{3\alpha}$ acquires a phase label $\exp\left(-i[\Omega_1 - \pi J_{12} - \pi J_{13}]t_1\right)$, which means that after Fourier transformation with respect to t_1 there will be a peak at $-[\Omega_1 - \pi J_{12} - \pi J_{13}]$ in the ω_1 dimension. In ω_2 the term $\hat{I}_{1\alpha}\hat{I}_{2-}\hat{I}_{3\alpha}$ gives rise to a peak at $[\Omega_2 - \pi J_{12} - \pi J_{23}]$, while the term $\hat{I}_{1\beta}\hat{I}_{2-}\hat{I}_{3\alpha}$ gives a peak at $[\Omega_2 + \pi J_{12} - \pi J_{23}]$.

Overall, the term $\hat{I}_{1+}\hat{I}_{2\alpha}\hat{I}_{3\alpha}$ present during t_1 gives rise to two components in the cross-peak multiplet:

$$\{\omega_1, \omega_2\} \;=\; \{-[\Omega_1 - \pi J_{12} - \pi J_{13}], [\Omega_2 - \pi J_{12} - \pi J_{23}]\} \quad \text{intensity: } +\tfrac{1}{4}\theta^2$$

$$\{\omega_1, \omega_2\} \;=\; \{-[\Omega_1 - \pi J_{12} - \pi J_{13}], [\Omega_2 + \pi J_{12} - \pi J_{23}]\} \quad \text{intensity: } -\tfrac{1}{4}\theta^2.$$

We now need to work through the remaining three operator products of the form $\hat{I}_{1+}\hat{I}_{1y}\hat{I}_{2y'}$ present during t_1. It turns out that these follow a very similar pattern, in which the final pulse causes each to split into two products which are observable on spin two. The results are summarized in the table below.

Down the side of the table are the operator products present at the start of t_1, and along the top are the four possible operator products which result in observable signals on spin two. The entries in the table give the coefficient for the transfer between the operator in the corresponding row and column.

	$\hat{I}_{1\alpha}\hat{I}_{2-}\hat{I}_{3\alpha}$	$\hat{I}_{1\beta}\hat{I}_{2-}\hat{I}_{3\alpha}$	$\hat{I}_{1\alpha}\hat{I}_{2-}\hat{I}_{3\beta}$	$\hat{I}_{1\beta}\hat{I}_{2-}\hat{I}_{3\beta}$	t_2 / ω_2
$\hat{I}_{1+}\hat{I}_{2\beta}\hat{I}_{3\beta}$	0	0	$-\tfrac{1}{4}\theta^2$	$+\tfrac{1}{4}\theta^2$	
$\hat{I}_{1+}\hat{I}_{2\alpha}\hat{I}_{3\beta}$	0	0	$+\tfrac{1}{4}\theta^2$	$-\tfrac{1}{4}\theta^2$	
$\hat{I}_{1+}\hat{I}_{2\beta}\hat{I}_{3\alpha}$	$-\tfrac{1}{4}\theta^2$	$+\tfrac{1}{4}\theta^2$	0	0	
$\hat{I}_{1+}\hat{I}_{2\alpha}\hat{I}_{3\alpha}$	$+\tfrac{1}{4}\theta^2$	$-\tfrac{1}{4}\theta^2$	0	0	
t_1 / ω_1					

Recall that each of the product operators down the side represents a line from the spin-one multiplet, whereas those along the top represent a line from the spin-two multiplet. Thus the table is in fact a direct picture of the 1–2 cross-peak multiplet.

The operators have been ordered in the table in a way which will match the line positions for the case where the active coupling is smaller than either of the passive couplings, which is the situation illustrated in Fig. 10.10 (c) on page 332. The pattern of intensities shown in the table therefore matches this diagram. What we have here is a reduced multiplet, with the two anti-phase square arrays appearing in the top right, and bottom left, quadrants of the table.

We saw that the transfer from $\hat{I}_{1+}\hat{I}_{2\alpha}\hat{I}_{3\alpha}$ to $\hat{I}_{1\beta}\hat{I}_{2-}\hat{I}_{3\beta}$ has an overall coefficient of

$$\left(-\tfrac{1}{2}i\theta\right)\left(-\tfrac{1}{2}i\theta\right)\left(\tfrac{1}{4}\theta^2\right) = -\tfrac{1}{16}\theta^4.$$

This transfer gives rise to a peak which is not part of the reduced multiplet. We can compare the intensity of this unwanted peak with that of the wanted peaks, which go as $\tfrac{1}{4}\theta^2$:

$$\frac{\text{unwanted}}{\text{wanted}} = \frac{\tfrac{1}{16}\theta^4}{\tfrac{1}{4}\theta^2}$$

$$= \tfrac{1}{4}\theta^2.$$

Thus if θ is 20°, which is 0.35 radians, this ratio is 0.03 i.e. the unwanted peaks will be 3% of the intensity of the wanted peaks. For most purposes, this degree of suppression of the unwanted peaks is probably sufficient.

Lineshapes for the cross-peak multiplet

We saw above that a typical component of the cross-peak multiplet arises from the transfer $\hat{I}_{1+}\hat{I}_{2\alpha}\hat{I}_{3\alpha} \longrightarrow \hat{I}_{1\alpha}\hat{I}_{2-}\hat{I}_{3\alpha}$. In t_1 the term $\hat{I}_{1+}\hat{I}_{2\alpha}\hat{I}_{3\alpha}$ has a phase modulation according to $\exp(-i[\Omega_1 - \pi J_{12} - \pi J_{13}]t_1)$, and during t_2 the term $\hat{I}_{1\alpha}\hat{I}_{2-}\hat{I}_{3\alpha}$ will acquire a phase modulation of the form $\exp(+i[\Omega_2 - \pi J_{12} - \pi J_{23}]t_2)$. Overall the time-domain function for this component of the cross-peak multiplet will be

$$S(t_1, t_2) = \exp(-i[\Omega_1 - \pi J_{12} - \pi J_{13}]t_1) \times \exp(+i[\Omega_2 - \pi J_{12} - \pi J_{23}]t_2).$$

Referring to Eq. 10.8 on page 340, we see that during t_1 there are, in addition to the terms of the form $\hat{I}_{1+}\hat{I}_{2\gamma}\hat{I}_{3\gamma'}$, an equivalent set of terms of the form $\hat{I}_{1-}\hat{I}_{2\gamma}\hat{I}_{3\gamma'}$. The only difference in the behaviour of these terms is that the *sign* of the evolution frequencies in t_1 are reversed. For example, the transfer $\hat{I}_{1-}\hat{I}_{2\alpha}\hat{I}_{3\alpha} \longrightarrow \hat{I}_{1\alpha}\hat{I}_{2-}\hat{I}_{3\alpha}$ gives rise to a contribution to the time-domain signal of the form

$$S'(t_1, t_2) = \exp(+i[\Omega_1 - \pi J_{12} - \pi J_{13}]t_1) \times \exp(+i[\Omega_2 - \pi J_{12} - \pi J_{23}]t_2).$$

The transfers $\hat{I}_{1+}\hat{I}_{2\alpha}\hat{I}_{3\alpha} \longrightarrow \hat{I}_{1\alpha}\hat{I}_{2-}\hat{I}_{3\alpha}$ and $\hat{I}_{1-}\hat{I}_{2\alpha}\hat{I}_{3\alpha} \longrightarrow \hat{I}_{1\alpha}\hat{I}_{2-}\hat{I}_{3\alpha}$ both contribute to the observed signal:

$$
\begin{aligned}
S_{\text{obs}}(t_1, t_2) &= S(t_1, t_2) + S'(t_1, t_2) \\
&= \{\exp(-i[\Omega_1 - \pi J_{12} - \pi J_{13}]t_1) + \exp(+i[\Omega_1 - \pi J_{12} - \pi J_{13}]t_1)\} \\
&\quad \times \exp(+i[\Omega_2 - \pi J_{12} - \pi J_{23}]t_2) \\
&= 2\cos([\Omega_1 - \pi J_{12} - \pi J_{13}]t_1)\exp(+i[\Omega_2 - \pi J_{12} - \pi J_{23}]t_2).
\end{aligned}
$$

Overall, we have cosine modulation, so the spectrum can be processed in the usual way to give absorption mode lineshapes (see section 8.12 on page 226).

The diagonal-peak multiplet

If the final small flip angle pulse does not transfer the operators $\hat{I}_{1\pm}\hat{I}_{2\gamma}\hat{I}_{3\gamma'}$ to another spin, but leaves them on spin one, the result will be a contribution to the diagonal-peak multiplet. As was the case for the cross peak, we will find that only a sub-set of the sixteen possible components of the diagonal peak multiplet have significant intensity.

If we start with the operator product $\hat{I}_{1+}\hat{I}_{2\alpha}\hat{I}_{3\alpha}$ present during t_1, we can see that contributions to the diagonal peak will arise from transfers of the form

$$\hat{I}_{1+}\hat{I}_{2\alpha}\hat{I}_{3\alpha} \longrightarrow \hat{I}_{1-}\hat{I}_{2\gamma}\hat{I}_{3\gamma'}.$$

Referring to Eq. 10.7 on page 340, the intensity of the four possible transfers of this type are:

$$
\begin{aligned}
\hat{I}_{1+}\hat{I}_{2\alpha}\hat{I}_{3\alpha} &\longrightarrow \left(\tfrac{1}{4}\theta^2\right)(1)(1)\hat{I}_{1-}\hat{I}_{2\alpha}\hat{I}_{3\alpha} \\
\hat{I}_{1+}\hat{I}_{2\alpha}\hat{I}_{3\alpha} &\longrightarrow \left(\tfrac{1}{4}\theta^2\right)(\tfrac{1}{4}\theta^2)(1)\hat{I}_{1-}\hat{I}_{2\beta}\hat{I}_{3\alpha} \\
\hat{I}_{1+}\hat{I}_{2\alpha}\hat{I}_{3\alpha} &\longrightarrow \left(\tfrac{1}{4}\theta^2\right)(1)(\tfrac{1}{4}\theta^2)\hat{I}_{1-}\hat{I}_{2\alpha}\hat{I}_{3\beta} \\
\hat{I}_{1+}\hat{I}_{2\alpha}\hat{I}_{3\alpha} &\longrightarrow \left(\tfrac{1}{4}\theta^2\right)(\tfrac{1}{4}\theta^2)(\tfrac{1}{4}\theta^2)\hat{I}_{1-}\hat{I}_{2\beta}\hat{I}_{3\beta}.
\end{aligned}
$$

Of these four transfers, the first goes as θ^2, and so will be of comparable intensity with the components in the cross peak. All of the other transfers go as θ^4 or θ^6, and so will be of negligible intensity.

Thus, starting from $\hat{I}_{1+}\hat{I}_{2\alpha}\hat{I}_{3\alpha}$, the only transfer of significant intensity is that to $\hat{I}_{1-}\hat{I}_{2\alpha}\hat{I}_{3\alpha}$; note that the spin states of spins two and three remain the same in this transfer. The result is a contribution to the time-domain signal of the form,

$$S(t_1, t_2) = \tfrac{1}{4}\theta^2 \exp\left(-i[\Omega_1 - \pi J_{12} - \pi J_{13}]t_1\right) \times \exp\left(+i[\Omega_1 - \pi J_{12} - \pi J_{13}]t_2\right).$$

Such a phase modulated signal gives rise to a phase-twist lineshape, intensity $\tfrac{1}{4}\theta^2$, at frequency

$$\{\omega_1, \omega_2\} = \{-[\Omega_1 - \pi J_{12} - \pi J_{13}], [\Omega_1 - \pi J_{12} - \pi J_{13}]\}.$$

Apart from the sign change, the ω_1 and ω_2 frequencies are the same, so this peak lies directly on the diagonal line $\omega_1 = -\omega_2$.

The other operators of the form $\hat{I}_{1+}\hat{I}_{2\gamma}\hat{I}_{3\gamma'}$ also give rise to peaks which have, apart from the sign change, identical frequencies in the two dimensions. Thus, the diagonal-peak multiplet consists of just four peaks, all of which lie on the diagonal line $\omega_1 = -\omega_2$, and whose frequencies are just the four lines of the spin-one multiplet. The four peaks are all positive.

These four components of the diagonal-peak multiplet are those in which the spin states of spins two and three, which are both passive, *remain the same*. Once again, the small flip angle pulse discriminates in favour of terms in which the spin states of the passive spins are preserved.

The terms of the type $\hat{I}_{1-}\hat{I}_{2\gamma}\hat{I}_{3\gamma'}$ present during t_1 do not need any transfer to make them observable in t_2. For example,

$$\hat{I}_{1-}\hat{I}_{2\alpha}\hat{I}_{3\alpha} \longrightarrow (1)(1)(1)\hat{I}_{1-}\hat{I}_{2\alpha}\hat{I}_{3\alpha}.$$

This transfer will give rise to a phase-twist peak of overall intensity 1, and at

$$\{\omega_1, \omega_2\} = \{[\Omega_1 - \pi J_{12} - \pi J_{13}], [\Omega_1 - \pi J_{12} - \pi J_{13}]\};$$

note that this lies on the diagonal line $\omega_1 = \omega_2$. The four operators of the type $\hat{I}_{1-}\hat{I}_{2\gamma}\hat{I}_{3\gamma'}$ thus give rise to four components of the diagonal peak, all of which lie on the diagonal line $\omega_1 = \omega_2$, as shown in Fig. 10.15.

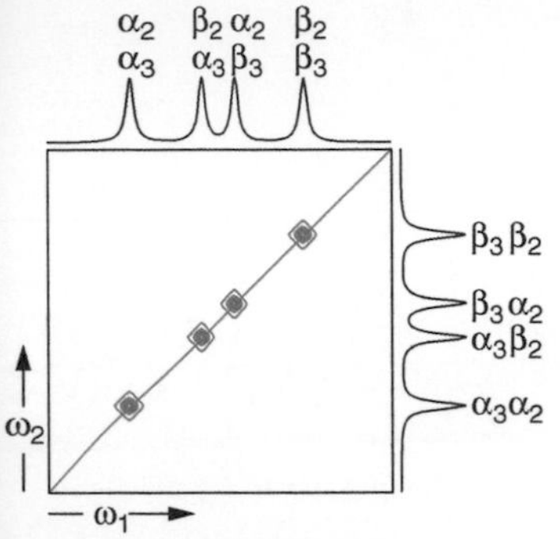

The key difference between the diagonal- and cross-peak multiplets is that for the cross-peak multiplet there are equal contributions from signals evolving at $+\Omega$ and $-\Omega$ during t_1, leading to overall cosine modulation and hence an absorption mode lineshape. In contrast, for the diagonal-peak multiplet, the two contributions are not equal, so we do not have cosine modulation, and hence cannot obtain an absorption mode lineshape.

The problem with small flip angle COSY

A small flip angle COSY does indeed generate reduced cross-peak multiplets, the individual peaks of which have the absorption mode lineshape. The overall intensity of the cross-peak components is $\tfrac{1}{4}\theta^2$, which makes them much less intense than the diagonal-peak components, which have intensity 1.

Recall as well that the diagonal-peak components have the unfavourable phase-twist lineshape. This, combined with the much greater intensity of the diagonal peaks, means that the cross-peak multiplets are

Fig. 10.15 Schematic diagonal-peak multiplet, for spin one, from a small flip angle COSY. These peaks arise from operators of the form $\hat{I}_{1-}\hat{I}_{2\gamma}\hat{I}_{3\gamma'}$ present during t_1. Out of the sixteen possible components of the multiplet, only four survive, and these are the peaks which lie directly on the diagonal line $\omega_1 = \omega_2$. These four peaks have the same spin state of the passive spins, spins two and three, in the two dimensions. As explained in the text, in practice the lines would have the phase twist lineshape.

easily swamped by the spreading diagonal. For these reasons, small flip angle COSY is not a particularly useful experiment.

However, the ZCOSY experiment, which we will describe next, gets around these problems in rather a neat way. In ZCOSY, the cross and diagonal peaks are of comparable intensity, and both have the absorption mode lineshape.

10.5   ZCOSY

The pulse sequence for the ZCOSY experiment is shown in Fig. 10.16. The mixing period consists of two small flip angle pulses separated by a small delay, and it is arranged that *only* population terms present between these two pulses contribute to the final spectrum. We will see that the resulting spectrum looks very similar to a small flip angle COSY, with the important difference that all peaks are in absorption mode.

The way in which this sequence works is easily appreciated using polarization operators. Let us start with the term $\hat{I}_{1+}\hat{I}_{2\alpha}\hat{I}_{3\alpha}$ present at the end of t_1 and see how it can be transferred to population terms by the first small flip angle pulse. Such population terms are represented by polarization operators of the form $\hat{I}_{1\gamma}\hat{I}_{2\gamma'}\hat{I}_{3\gamma''}$, where γ, γ' and γ'' can be α or β. Starting from $\hat{I}_{1+}\hat{I}_{2\alpha}\hat{I}_{3\alpha}$ and applying the transformations given in Eq. 10.7 on page 340, we can show that the pulse generates four population terms in which spin one is $\hat{I}_{1\alpha}$:

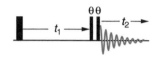

Fig. 10.16 Pulse sequence for the ZCOSY experiment. The final two pulses have small flip angles, typically of about 20°, and it is arranged that only population terms present between these two pulses contribute to the spectrum.

$$\hat{I}_{1+}\hat{I}_{2\alpha}\hat{I}_{3\alpha} \longrightarrow \left(+\tfrac{1}{2}\mathrm{i}\theta\right)(1)(1)\,\hat{I}_{1\alpha}\hat{I}_{2\alpha}\hat{I}_{3\alpha}$$

$$\hat{I}_{1+}\hat{I}_{2\alpha}\hat{I}_{3\alpha} \longrightarrow \left(+\tfrac{1}{2}\mathrm{i}\theta\right)\left(\tfrac{1}{4}\theta^2\right)(1)\,\hat{I}_{1\alpha}\hat{I}_{2\beta}\hat{I}_{3\alpha}$$

$$\hat{I}_{1+}\hat{I}_{2\alpha}\hat{I}_{3\alpha} \longrightarrow \left(+\tfrac{1}{2}\mathrm{i}\theta\right)(1)\left(\tfrac{1}{4}\theta^2\right)\,\hat{I}_{1\alpha}\hat{I}_{2\alpha}\hat{I}_{3\beta}$$

$$\hat{I}_{1+}\hat{I}_{2\alpha}\hat{I}_{3\alpha} \longrightarrow \left(+\tfrac{1}{2}\mathrm{i}\theta\right)\left(\tfrac{1}{4}\theta^2\right)\left(\tfrac{1}{4}\theta^2\right)\,\hat{I}_{1\alpha}\hat{I}_{2\beta}\hat{I}_{3\beta}.$$

We can see that, of these four possibilities, only the first will be significant if the flip angle is small. There are also four possibilities in which spin one is $\hat{I}_{1\beta}$, but again only one of these will be significant for small flip angles:

$$\hat{I}_{1+}\hat{I}_{2\alpha}\hat{I}_{3\alpha} \longrightarrow \left(-\tfrac{1}{2}\mathrm{i}\theta\right)(1)(1)\,\hat{I}_{1\beta}\hat{I}_{2\alpha}\hat{I}_{3\alpha}$$

The second small flip angle pulse has to generate observable terms from $\hat{I}_{1\alpha}\hat{I}_{2\alpha}\hat{I}_{3\alpha}$, so one of the operators must be rotated to $\hat{I}_{-}$. To contribute to the 1–2 cross peak, the spin-two operator must be transformed to $\hat{I}_{2-}$; the operators for spins one and three may also be changed from $\hat{I}_{1\alpha}$ to $\hat{I}_{1\beta}$, and from $\hat{I}_{3\alpha}$ to $\hat{I}_{3\beta}$:

$$\hat{I}_{1\alpha}\hat{I}_{2\alpha}\hat{I}_{3\alpha} \longrightarrow (1)\left(-\tfrac{1}{2}\mathrm{i}\theta\right)(1)\,\hat{I}_{1\alpha}\hat{I}_{2-}\hat{I}_{3\alpha}$$

$$\hat{I}_{1\alpha}\hat{I}_{2\alpha}\hat{I}_{3\alpha} \longrightarrow \left(\tfrac{1}{4}\theta^2\right)\left(-\tfrac{1}{2}\mathrm{i}\theta\right)(1)\,\hat{I}_{1\beta}\hat{I}_{2-}\hat{I}_{3\alpha}$$

$$\hat{I}_{1\alpha}\hat{I}_{2\alpha}\hat{I}_{3\alpha} \longrightarrow (1)\left(-\tfrac{1}{2}\mathrm{i}\theta\right)\left(\tfrac{1}{4}\theta^2\right)\,\hat{I}_{1\alpha}\hat{I}_{2-}\hat{I}_{3\beta}$$

$$\hat{I}_{1\alpha}\hat{I}_{2\alpha}\hat{I}_{3\alpha} \longrightarrow \left(\tfrac{1}{4}\theta^2\right)\left(-\tfrac{1}{2}\mathrm{i}\theta\right)\left(\tfrac{1}{4}\theta^2\right)\,\hat{I}_{1\beta}\hat{I}_{2-}\hat{I}_{3\beta}.$$

Again, of these only the first term will have significant amplitude if the flip angle is small. A corresponding transfer arises from the term $\hat{I}_{1\beta}\hat{I}_{2\alpha}\hat{I}_{3\alpha}$.

$$\hat{I}_{1\beta}\hat{I}_{2\alpha}\hat{I}_{3\alpha} \longrightarrow (1)\left(-\tfrac{1}{2}i\theta\right)(1)\,\hat{I}_{1\beta}\hat{I}_{2-}\hat{I}_{3\alpha}.$$

In summary, the result of the two small flip angle pulses acting on the term $\hat{I}_{1+}\hat{I}_{2\alpha}\hat{I}_{3\alpha}$ is the following contributions to the 1–2 cross peak:

$$\hat{I}_{1+}\hat{I}_{2\alpha}\hat{I}_{3\alpha} \longrightarrow \left(+\tfrac{1}{2}i\theta\right)\hat{I}_{1\alpha}\hat{I}_{2\alpha}\hat{I}_{3\alpha} \longrightarrow \left(+\tfrac{1}{2}i\theta\right)\left(-\tfrac{1}{2}i\theta\right)\hat{I}_{1\alpha}\hat{I}_{2-}\hat{I}_{3\alpha}$$

$$\hat{I}_{1+}\hat{I}_{2\alpha}\hat{I}_{3\alpha} \longrightarrow \left(-\tfrac{1}{2}i\theta\right)\hat{I}_{1\beta}\hat{I}_{2\alpha}\hat{I}_{3\alpha} \longrightarrow \left(-\tfrac{1}{2}i\theta\right)\left(-\tfrac{1}{2}i\theta\right)\hat{I}_{1\beta}\hat{I}_{2-}\hat{I}_{3\alpha}.$$

Leaving out the intermediate stage these transformations are

$$\hat{I}_{1+}\hat{I}_{2\alpha}\hat{I}_{3\alpha} \longrightarrow \left(+\tfrac{1}{4}\theta^2\right)\hat{I}_{1\alpha}\hat{I}_{2-}\hat{I}_{3\alpha}$$

$$\hat{I}_{1+}\hat{I}_{2\alpha}\hat{I}_{3\alpha} \longrightarrow \left(-\tfrac{1}{4}\theta^2\right)\hat{I}_{1\beta}\hat{I}_{2-}\hat{I}_{3\alpha}.$$

These are identical to the transfers we computed for the small flip angle COSY as shown in the table on page 342. Working through the rest of the operators, we will find exactly the same results as in the table, so the 1–2 cross peak will show the required reduced structure.

It is also easy to show that terms of the form $\hat{I}_{1-}\hat{I}_{2\gamma}\hat{I}_{3\gamma'}$ present during t_1 also give rise to a similar cross-peak multiplet, but with the sign of the ω_1 frequencies reversed. It will therefore be possible to process the spectra to give absorption mode lineshapes.

10.5.1   The diagonal-peak multiplet

We can go through the same procedure for the diagonal peak. For the first small flip angle pulse, the same considerations apply as for the cross peak, and the only two significant terms will be

$$\hat{I}_{1+}\hat{I}_{2\alpha}\hat{I}_{3\alpha} \longrightarrow \left(+\tfrac{1}{2}i\theta\right)\hat{I}_{1\alpha}\hat{I}_{2\alpha}\hat{I}_{3\alpha}$$

$$\hat{I}_{1+}\hat{I}_{2\alpha}\hat{I}_{3\alpha} \longrightarrow \left(-\tfrac{1}{2}i\theta\right)\hat{I}_{1\beta}\hat{I}_{2\alpha}\hat{I}_{3\alpha}.$$

The second pulse must make these observable on spin one. The transfers which are significant for a small flip angle pulse are

$$\left(+\tfrac{1}{2}i\theta\right)\hat{I}_{1\alpha}\hat{I}_{2\alpha}\hat{I}_{3\alpha} \longrightarrow \left(+\tfrac{1}{2}i\theta\right)\left(-\tfrac{1}{2}i\theta\right)\hat{I}_{1-}\hat{I}_{2\alpha}\hat{I}_{3\alpha}$$

$$\left(-\tfrac{1}{2}i\theta\right)\hat{I}_{1\beta}\hat{I}_{2\alpha}\hat{I}_{3\alpha} \longrightarrow \left(-\tfrac{1}{2}i\theta\right)\left(+\tfrac{1}{2}i\theta\right)\hat{I}_{1-}\hat{I}_{2\alpha}\hat{I}_{3\alpha}.$$

The overall transfers caused by the two small flip angle pulses are

$$\hat{I}_{1+}\hat{I}_{2\alpha}\hat{I}_{3\alpha} \longrightarrow \left(+\tfrac{1}{4}\theta^2\right)\hat{I}_{1-}\hat{I}_{2\alpha}\hat{I}_{3\alpha}$$

$$\hat{I}_{1+}\hat{I}_{2\alpha}\hat{I}_{3\alpha} \longrightarrow \left(+\tfrac{1}{4}\theta^2\right)\hat{I}_{1-}\hat{I}_{2\alpha}\hat{I}_{3\alpha}.$$

Both of the transfers give rise to a peak at

$$\{\omega_1, \omega_2\} = \{-[\Omega_1 - \pi J_{12} - \pi J_{13}], [\Omega_1 - \pi J_{12} - \pi J_{13}]\};$$

this is a component of the diagonal-peak multiplet which lies directly on the $\omega_1 = -\omega_2$ diagonal.

Working through the same calculation starting with the operator $\hat{I}_{1-}\hat{I}_{2\alpha}\hat{I}_{3\alpha}$ we find that it gives rise to the *same* contributions to both the cross- and diagonal-peak multiplets as did $\hat{I}_{1+}\hat{I}_{2\alpha}\hat{I}_{3\alpha}$, with the sole exception that the frequencies evolving during t_1 are of opposite sign; the amplitudes all go as $\frac{1}{4}\theta^2$. Therefore, all of the signals have cosine modulation in t_1, and so we can obtain an absorption mode lineshape.

ZCOSY is superior to small flip angle COSY first because the cross and diagonal peaks have similar intensities, and secondly because the spectra can be processed in such a way as to obtain absorption mode lineshapes. One practical difficulty is that it is vital to ensure that only population terms present between the two small flip angle pulses contribute to the observed signal; how this can be achieved effectively is discussed in section 11.15 on page 426.

10.6 HMBC

The HMBC experiment was discussed, for the case of two spins, in section 8.9 on page 215. In the present section we are going to investigate the form of the HMBC spectrum for a three-spin system containing two protons and one heteronucleus, such as ^{13}C. The two protons (I_1 and I_2) are coupled and, for simplicity, we will assume that only I_1 is long-range coupled to ^{13}C (the S spin). The topology of the spin system is shown in Fig. 10.17.

The pulse sequence for the HMBC experiment is shown in Fig. 10.18. As usual, we will start our analysis with equilibrium magnetization on spin I_1; we do not need to consider the magnetization on spin I_2 as this spin is not coupled to the heteronucleus. The 90° pulse generates $-\hat{I}_{1y}$, and then this term evolves for time τ under the influence of the offset of spin I_1, and the couplings of this spin to spins I_2 and S.

The result is rather a lot of terms, but based on the analysis for a two-spin system, we know that only those which are anti-phase with respect to the heteronucleus, the S spin, will be transformed into multiple quantum by the next S spin pulse. Therefore we will discard all of the other terms, leaving:

$$+ \sin(\pi J_{I_1 S}\tau)\cos(\pi J_{12}\tau)\cos(\Omega_{I_1}\tau)\, 2\hat{I}_{1x}\hat{S}_z$$
$$+ \sin(\pi J_{I_1 S}\tau)\cos(\pi J_{12}\tau)\sin(\Omega_{I_1}\tau)\, 2\hat{I}_{1y}\hat{S}_z$$
$$- \sin(\pi J_{I_1 S}\tau)\sin(\pi J_{12}\tau)\sin(\Omega_{I_1}\tau)\, 4\hat{I}_{1x}\hat{I}_{2z}\hat{S}_z$$
$$+ \sin(\pi J_{I_1 S}\tau)\sin(\pi J_{12}\tau)\cos(\Omega_{I_1}\tau)\, 4\hat{I}_{1y}\hat{I}_{2z}\hat{S}_z.$$

Note that all the terms include the factor $\sin(\pi J_{I_1 S}\tau)$ which comes from the evolution needed to create magnetization which is anti-phase with respect to the I_1–S coupling. For simplicity, these factors which depend on τ will be written $A_1, A_2 \ldots$, so that the situation at the end of τ is:

$$A_1\, 2\hat{I}_{1x}\hat{S}_z + A_2\, 2\hat{I}_{1y}\hat{S}_z + A_3\, 4\hat{I}_{1x}\hat{I}_{2z}\hat{S}_z + A_4\, 4\hat{I}_{1y}\hat{I}_{2z}\hat{S}_z.$$

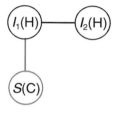

Fig. 10.17 The arrangement of spins considered in this discussion of the HMBC experiment. Spins I_1 and I_2 are protons, and are coupled, but only I_1 is (long-range) coupled to the S spin, ^{13}C. The coupling between the two I spins will be denoted J_{12}, and that between I_1 and S will be denoted $J_{I_1 S}$.

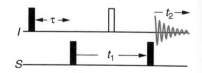

Fig. 10.18 Pulse sequence for the HMBC experiment.

The 90° pulse to the S spin simply changes $\hat{S}_z$ to $-\hat{S}_y$:

$$-A_1\,2\hat{I}_{1x}\hat{S}_y - A_2\,2\hat{I}_{1y}\hat{S}_y - A_3\,4\hat{I}_{1x}\hat{I}_{2z}\hat{S}_y - A_4\,4\hat{I}_{1y}\hat{I}_{2z}\hat{S}_y.$$

This brings us to the start of t_1. What we have here is a combination of various different types of heteronuclear multiple-quantum coherence between I_1 and S i.e. the long-range coupled proton and ^{13}C.

The 180° pulse placed in the middle of t_1, and applied to the I spins, will refocus the offset of I_1, so we do not need to consider its evolution. Furthermore, from the discussion in section 7.12.3 on page 176 we know that the evolution of multiple quantum between I_1 and S is not affected by the coupling between these spins, so we do not need to consider the evolution of J_{I_1S}.

The coupling between I_1 and I_2 (the two protons) *will* affect the evolution during t_1. However, we are going to ignore this evolution for two reasons. First, for the typical resolution that can be achieved in the ω_1 dimension of an HMBC spectrum (remember that it is the range of ^{13}C shifts which have to be covered in this dimension), it is unlikely that the splittings due to proton–proton couplings will be resolved. Secondly, ignoring the evolution of J_{12} simplifies the calculation considerably.

Therefore during t_1 we only need to consider the evolution due to the offset of the S spin (the heteronucleus). The result of this is

$$\cos\left(\Omega_S t_1\right)\left[-A_1\,2\hat{I}_{1x}\hat{S}_y - A_2\,2\hat{I}_{1y}\hat{S}_y - A_3\,4\hat{I}_{1x}\hat{I}_{2z}\hat{S}_y - A_4\,4\hat{I}_{1y}\hat{I}_{2z}\hat{S}_y\right]$$
$$+ \sin\left(\Omega_S t_1\right)\left[A_1\,2\hat{I}_{1x}\hat{S}_x + A_2\,2\hat{I}_{1y}\hat{S}_x + A_3\,4\hat{I}_{1x}\hat{I}_{2z}\hat{S}_x + A_4\,4\hat{I}_{1y}\hat{I}_{2z}\hat{I}_{3x}\right].$$

We must not forget the 180° pulse to the I spins, which will simply change the sign of the operators $\hat{I}_{1y}$ and $\hat{I}_{2z}$:

$$\cos\left(\Omega_S t_1\right)\left[-A_1\,2\hat{I}_{1x}\hat{S}_y + A_2\,2\hat{I}_{1y}\hat{S}_y + A_3\,4\hat{I}_{1x}\hat{I}_{2z}\hat{S}_y - A_4\,4\hat{I}_{1y}\hat{I}_{2z}\hat{S}_y\right]$$
$$+ \sin\left(\Omega_S t_1\right)\left[A_1\,2\hat{I}_{1x}\hat{S}_x - A_2\,2\hat{I}_{1y}\hat{S}_x - A_3\,4\hat{I}_{1x}\hat{I}_{2z}\hat{S}_x + A_4\,4\hat{I}_{1y}\hat{I}_{2z}\hat{S}_x\right].$$

The final pulse to the S spin makes some of these multiple-quantum terms observable on I_1. It is clear that as this pulse is about x, only the terms containing the operator $\hat{S}_y$ will become observable i.e. those in the first square brace:

$$\cos\left(\Omega_S t_1\right)\left[-A_1\,2\hat{I}_{1x}\hat{S}_z + A_2\,2\hat{I}_{1y}\hat{S}_z + A_3\,4\hat{I}_{1x}\hat{I}_{2z}\hat{S}_z - A_4\,4\hat{I}_{1y}\hat{I}_{2z}\hat{S}_z\right].$$

All four of these terms contribute to the spin-one multiplet, and they are all modulated at the offset of the S spin as a function of t_1. We will therefore see a two-dimensional multiplet centred at $\{\Omega_S, \Omega_1\}$ which, although only having one frequency in ω_1, has a complex multiplet structure in ω_2.

All of the contributions to the spin-one multiplet are anti-phase with respect to the coupling to the heteronucleus, the S spin. There are, in addition, contributions which are both in-phase and anti-phase with respect to the coupling to I_2, the other proton. Finally, all of these different contributions appear along both the x- and y-axes, in a complex mixture which depends on the factors A_i.

Figure 10.19 on the facing page shows examples of the I_1 spin multiplet structure that would be observed in the ω_2 dimension of an HMBC; the

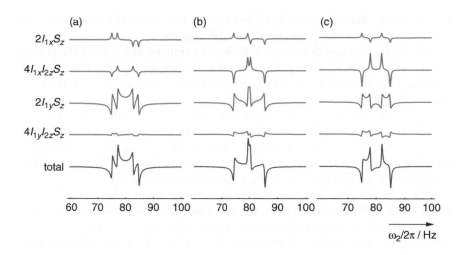

Fig. 10.19 Illustration of typical multiplet structures in the ω_2 dimension expected for HMBC cross-peak multiplets from the spin system shown in Fig. 10.17 on page 347. In each case, the contributions from the four product operators which contribute to the multiplet are shown separately, as well as their sum (marked 'total'), which is what would actually be observed in the spectrum. Note that the contribution from each operator depends on the values of the coupling constants and the offset. It has been assumed that x-magnetization will give the absorption lineshape. For all three multiplets, the offset of I_1 has been taken as 80 Hz, the linewidth is 0.5 Hz, and the delay τ is 40 ms. In (a) the coupling constants are $J_{12} = 2$ Hz, $J_{I_1S} = 7.5$ Hz; in (b) they are $J_{12} = 5$ Hz, $J_{I_1S} = 6$ Hz; and in (c) they are $J_{12} = 7$ Hz, $J_{I_1S} = 3$ Hz.

diagram also shows the individual contributions from the four operators. It is clear from this diagram that the multiplets have complex phase properties which will certainly defeat any attempt to measure the value of the long-range heteronuclear coupling, J_{I_1S}.

10.7 Sensitivity-enhanced experiments

Sensitivity is always at a premium in NMR spectroscopy, so those developing new multiple-pulse experiments must always pay close attention to ensuring that as much of the original equilibrium magnetization as possible ends up contributing to the observed signal. Inevitably, along the way magnetization will be lost due to relaxation, or due to delays not being at their optimum values, such as $1/(2J)$. It is also important to make sure that we are not losing magnetization by poor design of the pulse sequence.

It turns out that many two-dimensional experiments have a feature which results in a loss of sensitivity: this is that, at the end of t_1 only *one* component of the magnetization is transferred by the mixing period into observable signals. Generally, at the end of t_1 the magnetization will be somewhere in the xy-plane, such that it has a component along one axis (say x) which goes as $\cos(\Omega_1 t_1)$, and a component along the orthogonal axis (y) which goes as $\sin(\Omega_1 t_1)$. What usually happens is that only one of these components is transferred into observable magnetization; the other is discarded. Sensitivity is therefore lost as not all of the magnetization present at the end of t_1 leads to observable signals.

Of course, we can determine whether we transfer the sine or the cosine component by altering the pulse sequence, usually by shifting the phase of a pulse. However, this does not improve the situation as it is still the case that only *one* component is transferred.

For some types of experiments – principally heteronuclear ones – it turns out that a suitable modification of the pulse sequence will allow *both* components to be transferred. As a result, the signal-to-noise ratio of the spectrum can be increased by up to a factor of $\sqrt{2}$. Such modified experiments are described as being *sensitivity enhanced* (SE).

We will describe how this sensitivity-enhancement scheme can be applied to the HSQC experiment for the case of a two-spin system. The same approach is used in the more complex pulse sequences used to record three- and four-dimensional experiments for the study of labelled proteins and nucleic acids.

10.7.1   Sensitivity-enhanced HSQC

The HSQC experiment was described in section 8.7 on page 209, and its pulse sequence is given in Fig. 8.22 (b) on page 210. At the end of t_1 (period C), we showed that the following operators were present:

$$\cos(\Omega_S t_1)\sin(2\pi J_{IS}\tau_1)\,2\hat{I}_z\hat{S}_y - \sin(\Omega_S t_1)\sin(2\pi J_{IS}\tau_1)\,2\hat{I}_z\hat{S}_x.$$

We see that we have an anti-phase term along y, with cosine modulation as a function of t_1, and an anti-phase term along $-x$ with sine modulation. The subsequent $90°$ pulses (about x) transfer the first term to anti-phase on the I spin, but turn the second term into unobservable multiple-quantum coherence

$$-\cos(\Omega_S t_1)\sin(2\pi J_{IS}\tau_1)\,2\hat{I}_y\hat{S}_z + \sin(\Omega_S t_1)\sin(2\pi J_{IS}\tau_1)\,2\hat{I}_y\hat{S}_x.$$

As we commented on above, only one of the components (here the cosine modulated term) present at the end of t_1 ends up being observed. If we change the phase of the $90°$ pulse to the S spin at the end of t_1 from x to y, it will be the sine modulated component which is transferred to the I spin, but the cosine component will be transferred into unobservable muliple-quantum coherence.

The modified HSQC pulse sequence shown in Fig. 10.20 on the next page achieves transfer of *both* components present at the end of t_1. We will first describe what happens to each component, and once we have done this the general idea of how such a sequence works should become clear.

For simplicity we will assume that the delays τ_1 and τ_2 have their optimum values of $1/(4J_{IS})$; this means that during the spin echoes there will be complete conversion of in-phase to anti-phase, or vice versa. Up to the end of t_1 the pulse sequence is the same as a conventional HSQC, so from our previous analysis of that experiment we have, at point **a**, two anti-phase terms:

$$\text{point } \mathbf{a}:\quad \cos(\Omega_S t_1)\,2\hat{I}_z\hat{S}_y - \sin(\Omega_S t_1)\,2\hat{I}_z\hat{S}_x.$$

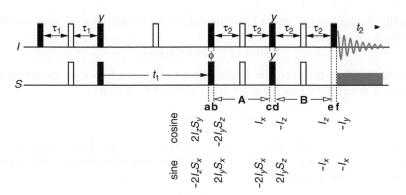

Fig. 10.20 Pulse sequence for sensitivity-enhanced HSQC. The operators present at each of the points **a–f** for both the cosine and sine modulated components are shown beneath the sequence. All of the pulses are of phase x unless otherwise noted; the second 90° pulse to the S spin has phase ϕ, which is $+x$ or $-x$ as described in the text.

First, we will follow the fate of the cosine component. Taking the phase ϕ of the second S spin 90° pulse to be $+x$, we find at point **b** that the anti-phase magnetization has been transferred to anti-phase on spin I:

$$\text{point } \mathbf{b}: \quad -\cos(\Omega_S t_1)\, 2\hat{I}_y \hat{S}_z.$$

The spin echo, period **A**, results in complete conversion of this anti-phase term to in phase. As usual, we can work out the effect of the echo by ignoring the offset, allowing the coupling to evolve for time $2\tau_2$, and then applying both 180° pulses. Recalling that $\tau_2 = 1/(4J_{IS})$, the result is, at point **c**,

$$\text{point } \mathbf{c}: \quad \cos(\Omega_S t_1)\, \hat{I}_x.$$

The 90°(y) pulse to the I spin rotates this in-phase term onto the z-axis:

$$\text{point } \mathbf{d}: \quad -\cos(\Omega_S t_1)\, \hat{I}_z.$$

During period **B** this z-magnetization does not evolve, but is just inverted by the 180° pulse. Therefore at the end of this period we have

$$\text{point } \mathbf{e}: \quad +\cos(\Omega_S t_1)\, \hat{I}_z.$$

Finally, the last 90° pulse to the I spin makes this term observable as an in-phase term along the y-axis:

$$\text{point } \mathbf{f}: \quad -\cos(\Omega_S t_1)\, \hat{I}_y.$$

Now let us turn to the sine modulated component. Once more assuming that $\phi = x$, we find that at point **b** there is a state of multiple-quantum coherence:

$$\text{point } \mathbf{b}: \quad \sin(\Omega_S t_1)\, 2\hat{I}_y \hat{S}_x.$$

During the spin echo, period **A**, the offsets are refocused and, as we have seen before, the multiple-quantum coherence between the I and S spins is unaffected by the coupling between these two spins (see section 7.12.3 on page 176). So, all that happens during this period is an inversion as a result of the 180° pulse to the I spin (the $\hat{S}_x$ term is unaffected by the 180°(x) pulse):

$$\text{point } \mathbf{c}: \quad -\sin(\Omega_S t_1)\, 2\hat{I}_y \hat{S}_x.$$

The 90°(y) pulse to the S spin rotates this multiple-quantum term to anti-phase y-magnetization on the I spin:

$$\text{point } \mathbf{d}: \quad \sin(\Omega_S t_1)\, 2\hat{I}_y \hat{S}_z.$$

During the spin echo, period **B**, this anti-phase magnetization evolves into in phase along x:

$$\text{point e:} \quad -\sin(\Omega_S t_1)\,\hat{I}_x.$$

The final $90°(x)$ pulse to the I spin has no effect on $\hat{I}_x$, so we are left with in-phase magnetization, on the I spin, along the $-x$-axis:

$$\text{point f:} \quad -\sin(\Omega_S t_1)\,\hat{I}_x.$$

The overall result is that both the cosine and sine modulated signals are transferred, in a single experiment, to in-phase magnetization on the I spin. In summary, the process is

$$\begin{aligned} \text{phase } \phi = x: \quad &\cos(\Omega_S t_1)\,2\hat{I}_z\hat{S}_y - \sin(\Omega_S t_1)\,2\hat{I}_z\hat{S}_x \longrightarrow \\ &\qquad\qquad -\cos(\Omega_S t_1)\,\hat{I}_y - \sin(\Omega_S t_1)\,\hat{I}_x. \end{aligned} \tag{10.9}$$

If we change the phase ϕ of the second S spin $90°$ pulse to $-x$ and work through the calculation again we will find that the cosine component changes sign, whereas the sine component does not. This is because the term $2\hat{I}_z\hat{S}_x$ present at point **a** is unaffected by this pulse.

$$\begin{aligned} \text{phase } \phi = -x: \quad &\cos(\Omega_S t_1)\,2\hat{I}_z\hat{S}_y - \sin(\Omega_S t_1)\,2\hat{I}_z\hat{S}_x \longrightarrow \\ &\qquad\qquad +\cos(\Omega_S t_1)\,\hat{I}_y - \sin(\Omega_S t_1)\,\hat{I}_x. \end{aligned} \tag{10.10}$$

Having analysed the sequence we can 'stand back' and see how it works. The cosine component is transferred to the I spin by the first pair of $90°$ pulses, the resulting anti-phase magnetization becomes in-phase during the spin echo **A**, and then this magnetization is rotated onto the z-axis. It remains there during period **B**, and is then made observable by the final pulse.

The sine component is first transferred into multiple-quantum coherence, which does not evolve during period **A**. The multiple-quantum coherence is then transferred to anti-phase on the I spin, and is finally rephased during period **B**, ending up along x so that it is unaffected by the final $90°$ pulse.

The key point is that for the cosine component the magnetization is 'stored' along z during period **B**, while the sine component rephases. Similarly, the sine component is stored as multiple quantum during period **A**, while the cosine component rephases. At the end, both components have been transferred from S to I and rephased.

In order to process the spectrum in the usual way we need to disentangle the sine and cosine modulated signals. We do this by repeating the experiment *twice* for each t_1 value, once with the phase ϕ set to $+x$, and once with the phase set to $-x$. The resulting data are kept separate.

Referring to Eqs 10.9 and 10.10, we can see that adding the result of the two experiments gives just the sine modulated term $-\sin(\Omega_S t_1)\,\hat{I}_x$, whereas subtracting the two gives just the cosine modulated term $-\cos(\Omega_S t_1)\,\hat{I}_y$. Note that these terms appear along different axes in the t_2 dimension; it will therefore be necessary to phase shift one of them by $90°$ to compensate for this. Having made all these manipulations, we have a cosine and a sine modulated data set which can be processed according to the SHR procedure, described in section 8.12.3 on page 230.

You might think, quite reasonably, that this sensitivity-enhanced experiment should improve the signal-to-noise ratio by a factor of *two*, since we are transferring *both* components of the magnetization present at the end of t_1, rather than just one of them. However, in order to be able to separate these components, we need to record two experiments, each of which comes with its own noise. When the data sets are combined, the signal increases by a factor of two, but the noise also increases, by a smaller factor of $\sqrt{2}$ (see section 2.4 on page 13). Overall, the signal-to-noise ratio therefore increases by a factor of $\sqrt{2}$.

10.7.2 Practical aspects of sensitivity-enhanced experiments

In practice we may not obtain the factor of $\sqrt{2}$ improvement in the signal-to-noise ratio: this is for two reasons. First, the sequence is longer and more complex. As a result, more magnetization will be lost due to relaxation, and there is also the possibility of further losses due to pulse imperfections.

Secondly, there will only be complete transfer of both components if the delays τ_2 are at their optimum values of $1/(4J_{IS})$. For a molecule containing a range of one-bond C–H or N–H couplings, a compromise value will have to be chosen, resulting in less that the full sensitivity gain for some resonances.

The sensitivity-enhanced HSQC experiment, and more complex sequences based on the same idea, have proved to be very popular in the NMR of large biomolecules. No doubt this is because in such systems there is such a premium on sensitivity. In addition, it turns out that selection using pulsed field gradients can be implemented into these sensitivity-enhanced sequences without further loss of sensitivity.

10.8 Constant time experiments

The 'constant time' method gives us a way of removing the splittings, in the ω_1 dimension, due to *homonuclear* couplings. Removing heteronuclear couplings is quite easy – we can use broadband decoupling, or a strategically placed $180°$ pulse to the heteronucleus – but removing homonuclear decouplings has proved to be difficult, and the constant time approach is one of the few practical ways of achieving this.

There are two reasons why we might want to remove the splittings due to homonuclear coupling. First, the collapse of multiplets to single lines will simplify the spectra, and therefore increase the effective resolution. Secondly, if all the intensity of a multiplet is concentrated into one line, the signal-to-noise ratio will increase. However, we will see that in practice although the first aim can be achieved, it is not always the case that the constant time method leads to an improvement in the signal-to-noise ratio.

The basic idea of a constant time period is illustrated in Fig. 10.21. The constant time is the period T between the dashed lines, and t_1 is the usual evolution time. Within this time T there is a $180°$ pulse, which is not placed centrally, but occurs a time $\frac{1}{2}(T - t_1)$ from the beginning. As a result, a spin echo forms at a time $(T - t_1)$ (indicated by the blue line). The remaining time until the end of the element is $T - (T - t_1) = t_1$.

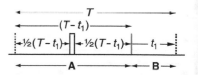

Fig. 10.21 The 'constant time' pulse sequence element, used to remove the splittings in the ω_1 dimension due to homonuclear coupling. The element occupies a fixed time period T, and contains a $180°$ pulse which forms a spin echo over the time $(T - t_1)$ (period A). As a result, the offset evolves only during period B (time t_1). In contrast, the coupling evolves for the whole time T, unaffected by the size of t_1. As a result, the evolution during t_1 depends only on the offset, so there are no splittings in the ω_1 dimension due to homonuclear coupling.

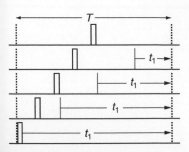

Fig. 10.22 Illustration of how the location of the 180° pulse in the constant time element changes as t_1 increases. The value $t_1 = 0$ is achieved by having the 180° pulse in the centre of the constant time T, as shown at the top. As t_1 increases, the 180° pulse moves to the left. When this pulse reaches the start of the constant time, t_1 is at its maximum value of T. An equivalent result is achieved by moving the 180° pulse to the right.

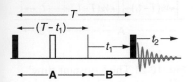

Fig. 10.23 Pulse sequence for the constant time COSY experiment.

The evolution of the offset (chemical shift) is refocused by the spin echo (period **A**), but during period **B** the offset evolves as normal. Therefore the overall result is that the offset evolves for time t_1, just as it would in a normal two-dimensional experiment.

Homonuclear coupling is *not* refocused by the spin echo, and so continues to evolve throughout the *whole* of the constant time period T. Thus, as t_1 is increased, the magnetization present at the end of the constant time is modulated by the offset, but the evolution of the coupling does not change. As a result, couplings do not modulate the signal as a function of t_1, and so there are no splittings in the ω_1 dimension.

Figure 10.22 illustrates the behaviour of the constant time element as t_1 increases. The value of $t_1 = 0$ is achieved by placing the 180° pulse in the middle of the constant time, and then t_1 is increased by moving the 180° pulse to the left (or, equivalently, to the right). The maximum value which t_1 can take is T.

10.8.1   Constant time COSY

The simplest experiment in which we can include this constant time element is COSY; the resulting pulse sequence is shown in Fig. 10.23. The sequence starts with a 90° pulse, and then the constant time period follows. At the end of the constant time we have the usual mixing pulse, followed by detection.

The analysis of this pulse sequence for a two-spin system is straightforward. Starting with equilibrium magnetization on spin one, $\hat{I}_{1z}$, the 90° pulse generates $-\hat{I}_{1y}$. As we have explained, the coupling evolves for time T, giving the following result:

$$- \cos\left(\pi J_{12} T\right) \hat{I}_{1y} + \sin\left(\pi J_{12} T\right) 2\hat{I}_{1x}\hat{I}_{2z}.$$

We need to take account of the 180°(x) pulse, which inverts the operators $\hat{I}_{1y}$ and $\hat{I}_{2z}$ to give

$$\cos\left(\pi J_{12} T\right) \hat{I}_{1y} - \sin\left(\pi J_{12} T\right) 2\hat{I}_{1x}\hat{I}_{2z}.$$

We now have to let these two terms evolve under the offset, which only acts for time t_1. This gives the following result:

$$\cos\left(\Omega_1 t_1\right) \cos\left(\pi J_{12} T\right) \hat{I}_{1y} - \sin\left(\Omega_1 t_1\right) \cos\left(\pi J_{12} T\right) \hat{I}_{1x}$$
$$- \cos\left(\Omega_1 t_1\right) \sin\left(\pi J_{12} T\right) 2\hat{I}_{1x}\hat{I}_{2z} - \sin\left(\Omega_1 t_1\right) \sin\left(\pi J_{12} T\right) 2\hat{I}_{1y}\hat{I}_{2z}.$$

The final 90° pulse results in the following two observable terms:

$$- \sin\left(\Omega_1 t_1\right) \cos\left(\pi J_{12} T\right) \hat{I}_{1x} \qquad [1]$$
$$+ \sin\left(\Omega_1 t_1\right) \sin\left(\pi J_{12} T\right) 2\hat{I}_{1z}\hat{I}_{2y}. \qquad [2]$$

Term [1] gives the diagonal peak, as it is modulated in t_1 at frequency Ω_1, and appears on spin one in t_2. In ω_2 we expect an in-phase doublet (arising from $\hat{I}_{1x}$), whereas in ω_1 there is just one modulating frequency. Therefore the diagonal-peak multiplet will consist of just two lines at $\{\Omega_1, \Omega_1 \pm \pi J_{12}\}$. Note that there is *no* splitting in the ω_1 dimension, which is just what we expect from a constant time experiment.

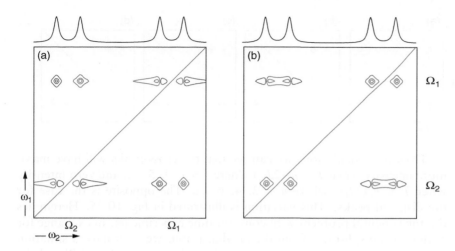

Fig. 10.24 Schematic constant time COSY spectra for a two-spin system; the absence of any splitting in the ω_1 dimension is immediately evident. In (a) the spectrum has been phased so that, in the ω_2 dimension, the cross peaks are in absorption and the diagonal peaks in dispersion. The anti-phase structure of the cross-peak multiplets is clear. In (b), the opposite phasing has been used, with the diagonal peaks in absorption and the cross peaks in dispersion. Now, the in-phase nature of the diagonal peaks is plain. The lineshape in the ω_1 dimension is the same for both diagonal and cross peaks. In these simulations, the constant time T has been set to a value which gives similar intensity for the cross and diagonal peaks; the diagonal is indicated by the blue line.

Term [2] gives the cross peak, as it appears on spin two. In contrast to the diagonal peak, the multiplet in ω_2 is in anti-phase (arising from $2\hat{I}_{1z}\hat{I}_{2y}$), but as for the diagonal peak there is one modulating frequency in ω_1. The result will be a multiplet with two anti-phase peaks at $\{\Omega_1, \Omega_2 \pm \pi J_{12}\}$.

Note that both the diagonal- and cross-peak terms have the *same* modulation, $\sin(\Omega_1 t_1)$, in t_1: they can therefore both be phased to absorption in ω_1. However, in t_2 the magnetization which gives rise to the diagonal peak appears along x, whereas that for the cross peak appears along y. The lineshapes will therefore be different in ω_2.

The equilibrium magnetization on spin two will give rise to an equivalent set of peaks at $\{\Omega_2, \Omega_2 \pm \pi J_{12}\}$ and $\{\Omega_2, \Omega_1 \pm \pi J_{12}\}$. The overall form of the spectrum is illustrated in Fig. 10.24. Note the lack of any splitting on the ω_1 dimension, and the different lineshapes of the cross and diagonal peaks in the ω_2 dimension.

Referring to terms [1] and [2] in our calculation we see that, although the value of the coupling J_{12} does not affect the frequency of the modulation in t_1, it does affect the *intensity* of the peaks via the factors $\cos(\pi J_{12}T)$ for the diagonal peak and $\sin(\pi J_{12}T)$ for the cross peak. The reason for this intensity effect is easy to see. Cross peaks only arise from magnetization which is anti-phase at the time of the final 90° pulse. The amount of anti-phase magnetization at this point depends on the evolution of the coupling prior to this pulse, which in this experiment has occurred for time T; this is the origin of the factor $\sin(\pi J_{12}T)$. Similarly, the diagonal peaks arise from in-phase magnetization present at the time of the final pulse, and the amount of this magnetization goes as $\cos(\pi J_{12}T)$.

Fig. 10.25 Schematic constant time COSY spectra of a two-spin system showing the effect of increasing the constant time T. For spectra (a)–(d) the value of T is $1/(8J_{12})$, $1/(4J_{12})$, $3/(8J_{12})$ and $1/(2J_{12})$, respectively.

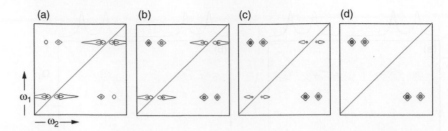

From our calculations we can see that the cross peaks will have maximum intensity when $T = n/(2J_{12})$ where $n = 1$, 3, 5 ..., and zero intensity when $T = n/(2J_{12})$ where $n = 2$, 4, 6 .... The opposite is the case for the diagonal peaks. This variation is illustrated in Fig. 10.25. Herein lies the fundamental problem with constant time experiments. In exchange for removing the splittings from the ω_1 dimension, we now have a situation where the *intensity* of the cross peak depends on the value of T and the size of the coupling constant. In a real molecule, there will inevitably be a range of coupling constants present, and so it will not be possible to choose a single value of T which will give the greatest intensity for all peaks. If we are unlucky, a cross peak could be entirely missing just because of an unfortunate choice of T.

The second problem with the constant time experiment is that the losses due to relaxation can be severe, since for *every* value of t_1 the time between the start of the experiment and the acquisition of the signal is always T. This is in contrast to normal COSY, where the time during which the magnetization decays due to relaxation starts from zero and increases steadily with t_1. Thus, the relaxation losses will be much greater in a constant time COSY than they are in a regular COSY experiment.

Linewidths in the ω_1 dimension

Constant time COSY experiments have the rather unusual feature that, in the ω_1 dimension, the linewidth is limited only by the *inhomogeneous* broadening. The way this comes about is as follows.

In the pulse sequence the presence of the spin echo means that, at the end of period **A**, the decay of the magnetization is due only to the homogeneous part of the linebroadening – any inhomogeneous decay will have been refocused (section 9.9 on page 300). Therefore, noting that the duration of period **A** is $(T - t_1)$, the magnetization present at the end of this period will be reduced by a factor

$$\exp(-R_{xy}[T - t_1]),$$

where R_{xy} is the rate constant for transverse relaxation i.e. the homogeneous decay.

During period **B**, which is of duration t_1, the magnetization will decay due to both the homogeneous and inhomogenous contributions. This introduces a further factor

$$\exp(-[R_{xy} + R_{\text{inhom}}]t_1),$$

where R_{inhom} is the rate constant for the inhomogeneous decay.

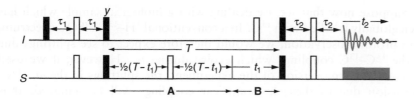

Fig. 10.26 Pulse sequence for the constant time HSQC experiment; this sequence should be compared with that for conventional HSQC shown in Fig. 8.22 (b) on page 210. The usual t_1 evolution has been replaced by the constant time period of Fig. 10.21 on page 353, so as a result in the ω_1 dimension there are no splittings due to homonuclear couplings. This is a useful feature for the case where the S spin is ^{13}C and where the sample is uniformly labelled with ^{13}C. Note that, as in conventional HSQC, a 180° pulse to the I spin is placed in the middle of t_1. This pulse is needed to refocus the evolution of the I–S coupling during t_1.

Thus, at the end of the constant time the magnetization will have been reduced by a factor

$$\exp\left(-R_{xy}[T - t_1]\right) \times \exp\left(-[R_{xy} + R_{\text{inhom}}]t_1\right) = \exp\left(-R_{xy}T\right)\exp\left(-R_{\text{inhom}}t_1\right).$$

The decay as a function of t_1, which is what will determine the linewidth in the ω_1 dimension, is therefore determined *only* by R_{inhom}. The homogeneous decay does affect the overall intensity, via the term $\exp\left(-R_{xy}T\right)$, but this decay does not affect the linewidth.

If well adjusted ('shimmed'), modern NMR magnets give exceptionally homogeneous fields, and so the inhomogeneous contribution to the linewidth is very small. As a result, the linewidth in the ω_1 dimension of a constant time experiment can be very small. In practice, though, the linewidth in this dimension is likely to be limited by the maximum value which t_1 can reach.

10.8.2 Constant time HSQC

Constant time COSY experiments have never proved to be particularly popular. However, in the area of biomolecular NMR, where the proteins and nucleic acids which are being studied have been globally labelled with ^{13}C and ^{15}N, the constant time element is often used as part of the complex three- and four-dimensional pulse sequences. We will illustrate such applications by describing a constant time HSQC experiment, the pulse sequence for which is shown in Fig. 10.26.

The constant time element is inserted between the two 90° pulses which transfer the magnetization to the S spin, and the two 90° pulses which transfer the magnetization back to the I spin. As before, the offset of the S spin is refocused over the period A, but evolves during period B, which is t_1.

In this heteronuclear experiment, the S spin 180° pulse in the middle of period A, and the I spin 180° pulse in the middle of period B, refocus the evolution of the heteronuclear coupling over these two times. Overall, for a two-spin system the appearance of the constant time HSQC experiment will be identical to that of the normal HSQC.

Suppose now that we are dealing with a biological sample which has been globally labelled with ^{13}C. In a conventional ^{1}H–^{13}C HSQC spectrum, with proton observation, we would therefore expect to see splittings due to the ^{13}C–^{13}C couplings which evolve during t_1. However, if we used the constant time version of the experiment, the splittings in the ω_1 (^{13}C) dimension due to these homonuclear couplings will be removed, thus simplifying the spectrum.

At the end of the constant time period, any ^{13}C magnetization which is anti-phase with respect to ^{13}C–^{13}C couplings will *not* be transferred back to proton, but will be transferred to other carbons or into multiple-quantum coherence by the final 90° pulse to the S spin (^{13}C). The magnetization which is in-phase with respect to the C–C couplings will be transferred to proton and, since the ^{13}C–^{13}C coupling has evolved for the whole of the constant time T, these in-phase terms will go as $\cos(\pi J_{CC} T)$. It is therefore important to choose T so as to maximize this term, which means that $\pi J_{CC} T = \pi, 2\pi \ldots$, i.e. $T = n / J_{CC}$ where $n = 1, 2, 3 \ldots$. The largest couplings present are the one-bond ^{13}C–^{13}C couplings. These do not vary that much with structure, so it is possible to find a value of T which is a reasonable compromise.

A constant time HSQC would be a completely pointless experiment for a natural abundance sample, in which there is a very small probability of finding two ^{13}C nuclei in one molecule.

10.9   TROSY

In section 9.11.2 on page 308 we described how cross correlation between CSA and dipolar relaxation can result in the two lines of a doublet having different linewidths. The effect is particularly pronounced for ^{15}N–^{1}H pairs in large molecules when the spectra are recorded at high field. In such cases, it is not uncommon for there to be a twenty-fold difference in the widths of the two lines of the doublet, both in the ^{15}N and ^{1}H spectra.

In conventional HSQC spectra it is usual to collapse, in both dimensions, the splittings due to the heteronuclear couplings. In ω_1 this is achieved by the 180° pulse applied to the I spins in the middle of t_1; in ω_2, the splittings are removed by observing the signal in the presence of broadband decoupling of the S spins.

As was explained in section 9.11.2 on page 308, collapsing the splittings in this way results in a line whose width is the average of the widths of the two lines of the doublet. In the case that one line is much broader than the other, the result will be a considerable reduction in peak height when the decoupled line is compared with the sharp line of the doublet (see Fig. 9.41 on page 309), and hence a reduction in the signal-to-noise ratio.

These observations lead to the idea that, in cases where cross-correlation effects are substantial, it is best *not* to remove the splittings due to the heteronuclear couplings. We will then obtain a higher signal-to-noise ratio on account of the greater peak height of the sharp line. This is the fundamental idea behind TROSY.

Experimentally, it is easy to modify the HSQC sequence so as to retain the splittings in each dimension. All we need to do is to remove (i) the

180° pulse to the I spin which is applied in the middle of t_1, and (ii) the broadband decoupling of the S spin during acquisition. The resulting pulse sequence is shown in Fig. 10.27.

Rather than giving a single peak at $\{\Omega_S, \Omega_I\}$, this modified sequence gives a multiplet centred at this frequency and split by J_{IS} in each dimension; all of the peaks are in phase. An example of such a multiplet is shown in Fig. 10.28 for two different cases.

In (a) the width of the two lines of the doublet are the same, and we see the familiar pattern of four peaks. However in (b) one line of the doublet (in both dimensions) has been made *ten* times broader than the other. The four components of the multiplet now all have different combinations of the linewidths in each dimension.

One peak (here the top right) is narrow in each dimension, two peaks are broad one way and sharp the other: they are just visible in the top left and bottom right positions. The final peak is broad in each dimension, and is invisible at the contour levels chosen. As the broad lines increase in width, the intensity of all but the top right-hand peak decreases, and in the limit only this peak is seen.

However, it is not always the case that the difference in the linewidths is such that only one out of the four peaks is seen. The presence of the other three peaks can cause confusion and crowding of the spectrum, so it is necessary to devise experiments in which all the unwanted peaks

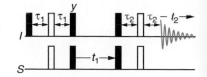

Fig. 10.27 Modified HSQC pulse sequence in which the IS coupling is retained in each dimension. Compared with conventional HSQC, Fig. 8.22 (b) on page 210, the changes are simply the removal of the 180° pulse to I in the middle of t_1, and the omission of broadband decoupling of the S spin during t_2.

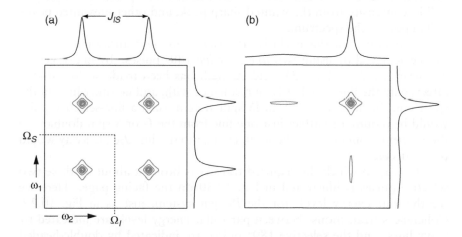

Fig. 10.28 Schematic multiplets, as would be recorded using the modified HSQC experiment of Fig. 10.27; note that the coupling is retained in both dimensions. In (a) the linewidths of all the peaks are the same, and we see the familiar square array of peaks. In (b) the width of one of the lines of each doublet has been increased by a factor of ten; the one-dimensional spectra plotted at the edges of the two-dimensional plots illustrate clearly the large reduction in the height of the broad peak. Of the four components of the multiplet, one – that at the top right – is unaffected as it is still narrow in each dimension. Two of the peaks are broad in one dimension and narrow in the other: they are just visible in the contour plot. The fourth peak is broad in each dimension, and is invisible in this contour plot. A cross peak of the type shown in (a) is characteristic of a small molecule, whereas that shown in (b) would be expected for a large molecule.

are suppressed deliberately. In the next section, we turn to how these experiments are designed.

10.9.1   Line-selective transfer

The key to designing a TROSY experiment is to understand the relationship between the four lines of the two-dimensional multiplet and the energy level diagram of a two-spin system: this is illustrated in Fig. 10.29. At the top of the figure are shown the four energy levels of a two-spin system, labelled with the spin states of each spin, the state of the I spin being given first. Transitions 1–3 and 2–4 involve flipping the I spin, and so are the two lines of the I-spin doublet which appear in the ω_2 dimension. Transitions 1–2 and 3–4 are the S-spin transitions, and correspond to the S-spin doublet which appears in the ω_1 dimension.

In the schematic two-dimensional multiplet, shown in the lower part of the figure, the ω_1 frequency of the sharp peak is that of the 3–4 transition, and in ω_2 the sharp peak is at the frequency of the 2–4 transition. Therefore we see that the sharp line arises from a transfer from a *specific* line of the S-spin doublet (here 3–4), to a specific line of the I-spin doublet (here 2–4).

If we want to devise an experiment in which *only* this sharp peak appears, what we need to do is cause the *selective* and *exclusive* transfer of coherence from the 3–4 transition to the 2–4 transition. It is important that the transfer is just between these two transitions: if it spreads elsewhere we will lose intensity from the wanted sharp peak, and other unwanted peaks will appear in the spectrum.

It turns out that this exclusive transfer from one transition to another can be achieved by using two *line-selective* 180° pulses. By line-selective, we mean a pulse whose RF field strength has been made so low that it affects only the line which it is on resonance with, and no other lines in the spectrum. In the context of an IS spin system, such a line-selective pulse would be required to affect just one line from the I- or S-spin doublet, so the RF field would have to be weak enough that a line J_{IS} Hz away would be unaffected.

The way in which the required transfer is brought about by these two selective pulses is illustrated in Fig. 10.30 on the facing page. Here we see the four energy levels for the IS spin system, just as in Fig. 10.29. Coherences (transitions) between particular energy levels are indicated by wavy lines, and the selective 180° pulses are indicated by double-headed arrows.

In Fig. 10.30 (a) we start out with coherence between levels 3 and 4 i.e. one of the lines of the S-spin doublet. The first 180° pulse is applied at the frequency of transition 1–3, which is one of the lines of the I-spin doublet: note that the transition (the wavy line), and the pulse (the double-headed arrow) share a common energy level, level 3. It turns out that the effect of this 180° pulse is to transfer the coherence from 3–4 to 1–4, as shown in the middle set of energy levels. We can think of this process as the pulse 'moving' the end of the curly line from level 3 to level 1. Just exactly why these line-selective 180° pulses work in this way is most readily appreciated using *single transition operators*; see the Further reading section at the end of the chapter for appropriate references.

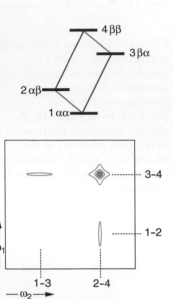

Fig. 10.29 At the top are shown the four energy levels of a two-spin system. Each level is labelled according to the spin states of the two spins I and S; for example, level 2 is labelled $\alpha\beta$, which means that spin I is in the α state, and spin S is in the β state. The two S-spin transitions, 1–2 and 3–4, are shown in blue: these correspond to the two lines of the S-spin doublet. Similarly, the two lines of the I-spin doublet correspond to the transitions 1–3 and 2–4, which are shown in dark grey. The lower part of the diagram shows a typical multiplet which would be observed in an I–S correlation spectrum. The ω_1 and ω_2 frequencies of the lines in this multiplet correspond to the transitions in the energy level diagram, as shown. In this case, the sharp peak in the multiplet arises from the transfer from the S-spin transition 3–4 to the I-spin transition 2–4.

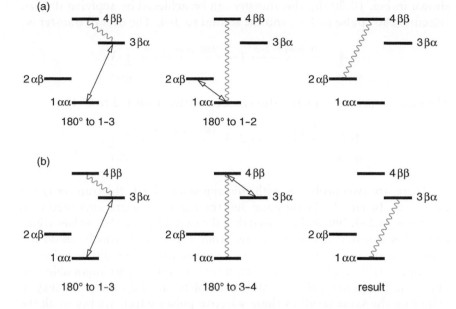

Fig. 10.30 Illustration of how coherence transfer from one transition to another can be brought about by the successive application of two line-selective 180° pulses. The diagram shows the four energy levels of an *IS* spin system, as in Fig. 10.29 on the preceding page. Coherences corresponding to particular transitions are indicated by wavy lines, and the selective 180° pulses are indicated by double-headed arrows. In (a) we see the transfer from 3–4 to 2–4 by application of selective pulses at the frequencies of transitions 1–3 and then 1–2. The effect of the first pulse is to 'move' the end of the curly line from energy level 3 down to level 1; in the same way, the second pulse moves the end of the curly line from level 1 to level 2. Shown in (b) is the transfer from 3–4 to 1–3 brought about by selective 180° pulses applied at the frequency of transition 1–3 and then at that of 3–4.

The second 180° pulse is applied at the frequency of transition 1–2, which is one of the lines of the *S*-spin doublet. As is shown in the diagram, this 'moves' the end of the curly line from level 1 to level 2, so the coherence ends up between levels 2 and 4 i.e. one of the lines of the *I*-spin doublet.

The overall effect can be summarized:

$$\underbrace{3\text{–}4}_{\text{spin } S} \xrightarrow{\;180° \text{ pulse to } 1\text{–}3\;} 1\text{–}4 \xrightarrow{\;180° \text{ pulse to } 1\text{–}2\;} \underbrace{2\text{–}4}_{\text{spin } I}.$$

Using this diagrammatic approach, we can show that this sequence of two selective 180° pulses will also cause the transfer

$$\underbrace{1\text{–}2}_{\text{spin } S} \xrightarrow{\;180° \text{ pulse to } 1\text{–}3\;} 2\text{–}3 \xrightarrow{\;180° \text{ pulse to } 1\text{–}2\;} \underbrace{1\text{–}3}_{\text{spin } I}.$$

Therefore, coherence associated with each line of the *S*-spin doublet ends up on a particular line of the *I*-spin doublet.

It may be that we wish the transfer to go to the other line of the *I*-spin doublet, so instead of 3–4 going to 2–4, we want it to go to 1–3. As is shown in Fig. 10.30 (b), this transfer can be achieved by applying the first selective 180° pulse to 1–3, and the second to 3–4. The overall transfer is

$$\underbrace{3\text{–}4}_{\text{spin } S} \xrightarrow{\text{180° pulse to 1–3}} 1\text{–}4 \xrightarrow{\text{180° pulse to 3–4}} \underbrace{1\text{–}3}_{\text{spin } I}.$$

The same sequence of pulses also causes transfer from 1–2 to 2–4

$$\underbrace{1\text{–}2}_{\text{spin } S} \xrightarrow{\text{180° pulse to 1–3}} 2\text{–}3 \xrightarrow{\text{180° pulse to 3–4}} \underbrace{2\text{–}4}_{\text{spin } I}.$$

There are two problems with this approach. First, the transfer is not exclusive. In Fig. 10.29 on page 360 the transfer we are interested in is from 3–4 to 2–4, but we have seen that the two selective 180° pulses which cause this transfer will also transfer from 1–2 to 1–3, which we do not want. The second problem is that in a sample with many different *IS* spin systems, it will be extremely inconvenient – if not next to impossible – to apply these selective pulses to all the doublets. Luckily, there is a way of achieving the same result as these selective pulses which applies to all the *IS* spin systems in the sample at the same time; we describe this approach in the next section.

10.9.2   Implementation of line-selective 180° pulses

Imagine that we start with the operator $\hat{I}_z$, apply a 90°(y) pulse to the *I* spin, and then observe the result. The pulse will generate the term $\hat{I}_x$, which gives an in-phase doublet, the two lines of which correspond to the transitions 1–3 and 2–4, as is shown in Fig. 10.31.

Now suppose we start with the operator $2\hat{I}_z\hat{S}_z$, and once again apply a 90°(y) pulse to the *I* spin. This time the pulse will generate $2\hat{I}_x\hat{S}_z$, which corresponds to an anti-phase doublet in which one of the lines is positive and one is negative, as is shown in the figure. We can say, therefore, that in going from $\hat{I}_z$ to $2\hat{I}_z\hat{S}_z$ one of the lines has been inverted, i.e. it has experienced a line-selective 180° pulse. Similarly, if we start with the state $-2\hat{I}_z\hat{S}_z$, we will also find an anti-phase doublet, but this time it is the other line which has been inverted.

Overall then, if we can find a pulse sequence which takes us from $\hat{I}_z$ to $2\hat{I}_z\hat{S}_z$, this will correspond to a 180° pulse to *one* of the lines of the *I*-spin doublet. Similarly, a sequence which takes us from $\hat{I}_z$ to $-2\hat{I}_z\hat{S}_z$ will correspond to a 180° pulse to the other line.

Such a pulse sequence is shown in Fig. 10.32 (a) on the facing page. We recognize that this sequence is a simple spin echo, flanked by a 90°(x) pulse to the *I* spin on one side, and a 90°($\pm y$) pulse to the *I* spin on the other side. The delay τ is set to $1/(4J_{IS})$, so the spin echo causes complete interconversion of in-phase and anti-phase magnetization.

Starting with $\hat{I}_z$, the first pulse generates $-\hat{I}_y$. During the spin echo of total duration $1/(2J_{IS})$, $-\hat{I}_y$ evolves into the anti-phase state $2\hat{I}_x\hat{S}_z$, and the 180° pulses change this to $-2\hat{I}_x\hat{S}_z$. The final 90° pulse, if it is about $+y$, generates $2\hat{I}_z\hat{S}_z$, or if it is about $-y$, generates $-2\hat{I}_z\hat{S}_z$.

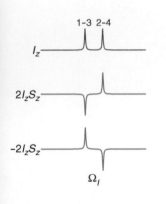

Fig. 10.31 Illustration of the form of the *I*-spin doublet which would arise from the application of an *I* spin 90°(y) pulse to the operators $\hat{I}_z$, $2\hat{I}_z\hat{S}_z$ and $-2\hat{I}_z\hat{S}_z$. For the latter two operators, one of the two lines of the doublet has been inverted. It therefore follows that the transformation $\hat{I}_z \rightarrow \pm 2\hat{I}_z\hat{S}_z$ can be thought of as being due to a selective 180° pulse to one of the lines of the *I*-spin doublet.

Thus, overall the sequence of Fig. 10.32 (a) achieves the transformation $\hat{I}_z \rightarrow \pm 2\hat{I}_z\hat{S}_z$, with the sign depending on the phase of the final pulse. The sequence is therefore equivalent to a selective 180° pulse to one of the lines of the I-spin doublet; which line is inverted depends on the phase of the last pulse. A similar analysis shows that sequence (b) has the same effect on the lines of the S-spin doublet.

These sequences are very useful as, due to the presence of the spin echo, their effect is *independent* of the offset. So, all IS spin systems in the sample, regardless of their offsets, experience the appropriate line-selective 180° pulses. The slight difficulty is that the sequences only work as described if $\tau = 1/(4J_{IS})$. If there is a range of values for the coupling constant J_{IS}, a compromise value for τ must be chosen, and as a result the inversion will not be perfect for all the spin systems.

10.9.3 A TROSY HSQC sequence

We can now include these two selective pulses into an HSQC-type sequence, to give the experiment whose pulse sequence is shown in Fig. 10.33. The sequence starts out, as in conventional HSQC, with magnetization being transferred from the I to the S spin. There is no I-spin 180° pulse placed in the middle of t_1 so that, as explained above, the splitting due to the I–S coupling is retained in ω_1.

At the end of t_1 the transfer back to the I spin is achieved by two selective 180° pulses, which are implemented using the sequences of Fig. 10.32. Period **A** is a 180° pulse to one of the I-spin transitions, either 1–3 or 2–4, depending on the phase ϕ_I. Period **B** is a 180° pulse to one of the S-spin transitions, either 1–2 or 3–4, depending on the phase ϕ_S. After these two periods, the magnetization is back on the I spin, where it is observed.

As was explained in section 10.9.1 on page 360, these line-selective 180° pulses do not achieve the exclusive transfer between two transitions. In addition, as it stands the pulse sequence will not produce data which

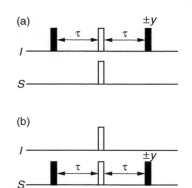

Fig. 10.32 Pulse sequences which achieve line-selective 180° pulses to: (a) one of the lines of the I-spin doublet; and (b) one of the lines of the S-spin doublet. Which line is inverted depends on the phase of the final pulse. The delay τ must be set to $1/(4J_{IS})$.

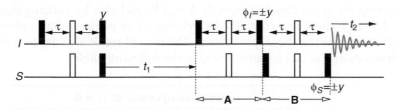

Fig. 10.33 HSQC-type pulse sequence using line-selective 180° pulses to implement the transfer from the S to the I spin. Up to the end of t_1, the sequence is the same as conventional HSQC, with the exception that there is no I-spin 180° pulse in the middle of t_1. Period **A** is a selective 180° pulse to one of the transitions of the I-spin doublet (i.e. 1–3 or 2–4, depending on the phase ϕ_I), implemented using the sequence of Fig. 10.32 (a). Period **B** is a 180° pulse to one of the transitions of the S-spin doublet (i.e. 1–2 or 3–4, depending on the phase ϕ_S); the pulse is implemented using the sequence of Fig. 10.32 (b). The overall effect of these two selective pulses is to transfer the magnetization to the I spin. The optimum value for τ is $1/(4J_{IS})$. Further processing, as described in the text, is needed to generate a spectrum in which only one peak is present in the two-dimensional multiplet.

can be processed to give an absorption mode spectrum. To get round these problems, it turns out that we need to repeat the experiment with different values of the phases ϕ_I and ϕ_S, and then combine the data in a fairly involved way. The details of exactly how this is done are described in the next section.

10.9.4   Processing the TROSY HSQC spectrum

Assuming that $\tau = 1/(4J_{IS})$, at the end of t_1 we have the following four terms

$$+\cos(\Omega_S t_1)\sin(\pi J_{IS} t_1)\,\hat{S}_x + \sin(\Omega_S t_1)\cos(\pi J_{IS} t_1)\,2\hat{I}_z\hat{S}_x$$
$$+\sin(\Omega_S t_1)\sin(\pi J_{IS} t_1)\,\hat{S}_y - \cos(\Omega_S t_1)\cos(\pi J_{IS} t_1)\,2\hat{I}_z\hat{S}_y.$$

To find the frequencies which are modulating t_1 we need to combine the trigonometric terms in the usual way. When we do this, four trigonometric factors keep appearing which, for brevity, we will replace with the following symbols:

$$c_+ = \cos([\Omega_S + \pi J_{IS}]t_1) \qquad c_- = \cos([\Omega_S - \pi J_{IS}]t_1)$$
$$s_+ = \sin([\Omega_S + \pi J_{IS}]t_1) \qquad s_- = \sin([\Omega_S - \pi J_{IS}]t_1).$$

The frequencies in the square brackets are just those of the two lines of the S-spin doublet.

Using these replacements, the four terms at the end of t_1 can be written

$$+\tfrac{1}{2}(s_+ - s_-)\,\hat{S}_x + \tfrac{1}{2}(s_+ + s_-)\,2\hat{I}_z\hat{S}_x$$
$$+\tfrac{1}{2}(-c_+ + c_-)\,\hat{S}_y + \tfrac{1}{2}(-c_+ - c_-)\,2\hat{I}_z\hat{S}_y. \tag{10.11}$$

We will drop the factor of $\tfrac{1}{2}$, as it makes no difference to the final result.

Each of these four terms gives rise to an observable I-spin operator. Working through all the details is rather tedious, but is made easier by recognizing the spin echoes in periods **A** and **B**, and also by noticing that, if $\tau = 1/(4J_{IS})$, there is a complete interchange of in-phase and anti-phase magnetization during these spin echoes. The overall result also depends on the phases ϕ_I and ϕ_S, as is summarized in the following table:

expt	ϕ_I	ϕ_S	observable operator at $t_2 = 0$			
			$\hat{I}_x$	$\hat{I}_y$	$2\hat{I}_x\hat{S}_z$	$2\hat{I}_y\hat{S}_z$
(a)	y	y	$(-c_+ - c_-)$	$(-s_+ + s_-)$	$(c_+ - c_-)$	$(s_+ + s_-)$
(b)	$-y$	y	$(-c_+ - c_-)$	$(s_+ - s_-)$	$(-c_+ + c_-)$	$(s_+ + s_-)$
(c)	y	$-y$	$(-c_+ - c_-)$	$(-s_+ + s_-)$	$(-c_+ + c_-)$	$(-s_+ - s_-)$
(d)	$-y$	$-y$	$(-c_+ - c_-)$	$(s_+ - s_-)$	$(c_+ - c_-)$	$(-s_+ - s_-)$

The entries in the table give the coefficients which multiply the operator heading the column for each of the four possible combinations of the phases ϕ_I and ϕ_S. Any one of the four experiments (a)–(d) creates a mixture of x- and y-magnetization, which will result in lines which have complex phase

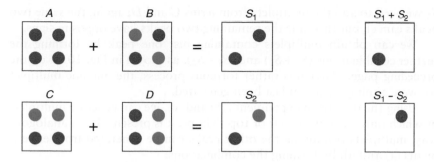

Fig. 10.34 Schematic form of the multiplets arising from terms A–D in Eqs 10.12 and 10.13. Adding together the multiplets from terms A and B results in two of the peaks cancelling, to give a multiplet containing just two negative peaks: this is the multiplet expected for combination S_1. In a similar way, combination S_2 contains the same two peaks as S_1, but in S_2 one peak is positive and one is negative. Multiplets with just one line can be created by further combining S_1 and S_2, as shown. Positive peaks are indicated by blue circles, and negative peaks by dark grey circles.

properties. Our first task is therefore to separate these two components of the magnetization.

After looking at the table for a while we realize that two useful combinations are

$$S_1 = \tfrac{1}{2}[(b) + (c)]$$
$$= \underbrace{(-c_+ - c_-)\,\hat{I}_x}_{A} + \underbrace{(-c_+ + c_-)\,2\hat{I}_x\hat{S}_z}_{B} \tag{10.12}$$

$$S_2 = \tfrac{1}{2}[(b) - (c)]$$
$$= \underbrace{(s_+ - s_-)\,\hat{I}_y}_{C} + \underbrace{(s_+ + s_-)\,2\hat{I}_y\hat{S}_z}_{D}. \tag{10.13}$$

Note that we now have a clean separation of x- and y-magnetization. Combination S_1 is processed to give a spectrum which is then phased to absorption. Combination S_2 is processed separately and also phased to absorption, which will require an additional phase correction on 90° in each dimension. This is on account of the observable magnetization appearing along y, rather than x, and the modulation in t_1 being of the form of a sine, rather than a cosine.

Each of the terms $A \ldots D$ will give rise to a two-dimensional multiplet centred at $\{\Omega_S, \Omega_I\}$ and split by J_{IS} in each dimension. However, the pattern of signs is different in each case. For example, term A is in phase in the ω_2 dimension, and both lines in the ω_1 dimension have the same sign, which happens to be negative. So, all four lines of the multiplet are negative.

In contrast, term B is anti-phase in both dimensions, and so gives rise to the familiar anti-phase square array. The multiplets arising from all of the four terms are shown schematically in Fig. 10.34.

The appearance of the spectrum from combination S_1 is found by adding together the multiplets from terms A and B. As shown in the diagram, the result is that two of the peaks cancel. Similarly, for combination

S_2 we need to add the multiplets from terms C and D; again, the same two peaks cancel, but this time the remaining two peaks have opposite signs.

We can obtain multiplets containing just one peak by forming the further combinations $(S_1 + S_2)$ and $(S_1 - S_2)$, as shown in Fig. 10.34 on the preceding page. After this rather tortuous process, the one-line multiplet we were aiming for has at last been generated.

Using the data from experiments (b) and (c), we can generate multiplets in which only the bottom left or top right peak is present. In an analogous way, multiplets containing the other peaks can be generated from experiments (a) and (d) by forming the combinations:

$$S_3 = \tfrac{1}{2}\left[(a) + (d)\right] \qquad S_4 = \tfrac{1}{2}\left[(a) - (d)\right].$$

Having processed these data separately, and phased them both to absorption, the further combinations $(S_3 + S_4)$ and $(S_3 - S_4)$ will give one-line multiplets, with the single line in either the top left or bottom right.

You may have noticed that this discussion has rather evaded the point as to *exactly* which combinations are needed to generate the multiplet containing just the sharp peak. With such a complex experiment and data processing, it is probably best to determine which combination is needed by trying them all and then picking the one that gives the sharp peak. Once the correct combination has been determined, it will remain the same for all spin systems of the same type e.g. all N–H pairs in a protein or nucleic acid.

The TROSY experiment, and the associated data processing, are somewhat involved, but the rewards, in terms of the gains in resolution and sensitivity, are very great in the case of large molecules at high field. The technique has proved to be very useful for ^{15}N–^{1}H pairs, and has also been used for ^{13}C–^{1}H pairs. Just like the sensitivity enhancement modification, TROSY-type transfer can be implemented into the more complex pulses sequences used to generate the three- and four-dimensional spectra used in biomolecular NMR.

10.10   Double-quantum spectroscopy of a three-spin system

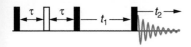

Fig. 10.35 The pulse sequence for two-dimensional double-quantum spectroscopy.

In section 8.5 on page 203 we showed that the double-quantum spectrum of a two-spin system consisted of two anti-phase multiplets in ω_2 which shared a common frequency, the double-quantum frequency, in the ω_1 domain. In such a spectrum the observation that the two multiplets share a common ω_1 frequency indicates that the two spins are coupled. The double-quantum spectrum from a system of three mutually coupled spins is rather more complex than for the two-spin case, and as a result its interpretation is less straightforward. Nevertheless such a spectrum does have some useful features, so it is worthwhile spending some time to unravel the details.

The relevant pulse sequence is shown in Fig. 10.35. To analyse the outcome of this sequence we recognize that the section $-\tau-180°-\tau-$ is a spin echo during which the offset is refocused but the coupling continues to evolve. Starting from equilibrium magnetization on spin one, $\hat{I}_{1z}$, we can therefore easily work out that the terms present at the end of the second delay τ are

$$\cos{(2\pi J_{12}\tau)}\cos{(2\pi J_{13}\tau)}\,\hat{I}_{1y}$$
$$-\sin{(2\pi J_{12}\tau)}\cos{(2\pi J_{13}\tau)}\,2\hat{I}_{1x}\hat{I}_{2z} - \cos{(2\pi J_{12}\tau)}\sin{(2\pi J_{13}\tau)}\,2\hat{I}_{1x}\hat{I}_{3z}$$
$$-\sin{(2\pi J_{12}\tau)}\sin{(2\pi J_{13}\tau)}\,4\hat{I}_{1y}\hat{I}_{2z}\hat{I}_{3z}.$$

As expected, the result is a mixture of in-phase, singly anti-phase and doubly anti-phase terms. The size of each of these terms depends on the values of the coupling constants and the delay τ.

The second 90° pulse rotates the singly anti-phase terms into a mixture of double- and zero-quantum coherence

$$2\hat{I}_{1x}\hat{I}_{2z} \xrightarrow{\;(\pi/2)(\hat{I}_{1x}+\hat{I}_{2x}+\hat{I}_{3x})\;} -2\hat{I}_{1x}\hat{I}_{2y} \quad [1]$$
$$2\hat{I}_{1x}\hat{I}_{3z} \xrightarrow{\;(\pi/2)(\hat{I}_{1x}+\hat{I}_{2x}+\hat{I}_{3x})\;} -2\hat{I}_{1x}\hat{I}_{3y}. \quad [2]$$

The doubly anti-phase term is rotated into a multiple-quantum term of a type which we have not encountered before

$$4\hat{I}_{1y}\hat{I}_{2z}\hat{I}_{3z} \xrightarrow{\;(\pi/2)(\hat{I}_{1x}+\hat{I}_{2x}+\hat{I}_{3x})\;} 4\hat{I}_{1z}\hat{I}_{2y}\hat{I}_{3y}. \quad [3]$$

Terms [1] and [2] give rise to what are known as *direct peaks* in the double-quantum spectrum whereas term [3] gives rise to what is known as a *remote peak*. These two types of peaks have rather different properties, so we will examine them separately.

10.10.1   Direct peaks

As described in section 7.12.1 on page 174, term [1] is a mixture of double- and zero-quantum coherence involving spins one and two. We now need to generalize the notation we introduced in that section in order to be able to deal with three spins. To do this, we define the following pure double- and zero-quantum operators

operator	definition
$\hat{DQ}_x^{(ij)}$	$(2\hat{I}_{ix}\hat{I}_{jx} - 2\hat{I}_{iy}\hat{I}_{jy})$
$\hat{DQ}_y^{(ij)}$	$(2\hat{I}_{ix}\hat{I}_{jy} + 2\hat{I}_{iy}\hat{I}_{jx})$
$\hat{ZQ}_x^{(ij)}$	$(2\hat{I}_{ix}\hat{I}_{jx} + 2\hat{I}_{iy}\hat{I}_{jy})$
$\hat{ZQ}_y^{(ij)}$	$(2\hat{I}_{iy}\hat{I}_{jx} - 2\hat{I}_{ix}\hat{I}_{jy})$

The superscript (ij) indicates that the double- or zero-quantum coherence involves spins i and j. Using this notation $2\hat{I}_{1x}\hat{I}_{2y}$ can be written as

$$\tfrac{1}{2}\left(\hat{DQ}_y^{(12)} - \hat{ZQ}_y^{(12)}\right),$$

therefore the pure double-quantum part of term [1] is $\tfrac{1}{2}\hat{DQ}_y^{(12)}$. We will assume from now on that only double-quantum operators contribute to the observed spectrum.

We now need to work out how this term evolves during t_1 under the influence of both the offset and the couplings. In section 7.12.3 on page 176 it was shown how double-quantum terms evolve under the influence of the offset, and this evolution was summarized in the form of the diagram given in Fig. 7.20 on page 178. Using this, we find that $\hat{DQ}_y^{(12)}$ evolves according to

$$\hat{DQ}_y^{(12)} \xrightarrow{\Omega_1 t_1 \hat{I}_{1z} + \Omega_2 t_1 \hat{I}_{2z} + \Omega_3 t_1 \hat{I}_{3z}} \cos(\Omega_{DQ}^{(12)} t_1)\, \hat{DQ}_y^{(12)} - \sin(\Omega_{DQ}^{(12)} t_1)\, \hat{DQ}_x^{(12)}, \quad (10.14)$$

where $\Omega_{DQ}^{(ij)}$ is the double quantum frequency between spins i and j:

$$\Omega_{DQ}^{(ij)} = \Omega_i + \Omega_j.$$

As we noted before, the double-quantum coherence between spins i and j does not evolve due to the coupling between these two spins. It does, however, evolve under the couplings to a third spin, k. For the double-quantum operators $\hat{DQ}_x^{(12)}$ and $\hat{DQ}_y^{(12)}$ evolution under this coupling results in the following transformations

$$\hat{DQ}_x^{(12)} \xrightarrow{2\pi J_{13} t \hat{I}_{1z}\hat{I}_{3z} + 2\pi J_{23} t \hat{I}_{2z}\hat{I}_{3z}} \cos(\pi\Sigma_{12,3} t)\, \hat{DQ}_x^{(12)} + \sin(\pi\Sigma_{12,3} t)\, 2\hat{DQ}_y^{(12)}\hat{I}_{3z}$$

$$\hat{DQ}_y^{(12)} \xrightarrow{2\pi J_{13} t \hat{I}_{1z}\hat{I}_{3z} + 2\pi J_{23} t \hat{I}_{2z}\hat{I}_{3z}} \cos(\pi\Sigma_{12,3} t)\, \hat{DQ}_y^{(12)} - \sin(\pi\Sigma_{12,3} t)\, 2\hat{DQ}_x^{(12)}\hat{I}_{3z},$$

where

$$\Sigma_{12,3} = J_{13} + J_{23}.$$

$\Sigma_{12,3}$ is the sum of the couplings between the passive third spin and the two spins (here 1 and 2) involved in the double-quantum coherence. This double-quantum evolution is analogous to evolution of the in-phase term $\hat{I}_{1x}$ into the anti-phase term $2\hat{I}_{1y}\hat{I}_{3z}$, and similarly of $\hat{I}_{1y}$ into $-2\hat{I}_{1x}\hat{I}_{3z}$. We can describe terms such as $2\hat{DQ}_y^{(12)}\hat{I}_{3z}$ and $2\hat{DQ}_x^{(12)}\hat{I}_{3z}$ as being 'anti-phase' double quantum.

The evolution during t_1 under the influence of the couplings can now be computed. The first term on the right of Eq. 10.14 evolves to give

$$\cos(\Omega_{DQ}^{(12)} t_1)\, \hat{DQ}_y^{(12)} \xrightarrow{2\pi J_{13} t_1 \hat{I}_{1z}\hat{I}_{3z} + 2\pi J_{23} t_1 \hat{I}_{2z}\hat{I}_{3z}}$$

$$\cos(\pi\Sigma_{12,3} t_1)\cos(\Omega_{DQ}^{(12)} t_1)\, \hat{DQ}_y^{(12)} - \sin(\pi\Sigma_{12,3} t_1)\cos(\Omega_{DQ}^{(12)} t_1)\, 2\hat{DQ}_x^{(12)}\hat{I}_{3z}.$$

Similarly, the second term gives

$$-\sin(\Omega_{DQ}^{(12)} t_1)\, \hat{DQ}_x^{(12)} \xrightarrow{2\pi J_{13} t_1 \hat{I}_{1z}\hat{I}_{3z} + 2\pi J_{23} t_1 \hat{I}_{2z}\hat{I}_{3z}}$$

$$-\cos(\pi\Sigma_{12,3} t_1)\sin(\Omega_{DQ}^{(12)} t_1)\, \hat{DQ}_x^{(12)} - \sin(\pi\Sigma_{12,3} t_1)\sin(\Omega_{DQ}^{(12)} t_1)\, 2\hat{DQ}_y^{(12)}\hat{I}_{3z}.$$

This brings us to the end of t_1.

The role of the final 90° pulse is to convert these double-quantum terms into observable signals. To determine the action of the pulse, we need to write out $\hat{DQ}_x^{(12)}$ and $\hat{DQ}_y^{(12)}$, and their anti-phase counterparts, in terms of the individual operators, and then act on these with the 90° pulse. It turns

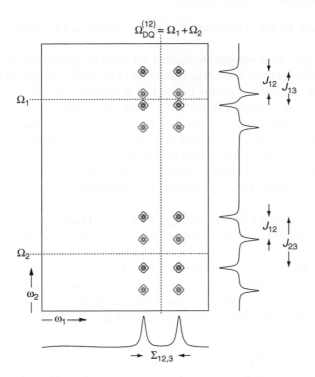

Fig. 10.36 Part of the double-quantum spectrum of a three-spin system showing the two multiplets which arise from the term $\hat{\mathrm{DQ}}_y^{(12)}$ present at the end of t_1. In ω_1 both multiplets share the same double-quantum frequency and have an in-phase splitting of $\Sigma_{12,3} = J_{13} + J_{23}$. The spin-one multiplet is centred at Ω_1 in ω_2, is anti-phase with respect to J_{12}, but in-phase with respect to J_{13}. The spin-two multiplet is centred at Ω_2 and is again anti-phase with respect to J_{12}, but in-phase with respect to J_{23}. The couplings have been chosen such that $J_{23} > J_{13} > J_{12}$; the frequency scales in the two dimensions are not the same.

out that, of all the operators present at the end of t_1, only $\hat{\mathrm{DQ}}_y^{(12)}$ is made observable by a 90° pulse:

$$\hat{\mathrm{DQ}}_y^{(12)} \xrightarrow{(\pi/2)(\hat{I}_{1x}+\hat{I}_{2x}+\hat{I}_{3x})} 2\hat{I}_{1x}\hat{I}_{2z} + 2\hat{I}_{1z}\hat{I}_{2x}.$$

Thus the final observable signal is

$$\tfrac{1}{2}[\sin(2\pi J_{12}\tau)\cos(2\pi J_{13}\tau)] \times$$
$$\cos(\pi\Sigma_{12,3}t_1)\cos(\Omega_{\mathrm{DQ}}^{(12)}t_1)\left(2\hat{I}_{1x}\hat{I}_{2z} + 2\hat{I}_{1z}\hat{I}_{2x}\right). \qquad (10.15)$$

The factor in square braces comes from the spin echo and determines the overall intensity of the peaks. The product of the two trigonometric terms which depend on t_1 can be expanded in the usual way to give

$$\tfrac{1}{2}\cos([\Omega_{\mathrm{DQ}}^{(12)} + \pi\Sigma_{12,3}]t_1) + \tfrac{1}{2}\cos([\Omega_{\mathrm{DQ}}^{(12)} - \pi\Sigma_{12,3}]t_1).$$

Each anti-phase term in Eq. 10.15 gives rise to a two-dimensional multiplet, as shown in Fig. 10.36. The first multiplet is centred as $\{\Omega_{\mathrm{DQ}}^{(12)}, \Omega_1\}$ and comes from the term $2\hat{I}_{1x}\hat{I}_{2z}$. The ω_2 frequencies are therefore those is the spin-one multiplet (a doublet of doublets), anti-phase with respect to the 1–2 coupling, but in phase with respect to the 1–3 coupling. The ω_1 frequencies are $\Omega_{\mathrm{DQ}}^{(12)} \pm \pi\Sigma_{12,3}$ i.e. 'doublet' centred on the double-quantum frequency and split by $2\pi\Sigma_{12,3}$; note that the splitting in this dimension is in phase.

The second two-dimensional multiplet comes from the term $2\hat{I}_{1z}\hat{I}_{2x}$, and is centred at $\{\Omega_{\mathrm{DQ}}^{(12)}, \Omega_2\}$. It has the same structure in ω_1 as the first multiplet, but in ω_2 the frequencies are those of the spin-two multiplet, anti-phase with respect to the 1–2 coupling, and in phase with respect to the 2–3 coupling.

These two-dimensional multiplets are directly analogous to those seen in the double-quantum spectrum, shown in Fig. 8.19 on page 205, of a two-spin system. In both cases the multiplets are centred at the same frequency – the double-quantum frequency $\Omega_{DQ}^{(12)}$ – in the ω_1 dimension and are anti-phase with respect to the 1–2 coupling in the ω_2 dimension. However, in the three-spin case in each dimension there is an additional in-phase splitting which can be attributed to the passive couplings to the third spin. These multiplets are described as *direct peaks* since they arise from double-quantum coherence between spins i and j which is observed on either spin i or spin j.

Term [2] on page 367 is analogous to term [1] except that it represents double-quantum coherence between spins one and three. Term [2] evolves in an analogous way and gives rise to two more multiplets centred at $\{\Omega_{DQ}^{(13)}, \Omega_1\}$ and $\{\Omega_{DQ}^{(13)}, \Omega_3\}$. These multiplets have an in-phase splitting in the ω_1 dimension of $\Sigma_{13,2} = J_{12} + J_{23}$, and in ω_2 have an in-phase splitting with respect to the coupling to spin two, which is the passive spin. These multiplets are also categorized as direct peaks.

10.10.2   Remote peaks

Term [3] on page 367 contains the operator product $4\hat{I}_{1z}\hat{I}_{2y}\hat{I}_{3y}$ which has a double-quantum part and so contributes to the spectrum. This term can be written as

$$\left[2\hat{I}_{1z}\right]\left[2\hat{I}_{2y}\hat{I}_{3y}\right],$$

and, using the table on page 367, we can rewrite the term $2\hat{I}_{2y}\hat{I}_{3y}$ in the following way

$$\left[2\hat{I}_{1z}\right]\tfrac{1}{2}\left[\hat{ZQ}_x^{(23)} - \hat{DQ}_x^{(23)}\right].$$

What we have here is double- and zero-quantum coherence between spins two and three, but the presence of the term $\hat{I}_{1z}$ indicates that the coherence is anti-phase with respect to the coupling to spin one. As before, we are just interested in the double-quantum part, $2\hat{I}_{1z}\hat{DQ}_x^{(23)}$. Note that although we started with equilibrium magnetization on spin *one* we have generated double quantum between spins two and three.

The evolution of this term under the influence of the offset is straightforward as the operator $\hat{I}_{1z}$ is unaffected. We can therefore use Fig. 7.20 on page 178 to give

$$2\hat{I}_{1z}\hat{DQ}_x^{(23)} \xrightarrow{\Omega_1 t_1 \hat{I}_{1z} + \Omega_2 t_1 \hat{I}_{2z} + \Omega_3 t_1 \hat{I}_{3z}} \tag{10.16}$$
$$\cos\left(\Omega_{DQ}^{(23)}t_1\right)2\hat{I}_{1z}\hat{DQ}_x^{(23)} + \sin\left(\Omega_{DQ}^{(23)}t_1\right)2\hat{I}_{1z}\hat{DQ}_y^{(23)},$$

where the double-quantum frequency is

$$\Omega_{DQ}^{(23)} = \Omega_2 + \Omega_3.$$

We saw before that under the influence of the couplings a term such as $\hat{DQ}_x^{(12)}$ evolves to an anti-phase term such as $2\hat{DQ}_y^{(12)}\hat{I}_{3z}$; this is closely analogous to $\hat{I}_{1x}$ evolving to $2\hat{I}_{1y}\hat{I}_{2z}$. In the same way, under the influence of the couplings, the anti-phase term $2\hat{I}_{1z}\hat{DQ}_x^{(23)}$ evolves into the in-phase term $\hat{DQ}_y^{(23)}$, and $2\hat{I}_{1z}\hat{DQ}_y^{(23)}$ evolves into $-\hat{DQ}_x^{(23)}$:

$$2\hat{I}_{1z}\hat{DQ}_x^{(23)} \xrightarrow{2\pi J_{12}t\hat{I}_{1z}\hat{I}_{2z}+2\pi J_{13}t\hat{I}_{1z}\hat{I}_{3z}} \cos{(\pi\Sigma_{23,1}t)}\,2\hat{I}_{1z}\hat{DQ}_x^{(23)} + \sin{(\pi\Sigma_{23,1}t)}\,\hat{DQ}_y^{(23)}$$

$$2\hat{I}_{1z}\hat{DQ}_y^{(23)} \xrightarrow{2\pi J_{12}t\hat{I}_{1z}\hat{I}_{2z}+2\pi J_{13}t\hat{I}_{1z}\hat{I}_{3z}} \cos{(\pi\Sigma_{23,1}t)}\,2\hat{I}_{1z}\hat{DQ}_y^{(23)} - \sin{(\pi\Sigma_{23,1}t)}\,\hat{DQ}_x^{(23)},$$

where

$$\Sigma_{23,1} = J_{12} + J_{13}.$$

As before, $\Sigma_{23,1}$ is the sum of the couplings to the passive spin, spin one, which is not involved in the double-quantum coherence.

If we allow the two terms on the right of Eq. 10.16 on the preceding page to evolve under the influence of the coupling, we find the following four terms present at the end of t_1:

$$+ \cos{(\pi\Sigma_{23,1}t_1)}\cos{(\Omega_{DQ}^{(23)}t_1)}\,2\hat{I}_{1z}\hat{DQ}_x^{(23)}$$

$$+ \sin{(\pi\Sigma_{23,1}t_1)}\cos{(\Omega_{DQ}^{(23)}t_1)}\,\hat{DQ}_y^{(23)}$$

$$+ \cos{(\pi\Sigma_{23,1}t_1)}\sin{(\Omega_{DQ}^{(23)}t_1)}\,2\hat{I}_{1z}\hat{DQ}_y^{(23)} \qquad (10.17)$$

$$- \sin{(\pi\Sigma_{23,1}t_1)}\sin{(\Omega_{DQ}^{(23)}t_1)}\,\hat{DQ}_x^{(23)}.$$

As before, we need to expand these terms out into their individual operators in order to work out the effect of the final 90° pulse. If we do this we find that only the terms $\hat{DQ}_y^{(23)}$ and $2\hat{I}_{1z}\hat{DQ}_x^{(23)}$ result in observable signals. We will consider each in turn.

The effect of the final 90° pulse on $\hat{DQ}_y^{(23)}$ is to give two anti-phase terms:

$$\hat{DQ}_y^{(23)} \xrightarrow{(\pi/2)(\hat{I}_{1x}+\hat{I}_{2x}+\hat{I}_{3x})} 2\hat{I}_{2x}\hat{I}_{3z} + 2\hat{I}_{2z}\hat{I}_{3x}.$$

The t_1 modulation for this term can be expanded to

$$\tfrac{1}{2}\sin{([\Omega_{DQ}^{(23)} + \pi\Sigma_{23,1}]t_1)} - \tfrac{1}{2}\sin{([\Omega_{DQ}^{(23)} - \pi\Sigma_{23,1}]t_1)};$$

note that minus sign. Thus $\hat{DQ}_y^{(23)}$ gives rise to two two-dimensional multiplets centred at $\{\Omega_{DQ}^{(23)}, \Omega_2\}$ and $\{\Omega_{DQ}^{(23)}, \Omega_3\}$. In ω_2 the multiplets are anti-phase with respect to the 2–3 coupling, and in phase with respect to the passive coupling to spin one. Since the double-quantum coherence leading to these peaks involved spins two and three, and the observable signals appear on spin two and on spin three, these multiplets are classified as direct peaks.

However, compared with the direct peaks which arose from terms [1] and [2] on page 367 there are some significant differences. First, the splitting in the ω_1 dimension is in anti-phase; secondly, the modulation is of the form of sine, rather than cosine, so the lineshape is different.

The overall intensity of the direct peaks arising from term [3] goes as

$$\tfrac{1}{2}\sin{(2\pi J_{12}\tau)}\sin{(2\pi J_{13}\tau)}.$$

Again this is in contrast to the direct peaks arising from terms [1] and [2] whose intensity goes as $\frac{1}{2}\sin(2\pi J_{12}\tau)\cos(2\pi J_{13}\tau)$.

Now we turn to the term $2\hat{I}_{1z}\hat{DQ}_x^{(23)}$ from Eq. 10.17 on the preceding page. This also leads to an observable term as a result of the effect of the final $90°$ pulse:

$$2\hat{I}_{1z}\hat{DQ}_x^{(23)} \xrightarrow{(\pi/2)(\hat{I}_{1x}+\hat{I}_{2x}+\hat{I}_{3x})} -4\hat{I}_{1y}\hat{I}_{2x}\hat{I}_{3x} + 4\hat{I}_{1y}\hat{I}_{2z}\hat{I}_{3z}.$$

The observable term, $4\hat{I}_{1y}\hat{I}_{2z}\hat{I}_{3z}$, appears on spin one as a doubly anti-phase multiplet. The modulation in t_1 can be expanded to

$$\tfrac{1}{2}\cos([\Omega_{DQ}^{(23)}+\pi\Sigma_{23,1}]t_1) + \tfrac{1}{2}\cos([\Omega_{DQ}^{(23)}-\pi\Sigma_{23,1}]t_1),$$

which means that the two-dimensional multiplet is centred at $\{\Omega_{DQ}^{(23)},\Omega_1\}$, and in ω_1 is split into an in-phase 'doublet' of separation $\Sigma_{23,1} = J_{12} + J_{13}$. The overall intensity of this peak is

$$\tfrac{1}{2}\sin(2\pi J_{12}\tau)\sin(2\pi J_{13}\tau).$$

This multiplet is described as a *remote peak* since the frequency in ω_1 is that of the double-quantum coherence involving spins two and three, but the frequency in ω_2 is that of spin one i.e. the spin which is *not* involved in the double-quantum coherence. Significantly, since the intensity of this term has no dependence on the coupling between spins two and three the remote peak will be present *even if J_{23} is zero*. These properties are in direct contrast to those shown by the direct peaks.

10.10.3   Summary

The double-quantum spectrum from three spins is therefore rather complex, with different kinds of peaks having subtly different dependences on the couplings, as well as different lineshapes. The following table summarizes the features of the five two-dimensional multiplets which arise from equilibrium magnetization on spin one.

In the table the following abbreviations are used: $S_{ij} = \sin(2\pi J_{ij}\tau)$, $C_{ij} = \cos(2\pi J_{ij}\tau)$. IP: $(\ldots)$ indicates that a multiplet is in-phase with respect to the coupling(s) given in the bracket, and AP: $(\ldots)$ indicates that a multiplet is in anti-phase.

	centre frequency in ω_1	centre frequency in ω_2	intensity	structure in ω_1	structure in ω_2
A	$(\Omega_1 + \Omega_2)$	Ω_1	$S_{12}C_{13}$	IP: $(J_{13} + J_{23})$	AP: (J_{12}), IP: (J_{13})
B	$(\Omega_1 + \Omega_2)$	Ω_2	$S_{12}C_{13}$	IP: $(J_{13} + J_{23})$	AP: (J_{12}), IP: (J_{23})
C	$(\Omega_1 + \Omega_3)$	Ω_1	$C_{12}S_{13}$	IP: $(J_{12} + J_{23})$	AP: (J_{13}), IP: (J_{12})
D	$(\Omega_1 + \Omega_3)$	Ω_3	$C_{12}S_{13}$	IP: $(J_{12} + J_{23})$	AP: (J_{13}), IP: (J_{23})
E	$(\Omega_2 + \Omega_3)$	Ω_2	$S_{12}S_{13}$	AP: $(J_{12} + J_{13})$	AP: (J_{23}), IP: (J_{12})
F	$(\Omega_2 + \Omega_3)$	Ω_3	$S_{12}S_{13}$	AP: $(J_{12} + J_{13})$	AP: (J_{23}), IP: (J_{13})
G	$(\Omega_2 + \Omega_3)$	Ω_1	$S_{12}S_{13}$	IP: $(J_{12} + J_{13})$	AP: (J_{12}), AP: (J_{13})

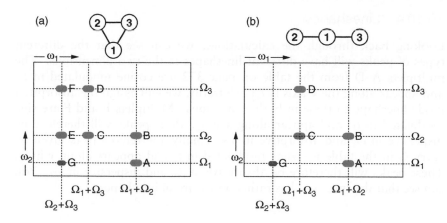

Fig. 10.37 Schematic two-dimensional double-quantum spectra for two different three-spin systems; only peaks which arise from equilibrium magnetization on spin one are shown. In (a) all three spins are coupled to one another, whereas in (b) there is no coupling between spins two and three. In both spectra the letters indicate the multiplets whose properties are listed in the table on the facing page. Direct peaks are shown in blue and indirect peaks in dark grey. In spectrum (b), the number of two-dimensional multiplets is reduced: there are no direct peaks between spins two and three, but the remote peak remains.

Multiplets A–F are all classed as direct peaks since in each case the frequency in ω_2 is that of one of the spins contributing to the double-quantum frequency in ω_2. Of these direct peaks, A and B arise from term [1] on page 367, C and D arise from term [2], and E and F arise from term [3]. Multiplet G is the remote peak which also arises from term [3].

There are an equivalent set of multiplets arising from $\hat{I}_{2z}$ which can be found by simply permuting the indices cyclically (e.g. $1 \rightarrow 2, 2 \rightarrow 3, 3 \rightarrow 1$). A further permutation will give the multiplets arising from $\hat{I}_{3z}$.

Figure 10.37 shows schematic double-quantum spectra for three-spin systems with two different coupling topologies; note that only peaks which arise from equilibrium magnetization on spin one are shown. The two-dimensional multiplets are labelled A–G, according to the entries in the table on the facing page. Spectrum (a) is for the case where all three spins are coupled to one another. In this case, all seven two-dimensional multiplets listed in the table are present in the spectrum.

If there is no coupling between spins two and three, the spectrum becomes much simpler, as is shown in (b). Referring to the table, we can see that the direct multiplets A–D will still occur, but multiplets E and F will vanish as in ω_2 they are anti-phase with respect to J_{23}, which is zero. Most interestingly multiplet G, the indirect peak, remains, as its intensity is independent of J_{23}.

The presence of the remote peak in spectrum (b) tells is that although spins two and three are not coupled, they must both have a coupling to spin one. This is more subtle information about the *topology* of the coupling network than we find from a simple COSY spectrum, and this feature of double-quantum spectra has proved to be useful in a number of contexts.

10.10.4   Lineshapes

Looking back through the calculations, we can see that the different types of peaks will have different lineshapes in the two dimensions. The multiplets A–D from the table on page 372 are cosine modulated in t_1, and appear along the x-axis in t_2: arbitrarily, we will assign the absorption mode lineshape to these in both dimensions. Multiplets E and F are sine modulated in t_1, and appear along x in t_2: these peaks will therefore be dispersive in ω_1 and absorptive in ω_2. Finally, the remote peak given by entry G in the table is cosine modulated in t_1 and appears along y in t_2. These peaks will therefore be absorptive in ω_1 and dispersive in ω_2. You can see that the spectrum has rather a mixture of lineshapes.

10.11   Further reading

Product operators for more extended spin systems:

Chapter 18 from M. H. Levitt, *Spin Dynamics* (2nd edition, John Wiley & Sons, Ltd, 2008).

A full account of the product operator method:

O. W. Sørensen, G. W. Eich, M. H. Levitt, G. Bodenhausen and R. R. Ernst, *Progress in Nuclear Magnetic Resonance Spectroscopy*, **16**, 163–192 (1983).

Polarization and single transition operators:

Chapter 2 from R. R. Ernst, G. Bodenhausen and A. Wokaun, *Principles of Nuclear Magnetic Resonance in One and Two Dimensions* (Oxford University Press, 1987).

Small flip angle COSY and ECOSY:

Chapter 6 from J. Cavanagh, W. J. Fairbrother, A. G. Palmer III, M. Rance and N. J. Skelton, *Protein NMR Spectroscopy* (2nd edition, Academic Press, 2007).

ECOSY:

C. Griesinger, O. W. Sørensen and R. R. Ernst, *J. Magn. Reson.*, **75**, 474–492 (1987).

ZCOSY:

H. Oschkinat, A. Pastore, P. Pfändler and G. Bodenhausen, *J. Magn. Reson.*, **69**, 559–566 (1986).

Constant time experiments and sensitivity-enhanced experiments:

Chapters 6 and 7 from J. Cavanagh, W. J. Fairbrother, A. G. Palmer III, M. Rance and N. J. Skelton, *Protein NMR Spectroscopy* (2nd edition, Academic Press, 2007).

TROSY:

C. Fernandez and G. Wider, *Current Opinion in Structural Biology*, **13**, 570–580 (2003).

10.12 Exercises

10.1 Draw sketches, roughly to scale, of the multiplets represented by the following operators: $\hat{I}_{2y}$, $2\hat{I}_{1z}\hat{I}_{2y}$, $2\hat{I}_{2y}\hat{I}_{3z}$ and $4\hat{I}_{1z}\hat{I}_{2y}\hat{I}_{3z}$, assuming that the couplings constants are: (a) $J_{12} = 2$ Hz, $J_{23} = 6$ Hz; and (b) $J_{12} = 6$ Hz, $J_{23} = 6$ Hz. Label each line of the multiplet according to the spin states of the passive spins. (You should assume that y-magnetization gives rise to an absorption mode lineshape.)

10.2 Starting with the operator $\hat{I}_{2y}$, work out the result of evolution of the 1–2, 1–3 and 2–3 couplings for a time t. Comment on the operators you obtain and the trigonometric factors associated with each.

Draw up a 'tree diagram', of the type shown in Fig. 10.2 on page 324, which represents the results of your calculation.

10.3 Starting with the operator $2\hat{I}_{2y}\hat{I}_{3z}$, draw up a tree diagram to show the evolution under first the 2–3 coupling, and then the 1–2 coupling. You do not need to work out all the sines and cosines to draw up the diagram, but just need to 'split' each operator into the appropriate pair which arise from evolution of the coupling.

By inspecting your tree diagram, write down the trigonometric factors which multiply the operator $4\hat{I}_{1z}\hat{I}_{2x}\hat{I}_{3z}$.

10.4 In a COSY of a three-spin system, the diagonal peak of spin one is represented by term [1] on page 326. From this term, work out the detailed form of the diagonal-peak multiplet, using the same approach as was used for the cross peak. Sketch the form of the multiplet and compare it with the cross peak.

You will need the identities

$$\cos A \cos B \equiv \tfrac{1}{2}\left[\cos\left(A + B\right) + \cos\left(A - B\right)\right]$$
$$\sin A \cos B \equiv \tfrac{1}{2}\left[\sin\left(A + B\right) + \sin\left(A - B\right)\right]$$
$$\sin\left(-A\right) \equiv -\sin\left(A\right) \qquad \cos\left(-A\right) \equiv \cos\left(A\right).$$

10.5 Identify all four anti-phase square arrays in each of the multiplets shown in Fig. 10.8 on page 329.

Sketch the form of the 1–2 cross peak for the following combinations of couplings: (a) $J_{12} = 8$ Hz, $J_{13} = 7$ Hz, $J_{23} = 4$ Hz; (b) $J_{12} = 8$ Hz, $J_{13} = 8$ Hz, $J_{23} = 4$ Hz; (c) $J_{12} = 8$ Hz, $J_{13} = 4$ Hz, $J_{23} = 0.5$ Hz; (d) $J_{12} = 8$ Hz, $J_{13} = 8$ Hz, $J_{23} = 0.5$ Hz. In each case, identify the four anti-phase square arrays.

10.6 Assuming that spin three is a heteronucleus, sketch the form of the reduced 1–2 cross-peak multiplets which would be expected for the cross peaks shown in Fig. 10.8 on page 329. Identify the two anti-phase square arrays in each case.

10.7 Explain why, whereas it is not usually possible to measure a value for the active coupling constant responsible for a cross peak, it is sometimes possible to measure values of the passive coupling constants.

10.8 In a three-spin system, give an interpretation of each of the following products of polarization operators. State, with reasons, which of the products are observable.

$$\hat{I}_{1\alpha}\hat{I}_{2-}\hat{I}_{3\alpha} \qquad \hat{I}_{1\alpha}\hat{I}_{2-}\hat{I}_{3-} \qquad \hat{I}_{1\beta}\hat{I}_{2\beta}\hat{I}_{3\beta} \qquad \hat{I}_{1\alpha}\hat{I}_{2\beta}\hat{I}_{3+}$$

Give the time evolution of the first and fourth operator products.

10.9 In a small flip angle COSY experiment, work out the contribution which the operator product $\hat{I}_{1+}\hat{I}_{2\beta}\hat{I}_{3\alpha}$, present during t_1, makes to the 1–2 cross peak. You should follow the same approach as used in section 10.4.4 on page 340 for the term $\hat{I}_{1+}\hat{I}_{1\alpha}\hat{I}_{3\alpha}$.

Without further detailed calculations, state the contribution to the 1–2 cross peak made by the term $\hat{I}_{1-}\hat{I}_{2\beta}\hat{I}_{3\alpha}$.

10.10 In the ZCOSY experiment, work out the contribution which the operator product $\hat{I}_{1+}\hat{I}_{2\beta}\hat{I}_{3\alpha}$, present during t_1, makes to the 1–2 cross peak. Compare your answer with that in the previous question.

10.11 The pulse sequence for a constant time version of double-quantum filtered COSY is shown below.

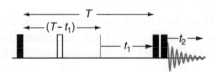

Show that, for a two-spin system and starting from $\hat{I}_{1z}$, the double quantum present between the final two pulses is given by

$$\tfrac{1}{2}\cos\left(\Omega_1 t_1\right)\sin\left(\pi J_{12}T\right)\left(2\hat{I}_{1x}\hat{I}_{2y} + 2\hat{I}_{1y}\hat{I}_{2x}\right).$$

Determine the effect of the final pulse on this double-quantum state, and hence predict the form of the diagonal- and cross-peak multiplets. Does double-quantum filtration have any practical benefits in this experiment?

10.12 Starting from $\hat{I}_{1z}$, determine the form of a constant time COSY spectrum for a three-spin system. Comment on the choice of the fixed time T for such a spin system.

10.13 Consider the three-spin system opposite which consists of a proton I coupled to a ^{13}C, S_1, with coupling constant J_{IS_1}; S_1 is further coupled to a second ^{13}C, S_2, with coupling constant $J_{S_1 S_2}$. Note that there is no coupling between I and S_2.

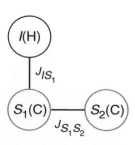

For the constant time HSQC pulse sequence shown in Fig. 10.26 on page 357, and starting with equilibrium magnetization in the I spin, work out the S-spin operators present at the end of the constant time period; assume that $\tau_1 = 1/(4J_{IS_1})$ and $\tau_2 = 1/(4J_{IS_1})$.

Determine which of these operators become observable on the I spin at the end of the sequence, and hence predict the form of the spectrum. How do the peaks vary in intensity as a function of the constant time T? What is the optimum value for this time?

10.14 Using the same approach as in Fig. 10.30 on page 361, verify that the following transfers take place:

$$\underbrace{1\text{--}2}_{\text{spin } S} \xrightarrow{\ 180° \text{ pulse to } 1\text{--}3\ } 2\text{--}3 \xrightarrow{\ 180° \text{ pulse to } 1\text{--}2\ } \underbrace{1\text{--}3}_{\text{spin } I}.$$

$$\underbrace{1\text{--}2}_{\text{spin } S} \xrightarrow{\ 180° \text{ pulse to } 1\text{--}3\ } 2\text{--}3 \xrightarrow{\ 180° \text{ pulse to } 3\text{--}4\ } \underbrace{2\text{--}4}_{\text{spin } I}.$$

Work out which two line-selective 180° pulses you could need to cause transfer from: (a) 1–3 to 3–4; (b) 1–3 to 1–2. How could these be implemented in a practical sequence?

10.15 Verify that the operators given at the start of section 10.9.4 on page 364 are indeed those present at the end of t_1. By combining the trigonometric functions in the usual way, also verify that Eq. 10.11 on page 364 is correct.

Follow the fate of the term $\hat{S}_x$, present at the end of t_1, through the rest of the TROSY HSQC sequence, for the case where both ϕ_I and ϕ_S are y. Check your answer against the table on page 364.

10.16 Using the same approach as in section 10.9.4 on page 364, determine the form of the two-dimensional multiplets arising from the combinations S_3 and S_4, and hence verify that by further combining these you can generate multiplets with one peak in the top left or bottom right.

10.17 Can you think of any disadvantages of the TROSY HSQC sequence, when applied to ^{15}N–^1H correlation?

10.18 In the pulse sequence for double-quantum spectroscopy (Fig. 10.35 on page 366), verify that, at the end of the second delay τ, the operators present are those stated on page 366.

10.19 Write out the following double-quantum operators in terms of the cartesian operators $\hat{I}_{ix}$, $\hat{I}_{iy}$ and $\hat{I}_{iz}$ (you will need to use the table on page 367):

$$\hat{DQ}_x^{(12)} \qquad \hat{DQ}_y^{(12)} \qquad 2\hat{DQ}_x^{(12)}\hat{I}_{3z} \qquad 2\hat{DQ}_y^{(12)}\hat{I}_{3z}.$$

Verify that, if a $90°(x)$ pulse is applied to all three spins, only the double-quantum term $\hat{DQ}_x^{(12)}$ results in observable magnetization.

10.20 Without performing detailed calculations, predict the form of the double-quantum spectrum of a three-spin system which would arise from equilibrium magnetization *on spin two* for the cases: (a) all three spins coupled; (b) no coupling between spins one and three. Make a sketch of the spectrum in each case. (You should be able to determine everything you need to know by cyclically permuting the indices in the table on page 372.)

Do the same for equilibrium magnetization starting on spin three and hence determine the form of the spectrum, in the two cases listed above, for a real experiment in which we start with equilibrium magnetization on all three spins.

11

Coherence selection: phase cycling and field gradient pulses

We now turn to a topic which has been rather glossed over up to this point, especially when discussing two-dimensional experiments, which is exactly *how* we can select different kinds of coherences at different points in our pulse sequences. The ability to make such a selection is very important, as is well illustrated by the three pulse sequences shown in Fig. 11.1. Each of these pulse sequences consists of three 90° pulses arranged in different ways, but what really makes them different is *our wish* to restrict the type of coherence or magnetization present at each stage.

Sequence (a) is DQF COSY, so we want single-quantum coherence to be present during t_1, and double quantum to be present between the final two pulses. Sequence (b) is double-quantum spectroscopy: here we want single quantum during the spin echo, and double quantum during t_1. Finally, (c) is NOESY, in which we want single quantum during t_1, and z-magnetization during τ.

The spins in the sample have no way of knowing, nor come to that do they care, what *we* want from the pulse sequence. They will just evolve according to the sequence of pulses and delays, and coherences other than the ones we want will certainly be generated. The result will be confused two-dimensional spectra, rather than the clean results which were presented in Chapter 8. It is therefore essential that, at each point, we have some way of selecting the coherences we want and rejecting all the others.

There are two ways of achieving this selection. The first is *phase cycling*. In this method, we repeat the pulse sequence (for the same value of t_1) several times over, each time shifting the phase of certain pulses in a predetermined way. The results from these separate experiments are then combined, probably using additional receiver phase shifts, in such a way that the signals arising from the wanted coherences add up, while those arising from unwanted coherences cancel. In this chapter, we will discover how to devise the sequence of phases, called *phase cycles*, that are needed to select particular coherences.

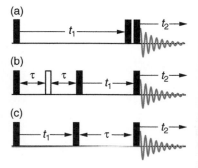

Fig. 11.1 Three different pulse sequences, all of which consist of three 90° pulses: (a) is DQF COSY, (b) is double-quantum spectroscopy, and (c) is NOESY. All that is really different between these experiments is the type of coherence or magnetization we *want* to be present at each stage.

Understanding NMR Spectroscopy, Second Edition James Keeler
© 2010 John Wiley & Sons, Ltd

We have, in fact, already come across a simple form of phase cycling, which we called difference spectroscopy. For example, in the HMQC experiment described in section 8.8 on page 212, we saw that we were able to suppress unwanted signals by repeating the experiment twice, once with the phase of one of the pulses shifted by 180°, and then subtracting the results. This difference experiment is essentially a two-step phase cycle. In this chapter we will develop longer phase cycles capable of greater discrimination than simple difference spectroscopy.

The second method for coherence selection is to use *pulsed field gradients*. During such a gradient the main magnetic field is deliberately made inhomogeneous for a short time. As a result, any coherences present dephase rapidly: this is exactly the same process as the inhomogeneous dephasing described in section 9.9 on page 300. However, it turns out that this dephasing can be reversed by applying a second field gradient, but – and this is the key part – by careful choice of the duration and position in the pulse sequence of the two gradients, we can make sure that *only* the required coherence is rephased. Later on in this chapter we will look in detail at exactly how these gradient pairs are designed.

Before getting into the details of how to construct phase cycles and gradient sequences, we will introduce the idea of *coherence order* and the *coherence transfer pathway*. These provide a convenient and unified framework for discussing both phase cycles and gradient sequences.

11.1 Coherence order

The idea of coherence order was introduced in section 7.12.1 on page 174, but this idea is so important for the present chapter that it is worth restating the key ideas once more. The coherence order, given the symbol p, is defined by what happens to an operator (or product of operators) when a z-rotation through an angle ϕ is applied. If, as a result of this rotation, the operator acquires a phase of $(-p \times \phi)$, the operator is classed as having order p:

$$\text{definition of order } p : \quad \hat{\rho}^{(p)} \xrightarrow{\text{rotate by } \phi \text{ about } z} \hat{\rho}^{(p)} \times \exp(-ip\phi). \quad (11.1)$$

In Eq. 11.1, $\hat{\rho}^{(p)}$ is an operator of order p.

The order can take both positive and negative integer values, including zero. An operator with $p = \pm 1$ is called single quantum, whereas one with $p = \pm 2$ is called double quantum, and so on. Operators with $p = 0$ are either zero quantum or z-magnetization.

Our usual product operators can be classified according to coherence order by expressing them in terms of the raising operator, $\hat{I}_{i+}$, and the lowering operator, $\hat{I}_{i-}$, for each spin i. These operators are defined as

$$\hat{I}_{i+} \equiv \hat{I}_{ix} + i\hat{I}_{iy} \qquad \hat{I}_{i-} \equiv \hat{I}_{ix} - i\hat{I}_{iy}, \quad (11.2)$$

where $\hat{I}_{ix}$ and $\hat{I}_{iy}$ are the usual x and y operators for spin i.

The operator $\hat{I}_{i+}$ has coherence order $p = +1$, something we can easily demonstrate by seeing how it is affected by a z-rotation:

$$\hat{I}_{i+} \equiv \hat{I}_{ix} + i\hat{I}_{iy} \xrightarrow{\ \phi\hat{I}_{iz}\ } \cos\phi\, \hat{I}_{ix} + \sin\phi\, \hat{I}_{iy} + i\left(\cos\phi\, \hat{I}_{iy} - \sin\phi\, \hat{I}_{ix}\right)$$

$$= \cos\phi\left(\hat{I}_{ix} + i\hat{I}_{iy}\right) - i\sin\phi\left(\hat{I}_{ix} + i\hat{I}_{iy}\right)$$

$$= \left(\cos\phi - i\sin\phi\right)\left(\hat{I}_{ix} + i\hat{I}_{iy}\right)$$

$$= \exp(-i\phi)\,\hat{I}_{i+}. \qquad (11.3)$$

This shows that under a z-rotation $\hat{I}_{i+}$ acquires a phase of $(-\phi)$. Therefore, from Eq. 11.1 on the preceding page, the operator must have coherence order $+1$. Using a similar approach we can show that $\hat{I}_{i-}$ has coherence order -1. The operator $\hat{I}_{iz}$ is unaffected by a z-rotation, and so has $p = 0$.

The definitions of the raising and lowering operators given in Eq. 11.2 on the facing page, can be turned round so as to express $\hat{I}_{ix}$ and $\hat{I}_{iy}$ in terms of $\hat{I}_{i+}$ and $\hat{I}_{i-}$:

$$\hat{I}_{ix} \equiv \tfrac{1}{2}\left(\hat{I}_{i+} + \hat{I}_{i-}\right) \qquad \hat{I}_{iy} \equiv \tfrac{1}{2i}\left(\hat{I}_{i+} - \hat{I}_{i-}\right). \qquad (11.4)$$

From these we see that both $\hat{I}_{ix}$ and $\hat{I}_{iy}$ are *equal* mixtures of coherence orders $+1$ and -1.

If we have a product operator, then each operator can be classified according to coherence order, and in addition we can define an overall order, which is the sum of the individual coherence orders. For example the operator $2\hat{I}_{1x}\hat{I}_{2z}$ can be expanded as

$$2\hat{I}_{1x}\hat{I}_{2z} \equiv (\ \underbrace{\hat{I}_{1+}}_{p_1=+1} + \underbrace{\hat{I}_{1-}}_{p_1=-1}\ )\ \underbrace{\hat{I}_{2z}}_{p_2=0}\ ,$$

from which we have $p_1 = +1$ or -1, and $p_2 = 0$. The overall order, $(p_1 + p_2)$, is therefore either $(+1 + 0) = +1$ or $(-1 + 0) = -1$, in equal measure.

We have already shown in section 7.12.1 on page 174 that products such as $2\hat{I}_{1x}\hat{I}_{2x}$ are mixtures of orders $p = 0$, $p = +2$ and $p = -2$:

$$2\hat{I}_{1x}\hat{I}_{2x} \equiv \tfrac{1}{2}(\underbrace{\hat{I}_{1+}\hat{I}_{2+}}_{p=+2} + \underbrace{\hat{I}_{1-}\hat{I}_{2-}}_{p=-2} + \underbrace{\hat{I}_{1+}\hat{I}_{2-}}_{p=0} + \underbrace{\hat{I}_{1-}\hat{I}_{2+}}_{p=0}).$$

Note that the operator contains an equal mixture of $p = +2$ and $p = -2$ terms.

11.1.1 Possible values of the overall coherence order

For any one spin the maximum coherence order is $+1$, so in a spin system composed of N spins the maximum overall coherence order that can be present is $+N$. For example, in a three-spin system the maximum coherence order is $+3$, such as would be represented by the operator product $\hat{I}_{1+}\hat{I}_{2+}\hat{I}_{3+}$; this is triple-quantum coherence.

In the same way, the most negative value of the overall coherence order is $-N$, which would be achieved by having all of the operators in the product of the type $\hat{I}_{i-}$. Overall a system of N spins can give rise to coherences of order $-p$ to $+p$, in integer steps.

These remarks apply only to systems of coupled spin-half nuclei; nuclei with $I > \tfrac{1}{2}$ can support higher orders of multiple quantum coherence than those described here.

We should just note here that in order to generate triple-quantum coherence one spin has to be coupled to two others, and to generate quadruple-quantum coherence, one spin needs to be coupled to three others. The generation of higher and higher orders of coherence therefore becomes increasingly unlikely as it requires the presence of more and more couplings to a single spin.

11.1.2   Evolution of operators of particular coherence orders

Under the influence of the offset term we know that an operator such as $\hat{I}_{ix}$ evolves into $\hat{I}_{iy}$ according to

$$\hat{I}_{ix} \xrightarrow{\;\Omega_i t \hat{I}_{iz}\;} \cos\left(\Omega_i t\right)\hat{I}_{ix} + \sin\left(\Omega_i t\right)\hat{I}_{iy},$$

where Ω_i is the offset of spin i. Evolution under the offset is just a z-rotation through an angle $(\Omega_i t)$, so from Eq. 11.3 on the preceding page, we can see that the operator $\hat{I}_{i+}$ evolves under the offset according to

$$\hat{I}_{i+} \xrightarrow{\;\Omega_i t \hat{I}_{iz}\;} \exp\left(-\mathrm{i}\Omega_i t\right)\hat{I}_{i+}.$$

Similarly, $\hat{I}_{i-}$ evolves as

$$\hat{I}_{i-} \xrightarrow{\;\Omega_i t \hat{I}_{iz}\;} \exp\left(+\mathrm{i}\Omega_i t\right)\hat{I}_{i-}.$$

From these it follows that a product such as $\hat{I}_{1+}\hat{I}_{2-}$, which has $p = 0$, evolves according to

$$\hat{I}_{1+}\hat{I}_{2-} \xrightarrow{\;\Omega_1 t \hat{I}_{1z}\;} \exp\left(-\mathrm{i}\Omega_1 t\right)\hat{I}_{1+}\hat{I}_{2-} \xrightarrow{\;\Omega_2 t \hat{I}_{2z}\;} \exp\left(+\mathrm{i}\Omega_2 t\right)\exp\left(-\mathrm{i}\Omega_1 t\right)\hat{I}_{1+}\hat{I}_{2-}.$$

The overall result is

$$\hat{I}_{1+}\hat{I}_{2-} \xrightarrow{\;\Omega_1 t \hat{I}_{1z}+\Omega_2 t \hat{I}_{2z}\;} \exp\left(+\mathrm{i}[-\Omega_1 + \Omega_2]t\right)\hat{I}_{1+}\hat{I}_{2-},$$

from which we recognize $[-\Omega_1 + \Omega_2]$ as the zero-quantum frequency. Similarly, for the $p = +2$ order term $\hat{I}_{1+}\hat{I}_{2+}$, the evolution is

$$\hat{I}_{1+}\hat{I}_{2+} \xrightarrow{\;\Omega_1 t \hat{I}_{1z}+\Omega_2 t \hat{I}_{2z}\;} \exp\left(-\mathrm{i}[\Omega_1 + \Omega_2]t\right)\hat{I}_{1+}\hat{I}_{2+},$$

where $[\Omega_1 + \Omega_2]$ is the double-quantum frequency.

In general, if in an operator product the coherence order of spin one is p_1, and that of spin two is p_2, the product will evolve over time according to

$$\hat{\rho}^{(p_1+p_2)} \xrightarrow{\;\text{free evolution for time } t\;} \hat{\rho}^{(p_1+p_2)} \times \exp\left(-\mathrm{i}\Omega^{(p_1+p_2)}t\right),$$

where

$$\Omega^{(p_1+p_2)} = p_1\Omega_1 + p_2\Omega_2.$$

If scalar coupling is taken into account the evolution becomes more complex, but for the present purposes we can ignore evolution of the coupling as it does not lead to a change in the coherence order, and so does not affect the process of coherence selection.

11.1.3   The effect of pulses

In principle, an RF pulse will cause any coherences present to be transferred to all possible coherence orders: there are no selection rules as to which transfers are allowed. However, at the detailed level, we know that whether or not a pulse will *actually* generate a certain coherence depends on the presence or otherwise of anti-phase states. For example, $\hat{I}_{1x}$ has coherence order ± 1, but a 90° pulse will not transfer this state to double quantum. On the other hand, $2\hat{I}_{1x}\hat{I}_{2z}$, which is also $p = \pm 1$, *will* be transferred into double- and zero-quantum coherence by a $90°(x)$ pulse.

A special case which will be of some interest to us is that a pulse applied to equilibrium magnetization, such as $\hat{I}_{1z}$ (coherence order zero), can *only* generate coherence orders ± 1. This is immediately clear as we know that a pulse can only generate $\hat{I}_{1x}$ or $\hat{I}_{1y}$ from $\hat{I}_{1z}$. It is not possible to generate multiple quantum directly from the equilibrium z-magnetization.

In some cases, the amount of a particular coherence order which is generated by transfer from another order depends on the flip angle of the pulse. Such effects turn out to be important in two-dimensional spectroscopy, so we will investigate them here.

Let us start with $\hat{I}_{iz}$, $p = 0$, and consider the effect of applying a pulse of flip angle θ about the x-axis:

$$\hat{I}_{iz} \xrightarrow{\theta \hat{I}_{ix}} \cos\theta\,\hat{I}_{iz} - \sin\theta\,\hat{I}_{iy}.$$

The operator $\hat{I}_{iy}$ can be rewritten in terms of $\hat{I}_{i+}$ and $\hat{I}_{i-}$ using Eq. 11.4 on page 383 to give

$$\hat{I}_{iz} \xrightarrow{\theta \hat{I}_{ix}} \cos\theta\,\hat{I}_{iz} - \tfrac{1}{2i}\sin\theta\left(\hat{I}_{i+} - \hat{I}_{i-}\right).$$

This equation tells us that, when applied to $\hat{I}_{iz}$, a pulse of any flip angle generates *equal* amounts of coherence orders ± 1.

If we apply a pulse to $\hat{I}_{i+}$ the situation is a little more complex. Writing $\hat{I}_{i+}$ as $\hat{I}_{ix} + i\hat{I}_{iy}$ allows us to work out the effect of the pulse in the usual way:

$$
\begin{aligned}
\hat{I}_{i+} \equiv \hat{I}_{ix} + i\hat{I}_{iy} \xrightarrow{\theta \hat{I}_{ix}} \ & \hat{I}_{ix} + i\cos\theta\,\hat{I}_{iy} + i\sin\theta\,\hat{I}_{iz} \\
= \ & \tfrac{1}{2}\left(\hat{I}_{i+} + \hat{I}_{i-}\right) + \tfrac{1}{2i}i\cos\theta\left(\hat{I}_{i+} - \hat{I}_{i-}\right) + i\sin\theta\,\hat{I}_{iz} \\
= \ & \tfrac{1}{2}\left(1 + \cos\theta\right)\hat{I}_{i+} + \tfrac{1}{2}\left(1 - \cos\theta\right)\hat{I}_{i-} + i\sin\theta\,\hat{I}_{iz} \\
= \ & \cos^2\left(\tfrac{1}{2}\theta\right)\hat{I}_{i+} + \sin^2\left(\tfrac{1}{2}\theta\right)\hat{I}_{i-} + i\sin\theta\,\hat{I}_{iz}. \qquad (11.5)
\end{aligned}
$$

In this calculation, to go to the second line we have used the identities of Eq. 11.4 on page 383, and to go to the last line we have used the identities

$$\cos^2\left(\tfrac{1}{2}\theta\right) \equiv \tfrac{1}{2}\left(1 + \cos\theta\right) \qquad \sin^2\left(\tfrac{1}{2}\theta\right) \equiv \tfrac{1}{2}\left(1 - \cos\theta\right).$$

Using a similar approach we can show that

$$\hat{I}_{i-} \equiv \hat{I}_{ix} - i\hat{I}_{iy} \xrightarrow{\theta \hat{I}_{ix}} \cos^2\left(\tfrac{1}{2}\theta\right)\hat{I}_{i-} + \sin^2\left(\tfrac{1}{2}\theta\right)\hat{I}_{i+} - i\sin\theta\,\hat{I}_{iz}. \qquad (11.6)$$

What Eqs 11.5 and 11.6 tell us is that the amount of transfer from $\hat{I}_{i+}$ to $\hat{I}_{i-}$ (and vice versa) depends on the flip angle of the pulse. For the special case that $\theta = \pi/2$ (a 90° pulse), $\sin(\pi/4) = 1/\sqrt{2}$, $\cos(\pi/4) = 1/\sqrt{2}$ and so $\sin^2(\pi/4) = \cos^2(\pi/4) = \frac{1}{2}$; there is therefore equal transfer to $\hat{I}_{i+}$ and $\hat{I}_{i-}$. As we will see, this special property of 90° pulses will turn out to be important in two-dimensional spectroscopy.

180° pulses

The effect of 180° pulses is rather simple, as in Eqs 11.5 and 11.6 for the case $\theta = \pi$, $\cos^2(\frac{1}{2}\theta) = 0$, $\sin^2(\frac{1}{2}\theta) = 1$ and $\sin\theta = 0$. The effect of such a 180° pulse is therefore:

$$\hat{I}_{i+} \xrightarrow{\pi\hat{I}_{ix}} \hat{I}_{i-} \qquad \hat{I}_{i-} \xrightarrow{\pi\hat{I}_{ix}} \hat{I}_{i+}.$$

We see that all a 180° pulse does is to *reverse the sign* of the coherence order.

An operator product with an overall order of +2 is thus changed to −2 by a 180° pulse:

$$\underbrace{\hat{I}_{1+}\hat{I}_{2+}}_{p=+2} \xrightarrow{\pi(\hat{I}_{1x}+\hat{I}_{2x})} \underbrace{\hat{I}_{1-}\hat{I}_{2-}}_{p=-2}$$

11.1.4   Observables

In an NMR experiment, it is the x- and y-magnetizations which we ultimately observe, and these magnetizations are represented by the operators $\hat{I}_{ix}$ and $\hat{I}_{iy}$. As we have seen, these operators both have coherence orders +1 and −1, so it is clear that $p = \pm 1$ are the *only* observable coherences. This is hardly a surprise, as we are used to the idea that we can only observe single-quantum coherence.

In section 5.2 on page 82 we saw that the usual procedure is to combine the observed signals from the x- and y-magnetizations into a complex time-domain signal, $S(t)$:

$$S(t) \propto \left[M_x(t) + iM_y(t)\right].$$

Constructing the observable signal $S(t)$ in this way can be shown to be equivalent to detecting *only* coherence order $p = -1$.

It is somewhat arbitrary as to whether we detect $p = -1$ or $p = +1$. However, if a complex signal $S(t)$ is constructed in the way described, it is definite that only *one* out of $p = \pm 1$ is observed. We will assume that it is $p = -1$ which is observable.

11.1.5   Summary

Coherence order is a key concept in this chapter, so let us summarize its properties.

- Coherence order, p, is defined by the response of an operator to a rotation about the z-axis. An operator of order p acquires a phase of $(-p\phi)$ when subject to a rotation of angle ϕ about the z-axis.

- The operators $\hat{I}_{iz}$, $\hat{I}_{i+}$ and $\hat{I}_{i-}$ have coherence orders 0, +1 and −1, respectively. The overall coherence order of a product of operators can be found by adding together the coherence orders of each operator in the product.

- The following identities are useful in assigning coherence orders:

$$\hat{I}_{ix} \equiv \tfrac{1}{2}\left(\hat{I}_{i+} + \hat{I}_{i-}\right) \qquad \hat{I}_{iy} \equiv \tfrac{1}{2i}\left(\hat{I}_{i+} - \hat{I}_{i-}\right).$$

- For a system of N coupled spins one-half, the coherence order can take all values between $-N$ and $+N$, in integer steps, including zero.

- Under free evolution, an operator (or product of operators) simply acquires a phase factor $\exp\left(-i\Omega^{(p_1+p_2+\ldots)}t\right)$, where the frequency $\Omega^{(p_1+p_2+\ldots)}$ is determined by the offsets and coherence orders of the individual operators in the product:

$$\Omega^{(p_1+p_2+\ldots)} = p_1\Omega_1 + p_2\Omega_2 + \ldots$$

- A pulse applied to equilibrium magnetization generates equal amounts of $p = +1$ and $p = -1$ coherence, and no higher orders.

- A 90° pulse has the special property of causing equal amounts of coherence to be transferred into order $+p$ as into order $-p$.

- A 180° pulse simply reverses the sign of the coherence order.

- Only coherence order −1 is observable.

11.2 Coherence transfer pathways

A convenient way of describing which coherences are desired at each stage in a pulse sequence is to draw a *coherence transfer pathway* (CTP) underneath the pulse sequence. Several examples of such pathways are shown in Fig. 11.2 on the following page.

The thick blue line in the CTP shows the coherence order, or orders, which we want to be present at each point in the sequence, and the transfers between these orders caused by the pulses. Note that during a delay the order remains constant, but that a pulse causes the orders to change. It is important to realize that the specified CTP is the one which we want to contribute to the spectrum. There will be many other pathways, not shown, which will also contribute. By using phase cycling or gradient pulses it is our task to select the required CTP and reject all others.

Sequence (a) is for DQF COSY. We start with coherence order $p = 0$, corresponding to equilibrium z-magnetization, from which the first 90° pulse generates $p = \pm 1$. Recall that this 90° pulse creates equal amounts of $p = +1$ and $p = -1$, and for reasons which will be discussed below it is important to retain both $p = +1$ and $p = -1$ during t_1.

The second 90° pulse transfers the single quantum into double quantum, $p = \pm 2$. Note that we have allowed for all possible transfers by this pulse: $+1 \rightarrow +2$, $+1 \rightarrow -2$, $-1 \rightarrow +2$ and $-1 \rightarrow -2$; not to allow all of

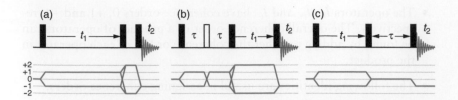

Fig. 11.2 Three pulse sequences with their corresponding coherence transfer pathways shown underneath. The grey 'tram lines' show the possible coherence orders, p, that can be present, which in this case we have restricted to the range -2 to $+2$, shown on the left. The thick blue line shows the desired order, or orders, of coherence present at each point in the sequence. During the delays the coherence order remains constant; in contrast, pulses cause a change in coherence order. The sequences are: (a) DQF COSY, (b) double-quantum spectroscopy, and (c) NOESY.

these transfers would result in a loss of signal. Finally, the third 90° pulse transfers $p = \pm 2$ to $p = -1$. Since only $p = -1$ is observable, we need only concern ourselves with the pathway that ends up with this order.

Sequence (b) is for double-quantum spectroscopy. Again, we start with $p = 0$, and the first pulse generates $p = \pm 1$. The 180° pulse just swaps the sign of the coherence order, so $+1 \to -1$ and $-1 \to +1$. The second 90° pulse generates double quantum and, as for DQF COSY, we allow all possible transfers between ± 1 and ± 2. As the pulse is 90°, equal amounts of $p = +2$ and $p = -2$ will be generated. The final pulse transfers the double quantum to observable magnetization, which has $p = -1$.

Finally, sequence (c) is NOESY. The first part of the sequence is as for DQF COSY, but this time the second 90° pulse is required to generate z-magnetization, which has $p = 0$. After the mixing time τ, the final 90° pulse generates observable $p = -1$. We can see that these coherence transfer pathways are convenient ways of expressing the desired outcome of an experiment.

11.2.1   Coherence transfer pathways in heteronuclear experiments

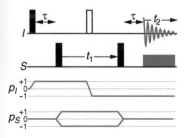

Fig. 11.3 Pulse sequence and coherence transfer pathways for the HMQC experiment. Note that separate coherence orders, p_I and p_S, are specified for the I and S spins, respectively.

In heteronuclear experiments it is sometimes convenient to write a separate coherence transfer pathway for each type of nucleus. As an example, Fig. 11.3 shows the pulse sequence and separate CTPs for the HMQC experiment. The coherence order for the I spin (typically proton) is given by the pathway labelled p_I, and that for the S spin (the heteronucleus) is given by the pathway labelled p_S.

There are several things to note about these pathways. The first is that the coherence order of the I spin only changes when pulses are applied to that spin: pulses to the S spin have no effect on the coherence order p_I. Secondly, as we observe the I-spin magnetization in this experiment, the CTP for this spin must end up at $p_I = -1$. In addition, the CTP for the S spin must end up at $p_S = 0$, so that the overall order, $(p_I + p_S)$, is -1. If p_S were not zero, then we would have a state of heteronuclear multiple-quantum coherence, which is not observable.

After the first pulse to the S spin, we have $p_I = +1$ and $p_S = \pm 1$; the overall order is therefore 0 or 2, which corresponds to heteronuclear double- and zero-quantum coherence. The 180° pulse to I simply inverts the sign of p_I, which in this case is equivalent to interchanging the double- and zero-quantum coherence. Note that the $p_I = -1$ coherence generated by the first pulse will be switched to $p_I = +1$ by the 180° pulse, and so will not be observable – this is why this pathway is not shown. The final 90° pulse to S transfers $p_S = \pm 1$ to $p_s = 0$, thus making the signal observable on the I spin.

Before we get down to the nitty-gritty of how these desired pathways are selected we need to explore the important topic of how a CTP is related to frequency discrimination and lineshape in two-dimensional spectra.

11.3   Frequency discrimination and lineshapes

This topic has been discussed before in section 8.12 on page 226, but we now need to revisit the topic in order to see how it is related to CTPs. It is helpful first to summarize the key results from section 8.12:

- Most two-dimensional experiments generate time-domain functions which are cosine or sine modulated as a function of t_1 i.e. data of the form

$$\cos(\Omega_A t_1)\exp(i\Omega_B t_2) \quad \text{or} \quad \sin(\Omega_A t_1)\exp(i\Omega_B t_2);$$

such data sets are said to be *amplitude modulated* in t_1. Fourier transformation of such data gives spectra which lack frequency discrimination in the ω_1 dimension, i.e. peaks at $\pm\Omega_A$ are not distinguished, which leads to confusion in the spectrum.

- It is usually possible, by simple modifications to the pulse sequence, to record both a cosine and a sine modulated data set in separate experiments.

- The cosine and sine modulated data sets can be combined to give a phase modulated signal of the form

$$\exp(i\Omega_A t_1)\exp(i\Omega_B t_2).$$

Such a signal gives a spectrum with frequency discrimination in the ω_1 dimension, but which has the unfavourable phase-twist lineshape.

- If cosine and sine modulated data sets are available, it is possible to process the data in such a way as to obtain *both* frequency discrimination and absorption mode lineshapes (i.e. avoiding the phase-twist lineshape).

The way in which these important ideas are related to CTPs is best illustrated by the example of a simple COSY spectrum. The COSY pulse sequence, along with three different CTPs, is shown in Fig. 11.4 on the next page. All pathways start with $p = 0$, and end with the $p = -1$, which is the observable coherence order. The difference between the pathways is in the coherence order(s) present during t_1: in (a) $p = +1$ is present, in (b)

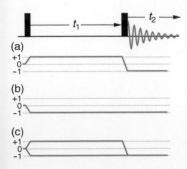

Fig. 11.4 The COSY pulse sequence, with three possible coherence transfer pathways. Pathway (a) gives the N-type or echo spectrum, (b) gives the P-type or anti-echo spectrum, and (c) gives amplitude modulated data.

$p = -1$ is present, and in (c) both $p = +1$ and $p = -1$ are present. As we shall see, these differences have a large influence on the detailed form of the spectrum.

Let us first consider CTP (a). From the discussion in section 11.1.2 on page 384, we know that the evolution of the $p = +1$ coherence during t_1 results in the signal acquiring a phase $\exp(-i\Omega_A t_1)$, where Ω_A is the modulating frequency in t_1.

The final pulse transfers the coherence to order -1, and then evolution during t_2 gives a signal of the form $\exp(+i\Omega_B t_2)$; we have written the frequency in t_2 as Ω_B, to allow for the possibility that it might be different to that in t_1. The really important point here is that the *sense* of the modulation in t_1 and t_2 is opposite, on account of the change in the sign of the coherence order between t_1 and t_2.

The overall form of the two-dimensional time-domain signal will therefore be

$$\text{CTP (a)}: \quad \exp(-i\Omega_A t_1)\,\exp(+i\Omega_B t_2).$$

This signal is *phase modulated* in t_1, and so the spectrum will be sensitive to the sign of Ω_A i.e. the spectrum will be frequency discriminated. From the discussion in section 8.12.2 on page 228, we can deduce that Fourier transformation of this time-domain signal will give rise to a peak at $\{\omega_1, \omega_2\} = \{-\Omega_A, \Omega_B\}$, and that this peak will have the phase-twist lineshape.

For CTP (b) all that is different from (a) is that the coherence order present during t_1 is -1 rather than $+1$. This simply alters the sense of the phase modulation in t_1, so that the overall two-dimensional signal is of the form:

$$\text{CTP (b)}: \quad \exp(+i\Omega_A t_1)\,\exp(+i\Omega_B t_2).$$

Fourier transformation of this gives a phase-twist line at $\{+\Omega_A, \Omega_B\}$.

11.3.1 N- and P-type spectra

CTP (a) gives rise to what is called the N-type or echo spectrum. The N stands for 'negative', which comes from the observation that the coherence order in t_1 is of *opposite* sign to that in t_2. The term 'echo' comes from the fact that the final 90° pulse causes a change in coherence order from $+p$ to $-p$. This is the same effect as a 180° pulse, which is associated with the formation of a spin echo.

CTP (b) gives rise to the P-type or anti-echo spectrum. The P stands for 'positive', the name coming from the fact that the coherence order in t_1 and in t_2 have the *same* sign. The term 'anti-echo' indicates that, with no change of the sign of the coherence order, an echo will not be formed.

11.3.2 Amplitude modulated spectra

We now turn to CTP (c) of Fig. 11.4. Here, both $p = +1$ and $p = -1$ are present during t_1, so the modulation in this dimension is the sum of two phase modulations

$$\underbrace{\exp(-i\Omega_A t_1)}_{p=+1} + \underbrace{\exp(i\Omega_A t_1)}_{p=-1},$$

where we have assumed that $p = +1$ and $p = -1$ contribute equally. Of course, the sum of these two is just $2\cos(\Omega_A t_1)$, so the overall two-dimensional time-domain function is of the form

$$\text{CTP (c)}: \quad \cos(\Omega_A t_1)\,\exp(+i\Omega_B t_2).$$

This signal is amplitude modulated in t_1, and so, as described in section 8.12 on page 226, the result will be a spectrum which lacks frequency discrimination.

However, assuming that we have access to the sine modulated data (which we usually do), the cosine and sine modulated data sets can be combined in such a way as to give frequency discrimination and at the same time retain absorption mode lineshapes. How this is done is described in section 8.12.3 on page 230 and in the following section 8.12.4 on page 231.

The important point here is that in order for the signals from the $p = +1$ and $p = -1$ pathways to combine to give cosine (or sine) modulation in t_1, the two pathways must have the *same* overall amplitude.

11.3.3   Summary

In summary, we have shown that if one coherence of order either $+p$ or $-p$ is present during t_1, the resulting spectrum will be frequency discriminated, but will have the unfavourable phase-twist lineshape.

If coherence of *both* orders $+p$ and $-p$ is present during t_1, and if the two contribute equally, the resulting data will be amplitude modulated. As a result the spectrum will lack frequency discrimination. However, assuming that both cosine and sine modulated data sets can be recorded, it is possible to process the data in such a way as to achieve frequency discrimination and retain absorption mode lineshapes.

Selecting a CTP in which both orders $+p$ and $-p$ are present and contribute equally during t_1 is referred to as *selecting symmetrical pathways*. You will note that all of the pathways shown in Fig. 11.2 on page 388 have this property, and so all the spectra can be processed to give frequency discriminated spectra with absorption mode lineshapes.

11.4   The receiver phase

The ability to change the phase of our RF pulses is very important in multiple-pulse NMR experiments, and is of course crucial in the whole process of phase cycling. There is a further phase which is under our control in an NMR spectrometer, the *receiver phase*, and this too plays a vital part in the implementation of phase cycles.

In section 5.2 on page 82 we saw that the spectrometer is capable of making simultaneous measurements of the x- and y-components of the magnetization, and that the resulting detected signals are used to form the real and imaginary parts of a complex time-domain signal. To avoid the confusion of having too many x's and y's, we will call the outputs of these two detectors A and B, rather than referring to them as the x and y detectors.

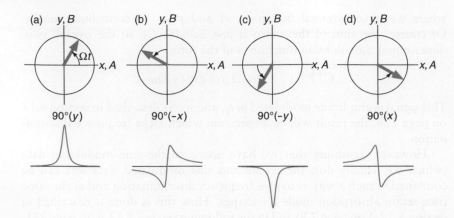

Fig. 11.5 Vector diagrams showing the result of a pulse–acquire experiment for different phases of the pulse. The position of the vector, which has offset Ω, is shown after free precession for time t, such that it has rotated through an angle (Ωt). The phase of the pulse is shown beneath each vector diagram. The magnetization along the x-axis gives rise to the detected signal A, and the magnetization along the y-axis gives rise to the signal B. Beneath each vector diagram is shown the *real* part of the spectrum which would result from Fourier transformation of the time-domain signal $(A + iB)$. Note how the phase of the line in the spectrum changes in response to the phase of the pulse.

Imagine a simple pulse–acquire experiment, in which the phase of the pulse is y, and where we have one line in the spectrum with offset Ω. The pulse rotates the magnetization onto the x-axis, and then free precession for time t results in the vector rotating through an angle (Ωt) towards the y-axis. The situation is depicted in Fig. 11.5 (a).

If for simplicity we assume that the equilibrium magnetization is of size one, then simple geometry tells us that $M_x = \cos(\Omega t)$ and $M_y = \sin(\Omega t)$. The detected signal A is thus $\cos(\Omega t)$, and B is $\sin(\Omega t)$. If we construct the time-domain signal from the combination $A + iB$, the result is

$$
\begin{aligned}
S_a(t) &= A + iB \\
&= \cos(\Omega t) + i\sin(\Omega t) \\
&= \exp(i\Omega t).
\end{aligned}
$$

Assuming the usual damping of this signal due to relaxation, Fourier transformation of this function will give rise to a spectrum in which the real part has an absorption mode line at frequency Ω. This is shown in the diagram, beneath the vector picture.

Now suppose that we shift the phase of the pulse from y to $-x$: the result is shown in Fig. 11.5 (b). It is now evident that the detected signal A will be $-\sin(\Omega t)$, and B will be $\cos(\Omega t)$. The combination $(A + iB)$, which we computed before, is now

$$
\begin{aligned}
S_{a'}(t) &= -\sin(\Omega t) + i\cos(\Omega t) \\
&= i(\cos\Omega t + i\sin\Omega t) \\
&= i\exp(i\Omega t) \\
&= \exp(i\pi/2)\exp(i\Omega t).
\end{aligned}
$$

The factor $\exp(i\pi/2)$ is a phase shift by $(\pi/2)$, and so the real part of the spectrum arising from this signal will show the dispersion mode lineshape (see section 5.3.2 on page 85); this is shown in the diagram. Not surprisingly, the phase of the line in the spectrum has been changed as a result of shifting the phase of the pulse.

However, we do not *have* to form the time domain from the combination $(A + iB)$, but can choose any combination we like. Suppose that for the case of the $90°(-x)$ pulse, (b) in the figure, we form the time-domain signal from the combination $(B - iA)$:

$$
\begin{aligned}
S_b(t) &= B - iA \\
&= \cos(\Omega t) - i[-\sin(\Omega t)] \\
&= \cos(\Omega t) + i\sin(\Omega t) \\
&= \exp(i\Omega t).
\end{aligned}
$$

Now we have exactly the same time-domain signal as we did for the $90°(y)$ pulse, and so will obtain an absorption mode lineshape in the real part of the spectrum.

In a similar way we can find a combination of signals A and B arising from the experiment with a $90°(-y)$ pulse, and with a $90°(x)$ pulse, which will give an absorption mode lineshape in the real part of the spectrum. All four combinations are summarized in the following table (the labels in the first column refer to the parts of Fig. 11.5 on the facing page):

	pulse phase	A	B	combination	rx phase
(a)	y	$\cos(\Omega t)$	$\sin(\Omega t)$	$(A + iB)$	0
(b)	$-x$	$-\sin(\Omega t)$	$\cos(\Omega t)$	$(B - iA)$	$-\pi/2$
(c)	$-y$	$-\cos(\Omega t)$	$-\sin(\Omega t)$	$(-A - iB)$	$-\pi$
(d)	x	$\sin(\Omega t)$	$-\cos(\Omega t)$	$(-B + iA)$	$-3\pi/2$

What we are seeing here is that, although changing the phase of the pulse changes the detected signals A and B, we can always find a combination of these signals which, after Fourier transformation, will give us an absorption mode spectrum. Changing the combinations in this way is usually described as changing the *phase of the receiver*.

The four combinations of A and B listed in the table are related to one another simply by multiplying the entry on the previous row by $-i$. For example, to go from the second to the third row we would simply compute:

$$
-i \times (B - iA) = -A - iB.
$$

Given that $\exp(-i\pi/2) = -i$, multiplying by $-i$ is the same thing as applying a phase shift of $(-\pi/2)$. Thus, the combinations of A and B in the table can be regarded as arising from a shift in the phase of the receiver by $(-\pi/2)$ each time we move from one row to the next. The total receiver phase shift, relative to that in the first experiment, is also shown in the table under the column headed 'rx phase'. Arbitrarily, we will assign a receiver phase of zero to the combination $A + iB$; such a phase can also be described as 'having the receiver along x'.

Fig. 11.6 Illustration of how both the phase of the pulse and of the receiver affects the lineshape in the spectrum; the black dot • indicates the phase of the receiver. The vector diagrams in (a) are the same as those in Fig. 11.5 on page 392. Since the receiver phase is fixed the four spectra have different lineshapes. In (b) the pulses go through the same phases as in (a), but this time the receiver phase is shifted by −90° each time the phase of the pulse advances (receiver phase shifts are measured *clockwise* from the x-axis). Consequently the magnetization is in a *fixed* position relative to the receiver phase and as a result each spectrum has the same lineshape.

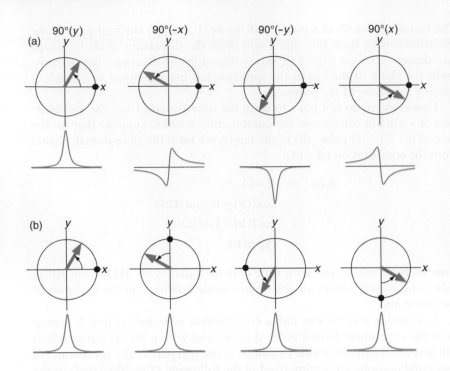

If we record four separate pulse–acquire experiments in which the phase of the pulse goes through the sequence [y, −x, −y, x], each will give a different lineshape, as shown in Fig. 11.5 on page 392. Adding up the results of these four experiments will result in complete cancellation, as the lines for pulse phases y and −y, and for phases −x and x, are equal and opposite.

However, if the receiver phase goes through the sequence of phases [0, −π/2, −π, −3π/2] in the four experiments, each will give the absorption mode lineshape, and so the spectra from all four experiments will add up. Figure 11.6 illustrates this idea. In (a) the phase of the receiver, indicated by the black dot •, remains fixed, so changing the phase of the pulse changes the phase of the spectrum. In (b), the receiver changes phase by −90° each time the pulse phase changes. As a result, the angle between the axis denoted by the black dot and the magnetization is fixed, and so the four spectra all have the same lineshape. We say that in (b) the receiver phase *follows* the phase shift of the magnetization, which in turn comes from the phase shift of the pulse.

It is rather inconvenient that, for the receiver phase to follow the magnetization, when the pulse phase *advances* in the sequence [y, −x, −y, x], we have to move the receiver phase in the *opposite* direction through the sequence of phases [0, −π/2, −π, −3π/2]. To get round this, the spectrometer software is configured so that requesting the receiver phases [0, π/2, π, 3π/2] will result in the signals adding up when the pulse goes through the sequence [y, −x, −y, x]. In other words, the software sorts out the minus sign for us.

11.4.1   Specifying the receiver phase

We can specify the receiver phase in any way we like, either in radians – as we have been doing up to now – or in degrees. So the sequence $[0, \pi/2, \pi, 3\pi/2]$, can be written $[0°, 90°, 180°, 270°]$. Equally well, rather than specifying the angle we can specify the axis, so that the sequence of phases can be written $[x, y, -x, -y]$.

The *absolute* receiver phase is not important, all that matters is by how much the phase advances in each step. Therefore in the pulse–acquire experiment we have been discussing, the sequence of receiver phases can just as well be $[-x, -y, x, y]$ or $[y, -x, -y, x]$ or $[270°, 0°, 90°, 180°]$. The absolute phase will affect the lineshape, but we can always adjust it to what we want by phasing the spectrum.

11.4.2   The take-home message

The key point to take away from this section is that if the signal generated by the pulse sequence acquires a phase shift, we can compensate for this by shifting the receiver phase by the *same* amount. If we do this, the signal will add up: this is a crucial part of how phase cycling works.

11.5   Introducing phase cycling

Coherence selection by phase cycling relies on the following property in relation to the general pathway illustrated in Fig. 11.7 (a):

If a pulse causes a change in coherence order from p_1 to p_2, then shifting the phase of the pulse by $\Delta\phi$ results in the coherence acquiring a phase shift of

$$-\Delta p \times \Delta\phi,$$

where Δp is the change in coherence order, given by

$$\Delta p = (p_2 - p_1).$$

This property means that pathways with different Δp acquire different phase shifts: as a result it is possible for us to differentiate between pathways. In practice, the way we do this is to repeat the experiment several times, with different values of $\Delta\phi$, and then combine the results in such a way that the signals which derive from the wanted pathway add up, whereas signals from all other pathways cancel.

It is important to realize that the phase which a coherence acquires as a result of following a particular pathway is carried forward with that coherence through the rest of the pulse sequence. So, if the coherence ends up contributing to the observed signal, then this signal will have the phase shift which the coherence acquired earlier in the sequence.

Our ability to alter the receiver phase is important in making sure that the wanted signals from different experiments add up. As we saw in section 11.4 on page 391, if we adjust the receiver phase so that it 'follows'

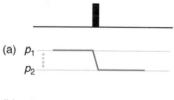

(a)
(b)

Fig. 11.7 Two examples of a single coherence transfer step brought about by a pulse. In (a) the change is from coherence order p_1 to order p_2, such that the change in coherence order, Δp, is $(p_2 - p_1)$. If the phase of the pulse is shifted by $\Delta\phi$, the coherence following the pathway shown acquires a phase $-\Delta p \times \Delta\phi$. Pathway (b) is for the particular case of $+2 \to -1$, which has $\Delta p = -3$, and is discussed in the text.

the phase shifts of the signal, then these will add up. On the other hand, if the receiver phase does not follow that of the signal, the latter will cancel.

How this all works is best seen using an example, which will be the selection of the pathway $+2 \rightarrow -1$, shown in Fig. 11.7 (b).

11.5.1   Selection of a single pathway

In this section we will concentrate on devising a phase cycle which selects the pathway $+2 \rightarrow -1$, which has $\Delta p = -3$. Imagine that we advance the phase of the pulse through the sequence x, y, $-x$, $-y$, or in degrees $0°$, $90°$, $180°$, $270°$. The phase shifts experienced by the pathway will be $-(-3 \times \Delta\phi) = 3\,\Delta\phi$; these are listed in the following table:

step	phase shift of pulse, $\Delta\phi$	$3\,\Delta\phi$	equiv($3\,\Delta\phi$)
1	0°	0°	0°
2	90°	270°	270°
3	180°	540°	180°
4	270°	810°	90°

The fourth column, headed 'equiv($3\,\Delta\phi$)', is just the phase in the third column reduced to the range 0–360°. The idea here is that a phase of $\phi + n \times 360°$, where n is an integer, is equivalent in every way to a phase of ϕ. We can therefore add or subtract multiples of 360° from any phase until it is in the range 0–360° without any loss of information. For example 540° is the same as 180° since $540° - 360° = 180°$, and 810° is the same as 90° since $810° - 2 \times 360° = 90°$. Reducing the phases in this way makes them a lot easier to comprehend.

What the table tells us is that as the phase of the pulse goes

$$\text{pulse phase:} \quad [0°, 90°, 180°, 270°]$$

the pathway with $\Delta p = -3$ acquires a phase which goes

$$\text{phase for } \Delta p = -3: \quad [0°, 270°, 180°, 90°].$$

If we want the signals in these four steps to add up, all we need to do is to make the *receiver* phase shift match that acquired by the $\Delta p = -3$ pathway on each step i.e. the sequence

$$\text{rx phase to select } \Delta p = -3: \quad [0°, 270°, 180°, 90°].$$

Now that we have seen how to select a particular pathway, we need to check that other pathways are rejected, as this is ultimately what we are trying to achieve. Let us consider the pathway with $\Delta p = -2$, for which the phase shift experienced will be $-(-2 \times \Delta\phi) = 2\,\Delta\phi$; the table below gives the result for each step in the phase cycle:

step	phase shift of pulse, $\Delta\phi$	$2\,\Delta\phi$	equiv($2\,\Delta\phi$)
1	0°	0°	0°
2	90°	180°	180°
3	180°	360°	0°
4	270°	540°	180°

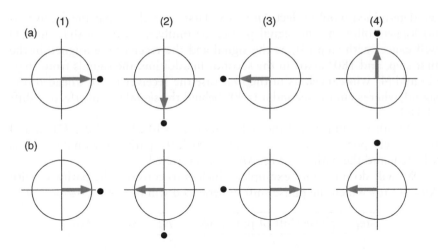

Fig. 11.8 Diagrammatic way of working out whether or not a particular set of receiver phase shifts will result in the selection or rejection of a particular pathway. The blue arrow indicates the phase shift acquired by the signal from a particular pathway as the pulse phase goes through the four steps [0°, 90°, 180°, 270°]. The phase of the receiver is shown by the black dot •, and both the receiver and signal phases are measured anti-clockwise from 3 o'clock. In (a) is shown the signal phases for $\Delta p = -3$, as given in the first table on page 396. The receiver phase goes [0°, 270°, 180°, 90°], so that the arrow and the black dot are always aligned: the signals from this pathway will therefore add up. In (b) the signal phases are for the pathway $\Delta p = -2$, as given in the second table on page 396, but the receiver phases are those used to select $\Delta p = -3$. Now the receiver phase no longer follows the phase of the signal, and in fact steps (1) and (3), and (2) and (4) cancel one another. The pathway with $\Delta p = -2$ is therefore rejected. It is important to realize that these are not vector diagrams in the NMR sense, but simply diagrammatic representations of various phase shifts.

The phase shifts given in the last column are, of course, different from those computed for the pathway $\Delta p = -3$. The question is, if we use the receiver phase shifts [0°, 270°, 180°, 90°] will the pathway with $\Delta p = -2$ cancel?

One way of determining this is shown in Fig. 11.8. In this figure, the blue vector represents the phase which the coherence has acquired as a result of shifting the phase of the pulse, and the phase of the receiver is indicated by the black dot. Both phases are measured anti-clockwise from 3 o'clock.

In (a) the arrows show the phases acquired by the pathway with $\Delta p = -3$ as the pulse is advanced through the four steps [0°, 90°, 180°, 270°] i.e. the phases from the final column in the first table on page 396. The receiver phases, indicated by the black dot, go through the sequence [0°, 270°, 180°, 90°], which is what we determined would result in the signals from the pathway with $\Delta p = -3$ adding up. It is clear from the diagram that, since the receiver phase follows the phase of the signal, the signals from all four steps will add up.

In Fig. 11.8 (b) we see the signal phases for the pathway with $\Delta p = -2$ (given in the second table on page 396), and the set of receiver phases which we determined would select $\Delta p = -3$. First, it is clear that the receiver is no longer following the signal phase. Secondly, we can see that step (1) will cancel with step (3), as the signal and the receiver are aligned in the first step, and 180° apart in the second. In addition, the signals from steps (2) and (4) will cancel, as in the first case the receiver is 90° ahead of the signal, whereas in the second it is 90° behind the signal i.e. an effective shift of 180°.

Overall we find that if the pulse phase goes [0°, 90°, 180°, 270°], and the receiver phase goes [0°, 270°, 180°, 90°], the pathway with $\Delta p = -3$ is selected, and the pathway with $\Delta p = -2$ is rejected.

We will do one more example, which is to consider the pathway with $\Delta p = +1$ for which the phase shift experienced will be $-(+1 \times \Delta\phi) = -\Delta\phi$.

step	phase shift of pulse, $\Delta\phi$	$-\Delta\phi$	equiv($-\Delta\phi$)
1	0°	0°	0°
2	90°	−90°	270°
3	180°	−180°	180°
4	270°	−270°	90°

This time the phases in the fourth column, headed 'equiv($-\Delta\phi$)', were obtained by *adding* 360° to the phases in the third column.

The phase shifts in the final column of the above table are *identical* to those in the first table on page 396, which were computed for the pathway $\Delta p = -3$. Therefore if we used the sequence of receiver phases [0°, 270°, 180°, 90°] the signals from *both* pathways will add up.

If we work through more values of Δp, we will find that this four-step cycle rejects $\Delta p = -2, -1, 0, 2, 3,$ and 4, but selects $\Delta p = -3, 1$ and 5. A convenient way of specifying this selectivity is to write out the possible values of Δp in a line, putting brackets around those which are rejected and emboldening those which are selected:

$$(-4) \quad \mathbf{-3} \quad (-2) \quad (-1) \quad (0) \quad \mathbf{1} \quad (2) \quad (3) \quad (4) \quad \mathbf{5}$$

Note that the selected values of Δp differ by four, which is no coincidence as this is the number of steps in the phase cycle. We can understand how this comes about in the following way.

Suppose that we have a pathway with a particular value of Δp, and that we then shift the phase of the pulse by $\frac{\pi}{2}$ (90°). This pathway will therefore experience a phase shift of $(-\Delta p \times \frac{\pi}{2})$. Now consider a second pathway which has a change in coherence order of $(\Delta p + 4)$. The effect of a $\frac{\pi}{2}$ phase shift of the pulse on such a pathway is:

$$
\begin{aligned}
-(\Delta p + 4) \times \tfrac{\pi}{2} &= -\Delta p \times \tfrac{\pi}{2} - 4 \times \tfrac{\pi}{2} \\
&= -\Delta p \times \tfrac{\pi}{2} - 2\pi \\
&= -\Delta p \times \tfrac{\pi}{2}.
\end{aligned}
$$

To go to the last line we have used the fact that a phase shift of 2π (360°) has no effect. What we see here is that a pathway with $(\Delta p + 4)$ experiences the *same* phase shift as one with Δp. Therefore, selecting one will automatically select the other.

Put more generally, this property of a phase cycle can be expressed as follows:

> If the phase of a pulse is cycled through a complete series of N steps
>
> $$0, \delta, 2\delta, \ldots (N-1)\delta,$$
>
> where the phase increment δ is $360°/N$, then selecting a pathway with a particular value of Δp will also result in the selection of pathways with $(\Delta p + nN)$, where n is a positive or negative integer.

This lack of selectivity of a phase cycle might at first appear to be something of a drawback, but we will see that it turns out to be quite useful to be able to select more than one pathway at the same time.

11.5.2 Combining phase cycles

Suppose that we have a pulse sequence containing two pulses, and we want to select $\Delta p = +1$ for the first pulse, and $\Delta p = -2$ for the second, as shown in Fig. 11.9. For the first pulse the phase acquired by the pathway is $-\Delta p \times \Delta\phi_1 = -\Delta\phi_1$. Using this we can construct a four-step cycle to select $\Delta p = +1$, as shown in the following table:

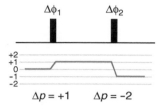

Fig. 11.9 A coherence transfer pathway involving two changes of coherence order, one with $\Delta p = +1$, and one with $\Delta p = -2$. The phase shifts of the two pulses are $\Delta\phi_1$ and $\Delta\phi_2$.

step	phase shift of pulse, $\Delta\phi_1$	$-\Delta\phi_1$	equiv($-\Delta\phi_1$)
1	$0°$	$0°$	$0°$
2	$90°$	$-90°$	$270°$
3	$180°$	$-180°$	$180°$
4	$270°$	$-270°$	$90°$

For the second pulse, the phase acquired by the pathway is $-\Delta p \times \Delta\phi_2 = 2\,\Delta\phi_2$. The table below gives the results for a four-step cycle:

step	phase shift of pulse, $\Delta\phi_2$	$2\,\Delta\phi_2$	equiv($2\,\Delta\phi_2$)
1	$0°$	$0°$	$0°$
2	$90°$	$180°$	$180°$
3	$180°$	$360°$	$0°$
4	$270°$	$540°$	$180°$

To select both pathways, we need to complete both four-step phase cycles *independently* of one another. This means that as $\Delta\phi_1$ goes through the sequence $[0°, 90°, 180°, 270°]$ and the receiver follows with $[0°, 270°, 180°, 90°]$, the phase of the second pulse must be held constant.

Having completed these four steps, the phase of the second pulse can be shifted by $90°$, and this is then held constant as the first pulse is again shifted through $[0°, 90°, 180°, 270°]$. However, this time the phase shift of $180°$ which results from shifting the phase of the *second* pulse, must be added to phase shifts which come from shifting the first pulse.

Table 11.1 Construction of the 16-step phase cycle needed to select the pathway shown in Fig. 11.9 on the preceding page.

step	$\Delta\phi_1$	$-\Delta\phi_1$	equiv($-\Delta\phi_1$)	$\Delta\phi_2$	$2\Delta\phi_2$	equiv($2\Delta\phi_2$)	total phase
1	0°	0°	0°	0°	0°	0°	0°
2	90°	−90°	270°	0°	0°	0°	270°
3	180°	−180°	180°	0°	0°	0°	180°
4	270°	−270°	90°	0°	0°	0°	90°
5	0°	0°	0°	90°	180°	180°	180°
6	90°	−90°	270°	90°	180°	180°	90°
7	180°	−180°	180°	90°	180°	180°	0°
8	270°	−270°	90°	90°	180°	180°	270°
9	0°	0°	0°	180°	360°	0°	0°
10	90°	−90°	270°	180°	360°	0°	270°
11	180°	−180°	180°	180°	360°	0°	180°
12	270°	−270°	90°	180°	360°	0°	90°
13	0°	0°	0°	270°	540°	180°	180°
14	90°	−90°	270°	270°	540°	180°	90°
15	180°	−180°	180°	270°	540°	180°	0°
16	270°	−270°	90°	270°	540°	180°	270°

The result of this is the following set of phases

$$[0° + 180°, 270° + 180°, 180° + 180°, 90° + 180°] \equiv [180°, 450°, 360°, 270°]$$

Reducing these to the range 0–360° gives [180°, 90°, 0°, 270°].

In the next four steps, the phase of the second pulse is shifted to 180°, and once again the first pulse is cycled through the four phases. Shifting the second pulse by 180° results in no shift in the phase of the signal, so for these four steps the total phase is [0°, 270°, 180°, 90°], just as it was for the first four steps.

Finally, the second pulse is shifted to 270°, which results in a phase shift of 180°, so the total phase for the last four steps is the same as for the second four: [180°, 90°, 0°, 270°]. Table 11.1 summarizes all of this for the 16 steps of the cycle.

In the table, the column headed 'total phase' is the phase acquired as a result of shifting both of the pulses. It is found by adding together the phases equiv($-\Delta\phi_1$) and equiv($2\Delta\phi_2$), and reducing the result to the range 0–360° in the usual way. To select these two pathways, the receiver phase would need to match the phases given in the right-hand column.

If we wanted to select three separate values of Δp, this would require $4^3 = 64$ independent steps: clearly, we want to avoid this if we can, as the experiment will become rather long. The next section explains how we can judiciously minimize the length of our phase cycles.

11.6 Some phase cycling 'tricks'

We need to be intelligent about how we go about applying phase cycling otherwise we will end up with absurdly long phase cycles which will be quite impractical. This section is about some 'tricks' which we can use in order to minimize the number of pulses which we need to cycle, and hence minimize the length of the phase cycle.

11.6.1 The first pulse

We noted in section 11.1.3 on page 385 that a pulse applied to equilibrium magnetization can only generate coherence orders ± 1. Often, this is exactly what we want – for example, it is a common feature of all the coherence transfer pathways in Fig. 11.2 on page 388. If this is the case, then there is no need to apply a phase cycle to the first pulse at all.

11.6.2 Grouping pulses together

In the previous section we described how a phase cycle can be constructed to select a particular value of Δp caused by a pulse. However, exactly the same considerations apply to the *overall* transformation brought about by a group of pulses: all that we have to do is to shift the phases of *all* the pulses at the same time. Any delays present between the pulses are not important, as the coherence order does not change during such periods.

A good example of the application of this idea is in the sequence used to generate double-quantum coherence, shown in Fig. 11.10. This sequence takes equilibrium magnetization, $p = 0$, and transforms part of it into double-quantum coherence which has $p = \pm 2$. The overall transformation caused by this group of three pulses therefore has $\Delta p = \pm 2$.

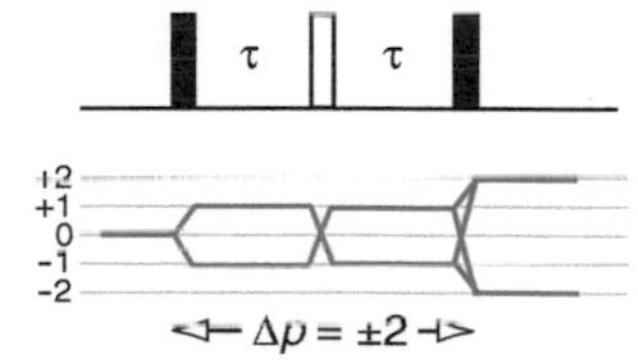

A four-step cycle in which all three pulses move through the steps $[0°, 90°, 180°, 270°]$ and the receiver phase goes $[0°, 180°, 0°, 180°]$ will select $\Delta p = +2$. In addition, since the cycle has four steps, we can deduce that it will also select $\Delta p = -2$, as this differs from $\Delta p = +2$ by four. Therefore the four-step cycle of the group of three pulses selects the transformation with $\Delta p = \pm 2$, which is what we require.

We do have to be cautious about one thing here. Just because we select $\Delta p = \pm 2$ overall, it does not mean that the detailed CTP drawn in Fig. 11.10 is selected. In fact, all coherence orders present during the spin echo will be selected, provided that they are transferred to $p = \pm 2$ by the final pulse.

In reality things are not as bad as they might seem. First, the initial pulse can only generate $p = \pm 1$, and secondly, the $180°$ pulse can only change the sign of the coherence order.[1] The pathways drawn during the spin echo are therefore the only possible ones, so we do in fact obtain the pathways specified.

Fig. 11.10 This three-pulse sequence can be used to generate double quantum, with $p = \pm 2$. The overall transformation caused by the group of three pulses is $\Delta p = \pm 2$, and they can be phase cycled together (as a unit) to select this change.

---

[1] If the $180°$ pulse is imperfect then other changes may occur; see section 11.6.5 on the next page.

11.6.3   The final pulse

The role of the final 90° pulse in the sequence is usually to generate observable coherence, $p = -1$, from whatever coherence orders are present just prior to this pulse. This is the case for all three sequences shown in Fig. 11.2 on page 388.

Suppose that, by appropriate phase cycling, we have *already* selected the coherence order(s) we want to be present just before this last pulse. If this is the case, then there is no need to select any particular pathway on the last pulse, as the only possible pathway is from the already selected coherence orders to $p = -1$. Pathways which end up with other orders of coherence are simply not observable, and so we do not need to worry about them.

11.6.4   High-order multiple-quantum terms

Look again at the sequence shown in Fig. 11.10 on the previous page which is used to generate $p = \pm 2$. If we select this pathway with a four-step cycle, we will also select $\Delta p = \pm 6$ i.e. the generation of sextuple-quantum coherence.

However, in practice we need not worry about this, as in order to generate such a coherence we would need to have one spin with a significant coupling to *five* others. This is just so unlikely in any real sample that we can discount it occurring.

Generally speaking, in devising phase cycles we do not need to consider the generation of high orders of multiple-quantum coherence. It is probably safe to assume that triple quantum is the highest order that is likely to be generated in any significant quantities unless we have some very exotic spin system.

11.6.5   Refocusing pulses

It was noted in section 11.1.3 on page 385 that a 180° pulse simply causes the coherence order to change sign. In fact we can say that the refocusing property which such a pulse has comes about *because* it causes such a change in the coherence order. A coherence of order p present before the 180° pulse acquires a phase $(-\Omega^{(p)}\tau_1)$ as a result of evolution for time τ_1. After the pulse, the coherence has order $-p$ and so acquires a phase $(-\Omega^{(-p)}\tau_2)$ as a result of evolution from time τ_2. However, since $\Omega^{(-p)} = -\Omega^{(p)}$ this latter phase can be written as $(+\Omega^{(p)}\tau_2)$. If $\tau_1 = \tau_2$, these two phases are equal and opposite, and so cancel one another out: this is the formation of an echo.

Although a 180° pulse simply changes the sign of the coherence order, if the pulse is in anyway imperfect – for example by being mis-calibrated so that it is not really a 180° pulse – then additional coherence transfers will be caused. These unwanted transfers can be suppressed using an appropriate phase cycle.

For example, if the 180° pulse is being used to refocus single-quantum coherence, the desired pathways from $p = +1 \rightarrow p = -1$ and $p = -1 \rightarrow p = +1$ have $\Delta p = \pm 2$, as shown in Fig. 11.11. The four-step phase cycle

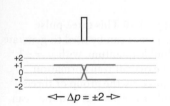

Fig. 11.11 A 180° pulse simply causes a change in the sign of the coherence order, here $p = \pm 1$ goes to $p = \mp 1$; the required pathways are therefore $\Delta p = \pm 2$. These pathways can be selected with a four-step phase cycle in which the pulse goes [0°, 90°, 180°, 270°] and the receiver phase goes [0°, 180°, 0°, 180°]. This is know as EXORCYCLE.

in which the pulse goes [0°, 90°, 180°, 270°] and the receiver phase goes [0°, 180°, 0°, 180°] selects both of these pathways. This phase cycle was one of the first to be used in multiple-pulse NMR, and is often called EXORCYCLE.

EXORCYCLE was originally developed to remove some artifacts, dubbed 'phantoms' and 'ghosts', from two-dimensional spectra – hence the name.

11.7   Axial peak suppression

In a two-dimensional experiment, our expectation is that it will be the equilibrium magnetization present before the first pulse which leads to the coherence present during t_1, and then finally to the observed signals during t_2. However, all through the pulse sequence the magnetization or coherences present are decaying due to relaxation, thus leading to the regeneration of z-magnetization. As this z-magnetization has been generated by relaxation (an essentially random process), it is not phase labelled i.e. it is not modulated by the evolution during t_1.

If this recovered z-magnetization is made observable by subsequent pulses in the sequence, it will give rise to peaks in the two-dimensional spectrum. However, as the magnetization is not modulated as a function of t_1, the peaks will appear at $\omega_1 = 0$ (their frequencies in ω_2 are just those of the usual spectrum). These peaks are usually called *axial peaks* on account of them lying on the $\omega_1 = 0$ axis. In section 9.7.4 on page 281 we came across a specific example of how such peaks arise in the NOESY experiment.

Axial peaks are easily suppressed by difference spectroscopy. All we do is repeat the experiment with the phase of the first pulse changed by 180°. This changes the sign of all of the wanted peaks, but leaves the axial peaks unaffected. Subtracting the signals from the two experiments will therefore cancel the axial peaks, but the wanted peaks will add up.

A convenient way of subtracting the results from the two experiments is simply to shift the phase of the receiver by 180°, in which case we have a two-step phase cycle:

axial peak suppression   first pulse: [0°, 180°]   rx phase: [0°, 180°].

Generally speaking it is convenient to include axial peak suppression as part of the phase cycling process which we are inevitably going to be using for coherence pathway selection.

11.8   CYCLOPS

As has been described previously, the spectrometer has detectors which measure both the x- and y-components of the magnetization, and we use the output of these detectors to construct a complex time-domain signal. Due to imperfections in the RF electronics in the spectrometer, the output of these two detectors may not correspond exactly to the x- and y-magnetizations, and this can lead to unwanted peaks in the spectrum.

Three kinds of imperfection have been identified. The first is where, rather than the output of the detectors corresponding to the x- and y-components of the magnetization, the detectors measure two components which are not quite at 90° to one another. Such an imperfection leads to the appearance of a small peak at minus the frequency of the real peak i.e. the peaks are symmetrically placed with respect to the centre of the spectrum, $\omega = 0$. Such peaks are called *quadrature artifacts* or *quadrature images*.

The second kind of imperfection is where the two detectors are not quite balanced, in the sense that the same amount of magnetization gives rise to a *different* output from the two detectors. This imperfection, like the first, gives rise to quadrature artifacts.

The third kind of imperfection is when, in the absence of any transverse magnetization, the output of either of the detectors is not zero but rather has a small steady value. This is sometimes called a *DC offset*: the 'DC' stands for direct current, which implies a steady, rather than fluctuating, value. Such a steady contribution to the time-domain signal will lead, after Fourier transformation, to a peak at zero frequency. This peak is sometimes called the *zero-frequency glitch* or *DC spike*.

On modern spectrometers with carefully constructed and adjusted receivers, the zero-frequency glitch and quadrature artifacts are likely to be reasonably small. Nevertheless, they can be inconvenient, especially if the spectrum has high dynamic range in which case the quadrature artifact from a large signal can obscure a genuine, weaker signal.

It turns out that, provided these artifacts are not too large, they can be suppressed in a simple pulse–acquire experiment using a four-step phase cycle, known as CYCLOPS:

CYCLOPS   pulse: [0°, 90°, 180°, 270°]   rx phase: [0°, 90°, 180°, 270°].

In terms of CTPs, we recognize that this cycle selects the pathway $\Delta p = -1$, which is exactly that required in a pulse–acquire experiment, since the magnetization starts on z, $p = 0$, and ends up as observable, $p = -1$.

In more complex pulse sequences, we can implement CYCLOPS by cycling the phases of *all* of the pulses together through the sequence [0°, 90°, 180°, 270°], with the receiver phase going [0°, 90°, 180°, 270°]. One way of thinking about this is that, taken together, the whole pulse sequence converts z-magnetization, $p = 0$, to observable magnetization, $p = -1$. This is a transfer with $\Delta p = -1$, which is the pathway selected by this cycle.

The problem with implementing CYCLOPS in more complex sequences is that it multiplies the length of the phase cycle by a factor of four, which may result in an unacceptably long cycle. However, it is sometimes the case that a phase cycle which is devised to select other pathways in the sequence turns out, almost as a by-product, to suppress quadrature artifacts and the zero frequency glitch. Clearly this is a desirable outcome.

11.9   Examples of practical phase cycles

In this section we will give examples of typical phase cycles for some of the most commonly used two-dimensional experiments.

11.9.1 COSY

Figure 11.12 shows the COSY pulse sequence with three different CTPs. In (a) and (b) only one coherence order is retained in t_1; as explained in section 11.3 on page 389, the resulting time-domain data will be phase modulated in t_1. The corresponding spectra will therefore be frequency discriminated, but will have the unfavourable phase-twist lineshape. As was discussed before, CTP (a) will give the N-type, and CTP (b) the P-type spectrum.

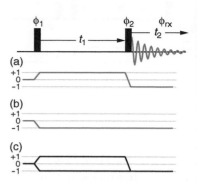

To select CTP (a) all we have to do is select $\Delta p = +1$ for the first pulse. If we do this, we do not need to select any pathway on the last pulse as we have already selected $p = +1$ prior to this pulse. A suitable four-step cycle is:

N-type: $\phi_1 = [0°, 90°, 180°, 270°]$ $\phi_{rx} = [0°, 270°, 180°, 90°]$.

Fig. 11.12 The COSY pulse sequence, together with three possible coherence transfer pathways: (a) gives the N-type spectrum, (b) gives the P-type spectrum and (c) retains symmetrical pathways in t_1. The phases of the two pulses are denoted ϕ_1 and ϕ_2, and the phase of the receiver is denoted ϕ_{rx}.

During this phase cycle, the phase of the second pulse, ϕ_2, is held constant.

Similarly, to select CTP (b) all we need to do is select $\Delta p = -1$ for the first pulse.

P-type: $\phi_1 = [0°, 90°, 180°, 270°]$ $\phi_{rx} = [0°, 90°, 180°, 270°]$.

An alternative approach would be to keep the phase of the first pulse fixed and cycle the phase of the second pulse. For the N-type spectrum, this means selecting $\Delta p = -2$, for which a suitable cycle is

N-type: $\phi_2 = [0°, 90°, 180°, 270°]$ $\phi_{rx} = [0°, 180°, 0°, 180°]$.

For the P-type spectrum, CTP (b), we need to select $\Delta p = 0$ on the last pulse, for which a suitable cycle is

P-type: $\phi_2 = [0°, 90°, 180°, 270°]$ $\phi_{rx} = [0°, 0°, 0°, 0°]$.

CTP (c) in Fig. 11.12 has symmetrical pathways in t_1, and so will lead to amplitude modulation as a function of t_1. Such data can be processed so as to obtain both frequency discrimination and absorption mode lineshapes.

The first pulse can only generate $p = \pm 1$, so we do not need to select this. As before, since the coherence orders we require in t_1 have already been selected, there is no need to cycle the final pulse. Thus, this CTP is the only one possible in this two-pulse experiment, so no phase cycling is needed. If necessary, a two-step cycle of the first pulse and the receiver can be used to suppress axial peaks.

11.9.2 DQF COSY

The pulse sequence, and two possible CTPs, for DQF COSY are shown in Fig. 11.13 on the following page. CTP (a) retains symmetrical pathways in t_1 and so can give an absorption mode spectrum; this pathway is also the simplest one to select.

The final pulse causes two transfers, one with $\Delta p = -3$, and one with $\Delta p = +1$. As these two required values of Δp differ by four, a

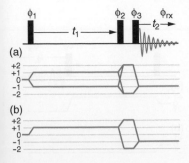

Fig. 11.13 The DQF COSY pulse sequence, together with two possible coherence transfer pathways: (a) retains symmetrical pathways in t_1; (b) gives an N-type spectrum with frequency discrimination.

four-step phase cycle will select *both* of them at the same time, which is very convenient for us. A suitable phase cycle is

CTP (a): $\phi_3 = [0°, 90°, 180°, 270°]$ $\phi_{rx} = [0°, 270°, 180°, 90°]$.

During this cycle, the phases of ϕ_1 and ϕ_2 are kept constant.

As only $p = -1$ is observable, selecting $\Delta p = -3$ and $\Delta p = +1$ on the last pulse unambiguously selects $p = \pm 2$ in the period between the second and third pulses. Since the first pulse can *only* generate $p = \pm 1$, there is no need for any further selection.

An alternative approach to selecting CTP (a) is to group the first two pulses together, and then cycle them as a unit to select $\Delta p = \pm 2$. Such a cycle would be

CTP (a): $(\phi_1, \phi_2) = [0°, 90°, 180°, 270°]$ $\phi_{rx} = [0°, 180°, 0°, 180°]$.

Having made the selection of $p = \pm 2$ between the second and third pulses, there is no need to cycle the last pulse as this can only generate the observable $p = -1$ from the already selected $p = \pm 2$ coherences.

Selecting CTP (b) from Fig. 11.13 needs a longer cycle. One way to approach this is to use the same four-step cycle as above to select $\Delta p = -3$ and $\Delta p = +1$ on the last pulse:

$$\phi_3 = [0°, 90°, 180°, 270°]   \phi_{rx} = [0°, 270°, 180°, 90°].$$

Then we need to select $\Delta p = +1$ on the first pulse, which will require the cycle

$$\phi_1 = [0°, 90°, 180°, 270°]   \phi_{rx} = [0°, 270°, 180°, 90°].$$

This is of course the same cycle as for the last pulse.

These two cycles need to be completed independently giving the following 16-step cycle:

step	ϕ_1	ϕ_3	ϕ_{rx}	step	ϕ_1	ϕ_3	ϕ_{rx}
1	0°	0°	0°	9	180°	0°	180°
2	0°	90°	270°	10	180°	90°	90°
3	0°	180°	180°	11	180°	180°	0°
4	0°	270°	90°	12	180°	270°	270°
5	90°	0°	270°	13	270°	0°	90°
6	90°	90°	180°	14	270°	90°	0°
7	90°	180°	90°	15	270°	180°	270°
8	90°	270°	0°	16	270°	270°	180°

11.9.3 Double-quantum spectroscopy

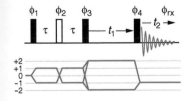

Fig. 11.14 Pulse sequence for double-quantum spectroscopy, together with a coherence transfer pathway in which symmetrical pathways are retained in t_1.

Figure 11.14 shows a pulse sequence for double-quantum spectroscopy, along with a CTP in which symmetrical pathways are retained in t_1; this CTP is similar in many ways to that for DQF COSY. As we did before, we can select $\Delta p = -3$ and $\Delta p = +1$ for the final pulse using the four-step cycle:

$$\phi_4 = [0°, 90°, 180°, 270°]   \phi_{rx} = [0°, 270°, 180°, 90°].$$

If necessary, the 180° pulse (phase ϕ_2) can be cycled to select $\Delta p = \pm 2$ according to the EXORCYCLE scheme. These two four-step cycles have to be completed independently, thus giving a 16-step cycle.

11.9.4 NOESY

The NOESY pulse sequence, along with a CTP which retains symmetrical pathways in t_1, is shown in Fig. 11.15. The final pulse causes the transformation $\Delta p = -1$, which can be selected with the four-step cycle:

$$\phi_3 = [0°, 90°, 180°, 270°] \qquad \phi_{rx} = [0°, 90°, 180°, 270°].$$

This selection ensures that $p = 0$ is present during the mixing time τ, and since the first pulse can only generate $p = \pm 1$, no further cycling is needed.

As well as selecting $\Delta p = -1$ for the last pulse, this four-step cycle selects $\Delta p = -5$ and $\Delta p = +3$. The first of these would correspond to the transfer $p = 4 \rightarrow p = -1$, and the second to the transfer $p = -4 \rightarrow p = -1$. This means that, in addition to selecting the z-magnetization present during τ, the four-step cycle would also select $p = \pm 4$. As was commented on above, for just about all practical situations there will be negligible amounts of quadruple quantum generated, so we do not need to worry about such coherences interfering with the NOESY cross peaks.

As was explained in section 9.7.4 on page 281, in a NOESY experiment it is important to suppress the axial peaks, so we need to add a simple two-step cycle in which the first pulse goes $[0°, 180°]$ and the receiver does the same. Combining this with the four-step cycle of the last pulse gives an overall eight-step cycle:

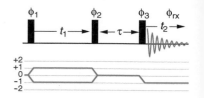

Fig. 11.15 Pulse sequence for NOESY, together with a coherence transfer pathway in which symmetrical pathways are retained in t_1.

step	1	2	3	4	5	6	7	8
ϕ_1	0°	0°	0°	0°	180°	180°	180°	180°
ϕ_3	0°	90°	180°	270°	0°	90°	180°	270°
ϕ_{rx}	0°	90°	180°	270°	180°	270°	0°	90°

11.9.5 HMQC

The pulse sequence and CTP for HMQC is shown in Fig. 11.16. As was explained in section 8.8 on page 212, a simple difference experiment is used to suppress the signal from the I spins which are not coupled to S. This involves repeating the experiment with the phase of the first S spin 90° pulse first set to x, and then to $-x$. Subtracting the data from these two experiments cancels the unwanted signals from the I spins.

In terms of the CTP this S-spin pulse causes the transfer $\Delta p_S = \pm 1$, both pathways which can be selected by the two-step cycle:

$$\phi_S = [0°, 180°] \qquad \phi_{rx} = [0°, 180°].$$

Note that as this is a two-step cycle, it selects both $\Delta p_S = +1$ and $\Delta p_S = -1$ as the Δp_S values differ by two.

The 180° pulse to the I spins is required to cause the transformation $\Delta p_I = -2$, and this can be selected using the usual four-step EXORCYCLE. Overall, we therefore have an eight-step cycle:

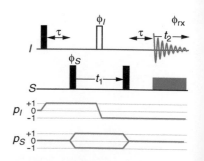

Fig. 11.16 Pulse sequence for HMQC, together with a coherence transfer pathway which leads to amplitude modulation as a function of t_1.

step	1	2	3	4	5	6	7	8
ϕ_S	0°	180°	0°	180°	0°	180°	0°	180°
ϕ_I	0°	0°	90°	90°	180°	180°	270°	270°
ϕ_{rx}	0°	180°	180°	0°	0°	180°	180°	0°

11.10   Concluding remarks about phase cycling

We will close this discussion of phase cycling with a summary of the key ideas, and then go on to comment on the deficiencies of the method.

11.10.1   Summary

- Shifting the phase of a pulse by $\Delta\phi$ results in a pathway which has a change in coherence order of Δp acquiring a phase $-\Delta p \times \Delta\phi$.

- A particular pathway can be selected by ensuring that, as the phase of the pulse is advanced, the receiver is phase shifted by an amount equal to the phase shift experienced by the desired pathway.

- If a phase cycle of N steps selects a pathway with a particular value of Δp, it will also select pathways with changes in coherence order $\Delta p \pm nN$, where n is an integer.

- Phase cycles designed to select particular values of Δp for different pulses must be completed independently.

- The length of phase cycles can be minimized by 'intelligent design', such as taking advantage of: (a) grouping pulses together; (b) recognizing that the first pulse can only generate $p = \pm 1$; (c) realizing that not *all* of the pulses need to be cycled in order to select a particular pathway unambiguously.

11.10.2   Deficiencies of phase cycling

The are two major problems with phase cycling as a method of coherence selection. The first is that, in order for the selection to work, we have to complete *all* of the steps in the cycle. Even if the signal-to-noise ratio is sufficient on one scan, we have to carry on and repeat all of the four, eight, or however many steps there are in the phase cycle. As a result, the experiment can end up taking far longer than is strictly necessary.

This problem is especially acute in two-dimensional spectra where we have to perform a separate experiment for each t_1 value. The need to complete a long phase cycle for each such value may limit the number of t_1 increments which can be recorded in the time available, and hence the resolution in the ω_1 dimension.

The second problem with phase cycling is that it relies on cancellation of the signals from unwanted pathways. For each step in the cycle, all possible pathways contribute to the observed signal. Then, as the signals from successive steps are combined, the unwanted signals will eventually be cancelled.

The problem is that the signals which are supposed to cancel one another are likely to have been recorded at times separated by many seconds, if not minutes. If anything has changed over this time, then the cancellation will not be perfect. The sorts of changes we are thinking of are fluctuations in the amplitude and phase of pulses due to imperfections in the RF electronics, changes in room temperature, changes in the static magnetic field caused by objects being moved near the magnet – in fact just about anything you can think of.

Modern spectrometers, and the environments in which they are housed, are very carefully designed so as to minimize any such sources of instability. However, there is a limit as to what can be achieved, and so inevitably the cancellation of unwanted pathways will be less than complete. This really shows up when the phase cycle has to suppress signals which are much stronger than the ones we are interested in. A good example of this is in inverse correlation experiments where we are trying to observe the weak signals from protons coupled to ^{13}C, and suppress the very much stronger signals from protons which are not coupled to ^{13}C. Such experiments really expose the limitations of the spectrometer, and hence of the phase cycling method.

Field gradient pulses, which are the topic of the rest of this chapter, provide an alternative to phase cycling for selecting a particular CTP. To a large extent, the use of gradient pulses avoids the difficulties which have been highlighted in relation to phase cycling. However, as we shall see, such pulses have other limitations.

11.11 Introducing field gradient pulses

Normally we take a great deal of trouble to make sure that the applied magnetic field is as homogeneous as possible across the sample, as this will give us the narrowest lines in the spectrum. However, we will see in this section that the ability to make the magnetic field inhomogeneous, for short periods and in a strictly controlled way, opens up an alternative way of selecting CTPs.

The usual arrangement for making the field inhomogeneous is to include, in the NMR probe, a small coil placed close to the RF coil used to excite and detect the NMR signal. The extra coil is designed so that when a current is passed through it a magnetic field is created which varies *linearly* along the z-axis i.e. along the direction of the main field B_0. Such a coil is said to create a *field gradient*.

The spectrometer is able to control both the size and direction of the flow of the current which passes through this field gradient coil. The greater the current flowing through the coil, the greater the field gradient i.e. the more rapidly the field changes with distance. If the direction of flow of the current is reversed, the field gradient is changed in *sign*. What this means is that if the current flowing one way results in the field increasing as we go along the positive z-direction, reversing the direction of flow of the current means that the field will decrease along the positive z-direction.

Figure 11.17 on the next page illustrates the relationship between the field gradient, the NMR sample and the spectrum. In (a) we see the sample

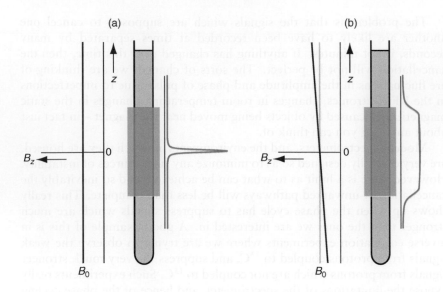

Fig. 11.17 Diagrammatic representation of the effect of a magnetic field gradient on the NMR spectrum. In (a) we see the usual NMR sample in a homogeneous magnetic field. The graph to the left of the tube shows a plot of the field along the z-direction, B_z, against z; in this case $B_z = B_0$ everywhere. The sensitive volume of the sample is shown by the blue rectangle. As the field is homogeneous across the sample, the spectrum expected for case (a) will have the usual narrow line, as shown to the right of the tube. When the gradient is applied, B_z varies linearly with z, as shown in (b); the variation in B_z has been greatly exaggerated. It is usual to arrange things so that the extra field due to the gradient is zero in the middle of the sample, $z = 0$. As a result of the variation in B_z, spins in different parts of the sample have different Larmor frequencies, and so we see a very broad line, as shown to the right of the tube. This line broadening is inhomogeneous.

in a homogeneous magnetic field B_0. To the left of the sample tube there is a graph of the magnetic field along the z-direction plotted against z: in this case, the graph is a straight line, as $B_z = B_0$ throughout.

Only part of the sample is actually excited and detected by the RF coil. Typically a region between 1 and 2 cm long forms this 'sensitive volume'; in the diagram this is shown by the blue rectangle. When the magnetic field is homogeneous, the spectrum will show a narrow line, as is illustrated schematically to the right of the tube.

If current is allowed to flow through the field gradient coil, the situation is changed to that shown in (b). Now we see that B_z varies linearly with z, as shown in the graph to the left of the tube. It is usual to arrange things such that the gradient coil produces no field in the middle of the sample, $z = 0$. As a result, B_z is greater than B_0 in one half of the sample, and less than B_0 in the other half.

A consequence of B_z varying along the tube is that different parts of the sample have different Larmor frequencies, so rather than there being one sharp line from the whole sample, at each z-coordinate there is a line with a different Larmor frequency. All of these lines merge together to give a very broad line, whose width is determined by the strength of the field gradient and the size of the sensitive volume.

Figure 11.17 (b) shows this broad line to the right of the tube. The intensity tails away at the edges of this line as a result of reaching the limits of the sensitive volume. This line is inhomogeneously broadened, in the sense described in section 9.9 on page 300.

The magnetic field, due to the combination of the gradient and the applied field B_0, can be written

$$B_z = B_0 + Gz, \tag{11.7}$$

where G is the *magnetic field gradient*, in units of T m^{-1}, and z is the coordinate along the field direction, measured (in m) from the centre of the sample. The sign of G can be reversed simply by changing the direction of the current flow through the gradient coil.

For historical reasons, the value of G is always quoted in 'Gauss per cm' (G cm^{-1}). One Gauss is 10^{-4} Tesla, so the conversion from G cm^{-1} to T m^{-1} is achieved simply by multiplying by 10^{-4} for the conversion Gauss to Tesla, and 10^2 for the conversion cm^{-1} to m^{-1}. So overall, we just multiply by 10^{-2}.

The upper limit on the field gradient which can be achieved by modern high-resolution spectrometers is about 60 G cm^{-1}, or 0.6 T m^{-1}. If we assume that the sensitive volume extends for about 1.5 cm, then from the top to the bottom of the sample the magnetic field due to such a gradient will vary by 9×10^{-3} T. With the aid of the usual relationship between the Larmor frequency and the magnetic field, $\omega = -\gamma B$, we can convert this range of magnetic field strength into a frequency range. Taking γ to be that of proton, we find that the frequency varies by 380 kHz from top to bottom. This range of frequencies produced by the gradient is very large indeed when compared with a typical NMR linewidth.

When a field gradient is applied, any transverse magnetization (or coherence) present will decay very quickly as a result of spread of Larmor frequencies across the sample. However, the crucial point is that this inhomogeneous decay can be reversed by a spin echo or, more generally, as a result of certain types of coherence transfer. How we can describe and exploit such processes are the topics of the next two sections.

11.11.1 The spatially dependent phase

As we have seen, when a gradient is applied the magnetic field becomes spatially dependent in the way described by Eq. 11.7. Consequently, the Larmor frequency also becomes spatially dependent in a way which we can find simply by multiplying this equation by the gyromagnetic ratio:

$$-\gamma B_z = -\gamma B_0 - \gamma Gz.$$

We have in fact multiplied by $-\gamma$ so that we can replace $-\gamma B_0$ with ω_0, the Larmor frequency when the field has its nominal value, B_0. Similarly, we will write $-\gamma B_z$ as $\omega(z)$, the Larmor frequency at position z:

$$\omega(z) = \omega_0 - \gamma Gz.$$

When a gradient is applied, the thing which is of interest to us is the *variation* in the Larmor frequency across the sample, i.e. the term $-\gamma Gz$.

The evolution at ω_0, due to the main magnetic field B_0, is the same in all parts of the sample and so does not cause any dephasing. We will therefore ignore it from now on, and simply write the spatially dependent part of the frequency, $\Omega(z)$, as

$$\Omega(z) = -\gamma G z.$$

If we have coherence of order $+1$ present, then it will evolve at frequency $\Omega(z)$ in the usual way:

$$\hat{I}_+ \xrightarrow{\Omega(z)t\hat{I}_z} \exp(-i\Omega(z)t)\,\hat{I}_+.$$

What this means is that the coherence acquires a phase $\phi(z) = -\Omega(z)t$. Coherence with order -1 acquires the opposite phase $\phi(z) = +\Omega(z)t$. The important point is that this phase is different in different parts of the sample: it is therefore called the *spatially dependent phase*.

If we have double-quantum coherence present, say with $p = +2$, the evolution due to the gradient will be

$$\hat{I}_{1+}\hat{I}_{2+} \xrightarrow{\Omega_1(z)t\hat{I}_{1z} + \Omega_2(z)t\hat{I}_{2z}} \exp(-i\Omega_1(z)t)\exp(-i\Omega_2(z)t)\,\hat{I}_{1+}\hat{I}_{2+}.$$

We have allowed for the possibility that the spatially dependent frequency will be different for the two spins, and so have written these two frequencies as $\Omega_1(z)$ and $\Omega_2(z)$. However, it is usually the case that the range of frequencies which the gradient causes across the sample is very much greater than the range of offsets in the normal spectrum. We saw an example of this above where the range of frequencies due to the gradient was 380 kHz, which should be compared with a range of offsets for proton spectra (at 500 MHz) of around 5 kHz. This being the case, we can safely assume that the spatially dependent frequency $\Omega(z)$ is the same for all spins. Thus, the evolution of the double quantum is just:

$$\hat{I}_{1+}\hat{I}_{2+} \xrightarrow{\Omega(z)t(\hat{I}_{1z} + \hat{I}_{2z})} \exp(-2i\Omega(z)t)\,\hat{I}_{1+}\hat{I}_{2+}.$$

The spatially dependent phase is therefore $\phi(z) = -2\Omega(z)t$.

These examples can be generalized to give the following expression for the spatially dependent phase acquired by a coherence of order p:

$$\phi(z) = -p \times \gamma G z t. \tag{11.8}$$

The crucial point is that the phase is proportional to the coherence order. It is this property which enables us to select CTPs using gradients, as explained in the following section.

Although the evolution of the offset in the presence of the gradient is not important in determining the spatially dependent phase, the evolution of this offset during the gradient does have some important consequences, which are discussed in section 11.12.5 on page 419.

11.11.2   Selection of a single pathway using two gradients

We are now in a position to explain how two gradients can be used to select a particular CTP using the arrangement shown in Fig. 11.18 on the next page. The basic idea is that during the first gradient G_1 coherences acquire a spatially dependent phase and are therefore dephased. The coherences are then transferred by the pulse, and acquire a further spatially dependent phase during the second gradient pulse G_2. If the phase acquired during the second gradient is equal and opposite to that acquired during the first, the coherence will be rephased at the end of the second gradient. On the other hand, if these phases are not equal and opposite, the coherence remains dephased and is effectively lost.

The process depicted in Fig. 11.18 can be described as follows. The phase acquired by coherence of order p_1 during the first gradient pulse is

$$\phi_1(z) = -p_1 \gamma G_1 z \tau_1.$$

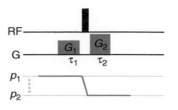

Fig. 11.18 Illustration of the use of field gradient pulses to select a coherence transfer pathway. The timing of the gradient pulses is given by the blue rectangles on the line marked 'G', whereas the RF pulses appear in the usual way on the line marked 'RF'. The duration of the first gradient pulse is τ_1 and the field gradient is of size G_1; the corresponding parameters for the second gradient pulse are τ_2 and G_2. Note that the gradients G_1 and G_2 can be positive or negative.

Similarly, the phase acquired by coherence of order p_2 during the second gradient pulse is

$$\phi_2(z) = -p_2 \gamma G_2 z \tau_2.$$

Therefore, at the end of the second gradient the total spatially dependent phase is

$$-p_1 \gamma G_1 z \tau_1 - p_2 \gamma G_2 z \tau_2.$$

For the CTP $p_1 \to p_2$ to be refocused at this point, the spatially dependent phase must be zero:

$$\text{refocusing condition:} \quad -p_1 \gamma G_1 z \tau_1 - p_2 \gamma G_2 z \tau_2 = 0.$$

What this means is that the dephasing due to the first gradient is exactly undone by the second. This expression can be rearranged to

$$\text{refocusing condition:} \quad \frac{G_1 \tau_1}{G_2 \tau_2} = -\frac{p_2}{p_1}.$$

By selecting the strengths and durations of the gradients such that this condition is satisfied, we can arrange for a particular pathway to be refocused. The hope is that coherences which have followed other pathways will be dephased.

For example, if we wish to select the pathway $+2 \to -1$, the refocusing condition is

$$p_1 = +2 \to p_2 = -1 \qquad \frac{G_1 \tau_1}{G_2 \tau_2} = -\frac{-1}{2}$$

$$= \frac{1}{2}.$$

If we make the two gradients the same length, $\tau_1 = \tau_2$, then to select this pathway the second gradient needs to be twice the strength of the first: $G_2 = 2G_1$.

As a second example, consider the pathway $-2 \to -1$; the refocusing condition is

$$p_1 = -2 \to p_2 = -1 \qquad \frac{G_1 \tau_1}{G_2 \tau_2} = -\frac{-1}{-2}$$

$$= -\frac{1}{2}.$$

For equal length gradients, this means that the second gradient needs to be twice the strength of the first *and* the gradients need to be applied in opposite senses: $G_2 = -2G_1$.

It is interesting to note that a pair of gradient pulses selects a particular *ratio* of coherence orders, whereas phase cycling selects a particular *change* in coherence order.

11.11.3   The spatially dependent phase in heteronuclear systems

The spatially dependent phase, given by Eq. 11.8 on page 412, depends on the gyromagnetic ratio of the nucleus in question, so if we are dealing with heteronuclear experiments we need to take this into account when devising our gradient selection schemes.

As was discussed in section 11.2.1 on page 388, we can assign separate coherence orders p_I and p_S to each type of nucleus, I and S. The spatially dependent phase arising from spin I depends on p_I and γ_I, and similarly for the S spin it depends on p_S and γ_S. Overall, the spatially dependent phase is given by

$$\phi(z) = -\left(p_I \gamma_I + p_S \gamma_S\right) G z t. \tag{11.9}$$

How this works out in practice is best illustrated using an example, such as the pathway shown in Fig. 11.19. Here we see that the coherence order on the S spin changes from +1 to 0, whereas the coherence order on the I spin remains unchanged at −1, simply because no pulse is applied to this spin. The spatially dependent phase caused by the first gradient pulse is

$$\phi_1(z) = -\left(-1 \times \gamma_I + 1 \times \gamma_S\right) G_1 z \tau_1,$$

and by the second gradient pulse is

$$\phi_2(z) = -\left(-1 \times \gamma_I + 0 \times \gamma_S\right) G_2 z \tau_2.$$

Therefore the refocusing condition, $\phi_1(z) + \phi_2(z) = 0$, is

$$\left(\gamma_I - \gamma_S\right) G_1 z \tau_1 + \gamma_I G_2 z \tau_2 = 0.$$

This condition rearranges to

$$\frac{G_1 \tau_1}{G_2 \tau_2} = \frac{\gamma_I}{\gamma_S - \gamma_I}$$
$$= \frac{1}{\left(\gamma_S / \gamma_I\right) - 1}.$$

If the I spin is ^{1}H and the S spin ^{13}C, then $(\gamma_S / \gamma_I) = 0.252$, and so $(G_1 \tau_1)/(G_2 \tau_2) = -1.34$. These numbers are a little easier to understand if we assume that γ for ^{1}H is four times that of ^{13}C, which is almost correct. With this assumption, $(\gamma_S / \gamma_I) = \frac{1}{4}$, and so $(G_1 \tau_1)/(G_2 \tau_2) = -\frac{4}{3}$.

11.11.4   Shaped gradient pulses

For technical reasons it is undesirable to switch the gradient pulse on and off suddenly. Rather, it is preferable to switch the gradient on and off in a smooth fashion. One approach which is commonly adopted is to make the envelope of the gradient pulse the first half of a sine wave, usually called a 'sine bell'. Mathematically, this is the function

$$\sin\left(\frac{\pi t}{\tau}\right).$$

When $t = 0$ and $t = \tau$ the function is zero, and it has its maximum value of 1 at $t = \frac{1}{2}\tau$.

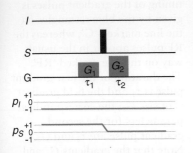

Fig. 11.19 Example of the selection of a coherence transfer pathway using gradients in a heteronuclear experiment. As no pulse is applied to the I spin, the coherence order on that spin does not change.

Looking back over the previous section, we can see that the spatially dependent phase from a rectangular shaped gradient pulse depends on the product $G \times \tau$. We can interpret this as the *area* under the envelope of the gradient pulse i.e. the width, τ, times the height, G.

The area under a sine bell shaped gradient is clearly less than the area under a rectangular gradient of the same height and duration, as is shown in Fig. 11.20. If the maximum of the sine bell is G, then the area under the gradient envelope is found from the integral

$$\int_0^\tau G \sin\left(\frac{\pi t}{\tau}\right) dt,$$

which has the value $(2G\tau/\pi)$. For a rectangular gradient, the area is simply $(G\tau)$, so the spatially dependent phase produced by the sine bell is $(2/\pi) = 0.64$ times that produced by a rectangular gradient.

It is common to define a shape factor, s, as

$$s = \frac{\text{area under the envelope of the shaped gradient}}{\text{area under a rectangular gradient of the same overall height and duration}},$$

and then to modify the expression for the spatially dependent phase to

$$\phi(z) = -s\,p\,\gamma G z\,t.$$

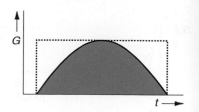

Fig. 11.20 The spatially dependent phase produced by a gradient depends on the area under its envelope, which is a plot of the gradient strength G against time. Here we see a comparison of the areas of a rectangular gradient, whose envelope is shown by the dashed line, and a sine bell gradient, whose envelope is shown by the solid line. Clearly the area under the sine bell shaped gradient, shown shaded in blue, is significantly less than that under the rectangular gradient.

11.11.5 Dephasing in a field gradient

In this section we will look at the details of how a coherence is dephased by a gradient. This will help us to understand how factors such as the length and strength of a gradient affect the rate of dephasing, and so how we might go about choosing these parameters in a particular experiment.

Let us consider the dephasing of observable coherence of order -1. If we start with the operator $\hat{I}_-$ at time zero, then after time t the operator will have acquired a phase $\gamma G z t$ at position z in the sample:

$$\exp(i\gamma G z t)\,\hat{I}_-. \tag{11.10}$$

The observable signal from the whole sample is found by adding up the contributions from all possible positions z, taking into account that the phase at each position is different.

If we assume that the sensitive volume of the sample extends from $-\frac{1}{2}z_m$ to $+\frac{1}{2}z_m$, then the signal from the whole sample is found by integrating the phase factor in Eq. 11.10 with respect to z, and in the range $-\frac{1}{2}z_m$ and $+\frac{1}{2}z_m$:

$$S(t) = \frac{1}{z_m} \int_{-\frac{1}{2}z_m}^{+\frac{1}{2}z_m} \exp(i\gamma G z t)\,dz.$$

We have divided by z_m, the size of the sensitive volume, in order to make $S(t)$ dimensionless, and so that it has the value 1 at $t = 0$; essentially this is a normalization factor.

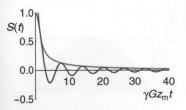

Fig. 11.21 The dark grey line is a plot of the function $S(t)$, given in Eq. 11.11, which shows how a coherence dephases during a gradient pulse. Note that the dephasing depends on the dimensionless parameter $\gamma G z_m t$. Also shown in blue is an approximation to the envelope of $S(t)$, as given in Eq. 11.12. After the first couple of oscillations, this envelope function is an excellent approximation to the envelope of $S(t)$.

The integral is straightforward to compute and gives us the rather neat result:

$$S(t) = \frac{\sin\left(\frac{1}{2}\gamma G z_m t\right)}{\frac{1}{2}\gamma G z_m t}. \tag{11.11}$$

Figure 11.21 shows a plot of this function against the dimensionless parameter $\gamma G z_m t$. The plot shows that $S(t)$ is an oscillating function, which decays steadily over time. In fact, it is this overall decay which we are more interested in than the oscillations, and it can be shown that once we are beyond the first couple of oscillations, the envelope of $S(t)$ is well approximated by

$$\text{envelope of } S(t) = \frac{1}{\frac{1}{2}|\gamma| G z_m t}. \tag{11.12}$$

This function is also shown in the plot in Fig. 11.21.

Not surprisingly, the dephasing goes as $1/(G\tau)$, so a stronger or longer gradient gives more dephasing. Also, the dephasing goes as $1/\gamma$ so, for a given field gradient, nuclei with higher gyromagnetic ratios are dephased more completely. This is not a surprise, as for a larger value of γ, the range of Larmor frequencies across the sample is greater for a given field gradient.

To take a specific example, suppose that $G = 20$ G cm^{-1} (0.2 T m^{-1}) and $z_m = 1$ cm (0.01 m). Then for the dephasing of protons, the envelope of $S(t)$ is

$$\text{envelope of } S(t) = \frac{3.7 \times 10^{-6}}{t}.$$

If we want the magnetization to be dephased to 1% of its initial value, i.e. $S(t) = 0.01$, then this last expression tells us that $t = 0.37$ ms, which is the length of the gradient pulse we would need. If we want more complete dephasing, say to 0.1%, then the gradient would need to be ten times longer, i.e. 3.7 ms.

To dephase coherences on ^{13}C, for which γ is one-quarter of that of proton, we would need gradients four times longer in order to achieve the same effect. For ^{15}N, for which γ is about one-tenth of that of the proton, we would need gradients which are *ten* times longer. Given that there is a practical limit on the length of a gradient pulse which can be applied, you can see that dephasing heteronuclei becomes progressively more difficult as their gyromagnetic ratios decrease.

11.12 Features of selection using gradients

Before we look at the way in which gradients can be used in some typical experiments, there are a number of features about the way gradients work, and the consequences of introducing them into pulse sequences, which we need to discuss.

11.12.1   Selection of multiple pathways

Consider the two pathways shown in Fig. 11.22 on the facing page, both of which involve the transfer of double quantum to single quantum. The conditions for refocusing these two pathways can easily be shown to be:

pathway (a): $\dfrac{G_1\tau_1}{G_2\tau_2} = \dfrac{1}{2}$ pathway (b): $\dfrac{G_1\tau_1}{G_2\tau_2} = -\dfrac{1}{2}.$

For pathway (a), the areas of the two gradients have to be in the ratio 1:2, and the gradients have to be in the same sense. For CTP (b), the areas are still in the ratio 1:2, but the gradients need to be in the opposite sense. There is *no combination* of two gradients which will select *both* pathways simultaneously. This is in contrast to phase cycling which, as was shown in section 11.9.2 on page 405, can select both of these pathways using a four-step phase cycle.

Looking back through the CTPs which we have specified for the common two-dimensional experiments discussed in section 11.9 on page 404, we see that it is very often the case that we wish more than one pathway to contribute. Unfortunately, if we use gradients for coherence selection, it is not possible (except in some special cases) to select more than one order of coherence at the point where the gradient is applied.

This feature of selection with gradients is unfortunate for two reasons. First, if we limit the number of desirable pathways which contribute, the signal which we observe will be reduced in size, and hence the signal-to-noise ratio will be reduced. For example, in Fig. 11.22 the consequence of selecting only pathway (a), or pathway (b), is a loss of *half* the signal when compared with phase cycling, which can select both pathways.

The second problem is that, if we apply a gradient during the evolution period t_1 of a two-dimensional experiment, we will inevitably select just *one* order of coherence during t_1: it will not be possible to retain symmetrical pathways. As was explained in section 11.3 on page 389, selecting one coherence order during t_1 results in a spectrum with the undesirable phase-twist lineshape. In order to be able to process the data to give absorption mode lineshapes, we must retain symmetrical pathways in t_1, which is incompatible with applying a gradient during t_1.

All is not lost, however, as there is a method, described in the next section, of regaining an absorption mode lineshape even when gradients have been used during t_1.

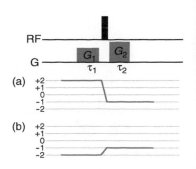

Fig. 11.22 Illustration of the two possible pathways by which a pulse can transfer double quantum, $p = \pm 2$, to single quantum, $p = -1$; such transfers take place, for example, on the last pulse of the DQF COSY and double-quantum spectroscopy experiments. Using gradients, it is possible to select either pathway (a) or pathway (b), but not both at once.

11.12.2 Obtaining absorption mode lineshapes when gradients are used in t_1

If we have used a gradient during t_1 it is possible to obtain an absorption mode spectrum using the following procedure. The experiment is repeated twice: once with the gradients set so as to select coherence order $+p$ during t_1, and once with the gradients set to select $-p$ during t_1. These two experiments give N- and P-type data sets, the time-domain signals for which are of the form

$$S_P(t_1, t_2) = \exp(+i\Omega_A t_1)\exp(i\Omega_B t_2) \quad S_N(t_1, t_2) = \exp(-i\Omega_A t_1)\exp(i\Omega_B t_2).$$

Note that the only difference between these is the sign of the modulation in t_1, which derives from the fact that signals come from either $+p$ or $-p$ order coherence during t_1.

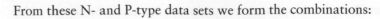

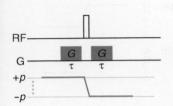

RF

G

+p
-p

Fig. 11.23 A 180° pulse simply causes the coherence order to change sign. Such a pathway, for *any* value of *p*, can be selected by two equal gradient pulses.

From these N- and P-type data sets we form the combinations:

$$S_c(t_1, t_2) = \frac{1}{2}[S_P(t_1, t_2) + S_N(t_1, t_2)]$$
$$= \cos(\Omega_A t_1)\exp(i\Omega_B t_2),$$

and

$$S_s(t_1, t_2) = \frac{1}{2i}[S_P(t_1, t_2) - S_N(t_1, t_2)]$$
$$= \sin(\Omega_A t_1)\exp(i\Omega_B t_2).$$

The resulting cosine and sine modulated data sets can then be processed to give an absorption mode spectrum using the SHR method (section 8.12.3 on page 230).

There is a cost to this method of obtaining absorption mode lineshapes. First, we have to record two separate experiments, with different gradient pulses, and secondly there is a reduction in signal-to-noise ratio by a factor of $\sqrt{2}$ compared with an experiment in which symmetrical pathways are retained during t_1.

11.12.3   Refocusing pulses

An ideal 180° pulse simply causes a change in the sign of the coherence order i.e. $p \rightarrow -p$, as is shown in Fig. 11.23. Such a change in coherence order is what leads to the formation of a spin echo, so when used in this way a 180° pulse is often called a refocusing pulse.

For *any* value of *p*, such a pathway is refocused by two equivalent gradients placed either side of the pulse. We can easily see how this works by noting that the spatially dependent phase from the first gradient pulse is $(-p\gamma Gz\tau)$, and that from the second gradient pulse is $(+p\gamma Gz\tau)$. Clearly, these are equal and opposite, so the total phase is zero and the pathway is refocused.

If the 180° pulse is imperfect, it will cause transfer to coherences other than $-p$: such pathways will not be refocused. So, this pair of gradients is a very good way of 'cleaning up' the results of an imperfect 180° pulse. What is more, because the selection of $p \rightarrow -p$ works for *any* value of *p*, we do not lose any signal.

11.12.4   180° pulses in heteronuclear experiments

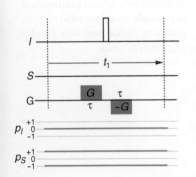

Fig. 11.24 A typical arrangement in which a centrally placed 180° pulse to the *I* spin is used to refocus the evolution of the heteronuclear coupling over the evolution time t_1. No coherence is present on the *I* spin, so the role of this 180° pulse is to invert the operators such as $\hat{I}_z$, rather than to act as a refocusing pulse in the way shown in Fig. 11.23. The 180° pulse to *I* causes no changes in the coherence order of the *S* spin. As explained in the text, two equal and opposite gradient pulses are useful for cleaning up any problems associated with an imperfect 180° pulse, while leaving the *S*-spin coherences unaffected.

In heteronuclear experiments, 180° pulses are often used to refocus the evolution of a heteronuclear coupling – for example during the evolution time of experiments such as HMQC and HSQC. Figure 11.24 illustrates a typical such arrangement.

There is no coherence present on the *I* spin, so the 180° pulse is not acting as a refocusing pulse in the sense described in the previous section. Rather, its role is to invert the $\hat{I}_z$ operators in product operators such as $2\hat{I}_z\hat{S}_x$ and $2\hat{I}_z\hat{S}_y$ which are present during t_1. This 180° pulse is therefore best described as an inversion pulse.

If the 180° pulse is perfect, it will cause the transformation $\hat{I}_z \rightarrow -\hat{I}_z$, and nothing else. However, if the pulse is imperfect, coherences may be generated (or transferred) by the pulse, and these may go on to give unwanted peaks in our spectrum. By placing a gradient *after* the 180° pulse, any coherences generated by the pulse will be dephased, and therefore will not contribute to the spectrum.

The problem with placing a gradient after the 180° pulse is that it will dephase the coherences present on the S spin – which is certainly not what we want to happen. To get round this, we place equal and opposite gradients either side of the 180° pulse. The S-spin coherences are dephased by the first gradient, and then promptly rephased by the second; this works for any value of p_S. Overall this pair of 'anti-phase' gradients cleans up any imperfections from the 180° pulse, and leaves the evolution of the S-spin coherences unaffected.

11.12.5   Phase errors due to gradient pulses

Up to now we have emphasized how, if the gradient pulses are correctly chosen, the spatially dependent phase due to the first gradient is equal and opposite to that of the second gradient, leading to refocusing at the end of the second gradient. However, this refocusing only applies to the phase which results *from the gradient pulse* itself: any phase due to the underlying evolution of offsets and couplings is *not* cancelled.

How this comes about is best illustrated by an example. We will use the DQF COSY pulse sequence, shown in Fig. 11.25, along with a gradient selection scheme and the associated CTP. The first gradient is placed during the double-quantum period, and the second just prior to t_2. By making the second gradient twice the area of the first, the pathway $+2 \rightarrow -1$ is refocused. We can therefore be sure that double-quantum filtration has taken place.

Now imagine recording a spectrum *without* the gradient pulses, but leaving in the delays τ where the gradients were, and using phase cycling to select the CTP shown. A typical value for these delays τ, the length of the gradient pulse, is 1 to 2 ms. So what we have is a pulse sequence with a significant delay between the second and third pulses, and a further significant delay between the last pulse and the start of acquisition.

During the first of these delays the double-quantum coherence will evolve at $(\Omega_1 + \Omega_2)$, and during the second delay the single-quantum coherence will evolve at its offset, as well as according to the couplings present. As a result of the evolution during these delays, phase errors will accrue, and these will affect the observed spectrum.

These phase errors are frequency dependent and can reach quite large values. For example, consider the evolution of the offset during the second delay τ. The offset is typically in the range 0–2500 Hz, and a typical value for the delay is 1.5 ms; this results in a frequency-dependent phase which reaches 1350° at the edge of the spectrum. It is simply not possible to correct such large phase errors by the usual phasing procedures. Furthermore, there will be additional frequency-dependent phase errors due to the evolution of the double quantum during the first delay τ. The overall result will be a spectrum which simply cannot be phased.

Putting the gradients back into the sequence does not eliminate these phase errors. Of course, the second gradient refocuses the spatially dependent phase caused by the first, but this refocusing effect does not extend

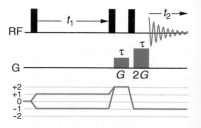

Fig. 11.25 A DQF COSY pulse sequence, along with a suitable pair of gradients to select the pathway $+2 \rightarrow -1$ caused by the final pulse. To select this pathway, the second gradient must have twice the area of the first; here we have chosen to achieve this by keeping the gradients the same length and doubling the strength of the second one. As explained in the text, the evolution of the offset during the two periods τ occupied by the gradient pulses leads to very large phase errors in the spectrum.

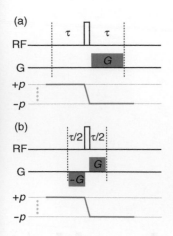

Fig. 11.26 Illustration of how the phase errors associated with the underlying evolution of the offsets during a gradient can be refocused. In (a) a 180° pulse forms a spin echo such that the evolution during the first time τ is refocused at the end of the second time τ. As a result, there is no net evolution of the offset over the entire period 2τ. The gradient, of duration τ, is placed *after* the 180° pulse so that the spatially dependent phase produced by the gradient is unaffected by the pulse. Sequence (b) also refocuses the evolution of the underlying offset, but is more time efficient than (a), giving the same spatially dependent phase in half the time.

to the underlying evolution of the offsets, which continues regardless of the gradients. What we have discovered is that we cannot simply insert gradients into our existing pulse sequences, without considering the effect that the *time* occupied by the gradient will have on the phase in the spectrum.

The solution to this problem is to place the gradient within a spin echo, as shown in Fig. 11.26 (a). The gradient pulse (duration τ) generates a spatially dependent phase in the usual way, but by containing the pulse in the second half of a spin echo, the evolution of the underlying offset over the first delay τ is refocused during the second delay τ. As a result, there is no net evolution of the offset over the total time 2τ. We should note that the spin echo *does not* refocus the evolution of the (homonuclear) coupling, and that the spatially dependent phase generated by the gradient is ($p\gamma zG\tau$).

Sequence (b) is slightly more time efficient, as it achieves the same effect as (a) but in half the time. In (b) the gradient has been split into two equal and opposite parts, but as before the spin echo ensures that there is no net evolution of the underlying offset over the total time τ. The first gradient generates a spatially dependent phase of $\frac{1}{2}(p\gamma zG\tau)$, and the second generates the same, so overall the phase is the same as in sequence (a). The reason that the gradients in (b) have to be applied in the opposite sense to one another is that the 180° pulse changes the sign of the coherence order.

In principle, when we want to insert a gradient into a pulse sequence, we should use sequence (a) or (b) in order to refocus the evolution of the underlying offsets. Unfortunately, this complicates the sequences by adding extra delays and extra refocusing pulses, which can themselves be a source of imperfections.

In many pulse sequences, especially heteronuclear ones, there are already spin echoes present as part of the sequence. It may be – if we are lucky – that we can insert our gradient into one of these existing echoes, and thereby avoid the need to introduce extra 180° pulses.

11.12.6   Selection of z-magnetization

Magnetization along the z-axis does not evolve during a delay, and so is unaffected by a field gradient. Another way of looking at this is to say that such magnetization has coherence order zero, and so does not acquire any spatially dependent phase during a gradient. Therefore, if we wish to *retain* z-magnetization and *reject* all other coherences, all we need to do is apply a single gradient. This is in contrast to the way we select other coherences, where we always need two gradients: one to dephase the coherence, and one to rephase it.

A gradient which is just used to destroy unwanted coherences is sometimes called a *purge gradient* or a *homospoil pulse*. In the following section we will see a number of cases where such gradients can be used to advantage in practical pulse sequences.

Zero-quantum coherence also has $p = 0$, and so, like z-magnetization, is not dephased by a gradient; therefore, the two cannot be separated. In some experiments, this turns out to be rather a problem as the presence of unwanted zero-quantum coherence leads to phase distortions. In section 11.15 on page 426 we will discuss how the contribution from zero quantum can be suppressed.

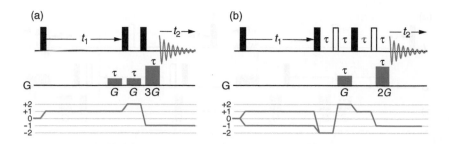

Fig. 11.27 Two different versions of the DQF COSY experiment which utilize gradients for CTP selection. Sequence (a) retains just $p = +1$ during t_1, and so leads to a frequency discriminated spectrum with the phase-twist lineshape. No attempt is made to control the phase errors which will accrue due to evolution of the underlying offsets during the three gradient pulses, so the spectrum has to be displayed in the absolute value mode. The areas of the gradients need to be in the ratio 1:1:3 to select the pathway shown. Sequence (b) retains symmetrical pathways in t_1, and so will give rise to data which can be processed to give absorption mode lineshapes. Both of the gradients are contained within spin echoes so that the phase errors due to the evolution of the underlying offsets are removed.

11.13 Examples of using gradient pulses

In this section we will look at how gradient pulses can be implemented into a number of the commonly used two-dimensional experiments.

11.13.1 DQF COSY

Figure 11.27 shows two different versions of the DQF COSY experiment in which gradient pulses are used for CTP selection. In sequence (a) only $p = +1$ is retained during t_1, so the resulting data set will be phase modulated as a function of t_1. Processing this data set will give a frequency discriminated spectrum, with the phase-twist lineshape. The areas under three gradient pulses need to be in the ratio 1:1:3 to select the pathway shown.

In this sequence, the gradients are not placed within spin echoes so, as described in section 11.12.5 on page 419, the resulting spectrum will show large frequency-dependent phase errors due to the evolution of the underlying offsets during the gradients. These phase errors, combined with the fact that the spectrum has the phase-twist lineshape, mean that the only feasible thing to do is to display the spectrum in the absolute value mode, as described in section 8.12.6 on page 234. While such a display does not give such high resolution as an absorption mode spectrum, it is convenient for routine spectroscopy where resolution is not at a premium.

If we want an absorption mode spectrum, then we will need to use the sequence shown in Fig. 11.27 (b) in which, since no gradient is applied during t_1, symmetrical pathways are retained. The resulting data set will be amplitude modulated in t_1, and so we will need to use the SHR or TPPI procedure to achieve frequency discrimination (see section 8.12 on page 226).

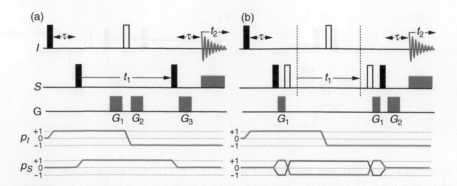

Fig. 11.28 Two alternative HMQC pulse sequences using gradients for selection. Sequence (a) is only suitable for generating spectra to be displayed in the absolute value mode since large phase errors accrue due to the evolution of the underlying offsets during gradients G_1 and G_2. In contrast, in sequence (b) this evolution is refocused by placing the gradients within spin echoes (the two gradients G_1) or in existing delays in the sequence (gradient G_2). Two alternative coherence transfer pathways are shown for this sequence: the blue pathway gives rise to the P-type spectrum, and the grey pathway gives rise to the N-type spectrum; as explained in the text, different gradient strengths are needed to select these two pathways. An absorption mode spectrum can be obtained by combining the P- and N-type data in the manner described in section 11.12.2 on page 417. The exact strengths and lengths needed for the gradients G_1, G_2 and G_3 depend on the gyromagnetic ratios of the I and S spin, as described in the text.

In this sequence both gradients appear within spin echoes, so the evolution of the underlying offsets during the gradients is refocused. As a result, the large frequency-dependent phase errors found in the spectra from sequence (a) are avoided. The CTP looks rather tortuous as a result of the fact that the 180° pulses change the sign of the coherence order. When working out the desired pathway it is sometimes useful to work back from the end, since we know that the pathway must finish at −1.

11.13.2   HMQC

Figure 11.28 shows two different HMQC pulse sequences using gradient selection. Sequence (a) retains only one pathway, $p_S = +1$, during t_1, and so gives rise to a frequency-discriminated spectrum with phase-twist lineshapes. The final gradient G_3 is placed within an existing delay in the pulse sequence, but the two gradients G_1 and G_2 are not, and so will give rise to large frequency-dependent phase errors in ω_1. The resulting spectrum is therefore only suitable for an absolute value display.

It is not strictly necessary to use two gradients during t_1. However, by placing one on either side of the 180° pulse we can select the pathway $p_I = +1 \rightarrow p_I = -1$, which this pulse is required to bring about. If the pulse is imperfect, then the gradients will make sure that any unwanted transfers are dephased.

At the end of the sequence the spatially dependent phase for the pathway shown is

$$(-\gamma_I G_1 z \tau_1 - \gamma_S G_1 z \tau_1) + (\gamma_I G_2 z \tau_2 - \gamma_S G_2 z \tau_2) + (\gamma_I G_3 z \tau_3),$$

where we have assumed that gradient G_1 is of duration τ_1, and so on. Rearranging this so that the $(G_i z \tau_i)$ are factors gives

$$G_1 z \tau_1 (-\gamma_I - \gamma_S) + G_2 z \tau_2 (\gamma_I - \gamma_S) + G_3 z \tau_3 (\gamma_I).$$

There are many combinations $G_1 \tau_1$, $G_2 \tau_2$ and $G_3 \tau_3$ which will make this spatially dependent phase zero i.e. refocus the pathway.

To get a handle on one of the possibilities it is easier to think of a specific case. Imagine that I is ^{1}H and S is ^{13}C, and let us also assume that $\gamma_H = 4\gamma_C$. To simplify things further we will also let all of the gradients have the same length, so that $\tau_2 = \tau_1$ and $\tau_3 = \tau_1$. With these simplifications and assumptions the refocusing condition becomes:

$$G_1 z \tau_1 (-4\gamma_C - \gamma_C) + G_2 z \tau_1 (4\gamma_C - \gamma_C) + G_3 z \tau_1 (4\gamma_C) = 0.$$

Cancelling a factor of $z \tau_1 \gamma_C$ simplifies this to

$$-5G_1 + 3G_2 + 4G_3 = 0.$$

One solution to this is for the gradient strengths to be in the ratio

$$G_1 : G_2 : G_3 = 5 : 3 : 4.$$

In the sequence shown in Fig. 11.28 (b) on the preceding page the two gradients G_1 are placed within spin echoes, and the final gradient G_2 is placed in an existing delay. As a result, the evolution due to the underlying offsets will be refocused, and so it should be possible to phase the spectrum in a straightforward manner.

The evolution of the S spin offset during the first gradient G_1 is refocused by the first S-spin 180° pulse. The second such 180° pulse refocuses the evolution during the second gradient G_1. For the I spin, the evolution during the first gradient G_1 is refocused during the second gradient G_1 by the I-spin 180° pulse which appears between these two gradients; this pulse was part of the original sequence.

Since gradients are applied during t_1, the resulting spectra will be phase modulated as a function of t_1. However, by recording both the P- and N-type spectra (the blue and grey lines on the CTP) it will be possible to obtain an absorption mode spectrum using the method described in section 11.12.2 on page 417.

For the solid CTP shown in sequence (b) (the P-type spectrum), the spatially dependent phase at the end of the sequence is:

$$(-\gamma_I G_1 z \tau_1 - \gamma_S G_1 z \tau_1) + (\gamma_I G_1 z \tau_1 - \gamma_S G_1 z \tau_1) + (\gamma_I G_2 z \tau_2).$$

The phase accrued by the I spin due to the first gradient pulse G_1 is equal and opposite to that accrued during the second gradient pulse G_1 on account of the change in sign of p_I caused by the I-spin 180° pulse. Thus, the first and third terms in the above expression cancel. The refocusing condition is

$$-2\gamma_S G_1 z\tau_1 + \gamma_I G_2 z\tau_2 = 0$$

which rearranges to

$$\frac{G_1\tau_1}{G_2\tau_2} = \frac{\gamma_I}{2\gamma_S}.$$

In the case that I and S are ^{1}H and ^{13}C, respectively, the refocusing condition is $(G_1\tau_1) = 2(G_2\tau_2)$.

We need different gradients to select the grey CTP which corresponds to the N-type spectrum. A similar calculation to the above shows that for this pathway the refocusing condition is

$$2\gamma_S G_1 z\tau_1 + \gamma_I G_2 z\tau_2 = 0$$

which rearranges to

$$\frac{G_1\tau_1}{G_2\tau_2} = -\frac{\gamma_I}{2\gamma_S}.$$

So for the case of a ^{1}H–^{13}C HMQC, the gradient ratio is $(G_1\tau_1) = -2(G_2\tau_2)$ i.e. one of the gradients needs to be applied in the opposite direction to those needed for the P-type experiment.

Suppressing the unwanted I-spin magnetization

From our original discussion of the HMQC experiment you will recall that a major difficulty is suppressing the signal from the I spins which are *not* coupled to S. In the case where I is ^{1}H, and S is ^{13}C, the unwanted signal is around a hundred times stronger than the wanted signal, so we have to suppress the former very effectively if we are to be able to see the latter.

Looking at the sequences in Fig. 11.28 on page 422 we can see that, with the gradient combinations we chose, the uncoupled I-spin magnetization remains dephased at the end of the sequence. For example, in sequence (b) this I-spin magnetization is dephased by the first gradient G_1, but is then rephased by the second gradient G_1 on account of the change in sign of the coherence order caused by the I-spin 180° pulse. However, the magnetization is once again dephased by gradient G_2, and so does not contribute to the observed signal.

The fact that the unwanted pathways are dephased at the end of the sequence, and so *never* contribute to the observed signal, is one of the very attractive features of selecting a pathway with gradients. This is in contrast to phase cycling where, for each step of the cycle, all pathways contribute to the signal, but it is arranged that the unwanted contributions will cancel when the signals from all the steps are combined. As was mentioned above, the effectiveness of such cancellation is very much dependent on the stability of the spectrometer. Generally speaking, it has been found that selection with gradients is far more effective than phase cycling when it comes to suppressing intense unwanted signals.

11.13.3   HSQC

Figure 11.29 on the facing page shows how gradients can be used to select the required pathway in an HSQC pulse sequence. Like the HMQC

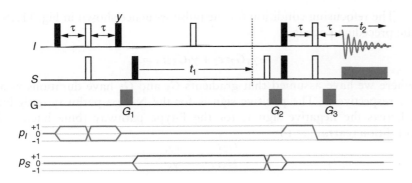

Fig. 11.29 Pulse sequence for an HSQC experiment utilizing gradient pulses for coherence selection. Gradient G_1 is a purge gradient as, when it is applied, the wanted magnetization is along z, whereas the unwanted magnetization (from I spins not coupled to S) is transverse. The gradients G_2 and G_3 can be chosen to select either the P-type pathway (blue line) or the N-type pathway (grey line); by combining these two data sets it is possible to obtain an absorption mode spectrum. Gradient G_2 is contained within a spin echo, so the evolution of the underlying S spin offset during the gradient is refocused. Gradient G_3 is placed within a spin echo present in the original HSQC sequence, and so the evolution of the I spin offset is therefore refocused.

sequence shown in Fig. 11.28 (b) on page 422, the presence of a gradient during t_1 means that the sequence will give a P- or N-type data set, depending on the choice of the gradients. By recombining these data sets, an absorption mode spectrum can be obtained.

Gradient G_2 is placed in a spin echo and therefore the evolution of the underlying S spin offset is refocused. Similarly, the evolution of the I-spin offset during gradient G_3 is refocused as this gradient is placed in the final spin echo which forms a part of the original HSQC sequence. We therefore expect to be able to phase the spectrum.

Gradient G_1, which is inserted between the second I-spin 90° pulse and the first S-spin 90° pulse, plays a role which we have not encountered before. If we work through the product operator analysis of this pulse sequence, we will find that the anti-phase term $2\hat{I}_x\hat{S}_z$, present at the end of the second delay τ, is transformed into the term $2\hat{I}_z\hat{S}_z$ by the second I-spin 90° pulse. This term is along z, and therefore is unaffected by the gradient.

In contrast, if we consider the magnetization from the I spins which are *not* coupled to S, at the end of the second τ delay this magnetization appears along y. It is therefore unaffected by the 90°(y) pulse, leaving the magnetization to be dephased by the gradient G_1. The role that this gradient plays is therefore to dephase the unwanted magnetization, while leaving the wanted z-magnetization unaffected. In other words, G_1 is a purge gradient.

It may be that this purge gradient, in conjunction with phase cycling, will give acceptable suppression of the unwanted I-spin magnetization. If this is the case, then there is no need to go to the complication of introducing the further gradients G_2 and G_3. For biological samples which have been globally labelled in ^{13}C or ^{15}N, it is generally found that perfectly adequate suppression of the unwanted signals can be obtained using the purge gradient G_1 and some limited phase cycling.

The refocusing condition for the pulse sequence shown in Fig. 11.29 on the preceding page is

$$\pm\gamma_S G_2 z\tau_2 + \gamma_I G_3 z\tau_3 = 0,$$

where we have assumed that gradients G_2 and G_2 have durations τ_2 and τ_3, respectively. The positive sign is for the N-type pathway (grey line), whereas the negative sign is for the P-type pathway (blue line). This condition rearranges to

$$\frac{G_2\tau_2}{G_3\tau_3} = -\frac{\pm\gamma_I}{\gamma_S}.$$

11.14   Advantages and disadvantages of coherence selection with gradients

We have already mentioned, in the context of the HMQC experiment, that the big advantage of using gradient pulses is that the unwanted pathways simply do not contribute to the observed signal. In contrast to phase cycling, therefore, the stability of the spectrometer is not so crucial.

Selection made using gradients is immediate, which is quite the opposite to phase cycling, where the selection process is only complete once the phase cycle has been finished. If the sample gives spectra with good signal-to-noise ratio in a single transient, completing all the steps of the phase cycle will just make the experiment unnecessarily long, but we have no option but to complete all the steps. In contrast, using gradient selection we can keep the experiment time to the absolute minimum needed to achieve the required signal-to-noise ratio. Experiments using gradient selection can therefore be recorded in the shortest possible times consistent with achieving the required signal-to-noise ratio.

However, these advantages of gradient selection come with a price. As we have seen, the inability of gradients to select symmetrical pathways leads to a loss of signal. Also, if we wish to retain absorption mode lineshapes, special steps need to be taken to control the evolution of the offset during the gradients. Usually, this will involve adding extra delays and extra refocusing pulses to the sequence. Finally, if gradients are introduced into the evolution time t_1, we will have to record separate P- and N-type data sets in order to be able to obtain an absorption mode spectrum.

Generally, the advantages of coherence selection with gradients outweigh the disadvantages, and as a result the use of gradients has become a matter of routine. Inverse correlation experiments, such as HMQC, on natural abundance samples benefit very much from selection with gradients, and it is fair to say that such experiments have only become routine as a result of the use of gradients. In the complex pulse sequences devised for biomolecular NMR selection using gradient pulses also plays a key role.

11.15   Suppression of zero-quantum coherence

We have already noted that z-magnetization is not dephased by a gradient simply because this type of magnetization does not evolve when a field is

(a) 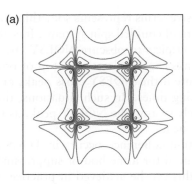   (b)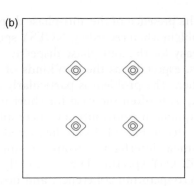

Fig. 11.30 Illustration of how a NOESY cross peak can be swamped by zero-quantum interference. The cross-peak multiplet shown in (a) is a mixture of an NOE contribution (about 20%) and a contribution from zero-quantum coherence present during the mixing time (about 80%). The NOESY contribution is an in-phase absorption mode doublet in each dimension, whereas the zero-quantum contribution is anti-phase in each dimension and has the dispersion mode lineshape. In multiplet (a) the NOE contribution is simply not visible, but is swamped by the anti-phase dispersive multiplet. If this contribution from the zero quantum is suppressed, the NOE multiplet becomes clearly visible, as shown in (b).

applied along the z-axis. In terms of coherence orders, z-magnetization is classified as having $p = 0$ and so it follows from Eq. 11.8 on page 412 that the spatially dependent phase is zero.

Zero-quantum coherence also has $p = 0$ and therefore, like z-magnetization, it does not acquire a spatially dependent phase. As a result, it is not straightforwardly possible to separate zero-quantum coherence and z-magnetization using gradients. The same is true for phase cycling, since any pathway which selects for $p = 0$ will necessarily select both zero-quantum coherence and z-magnetization.

This inability to separate zero quantum and z-magnetization causes difficulties in many two-dimensional experiments. For example, in NOESY the desired cross peaks arise from z-magnetization i.e. operators such as $\hat{I}_{1z}$ and $\hat{I}_{2z}$, present during the mixing time. The final pulse transforms these operators into $-\hat{I}_{1y}$ and $-\hat{I}_{2y}$, which give in-phase multiplets.

In a coupled two-spin system, anti-phase terms will develop during t_1, and some of these will be turned into zero-quantum coherence by the 90° pulse placed at the end of t_1. Part of this zero-quantum coherence, specifically $\hat{ZQ}_y$, is transformed by the final 90° pulse into anti-phase magnetization along the x-axis:

$$\hat{ZQ}_y \equiv 2\hat{I}_{1y}\hat{I}_{2x} - 2\hat{I}_{1x}\hat{I}_{2y} \xrightarrow{(\pi/2)(\hat{I}_{1x}+\hat{I}_{2x})} 2\hat{I}_{1z}\hat{I}_{2x} - 2\hat{I}_{1x}\hat{I}_{2z}.$$

The important point to note here is that the wanted NOESY peaks are in-phase and appear along the y-axis, whereas the unwanted peaks, which arise from zero-quantum coherence, are anti-phase and along the x-axis. Therefore, if the NOESY peaks are phased to absorption in ω_2, the peaks from the zero quantum will be in anti-phase dispersion. The same is true for the ω_1 dimension: the NOESY peaks are in phase and in absorption, whereas the zero-quantum derived peaks are anti-phase and dispersive.

The presence of the anti-phase dispersive terms represents something of a problem when recording NOESY spectra of coupled spin systems. It is all too easy for the anti-phase dispersive multiplets to obscure NOESY cross peaks, especially as the two kinds of peaks can lie directly on top of one another. The problem is particularly acute if the NOESY cross peaks are weak, as is often the case for short mixing times. In such situations, the zero-quantum contribution can completely swamp the NOESY peak, as is illustrated in Fig. 11.30 on the preceding page.

Similar interference from zero-quantum coherence occurs in TOCSY and ZCOSY spectra. There is clearly a need for a method of suppressing the zero-quantum coherence, and how this can be achieved in practice is described in the following sections. However, before doing this we will explore in a little more detail just why it is that zero-quantum coherence does not dephase in a gradient, since an understanding of this process is relevant to how we might go about suppressing this kind of coherence.

Zero-quantum coherence is represented by operator products of the type $\hat{I}_{1+}\hat{I}_{2-}$. In the absence of a gradient, the $\hat{I}_{1+}$ term will evolve in the usual way, acquiring a phase factor $\exp(-i\Omega_1 t)$ which depends on the offset of spin one. $\hat{I}_{2-}$ will acquire a phase factor $\exp(i\Omega_2 t)$, which is in the opposite sense and depends on the offset of spin two. Overall, the operator product acquires a phase which depends on the *difference* of the two offsets:

$$\exp(i[\Omega_2 - \Omega_1]t)\,\hat{I}_{1+}\hat{I}_{2-}.$$

As we saw above, in the presence of a gradient, the spatial dependence of the magnetic field results in the evolution frequency having an extra term $\Omega(z)$, where $\Omega(z) = -\gamma G z$. This term is simply added to the offsets of the spins, so that the evolution of $\hat{I}_{1+}$ and $\hat{I}_{2-}$ is now

$$\exp(-i[\Omega_1 + \Omega(z)]t)\hat{I}_{1+} \qquad \exp(i[\Omega_2 + \Omega(z)]t)\hat{I}_{2-}.$$

As a result, the evolution of the zero-quantum term is

$$\exp(i[\Omega_2 + \Omega(z) - \Omega_1 - \Omega(z)]t)\,\hat{I}_{1+}\hat{I}_{2-}.$$

Clearly, the two offset terms $\Omega(z)$ which derive from the gradient cancel one another. Therefore, the evolution of the zero-quantum coherence is *unaffected* by the presence of the gradient. What is happening here is that the gradient is affecting the two spins involved in the zero-quantum coherence in an equal and opposite way, such that overall there is no net effect.

11.15.1   The z-filter

A convenient framework for thinking about this problem of suppressing zero-quantum coherence is to consider the pulse sequence element known as a *z-filter*, shown in Fig. 11.31 on the next page. The idea of this element is that in-phase y-magnetization present at point **A** will reappear as in-phase y-magnetization at point **D**, but that *all* other magnetization will be suppressed.

The element works by the first pulse rotating $\hat{I}_{1y}$ to $\hat{I}_{1z}$; the gradient does not affect this z-magnetization, and it is then simply rotated back to $\hat{I}_{1y}$ by the second 90° pulse, which is about $-x$. Any in-phase x-magnetization present at point **A** is unaffected by the first pulse, and so is dephased by the gradient. Similarly, anti-phase terms along y are transferred to the coupled spin by the first pulse; as they are still single-quantum coherence, they are dephased by the gradient. For example, $2\hat{I}_{1y}\hat{I}_{2z}$ is transferred to $-2\hat{I}_{1z}\hat{I}_{2y}$, which is dephased.

The only problem with this sequence comes from anti-phase terms along x. For example, $2\hat{I}_{1x}\hat{I}_{2z}$ present at point **A** is transferred to $-2\hat{I}_{1x}\hat{I}_{2y}$ at point **B** by the first pulse. This state is a mixture of double- and zero-quantum coherence, of which only the zero-quantum part, $\frac{1}{2}(2\hat{I}_{1y}\hat{I}_{2x}-2\hat{I}_{1x}\hat{I}_{2y})$, survives the gradient to point **C**. This zero-quantum term is transformed back to anti-phase magnetization along x, $\frac{1}{2}(-2\hat{I}_{1z}\hat{I}_{2x} + 2\hat{I}_{1x}\hat{I}_{2z})$, by the final pulse.

The z-filter is therefore not entirely effective at selecting just the in-phase component along y. An anti-phase component along x passes through the filter because this anti-phase term becomes zero-quantum coherence in the interval between the two pulses. If we had a way of suppressing the zero-quantum coherence, we would have a perfect z-filter.

The key to achieving this suppression is the realization that, during the delay τ_z between the two pulses of the z-filter, the zero-quantum coherence evolves and acquires a phase, whereas the z-magnetization does not. It is easy to work out the details of how the zero quantum evolves using the information given in section 7.12.3 on page 176.

At point **C** we have the term $\frac{1}{2}(2\hat{I}_{1y}\hat{I}_{2x} - 2\hat{I}_{1x}\hat{I}_{2y})$ or $\frac{1}{2}\hat{ZQ}_y$. During τ_z this coherence evolves according to

$$\frac{1}{2}\cos([\Omega_1 - \Omega_2]\tau_z)\,\hat{ZQ}_y - \frac{1}{2}\sin([\Omega_1 - \Omega_2]\tau_z)\,\hat{ZQ}_x.$$

The final pulse only creates observable magnetization from the $\hat{ZQ}_y$ term, giving

$$\frac{1}{2}\cos([\Omega_1 - \Omega_2]\tau_z)(-2\hat{I}_{1z}\hat{I}_{2x} + 2\hat{I}_{1x}\hat{I}_{2z}).$$

The key thing here is that the *amplitude* and *sign* of this unwanted term depends on the zero-quantum frequency $[\Omega_1 - \Omega_2]$ and the time τ_z.

The trick to suppressing this contribution is to repeat the experiment for two different values of the z-filter delay, $\tau_{z,1}$ and $\tau_{z,2}$, such that $\cos([\Omega_1 - \Omega_2]\tau_{z,1})$ is equal and opposite to $\cos([\Omega_1 - \Omega_2]\tau_{z,2})$. If the results of the two experiments are then added together, the anti-phase terms will cancel completely.

A simple solution is to choose $\tau_{z,1}$ such that $\cos([\Omega_1 - \Omega_2]\tau_{z,1}) = 1$, and $\tau_{z,2}$ such that $\cos([\Omega_1 - \Omega_2]\tau_{z,2}) = -1$. In other words:

$$|\Omega_1 - \Omega_2|\,\tau_{z,1} = 0,\ 2\pi,\ 4\pi \ldots \quad \text{and} \quad |\Omega_1 - \Omega_2|\,\tau_{z,2} = \pi,\ 3\pi,\ 5\pi \ldots$$

Taking the first option from the list in each case gives the following values

$$\tau_{z,1} = 0 \quad \text{and} \quad \tau_{z,2} = \frac{\pi}{|\Omega_1 - \Omega_2|}.$$

If we express the offsets in terms of Hz, rather than rad s^{-1}, the value of $\tau_{z,2}$ looks more familiar:

$$\tau_{z,1} = 0 \quad \text{and} \quad \tau_{z,2} = \frac{1}{2(|\nu_1 - \nu_2|)}.$$

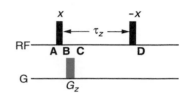

Fig. 11.31 The z-filter pulse sequence element, in which the gradient ensures that only coherence order $p = 0$ is present between the two 90° pulses. In its ideal form, only in-phase y-magnetization passes from point **A** to point **D**. However, the presence of zero quantum during τ_z results in anti-phase magnetization along x passing through the sequence. These contributions from zero quantum can be eliminated by repeating the sequence for a range of values for the delay τ_z.

It is clear that $\tau_{z,2}$ is the time needed for the zero quantum to precess through half a revolution.

In a real molecule there will be more than one zero-quantum frequency, so the choice of the z-filter delays is not so straightforward. However, it turns out that there is a systematic way of choosing a set of these delays such that coherences with a certain range of zero-quantum frequencies are suppressed: the details can be found in the publications listed under *Further reading*.

The difficulty with this approach is that it is necessary to repeat the experiment several times. This can result in an unacceptably long experiment, just as can be the case where a long phase cycle is used.

Implementation of z-filters in two-dimensional experiments

The z-filter element occurs in a number of important two-dimensional experiments, such as NOESY, ZCOSY and TOCSY, whose pulse sequences are shown in Fig. 11.32. In each case, the presence of zero-quantum coherence between the two pulses of the filter results in unwanted dispersive contributions to the spectrum. Just as was described in the previous section, it is possible to suppress this contribution from zero-quantum coherence by repeating the experiment with a set of carefully chosen values for the delay between the two pulses of the filter.

In the case of ZCOSY, the two pulses which form the filter have small flip angles, rather than being 90° pulses. However, the issue remains the same, regardless of the flip angle of the pulses, as in this experiment we wish to retain only the contribution due to populations (z-magnetization) between the two pulses, and reject all coherences, including zero-quantum.

In TOCSY, the mixing sequence is placed within a z-filter, and to suppress the zero-quantum contributions we need to vary both the delay between the first 90° pulse and the mixing sequence, as well as the delay between the mixing sequence and the second 90° pulse. This is necessary because the mixing sequence also generates zero-quantum coherence (see section 8.11.1 on page 222).

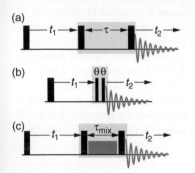

Fig. 11.32 Three pulse sequences for two-dimensional experiments, all of which contain a z-filter element, which is highlighted by the blue box. The sequences (a), (b) and (c) are NOESY, ZCOSY and TOCSY, respectively. In each case, a gradient is applied in the z-filter delay in order to dephase coherences other than those with $p = 0$.

11.15.2   Zero-quantum dephasing

In this last section, we turn to a modification of the z-filter which makes it possible to eliminate the zero-quantum coherence *in a single experiment*. To understand how the modification works, it is helpful to follow through a sequence of 'thought' experiments.

First, imagine taking the z-filter shown in Fig. 11.33 (a) on the next page, and then inserting a 180° pulse during τ_z, to give the sequence shown in (b). The 180° pulse is placed at time τ from the start of the filter, so will create a spin echo at time 2τ (indicated by the blue line). As a result, the zero quantum will only evolve for time ($\tau_z - 2\tau$). By moving the 180° pulse around in the filter (i.e. by changing τ), we can vary the time over which the zero quantum will evolve. The result is entirely equivalent to changing the delay τ_z in the original z-filter, sequence (a).

Now suppose that we could arrange things so that, in *different parts of the sample*, the 180° pulse appears at *different times* τ. What would happen

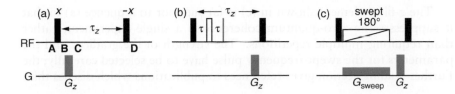

Fig. 11.33 Different versions of the z-filter element. The original sequence, which has already been discussed, is shown in (a). Sequence (b) gives an alternative way of changing the time for which the zero quantum evolves; as explained in the text, the zero quantum evolves for time $(\tau_z - 2\tau)$. Sequence (c) results in dephasing of the zero-quantum coherence in a single experiment. The key feature of the sequence is a swept-frequency 180° pulse (indicated by the rectangle with the diagonal line) applied during a gradient. See text for further details.

is that the zero quantum would evolve for different times in different parts of the sample. For example, we could arrange things so that at the top of the sample $\tau = 0$, so the zero quantum evolves for τ_z, and that at the bottom of the sample $\tau = \frac{1}{2}\tau_z$, so that the zero quantum does not evolve at all. As we go from the top to the bottom we arrange things such that τ increases steadily, and so the time for evolution of the zero quantum goes from τ_z at the top, steadily down to zero at the bottom. The idea is illustrated in Fig. 11.34.

When we observe our NMR signal, it is from the *whole* sample at once. So if we were able to arrange for the zero-quantum evolution time to vary along the sample, what we would actually observe would be the *sum* of the signals from all these filters with different zero-quantum evolution times. This would be the same as adding up the results from a large number of experiments with different z-filter delays, and so the contribution from the zero-quantum coherence would be suppressed. However, in our thought experiment we achieve this suppression in a *single experiment*, which is a great advantage. In effect, the zero quantum is dephased as a result of its evolution becoming spatially dependent, just as in a normal gradient.

The final part of the story is turning this thought experiment into a practical pulse sequence, which we do by using a combination of a gradient with a swept-frequency 180° pulse, as shown in Fig. 11.33 (c). As is illustrated in Fig. 11.17 on page 410, when a gradient is applied the NMR line becomes very broad and – most importantly – different parts of the line correspond to different *positions* in the sample. So, as shown in the diagram, the high frequency part of the line corresponds to the top of the sample, and the low frequency part to the bottom.

With the gradient switched on, we then apply a swept-frequency 180° pulse. Such pulses are rather different from the ones we have encountered so far, in that the frequency of the RF used to generate them is *not* constant, but is swept steadily from one edge of the spectrum to the other. As a result, lines at different *offsets* experience the 180° pulse at different *times*.

If we apply such a swept-frequency pulse when the gradient is on, and sweep the frequency from one end of the broad line to the other, the result will be a 180° pulse appearing at different *times* in different *parts* of the sample. This is exactly what we imagined in our thought experiment.

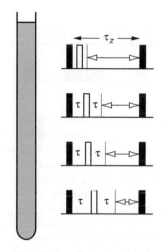

Fig. 11.34 Illustration of a thought experiment in which we imagine that the timing of the 180° pulse in the z-filter varies as we move along the sample. As a result, the time for which the zero-quantum coherence evolves, indicated by the double-headed open arrow, varies along the sample.

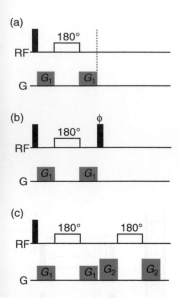

Fig. 11.35 Illustration of how gradients can be employed to advantage in selective excitation. In sequence (a), magnetization which experiences the selective 180° refocusing pulse is rephased at the end of the spin echo, but other magnetization is dephased by both gradient pulses. This sequence therefore gives very clean excitation in which the *only* magnetization present is from the selectively excited resonances. By adding a 90° pulse of appropriate phase ϕ to the end of the sequence, the selected magnetization will be rotated to $-z$; the result is a selective inversion sequence, shown in (b). Sequence (c) is simply a development of (a) in which two, rather than one, gradient echoes are used; the sequence is called the double pulsed field gradient spin echo, DPFGSE. As explained in the text, this sequence has more desirable phase properties than the simple gradient echo.

The z-filter element shown in (c) is far superior to sequence (a) in that it suppresses the zero-quantum coherence in a single experiment, rather than requiring multiple repetitions. The strength of the gradient and the parameters for the swept-frequency pulse have to be selected carefully; the Further reading section gives references to publications which discuss this.

11.16   Selective excitation with the aid of gradients

In section 4.11.2 on page 69 it was described how, by reducing the RF field strength of a pulse, only lines which are on resonance, or close to resonance, are excited. Such selective excitation is used quite often in NMR experiments, such as the transient NOE experiment described in section 9.7.1 on page 274. In this section we will describe how field gradient pulses can be used to improve the quality of selective excitation, and go on to show how this can be used to great advantage in one-dimensional NOE experiments.

The key idea of how gradients can be used to improve selective excitation is shown in Fig. 11.35 (a). The sequence starts with a non-selective 90° pulse which excites transverse magnetization from all of the spins; this magnetization is then dephased by gradient G_1. Next comes a selective 180° pulse whose frequency is set in the middle of the resonances which we want to excite, and whose field strength has been chosen so that only the resonances over the required range will be affected (e.g. one line or one multiplet).

The sequence ends with the second gradient G_1, which is identical to the first. For magnetization which has *not* experienced the selective 180° pulse, this second gradient simply causes further dephasing. On the other hand, the magnetization which *does* experience the selective 180° pulse is rephased by the second gradient, as the pulse sequence is a simple spin echo.

The overall result is that magnetization from resonances which experience the selective 180° pulse is refocused at the end of the second gradient, whereas *all* other magnetization is dephased. In an experiment which uses selective excitation we are only interested in the fate of the magnetization which has been excited. The advantage of this selective gradient echo method is that the magnetization from all of the spins other than the selectively excited ones is dephased, and is therefore unobservable, even if further pulses are applied.

Typically the excitation sequence of Fig. 11.35 (a) is used at the start of a more complex pulse sequence in which the selectively excited magnetization is manipulated further. For example, in the selective COSY experiment shown in Fig. 7.18 on page 172, the initial selective 90° pulse can be replaced by this gradient echo sequence. If we do this, it is not necessary to compute a difference spectrum as all but the magnetization from the excited spin is dephased. The magnetization generated by the final non-selective pulse must therefore come from the excited spin.

In the transient NOE experiment, whose pulse sequence is shown in Fig. 9.20 on page 274, we start out with selective inversion of the resonances of one spin. This can be achieved simply by adding a 90° pulse, of the appropriate phase, at the end of the gradient echo sequence, as is shown in Fig. 11.35 (b). Applying the pulse about $-x$ rotates the

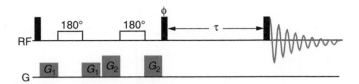

Fig. 11.36 The pulse sequence for the DPFGSE NOE experiment. This is essentially a one-dimensional transient NOE experiment in which the selective 180° pulse has been replaced by the combination of a DPFGSE sequence followed by a 90° pulse, of phase ϕ. If $\phi = +x$ the selectively excited magnetization is inverted, whereas if the phase is $-x$ the magnetization is returned to the z-axis. Cross relaxation takes place during τ, and the final pulse makes the z-magnetization observable. It is necessary to compute a difference spectrum in order to suppress the signals arising from z-magnetization which recovers due to relaxation during τ. Two experiments are recorded, with the phase ϕ set to $+x$ and then to $-x$: taking the difference between these eliminates the unwanted signals.

selectively excited magnetization onto $-z$, whereas applying the pulse about $+x$, puts the magnetization onto $+z$. In either case, all other magnetization is dephased.

11.16.1   The double pulsed field gradient spin echo

One of the difficulties with the gradient echo, Fig. 11.35 (a) on the facing page, is that the axis along which the selected magnetization rephases (i.e. the phase of the magnetization) is affected by the phase properties of the 180° pulse. For example, some shaped selective pulses (section 4.11.2 on page 69) give rise to phase shifts which vary across the range of offsets which are excited.

These difficulties with the phase of the resulting magnetization are all neatly side-stepped by using two gradient echoes, as shown in Fig. 11.35 (c). It turns out that the phase of the magnetization excited by this sequence is *independent* of the phase properties of the selective 180° pulses (provided the two pulses are the same). An important proviso is that the gradients must be chosen such that *only* magnetization which experiences *both* 180° pulses is refocused i.e. we must select only the pathway $-1 \rightarrow +1 \rightarrow -1$.

This double pulsed field gradient spin echo, or DPFGSE, sequence can be turned into a selective inversion pulse simply by adding a 90° pulse of the appropriate phase at the end of the sequence, just as we did for the single echo. Such an inversion pulse has found an important application in one-dimensional NOE experiments, which are described in the next section.

11.16.2   The DPFGSE NOE experiment

The DPFGSE NOE experiment is essentially a modification of the one-dimensional transient NOE experiment, shown in Fig. 9.20 on page 274, in which the selective inversion pulse has been replaced by a DPFGSE inversion sequence: the pulse sequence is shown in Fig. 11.36.

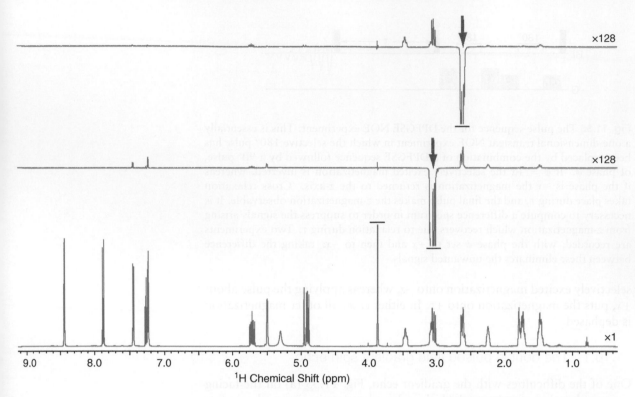

Fig. 11.37 Experimental DPFGSE NOE spectra for quinine, recorded at 500 MHz. The normal ^{1}H spectrum is shown at the bottom, along with two NOE spectra in which different multiplets have been inverted (indicated by the dark grey arrow). The mixing time was 0.5 s, and the NOE spectra are shown on an expanded vertical scale. Several NOE enhancements, including some rather small ones, are clearly visible against the clean baseline of the spectrum. Note the excellent suppression of the very strong peak at 3.85 ppm.

The initial 90° pulse generates magnetization along $-y$, and that part of the magnetization which experiences both of the selective 180° pulses refocuses along $-y$ at the end of the second gradient G_2. If the following 90° pulse has phase $\phi = +x$ the magnetization will be rotated onto $-z$ i.e. inverted, whereas if $\phi = -x$, the magnetization will be returned to $+z$. During the delay τ, cross relaxation takes place, and then the final 90° pulse makes the result observable.

Although the magnetization from all but the selectively excited spin is dephased at the end of the DPFGSE sequence, relaxation during the rather long delay τ allows z-magnetization to recover. This recovered magnetization will be made observable by the final 90° pulse. As a result, the spectrum will *not* simply show peaks from the initially excited spin and those spins which are cross relaxing with it.

The unwanted peaks from this recovered z-magnetization can easily be eliminated by a difference experiment. All we do is record the spectrum twice, once with the phase ϕ set to $+x$, and once with it set to $-x$. The signals due to the recovered magnetization will be the *same* in the two experiments, so taking the difference will eliminate them. This is, of course, exactly the same difference procedure we used to reveal the NOE enhancements in the simple transient NOE experiment.

The big advantage of the DPFGSE NOE experiment over the simple transient NOE experiment of Fig. 9.20 on page 274 is that, in the former, the difference step is only needed to suppress the signals arising from the z-magnetization which has recovered during τ. In contrast, in the simple transient experiment, the difference step has to suppress the much larger signals from the equilibrium z-magnetization of all the spins which are not affected by the 180° pulse. In practical use, the DPFGSE NOE gives much higher quality spectra, enabling smaller NOE enhancements to be detected with greater confidence.

Figure 11.37 on the preceding page shows experimental DPFGSE NOE spectra of quinine. The excellent suppression of the generality of signals makes it possible to observe even small NOE enhancements.

11.17 Further reading

Phase cycling and coherence transfer pathways:

G. Bodenhausen, H. Kogler and R. R. Ernst, *J. Magn. Reson.*, **58**, 370–388 (1984).

Appendix A.11 from M. H. Levitt, *Spin Dynamics* (2nd edition, John Wiley & Sons, Ltd, 2008).

CYCLOPS:

Chapter 6 from R. Freeman, *Spin Choreography* (Spektrum, 1997).

Sensitivity of experiments using gradient selection:

G. Kontaxis, J. Stonehouse, E. D. Laue and J. Keeler, *J. Magn. Reson. Ser. A*, **111**, 70–76 (1994).

The z-filter:

O. W. Sørensen, M. Rance and R. R. Ernst, *J. Magn. Reson.*, **56**, 527–534 (1984).

Zero-quantum dephasing:

M. J. Thrippleton and J. Keeler, *Angew. Chem., Int. Ed. Engl.*, **42**, 3938–3941 (2003).

Selective excitation using the DPFGSE, and its application to NOE experiments:

T. L. Hwang and A. J. Shaka, *J. Magn. Reson. Ser. A*, **112**, 275–279 (1995)

K. Stott, J. Keeler, Q. N. Van and A. J. Shaka, *J. Magn. Reson.*, **125**, 302–324 (1997).

11.18 Exercises

11.1 Using the same approach as was used to derive Eq. 11.3 on page 383, show that a z-rotation through an angle ϕ of the operator $\hat{I}_{i-}$ generates a phase factor of $\exp(+i\phi)$.

State the overall coherence order, or orders, of each of the following operators:

$$\hat{I}_{1+}\hat{I}_{2-} \quad 2\hat{I}_{1+}\hat{I}_{2+}\hat{I}_{3z} \quad \hat{I}_{1x} \quad \hat{I}_{2y} \quad 2\hat{I}_{1z}\hat{I}_{2y} \quad (2\hat{I}_{1x}\hat{I}_{2x} + 2\hat{I}_{1y}\hat{I}_{2y})$$

You may need to express the operators $\hat{I}_x$ and $\hat{I}_y$ in terms of $\hat{I}_+$ and $\hat{I}_-$ using Eq. 11.4 on page 383.

In a heteronuclear spin system a coherence order can be assigned to each spin, I and S. Assign such orders for the following operators:

$$\hat{I}_x \quad \hat{S}_y \quad 2\hat{I}_x\hat{S}_z \quad 2\hat{I}_x\hat{S}_x \quad 2\hat{I}_x\hat{S}_y.$$

Following the discussion in section 11.1.2 on page 384, write down the result of allowing each of the following operators to evolve freely for a time t:

$$\hat{I}_{1+} \quad \hat{I}_{2-} \quad \hat{I}_{1+}\hat{I}_{2+} \quad \hat{I}_+\hat{S}_- \quad \hat{I}_{1-}\hat{I}_{2-}\hat{I}_{3-}.$$

11.2 Draw up coherence transfer pathways for the following experiments: (a) triple-quantum filtered COSY (which is identical to DQF COSY, except that $p = \pm 3$ between the last two pulses); (b) zero-quantum spectroscopy (which is identical to double-quantum spectroscopy, except that we have $p = 0$ during t_1); (c) ZCOSY; (d) HSQC. In all cases, retain symmetrical pathways in t_1.

For the HMQC experiment, draw up coherence transfer pathways which will give: (a) a P-type spectrum; (b) an N-type spectrum; (c) a spectrum which can be processed to give absorption mode lineshapes. Which of these spectra will be frequency discriminated in the ω_1 dimension?

11.3 Confirm that each of the combinations of A and B given in the table on p. 393 does indeed give rise to a spectrum with the same lineshape.

Draw up a diagram, similar to that of Fig. 11.6 on page 394, to illustrate that, in a pulse–acquire experiment where the pulse phase goes through the sequence $[x, y, -x, -y]$ and the receiver phase goes through the sequence $[-180°, -270°, 0°, -90°]$, each spectrum has the same lineshape.

11.4 In section 11.5.1 on page 396 it was shown that a pathway with $\Delta p = -3$ could be selected with the following four-step cycle.

pulse phase: $[0°, 90°, 180°, 270°]$ rx phase: $[0°, 270°, 180°, 90°]$.

Show that this cycle rejects pathways with $\Delta p = -1$ and $\Delta p = 0$, but selects a pathway with $\Delta p = +5$.

11.5 In section 11.5.2 on page 399 it was shown how the independent completion of two four-step cycles, the first selecting $\Delta p = +1$ and the second selecting $\Delta p = -2$, generates a 16-step cycle. Draw up a table, similar to Table 11.1 on page 400, in which the required receiver phases are shown for the case where the *second* pulse is cycled first.

Draw up a 16-step phase cycle for a two-pulse sequence in which Δp is -1 for the first pulse and $+3$ for the second; you should determine the sequence of receiver phases needed to select this pathway.

11.6 Consider a phase cycle in which the pulse phase goes through the three steps $[0°, 120°, 240°]$. Devise a set of accompanying receiver phase shifts (which will not be multiples of 90°) which will select a pathway with $\Delta p = -2$. Without further detailed calculations, explain which other values of Δp will be selected or rejected by this three-step sequence.

Explain how this three-step sequence could be used to select the appropriate pathway for N-type COSY, and devise another three-step cycle to select the pathway for P-type COSY.

11.7 Shown below are the pulse sequence and coherence transfer pathway for triple-quantum filtered COSY.

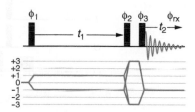

Grouping the first two pulses together, devise a *six*-step phase cycle which will select the required pathway (i.e. step the pulse phase in increments of 60° and determine the correct receiver phase shifts). Once this pathway has been selected, is any further phase cycling necessary?

Without further detailed calculations explain what other pathways are selected by your phase cycle, and comment on whether or not they are a matter for concern in practical spectroscopy.

Write an alternative six-step phase cycle for the above experiment in which just the phase of the last pulse, and the receiver, are shifted; you will need receiver phase shifts which are not multiples of 90°.

11.8 For the NOESY experiment, group the first two pulses together and devise a four-step phase cycle which selects the pathway shown in Fig. 11.15 on page 407.

Add axial peak suppression to your cycle (write out all eight steps).

Write a coherence transfer pathway for N-type NOESY, and devise a suitable phase cycle to select this pathway (16 steps are needed). Is is necessary to add axial peak suppression to your cycle?

11.9 Consider the pathway shown below.

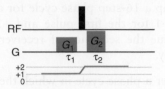

Write down the spatially dependent phase which accrues during the two gradient pulses, and hence determine the ratio $(G_2\tau_2)/(G_1\tau_1)$ which will refocus the pathway.

Determine: (a) the values of G_1 and G_2 needed if the gradients are of the same length; (b) the values of τ_1 and τ_2 needed if the two gradients have the same absolute strength.

11.10 Consider the pathway shown below for a heteronuclear experiment.

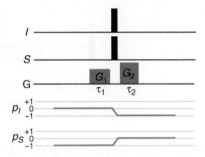

Write down the spatially dependent phase which accrues during the two gradient pulses, and hence determine the ratio $(G_1\tau_1)/(G_2\tau_2)$ which will refocus the pathway.

Assuming that I is ^1H and S is ^{15}N, and that the gradients have the same duration, work out the ratio of gradient strengths needed to select this pathway, given that $\gamma(^1H) \approx -10\,\gamma(^{15}N)$.

11.11   Draw up coherence transfer pathways for the following experiments, and explain how gradients could be used to select the specified pathways. In some cases you may need to modify the pulse sequences, for example, by the inclusion of extra refocusing pulses. Determine the relative strengths of any gradient pulses you propose to include.

(a)   P-type DQF COSY (i.e. $p = -1$ during t_1) (absolute value display).

(b)   N-type triple-quantum filtered COSY (i.e. $p = +1$ during t_1) (absolute value display).

(c)   N-type COSY (absolute value display).

(d)   Double-quantum spectroscopy, intended to give an absorption mode spectrum (i.e. refocusing phase errors due to gradients, and recording separate P- and N-type spectra).

(e)   N-type HSQC (absolute value display) for ^{1}H–^{13}C correlation.

12

Equivalent spins and spin system analysis

This chapter is concerned mainly with understanding how NMR spectra and the outcome of multiple-pulse experiments are affected by the presence of *equivalent spins*, such as the protons in a $-CH_3$ group. We will start out by looking at how simple spectra are affected by the presence of equivalent spins, and then go on to adapt the product operator method to deal with such spin systems so that we can understand the behaviour of CH_n groups in the INEPT experiment. The important APT and DEPT experiments, which are used to distinguish between resonances from CH, CH_2 and CH_3 groups in ^{13}C spectra, will also be discussed.

The chapter closes by considering the spectra which arise from strongly coupled spin systems and how these can be analysed. We will see that such spin systems can sometimes give rise to rather misleading spectral features which it is as well to be aware of.

Equivalent spins

When we learn about coupling in proton NMR, one of the first things we are told is that 'the couplings between equivalent spins do not give rise to any splittings'. For example, in chloroethane, CH_3CH_2Cl, the three protons from the methyl group appear as a triplet due to their coupling to the CH_2 protons, and the CH_2 protons appear as a quartet due to their coupling to the methyl protons. What we do *not* see is any splittings caused by the coupling between the methyl protons or by the coupling between the CH_2 protons.

Of course there *must* be a coupling between the protons in the methyl group, as they are only separated by two bonds, but for some reason this coupling does not affect the spectrum. Exactly why this is so is rather a subtle matter which is best addressed by first thinking about the spectrum of two *strongly coupled* spins, so this is the point from which we start.

Understanding NMR Spectroscopy, Second Edition James Keeler
© 2010 John Wiley & Sons, Ltd

12.1   Strong coupling in a two-spin system

Up to this point, all of our discussion applies only to *weakly coupled* spins. As was explained in section 2.3.2 on page 12, this is the limit in which the separation between the Larmor frequencies of any two coupled spins is much larger in magnitude than the scalar coupling between them. The form of the multiplets arising from weakly coupled spins can be predicted using the familiar 'tree diagrams' discussed briefly in section 2.3 on page 10. Note that, except in the case where the couplings are such that lines overlap, the intensity of each line within a given multiplet is the same.

If the weak-coupling condition does not apply i.e. if the frequency separation is comparable with or less than the coupling, then the system is said to be *strongly coupled*. Under these circumstances, both the *positions* and *intensities* of the lines in a multiplet are different from those expected in the weakly coupled limit. It is not possible to predict the frequencies and intensities of the lines without resorting to explicit quantum mechanical calculations.

In this section we are going to explore in some detail the spectrum from two strongly coupled spins. This discussion will reveal many of the general features of such strongly coupled systems, and will allow us to introduce a number of ideas which will be very useful when it comes to discussing equivalent spins.

12.1.1   The Hamiltonian and its energy levels

The spectrum of two strongly coupled spins can be predicted using the same approach as in section 3.5 on page 35. First, we construct the appropriate Hamiltonian and then find the associated energy levels and wavefunctions; the form of the spectrum is found by looking at the allowed transitions between the energy levels.

The Hamiltonian we used before is appropriate for two weakly coupled spins. When written in frequency units (Hz) this Hamiltonian is

$$\hat{H}_{\text{weak}} = \upsilon_{0,1}\hat{I}_{1z} + \upsilon_{0,2}\hat{I}_{2z} + J_{12}\hat{I}_{1z}\hat{I}_{2z}.$$

In the case of strong coupling the term which represents the coupling has to be modified to $J_{12}\,\hat{\mathbf{I}}_1 \cdot \hat{\mathbf{I}}_2$

$$\hat{H}_{\text{strong}} = \upsilon_{0,1}\hat{I}_{1z} + \upsilon_{0,2}\hat{I}_{2z} + J_{12}\hat{\mathbf{I}}_1 \cdot \hat{\mathbf{I}}_2,$$

where

$$\hat{\mathbf{I}}_1 = \left(\hat{I}_{1x}, \hat{I}_{1y}, \hat{I}_{1z}\right) \quad \text{and} \quad \hat{\mathbf{I}}_2 = \left(\hat{I}_{2x}, \hat{I}_{2y}, \hat{I}_{2z}\right).$$

$\hat{\mathbf{I}}_1$ and $\hat{\mathbf{I}}_2$ are vectors comprising the operators which represent the x-, y- and z-components of the angular momentum. $\hat{\mathbf{I}}_1 \cdot \hat{\mathbf{I}}_2$ is the *scalar product* between these two vectors, which is defined in the following way

$$\hat{\mathbf{I}}_1 \cdot \hat{\mathbf{I}}_2 = \hat{I}_{1x}\hat{I}_{2x} + \hat{I}_{1y}\hat{I}_{2y} + \hat{I}_{1z}\hat{I}_{2z}.$$

The product functions $\psi_{\alpha,1}\psi_{\alpha,2}$, $\psi_{\alpha,1}\psi_{\beta,2}$, $\psi_{\beta,1}\psi_{\alpha,2}$ and $\psi_{\beta,1}\psi_{\beta,2}$, are eigenfunctions of $\hat{H}_{\text{weak}}$ but it turns out that, on account of the modified coupling term, these functions are *not* eigenfunctions of $\hat{H}_{\text{strong}}$. However, given the

Table 12.1 Eigenfunctions and corresponding eigenvalues (energy levels) for two strongly coupled spins.

	M	eigenfunction	eigenvalue (energy / Hz)
a	+1	$\psi_{\alpha,1}\psi_{\alpha,2}$	$+\frac{1}{2}\Sigma + \frac{1}{4}J_{12}$
b	0	$\cos\left(\frac{1}{2}\xi\right)\psi_{\alpha,1}\psi_{\beta,2} + \sin\left(\frac{1}{2}\xi\right)\psi_{\beta,1}\psi_{\alpha,2}$	$+\frac{1}{2}D - \frac{1}{4}J_{12}$
c	0	$\cos\left(\frac{1}{2}\xi\right)\psi_{\beta,1}\psi_{\alpha,2} - \sin\left(\frac{1}{2}\xi\right)\psi_{\alpha,1}\psi_{\beta,2}$	$-\frac{1}{2}D - \frac{1}{4}J_{12}$
d	−1	$\psi_{\beta,1}\psi_{\beta,2}$	$-\frac{1}{2}\Sigma + \frac{1}{4}J_{12}$

fact that the two Hamiltonians $\hat{H}_{\text{weak}}$ and $\hat{H}_{\text{strong}}$ are not that different, we can use a well-established strategy in quantum mechanics which is to assume that the eigenfunctions of $\hat{H}_{\text{strong}}$ are linear combinations of the known eigenfunctions of $\hat{H}_{\text{weak}}$. This strategy works perfectly in the present case, although the details of how the correct linear combinations are chosen are beyond the level of the present discussion.

See M. H. Levitt, *Spin Dynamics* (2nd edition, John Wiley & Sons, Ltd, 2008) pages 615–621 for a detailed explanation of how these eigenfunctions for a two-spin system can be found.

The eigenfunctions and the corresponding energy levels of $\hat{H}_{\text{strong}}$ are given in Table 12.1, where the following definitions are used

$$\Sigma = \upsilon_{0,1} + \upsilon_{0,2} \qquad D = \sqrt{(\upsilon_{0,1} - \upsilon_{0,2})^2 + J_{12}^2}. \qquad (12.1)$$

Note that D is defined to be a *positive* quantity. The angle ξ is given by

$$\tan(\xi) = \frac{J_{12}}{\upsilon_{0,1} - \upsilon_{0,2}}.$$

In computing the angle ξ it is important to place it in the correct quadrant; this point is explored further on page 469.

Comparison of these eigenfunctions and eigenvalues with those given in Table 3.2 on page 38 for two weakly coupled spins shows that eigenfunctions a and d are the same as the weakly coupled eigenfunctions 1 and 4. However, eigenfunction b is a mixture of the weakly coupled eigenfunctions 2 ($\psi_{\alpha,1}\psi_{\beta,2}$) and 3 ($\psi_{\beta,1}\psi_{\alpha,2}$). Similarly, eigenfunction c is also a mixture of these two functions.

The degree of mixing of the wavefunctions depends on the angle ξ which is often called the *strong coupling parameter*. In the limit of weak coupling the difference in the Larmor frequencies is much greater than the coupling constant, $|\upsilon_{0,1} - \upsilon_{0,2}| \gg |J_{12}|$, so that $\tan(\xi)$, and hence ξ, is very close to zero. As a result $\cos\left(\frac{1}{2}\xi\right) = 1$ and $\sin\left(\frac{1}{2}\xi\right) = 0$, so there is no mixing of the wavefunctions. In this limit eigenfunctions b and c are then the same as the weakly coupled eigenfunctions 2 and 3.

The eigenvalues (energies) of a and d are the same as for the weakly coupled cases, since the eigenfunctions are the same. However, the eigenvalues of b and c are changed as a result of the mixing. In the weak-coupling limit $|\upsilon_{0,1} - \upsilon_{0,2}| \gg |J_{12}|$, so it follows that $D = |\upsilon_{0,1} - \upsilon_{0,2}|$. If this is the case, the eigenvalues in Table 12.1 are the same as for the weakly coupled case.

12.1.2 Form of the spectrum

In section 3.6 on page 38 it was explained that the allowed transitions were ones in which the quantum number m of *one* of the spins changes by ±1.

Table 12.2 Frequencies and intensities of the allowed transitions in a strongly coupled two-spin system, along with corresponding expressions for a weakly coupled system.

	strong coupling			weak coupling	
transition	frequency	intensity		transition	frequency
$a \to b$	$+\frac{1}{2}D - \frac{1}{2}\Sigma - \frac{1}{2}J_{12}$	$\frac{1}{2}(1 + \sin\xi)$		$1 \to 2$	$-v_{0,2} - \frac{1}{2}J_{12}$
$c \to d$	$+\frac{1}{2}D - \frac{1}{2}\Sigma + \frac{1}{2}J_{12}$	$\frac{1}{2}(1 - \sin\xi)$		$3 \to 4$	$-v_{0,2} + \frac{1}{2}J_{12}$
$a \to c$	$-\frac{1}{2}D - \frac{1}{2}\Sigma - \frac{1}{2}J_{12}$	$\frac{1}{2}(1 - \sin\xi)$		$1 \to 3$	$-v_{0,1} - \frac{1}{2}J_{12}$
$b \to d$	$-\frac{1}{2}D - \frac{1}{2}\Sigma + \frac{1}{2}J_{12}$	$\frac{1}{2}(1 + \sin\xi)$		$2 \to 4$	$-v_{0,1} + \frac{1}{2}J_{12}$

In a strongly coupled system the selection rule is expressed in terms of the quantum number M, introduced in section 3.6.1 on page 39, which is the sum of the m values for the two spins. The values of M for each of the eigenfunctions are given in Table 12.1 on the previous page.

Eigenfunction b is a mixture of the product functions $\psi_{\alpha,1}\psi_{\beta,2}$ and $\psi_{\beta,1}\psi_{\alpha,2}$, which have M values of $+\frac{1}{2} - \frac{1}{2} = 0$ and $-\frac{1}{2} + \frac{1}{2} = 0$, respectively. Thus this eigenfunction has $M = 0$, and the same is true for eigenfunction c. Eigenfunctions a and d have $M = +\frac{1}{2} + \frac{1}{2} = +1$ and $M = -\frac{1}{2} - \frac{1}{2} = -1$, respectively.

It is no coincidence that the two product functions which are mixed together to form eigenfunction b have the same value of M. This arises from the form of the coupling term in the Hamiltonian.

The allowed transitions are the ones in which the quantum number M changes by ± 1 i.e. transitions a–b, c–d, a–c and a–d. In the case of a weakly coupled spin system all the allowed transitions have the same intensity, however this is no longer true in a strongly coupled system and we have to use further quantum mechanical techniques, which are beyond the scope of this discussion, to predict the intensities. Table 12.2 gives the frequencies and intensities of the four transitions, along with the corresponding values in the weakly coupled case; note that the intensities are expressed in terms of the angle ξ.

In the limit of weak coupling ξ goes to zero so that $\sin\xi$ is also zero. All of the lines then have the same intensity. The expressions for the frequencies can be taken to the weak coupling limit by setting $D = |v_{0,1} - v_{0,2}|$, as before.

Figure 12.1 on the next page shows a set of spectra computed from the expressions given in the table and for progressively increasing degrees of strong coupling. In spectrum (a) the separation of the Larmor frequencies is sixteen times greater than the coupling constant, so the spectrum is very close to the weakly coupled limit. We see the expected two doublets, but the intensity of the lines within each of the doublets is not quite the same. As the Larmor frequencies move closer together in (b) and then (c), these intensity perturbations become more pronounced. The two 'outer' lines, transitions cd and ac, become progressively weaker, whereas the two 'inner' lines, ab and bd, become stronger. This leads to what is called *roofing*, since the profile of the intensities is reminiscent of the slope of a roof, as indicated by the dashed lines in spectrum (c).

In the weakly coupled case the two doublets are centred at the Larmor frequencies of the two spins. However, as the coupling becomes stronger the two lines are no longer symmetrically disposed about the Larmor

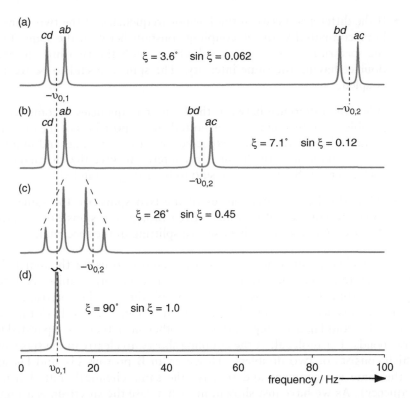

Fig. 12.1 Computed spectra from two strongly coupled spins. In all cases $J_{12} = 5.0$ Hz and the Larmor frequency of spin one is -10 Hz. As we go from (a) to (d) the Larmor frequency of spin two is moved progressively closer to spin one, with $v_{0,2}$ taking the values -90, -50, -20 and -10 Hz, respectively; the Larmor frequencies are indicated by the dashed vertical lines. Also given are the values (in degrees) of the angle ξ and the value of $\sin\xi$. The transitions are labelled according to Table 12.2 on the preceding page. Note particularly that when the two Larmor frequencies are equal the spectrum, shown in (d), consists of a single line.

frequencies. In fact, the stronger of the two lines (ab and bd) move closer to the Larmor frequencies, whereas the weaker lines move away – this is particularly evident in spectrum (c). Rather strangely, the separation of the two lines in each doublet is always J_{12}, regardless of the strength of the coupling.

If the two Larmor frequencies are the same, then ($v_{0,1} - v_{0,2}$) goes to zero and $\tan\xi$ goes to infinity i.e. the angle ξ is $\pi/2$ radians or $90°$. In this limit $\sin\xi = 1$, so the intensity of transitions cd and ac, which are given by $\frac{1}{2}(1 - \sin\xi)$, go to zero. In contrast the other two transitions have intensity 1. In other words, only two lines remain in the spectrum.

The frequency of transition ab is, from the table, $+\frac{1}{2}D - \frac{1}{2}\Sigma - \frac{1}{2}J_{12}$. However, if $v_{0,1} = v_{0,2}$ it follows from their definitions, Eq. 12.1 on page 443, that $D = |J_{12}|$ and $\Sigma = 2v_{0,1}$. Therefore the frequency of the transition ab is $-v_{0,1}$; a similar line of argument shows that the frequency of bd is also $-v_{0,1}$. Note carefully that the frequencies of these two lines are the same and *do not* depend on the value of the coupling constant.

The conclusion is that if we have two coupled spins whose Larmor frequencies (i.e. chemical shifts) are the same, then in the spectrum we simply see *one* line at the Larmor frequency. The value of the coupling constant has *no effect* on the position of the line, and although there is a coupling between the two nuclei this does not give rise to any observable splittings in the spectrum.

12.1.3   Summary

It is useful at this point to summarize what we have found about the spectrum of two coupled spins.

- If the difference between the Larmor frequencies of the two spins is large compared with the coupling constant between the spins, then the spectrum consists of two doublets, with the two lines in each doublet having the same intensity. The spins are said to be weakly coupled.

- When the difference between the Larmor frequencies becomes comparable with the coupling constant, then the positions of the lines are perturbed from the values expected for weak coupling, and in addition the intensities of the lines are altered to give the characteristic roofing effect. Such spins are said to be strongly coupled.

- When the Larmor frequencies of the two spins are the same, the spectrum consists of one line. Even though the spins are coupled, this coupling does not give rise to a splitting in the spectrum.

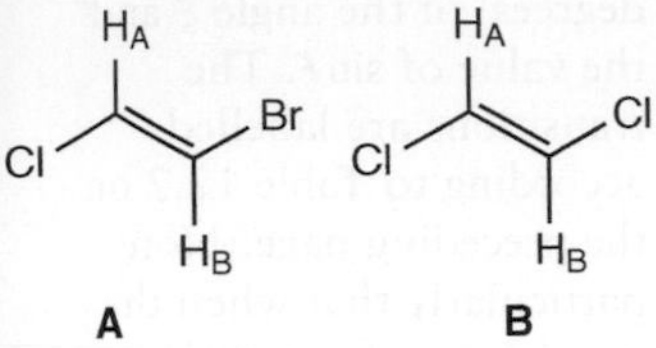

It is this last point which is of greatest interest to us for the remainder of this chapter, since it lies at the root of the reason why coupling between equivalent spins does not give rise to any splittings in the spectrum. For example, consider the two molecules shown in Fig. 12.2: in both the two protons H_A and H_B are coupled to one another since they are separated by three bonds. For molecule **A** the coupling shows up clearly in the spectrum which consists of two doublets. However, in **B** protons H_A and H_B are in the same environment and so have the same chemical shift (Larmor frequency). As we have just shown, in such a case the spectrum is a single line and there are no splittings due to the coupling, despite it being present.

Although we have illustrated this point for just two spins, the same idea applies to larger spin systems, and it is found that the couplings between spins which have the same Larmor frequency (same chemical shift) have no effect on the spectrum. Such spins are said to be *equivalent*, but as we shall see in the next section we need to distinguish between different kinds of equivalent spins.

Fig. 12.2 The two protons in **A** are in different chemical environments and are coupled to one another, so the spectrum consists of two doublets. In **B** the two protons are in identical environments and so have the same Larmor frequency. Thus, although they are still coupled, no splittings are seen in the spectrum, which consists of a single line.

12.2 Chemical and magnetic equivalence

In this section we will explore the difference between *magnetic* and *chemical* equivalence. Spins which are chemically equivalent have the same chemical shift, but for spins to be magnetically equivalent not only do the shifts need to be the same but also the couplings to all of the spins need to be the same. The reason why this distinction is important is that the couplings *between* magnetically equivalent spins have no effect on the spectrum, but the same is *not* true for spins which are simply chemically equivalent.

12.2.1 Chemical equivalence

The two protons in molecule **B** (shown in Fig. 12.2) are in the same chemical environment and hence have the same chemical shift (Larmor frequency): the two protons are said to be *chemically equivalent*. Likewise, the six protons in benzene are all in the same environment and so are also chemically equivalent.

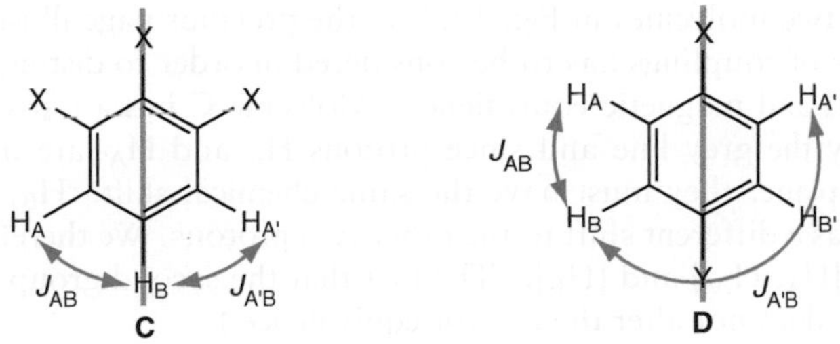

Fig. 12.3 Illustration of the difference between chemical and magnetic equivalence. In C the presence of the mirror plane (indicated by the grey line) makes H_A and $H_{A'}$ chemically equivalent (i.e. they have the same shift). The couplings J_{AB} and $J_{A'B}$ are clearly the same, so H_A and $H_{A'}$ are magnetically equivalent. In D, H_A and $H_{A'}$ are chemically equivalent, as are H_B and $H_{B'}$. However, $J_{AB} \neq J_{A'B}$ so H_A and $H_{A'}$ are *not* magnetically equivalent.

Chemical equivalence is often the result of symmetry. For example, molecule **B** has a two-fold rotation axis located in the middle of the double bond and coming out of the paper. Rotation by 180° about this axis interconverts H_A and H_B, and so the two protons are chemically equivalent. Similarly, the six protons in benzene are interconverted by rotation through 60° about the six-fold axis which lies perpendicular to the plane of the ring, and as a result all six protons are equivalent.

A two-fold axis is an axis about which rotation by 180° (i.e. half a full rotation) maps the molecule onto itself. Similarly, a rotation about a six-fold axis by 360° / 6 = 60° maps the molecule onto itself.

12.2.2 Magnetic equivalence

Groups of spins which have the same shift may also be *magnetically equivalent*, which is a more subtle form of equivalence involving the couplings as well as the shifts. Imagine that in a molecule we have a group of three spins A_1, A_2 and A_3 which all have the same shift, and a group of two spins B_1 and B_2 which have the same shift as one another, but different to that of the A spins.

If the coupling between each of the A spins and each of the B spins is identical, then the three A spins are magnetically equivalent, and the two B spins are magnetically equivalent. To be explicit, the couplings A_1–B_1, A_1–B_2, A_2–B_1, A_2–B_2, A_3–B_1 and A_3–B_2 must all be *identical* for the three A spins and the two B spins to form magnetically equivalent groups.

Expressed more formally, a group of spins is magnetically equivalent if: (1) they have the same chemical shift *and* (2) if each spin in the group has identical couplings to any other magnetically equivalent group of spins in the molecule. There is one exception to this rule which is that if there is only one group of spins in the molecule, then these are magnetically equivalent.

12.2.3 Examples

The test for magnetic equivalence is best illustrated by some examples. First consider the benzene molecule, in which all six protons have the same shift. Since there is only one group of spins in this molecule, and these have the same shift, they are magnetically equivalent. Other examples of spins which are magnetically equivalent for the same reason are the two protons in water and the four protons in methane.

The two molecules in Fig. 12.3 on the previous page illustrate how the network of couplings has to be considered in order to distinguish between chemical and magnetic equivalence. Molecule **C** has a mirror plane indicated by the grey line and since protons H_A and $H_{A'}$ are interconverted by this plane, they must have the same chemical shift. H_B, on the other hand, has a different shift to the other two protons. We therefore have two groups $[H_A, H_{A'}]$ and $[H_B]$. (The fact that the second group only has one member does not alter the test for equivalence.)

The size of the coupling between H_A and H_B is identical to that between $H_{A'}$ and H_B, since both pass through an identical set of bonds. The test for magnetic equivalence is therefore satisfied and thus H_A and $H_{A'}$ are magnetically equivalent.

Now turn to molecule **D**, which also has a mirror plane. Clearly H_A and $H_{A'}$ have the same shift since they are interconverted by the mirror plane. The same is true of H_B and $H_{B'}$, so we have two groups $[H_A, H_{A'}]$ and $[H_B, H_{B'}]$. The coupling of H_B to H_A is *not* the same as the coupling of H_B to $H_{A'}$, since the former is through three bonds and the latter through five bonds. Thus the condition for magnetic equivalence is *not* satisfied as the couplings of each member of the first group are not the same to each member of the second group. It follows that neither the group $[H_A, H_{A'}]$ nor the group $[H_B, H_{B'}]$ are magnetically equivalent.

The molecule PF_5, illustrated in Fig. 12.4, has a trigonal bipyramidal geometry. There are two environments for the fluorine atoms: the two axial positions (denoted a) and the three equatorial positions (denoted e). From the geometry, it is clear that each axial fluorine has an identical coupling to each of the equatorial fluorine atoms. As a result the two axial fluorines are magnetically equivalent, as are the three equatorial fluorines.

In molecules which have some degree of conformational flexibility, such as the rotation about single bonds, it is rather more difficult to decide whether or not a group of spins is magnetically equivalent. For example, consider the case of chloroethane, CH_3CH_2Cl. In this molecule the barrier to rotation about the C–C bond is not very high, but the staggered arrangements, illustrated as Newman projections in Fig. 12.5, are the lowest energy rotamers and therefore the ones which will be most populated.

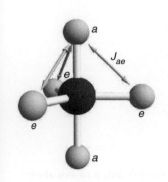

Fig. 12.4 PF_5 is trigonal pyramidal and so has two fluorine environments: axial (*a*) and equatorial (*e*). As indicated by the arrows, each equatorial fluorine has an identical coupling to each axial fluorine. The three equatorial fluorines are thus magnetically equivalent, and the two axial fluorines are also magnetically equivalent.

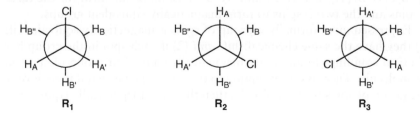

Fig. 12.5 The three low-energy rotamers of chloroethane depicted as Newman projections. The two methylene protons are labelled A and A', and the three methyl protons are labelled B, B' and B''. In any one rotamer the couplings between a particular A proton and the three B protons are not all the same. However, if there is rapid exchange between the three rotamers, and each is populated equally, the *averaged* values of these couplings are identical. As a result the CH_2 protons are magnetically equivalent, as are the CH_3 protons.

In rotamer R_1 the coupling between H_A and H_B is *not* the same as that between H_A and $H_{B'}$. Therefore, neither the three CH_3 protons nor the two CH_2 protons are magnetically equivalent. The same is true in the other two rotamers.

However, if the molecule is jumping between rotamers at a rate fast compared with the range of couplings involved, which is certainly the case for this molecule under normal conditions, we need to consider the *average* value of the couplings. If the three rotamers are equally populated, then the average value of the coupling from any one of the CH_2 protons to any one of the CH_3 protons is the same. Thus, the CH_2 protons form a magnetically equivalent group, as do the CH_3 protons. For chloroethane the interactions are the same in all three rotamers, so we expect them to be equally populated.

In more complex examples it may not be so immediately obvious whether or not a group of protons are magnetically equivalent, for example in situations in which the populations of the rotamers are not equal.

12.2.4 Consequences of magnetic equivalence

The most important consequence of magnetic equivalence is that the couplings between a group of magnetically equivalent spins have *no effect* on the spectrum i.e. these couplings do not give rise to any observable splittings in the spectrum. In section 12.1 on page 442 we saw an example of this for the case of two spins: when the shift of the two spins is the same then we see just one line in the spectrum, despite the fact that the two spins are coupled.

In fact, it can also be shown that the couplings between equivalent nuclei have no effect on the outcome of any NMR experiment, regardless of the pulse sequence. It therefore follows that in making product operator calculations on such spin systems we can simply ignore the couplings between equivalent spins. This greatly simplifies the calculations, and is a feature we will make use of in the remainder of this chapter.

The proof that the couplings between magnetically equivalent spins have no effect on the spectrum or the outcome of any experiment requires the use of quantum mechanical techniques which are well beyond the scope of this text. An elegant proof of this property is given in Appendix 9 of M. H. Levitt, *Spin Dynamics* (2nd edition, John Wiley & Sons, Ltd, 2008).

For the remainder of this chapter we will be focusing on heteronuclear NMR experiments involving the spins systems $^{13}C^1H$, $^{13}C(^1H)_2$ and $^{13}C(^1H)_3$. For the latter two we can reasonably assume that the protons have the same shift and, since they also have the same (one-bond) coupling to the ^{13}C, it follows that the protons are magnetically equivalent. We are therefore able to ignore the coupling between these spins thus greatly simplifying the calculations.

12.2.5 Notation for spin systems

There is a commonly used notation for spin systems which is especially useful when dealing with chemically or magnetically equivalent spins. In this notation a different capital letter is used for each spin with a distinct

The argument as to why all the averaged couplings are the same is as follows. The three-bond coupling between the A and B protons depends on the dihedral angle. Choose any one of the A protons, say H_A, and any one of the B protons, say H_B. In R_1 the dihedral angle between these protons is 180°, in R_2 the angle is 60°, and in R_3 it is 60°. We therefore have two rotamers in which the dihedral angle is 60° and one in which the angle is 180°. The same is true for any pair of protons we select from groups A and B. Thus, the average coupling, assuming that all the rotamers are equally populated, is the same for any A–B pair.

chemical shift. If the spins are strongly coupled (i.e. their shift separation is comparable with the coupling constant between them), then two letters close in the alphabet are used, whereas weakly coupled spins are denoted by letters far apart in the alphabet.

Therefore two weakly coupled spins are described as an 'AX spin system' whereas two strongly coupled spins are called an 'AB spin system'. Three weakly coupled spins would be denoted AMX and three strongly coupled spins would be denoted ABC. If two of the spins are strongly coupled and one is weakly coupled we have an ABX spin system.

In this notation groups of magnetically equivalent spins are denoted by adding a subscript, so AX_2 indicates a magnetically equivalent group of two spins (the X_2) weakly coupled to the first spin A. Likewise, AX_3 denotes a set of three magnetically equivalent nuclei coupled to the A spin.

Groups of spins which are chemically equivalent (as distinct from being magnetically equivalent) are denoted by adding primes, so AA'X indicates two chemically equivalent spins, A and A', weakly coupled to a third spin X. It is necessary to distinguish the chemically equivalent A spins using the prime as the coupling of X to A will not be the same as that to A'.

Using this notation the spin system of molecule **C** in Fig. 12.3 on page 447 would be denoted A_2X, whereas that of **D** would be denoted AA'XX', assuming in both cases that the spins are weakly coupled.

12.3   Product operators for AX_n (I_nS) spin systems

In this section we are going to explore how the product operator method can be extended to compute the evolution of AX_2 and AX_3 spin systems. The main reason for doing this is that we are interested in the behaviour of $^{13}CH_2$ and $^{13}CH_3$ fragments in certain heteronuclear experiments. Following on from the notation used in earlier chapters we will denote the operators of the heteronucleus using S, and those of the protons as I_1, I_2 and I_3. To be consistent with this notation we will therefore describe the spin systems as I_2S and I_3S.

Recall from the foregoing discussion that because the I spins are magnetically equivalent the coupling between them can be ignored. Thus, in the I_nS spin system we only have to consider the evolution of the I–S coupling, which is the same for each pair of spins. For brevity this coupling will be denoted J.

In section 10.1 on page 320 we described the product operators that arose for three coupled spins and how these evolved under free precession and pulses. As is explained in the following sections, the approach taken there is readily extended to cope with I_2S and I_3S spin systems.

12.3.1   Hamiltonians for free precession and pulses

Free precession

The free precession Hamiltonian has a term for the offset of each spin, along with terms describing the couplings. For the I_3S spin system it will be

$$\hat{H}_{\text{free}} = \underbrace{\Omega_I\hat{I}_{1z} + \Omega_I\hat{I}_{2z} + \Omega_I\hat{I}_{3z}}_{\text{offset of } I \text{ spins}} + \underbrace{\Omega_S\hat{S}_z}_{\text{offset of } S \text{ spin}}$$

$$\underbrace{+2\pi J\hat{I}_{1z}\hat{S}_z + 2\pi J\hat{I}_{2z}\hat{S}_z + 2\pi J\hat{I}_{3z}\hat{S}_z,}_{I\text{-}S \text{ coupling terms}}$$

where the offset of the I spins is Ω_I (all the same), and that of the S spin is Ω_S. For the I_2S spin system we simply drop the terms in $\hat{I}_{3z}$, and for the IS spin system we also drop the terms in $\hat{I}_{2z}$. Note that since the I spins are magnetically equivalent it is not necessary to include any terms for the coupling between them.

Each of the terms in the free precession Hamiltonians commutes with all the others, so the evolution can be computed by considering the effect of each in turn, in any order. Using the arrow notation this becomes, for the offset terms

$$\rho(0) \xrightarrow{\Omega_I t\,\hat{I}_{1z}} \xrightarrow{\Omega_I t\,\hat{I}_{2z}} \xrightarrow{\Omega_I t\,\hat{I}_{3z}} \xrightarrow{\Omega_S t\,\hat{S}_z},$$

and for the coupling terms

$$\rho(0) \xrightarrow{2\pi Jt\,\hat{I}_{1z}\hat{S}_z} \xrightarrow{2\pi Jt\,\hat{I}_{2z}\hat{S}_z} \xrightarrow{2\pi Jt\,\hat{I}_{3z}\hat{S}_z}.$$

Pulses

Since the I and S spins are different types of nucleus (e.g. ^{1}H and ^{13}C) pulses are applied to each separately. For a (strong) x-pulse to the S spins the Hamiltonian is

$$\hat{H}_{x,S} = \omega_{1,S}\hat{S}_x,$$

and for the I spins it is

$$\hat{H}_{x,I} = \omega_{1,I}\hat{I}_{1x} + \omega_{1,I}\hat{I}_{2x} + \omega_{1,I}\hat{I}_{3x}.$$

As with the free precession Hamiltonian, each of these terms commutes and so their effect can be considered separately and in any order. In the arrow notation we have, for a pulse to the I spins,

$$\rho(0) \xrightarrow{\omega_1 t\,\hat{I}_{1x}} \xrightarrow{\omega_1 t\,\hat{I}_{2x}} \xrightarrow{\omega_1 t\,\hat{I}_{3x}}.$$

As before, for the I_2S spin system we simply drop the terms in $\hat{I}_{3x}$, and for the IS spin system we also drop the terms in $\hat{I}_{2x}$.

12.3.2   Observable terms

As we have seen for two and three spins, only those operator products containing one $\hat{I}_x$ or $\hat{I}_y$ operator combined with any number of $\hat{I}_z$ operators are observable. For operator products which are observable on the S spin, the presence of an $\hat{I}_{i,z}$ operator indicates that the multiplet will be anti-phase with respect to the coupling to spin i. The appearance of the multiplet depends on the number of $\hat{I}_z$ operators present and on the number of coupled spins, as is described below.

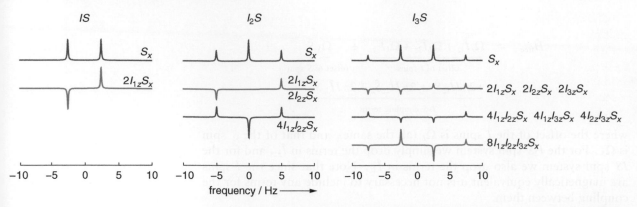

Fig. 12.6 Illustration of the multiplets arising from various operator products in IS, I_2S and I_3S spin systems. In each case $\hat{S}_x$ gives rise to the familiar multiplet with intensity patterns 1:1, 1:2:1 and 1:3:3:1. Introduction of $\hat{I}_{iz}$ operators into the product gives the more complex anti-phase multiplets shown. Note that in the I_2S spin system the operators $\hat{I}_{1z}$ and $\hat{I}_{2z}$ are interchangeable as they represent equivalent spins; similarly in the case of I_3S all three I spin operators are interchangeable. The offset of the S spin has been set to 0 Hz and the I–S coupling has been set to 5 Hz. The in-phase multiplets are drawn such that each has the same integral.

S spin observables

The operator $\hat{S}_x$ represents a fully in-phase multiplet on the S spin. For the IS spin system this means a 1:1 doublet, for the I_2S system it represents a 1:2:1 triplet and for the I_3S system it represents a 1:3:3:1 quartet. These are of course the multiplets expected for spin S in the conventional NMR spectrum of an I_nS spin system, with the intensity patterns that can be predicted using the usual tree diagram. These multiplets are illustrated along the top row of Fig. 12.6.

The term $2\hat{I}_{1z}\hat{S}_x$ indicates magnetization which is anti-phase with respect to the I_1–S coupling. For the IS spin system it gives rise to the familiar anti-phase multiplet with intensity -1:1. In the I_2S system the intensity pattern is -1:0:1 i.e. the middle line is absent, and in the I_3S system the pattern is -1:-1:1:1. These multiplets are illustrated in the second row of Fig. 12.6.

For the I_2S spin system the operator $2\hat{I}_{2z}\hat{S}_x$ gives the same pattern as $2\hat{I}_{1z}\hat{S}_x$. Similarly, for the I_3S system the operators $2\hat{I}_{1z}\hat{S}_x$, $2\hat{I}_{2z}\hat{S}_x$ and $2\hat{I}_{3z}\hat{S}_x$ all give the same multiplets. This property arises because the I spins are magnetically equivalent, and are thus interchangeable.

For the I_2S and I_3S spin systems we can have doubly anti-phase operators such as $4\hat{I}_{1z}\hat{I}_{2z}\hat{S}_x$; note that the normalizing factor is now 4 on account of there being three operators in the product. This operator gives rise to a multiplet with intensity pattern 1:-2:1 in the I_2S system, and a multiplet with intensity pattern 1:-1:-1:1 in the I_3S system. In the latter, the operators $4\hat{I}_{1z}\hat{I}_{3z}\hat{S}_x$ and $4\hat{I}_{2z}\hat{I}_{3z}\hat{S}_x$ give rise to identical multiplets. These doubly anti-phase multiplets are illustrated in the third row of the figure.

Finally, in the I_3S system we can have the triply anti-phase operator $8\hat{I}_{1z}\hat{I}_{2z}\hat{I}_{3z}\hat{S}_x$ which gives rise to a -1:3:-3:1 multiplet, also illustrated in the figure. As there are four operators in the product, the normalizing factor is now 8.

Changing the S spin operator from $\hat{S}_x$ to $\hat{S}_y$ gives an identical set of multiplets except that the phase is shifted by 90° i.e. the lines would all be in dispersion. As usual, a phase correction by 90° will return the lines to absorption. The form of the multiplets arising for these various operator products can be worked out by following an analogous process to that used in section 7.5 on page 152, a topic which is explored in more detail in one of the Exercises at the end of this chapter.

I spin observables

For all of these I_nS spin systems the I spin spectrum is simply a doublet due to the coupling to the S spin (recall that the couplings between the magnetically equivalent I spins have no effect on the spectrum). Thus the operators $\hat{I}_{i,x}$, where i is 1, 2 or 3, all give rise to an in-phase doublet on the I spin, and the operators $2\hat{I}_{i,x}\hat{S}_z$ all give rise to anti-phase doublets. Along with their counterparts $\hat{I}_{i,y}$ and $2\hat{I}_{i,y}\hat{S}_z$, these are the only observable operators on the I spins.

12.3.3 Evolution due to coupling

The anti-phase terms arise due to the evolution of coupling which follows the same rules as for a two-spin system, as illustrated in Fig. 7.6 on page 152. Starting with $\hat{S}_x$ the effect of the coupling to spin I_1 is

$$\hat{S}_x \xrightarrow{2\pi Jt\,\hat{I}_{1z}\hat{S}_z} \cos(\pi Jt)\,\hat{S}_x + \sin(\pi Jt)\,2\hat{I}_{1z}\hat{S}_y.$$

Note that the evolution of the coupling to I_1 gives the term $\hat{I}_{1z}$ in the operator product. By analogy, evolution of the coupling to spin I_2 gives the term $\hat{I}_{2z}$ in the product:

$$\hat{S}_x \xrightarrow{2\pi Jt\,\hat{I}_{2z}\hat{S}_z} \cos(\pi Jt)\,\hat{S}_x + \sin(\pi Jt)\,2\hat{I}_{2z}\hat{S}_y.$$

In the I_2S and I_3S spin systems the singly anti-phase term $2\hat{I}_{1z}\hat{S}_y$ can evolve further under the coupling to spin I_2. The relevant term in the Hamiltonian is $2\pi J\hat{I}_{2z}\hat{S}_z$, and since this does not include any operators for spin I_1 the operator $\hat{I}_{1z}$ in the product $2\hat{I}_{1z}\hat{S}_y$ will be unaffected. We can therefore write $2\hat{I}_{1z}\hat{S}_y$ as $A\hat{S}_y$, where A is a constant. The evolution of this term under the coupling to spin I_2 follows the usual rule

$$A\hat{S}_y \xrightarrow{2\pi Jt\,\hat{I}_{2z}\hat{S}_z} \cos(\pi Jt)\,A\hat{S}_y - \sin(\pi Jt)\,2A\hat{I}_{2z}\hat{S}_x.$$

Replacing A with $2\hat{I}_{1z}$ gives

$$2\hat{I}_{1z}\hat{S}_y \xrightarrow{2\pi Jt\,\hat{I}_{2z}\hat{S}_z} \cos(\pi Jt)\,2\hat{I}_{1z}\hat{S}_y - \sin(\pi Jt)\,4\hat{I}_{1z}\hat{I}_{2z}\hat{S}_x.$$

We see that the singly anti-phase term $2\hat{I}_{1z}\hat{S}_y$ evolves into the doubly anti-phase term $4\hat{I}_{1z}\hat{I}_{2z}\hat{S}_x$. Figure 12.7 illustrates the operators which arise from $\hat{S}_x$ due to the effect of coupling to spin I_1 and then to spin I_2.

In the I_3S spin system an analogous term arises due to the evolution under the coupling to spin I_3

$$2\hat{I}_{1z}\hat{S}_y \xrightarrow{2\pi Jt\,\hat{I}_{3z}\hat{S}_z} \cos(\pi Jt)\,2\hat{I}_{1z}\hat{S}_y - \sin(\pi Jt)\,4\hat{I}_{1z}\hat{I}_{3z}\hat{S}_x.$$

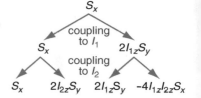

Fig. 12.7 Illustration of the operators arising from $\hat{S}_x$ under the evolution of first the coupling to spin I_1 and then the coupling to spin I_2. An arrow to the left is associated with a factor $\cos(\pi Jt)$, and an arrow to the right is associated with a factor $\sin(\pi Jt)$.

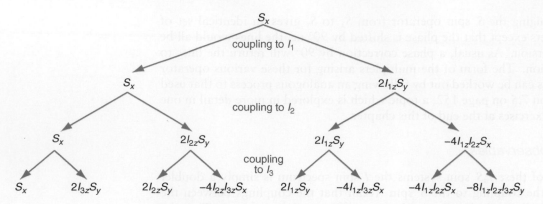

Fig. 12.8 Illustration of the operators arising from $\hat{S}_x$ under the evolution of the couplings to spins I_1, I_2 and I_3. An arrow to the left is associated with a factor $\cos(\pi Jt)$, and an arrow to the right is associated with a factor $\sin(\pi Jt)$.

In this spin system the doubly anti-phase term $4\hat{I}_{1z}\hat{I}_{2z}\hat{S}_x$ can evolve further under the influence of the coupling to spin I_3. As before, the terms $\hat{I}_{1z}$ and $\hat{I}_{2z}$ are unaffected by this coupling so the doubly anti-phase term can be written $B\hat{S}_x$, where B is a constant. The evolution under the coupling to spin I_3 follows the usual rule:

$$B\hat{S}_x \xrightarrow{\;2\pi Jt\,\hat{I}_{3z}\hat{S}_z\;} \cos(\pi Jt)\,B\hat{S}_x + \sin(\pi Jt)\,2B\hat{I}_{3z}\hat{S}_y.$$

Replacing B with $4\hat{I}_{1z}\hat{I}_{2z}$ gives

$$4\hat{I}_{1z}\hat{I}_{2z}\hat{S}_x \xrightarrow{\;2\pi Jt\,\hat{I}_{3z}\hat{S}_z\;} \cos(\pi Jt)\,4\hat{I}_{1z}\hat{I}_{2z}\hat{S}_x + \sin(\pi Jt)\,8\hat{I}_{1z}\hat{I}_{2z}\hat{I}_{3z}\hat{S}_y.$$

We see that evolution of the doubly anti-phase term $4\hat{I}_{1z}\hat{I}_{2z}\hat{S}_x$ gives the triply anti-phase term $8\hat{I}_{1z}\hat{I}_{2z}\hat{I}_{3z}\hat{S}_y$ (note that this is only possible in the I_3S spin system). Figure 12.8 illustrates the evolution of $\hat{S}_x$ under the coupling to three I spins.

These anti-phase terms evolve back into in-phase terms in an analogous way to that we have seen before. For example, when the term $4\hat{I}_{1z}\hat{I}_{2z}\hat{S}_x$ evolves under the coupling to spin I_2 we recognize that the term $\hat{I}_{1z}$ will be unaffected, so the problem boils down to the evolution of $2\hat{I}_{2z}\hat{S}_x$ under the coupling to I_2, which simply gives rise to $\hat{S}_y$. The full result is

$$4\{\hat{I}_{1z}\}\hat{I}_{2z}\hat{S}_x \xrightarrow{\;2\pi Jt\,\hat{I}_{2z}\hat{S}_z\;} \cos(\pi Jt)\,4\{\hat{I}_{1z}\}\hat{I}_{2z}\hat{S}_x + \sin(\pi Jt)\,2\{\hat{I}_{1z}\}\hat{S}_y,$$

where the unaffected operator has been placed in curly braces. Similarly, in the following transformation both $\hat{I}_{1z}$ and $\hat{I}_{2z}$ are unaffected

$$8\{\hat{I}_{1z}\hat{I}_{2z}\}\hat{I}_{3z}\hat{S}_x \xrightarrow{\;2\pi Jt\,\hat{I}_{3z}\hat{S}_z\;} \cos(\pi Jt)\,8\{\hat{I}_{1z}\hat{I}_{2z}\}\hat{I}_{3z}\hat{S}_x + \sin(\pi Jt)\,4\{\hat{I}_{1z}\hat{I}_{2z}\}\hat{S}_y.$$

12.4 Spin echoes in I_nS spin systems

The first experiment we will look at is the straightforward spin echo on the S spin, the pulse sequence for which is shown in Fig. 12.9. Following the discussion in section 7.8.4 on page 164 we note that since 180° pulses are applied to both the I and S spins we expect that the offset of both will be refocused, but that the coupling between them will continue to evolve for the entire time 2τ. Furthermore, the outcome of the $-\tau$–180°–τ– segment can be computed by allowing the coupling to evolve for time 2τ and then applying the two 180° pulses.

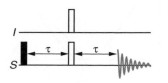

Fig. 12.9 A simple spin echo sequence applied to the S spin. The offsets of the I and S spins will be refocused, but the I–S coupling will continue to evolve for the whole time 2τ.

12.4.1 *IS* spin system

The initial 90° pulse generates the operator $-\hat{S}_y$. Using the short cut just discussed, the result of the spin echo for an IS spin system is

$$-\hat{S}_y \xrightarrow{\ 2\pi J(2\tau)\hat{I}_{1z}\hat{S}_z\ } -\cos{(2\pi J\tau)}\,\hat{S}_y + \sin{(2\pi J\tau)}\,2\hat{I}_{1z}\hat{S}_x$$

$$\xrightarrow{\ \pi(\hat{I}_{1x}+\hat{S}_x)\ } \cos{(2\pi J\tau)}\,\hat{S}_y - \sin{(2\pi J\tau)}\,2\hat{I}_{1z}\hat{S}_x.$$

If the delay τ is zero we have the in-phase term $\hat{S}_y$. Setting the delay to $1/(4J)$ gives complete conversion to the anti-phase term $-2\hat{I}_{1z}\hat{S}_x$, and a delay of $1/(2J)$ returns the in-phase term $-\hat{S}_y$, but with a sign change. A delay of $1/J$ gives the same result as a delay of zero.

12.4.2 *I_2S* spin system

For the I_2S spin system we need to consider the effect of the coupling first to spin I_1 and then to spin I_2. The first of these couplings gives the same result as above, and the second gives rise to additional terms

$$-\hat{S}_y \xrightarrow{\ 2\pi J(2\tau)\hat{I}_{1z}\hat{S}_z\ } -\cos{(2\pi J\tau)}\,\hat{S}_y + \sin{(2\pi J\tau)}\,2\hat{I}_{1z}\hat{S}_x$$

$$\xrightarrow{\ 2\pi J(2\tau)\hat{I}_{2z}\hat{S}_z\ } -\cos{(2\pi J\tau)}\cos{(2\pi J\tau)}\,\hat{S}_y + \sin{(2\pi J\tau)}\cos{(2\pi J\tau)}\,2\hat{I}_{2z}\hat{S}_x$$
$$+ \cos{(2\pi J\tau)}\sin{(2\pi J\tau)}\,2\hat{I}_{1z}\hat{S}_x + \sin{(2\pi J\tau)}\sin{(2\pi J\tau)}\,4\hat{I}_{1z}\hat{I}_{2z}\hat{S}_y$$

$$\xrightarrow{\ \pi(\hat{I}_{1x}+\hat{I}_{2x}+\hat{S}_x)\ } + \cos{(2\pi J\tau)}\cos{(2\pi J\tau)}\,\hat{S}_y - \sin{(2\pi J\tau)}\cos{(2\pi J\tau)}\,2\hat{I}_{2z}\hat{S}_x$$
$$- \cos{(2\pi J\tau)}\sin{(2\pi J\tau)}\,2\hat{I}_{1z}\hat{S}_x - \sin{(2\pi J\tau)}\sin{(2\pi J\tau)}\,4\hat{I}_{1z}\hat{I}_{2z}\hat{S}_y$$

Using the short-hand notation $c = \cos{(2\pi J\tau)}$ and $s = \sin{(2\pi J\tau)}$ the final result is written more compactly as

$$c^2\,\hat{S}_y - cs\left(2\hat{I}_{2z}\hat{S}_x + 2\hat{I}_{1z}\hat{S}_x\right) - s^2\,4\hat{I}_{1z}\hat{I}_{2z}\hat{S}_y. \tag{12.2}$$

There is a clear pattern to the trigonometric coefficients in front of each operator: each $\hat{I}_{iz}$ operator in the product leads to a cosine term being replaced by a sine term. Furthermore, the fact that the coefficients for $2\hat{I}_{2z}\hat{S}_x$ and $2\hat{I}_{1z}\hat{S}_x$ are the same is simply a result of the I–S couplings all being the same.

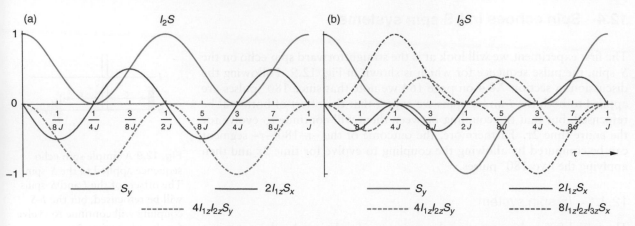

Fig. 12.10 Plots showing how the amounts of the operator products generated in a spin echo sequence vary with the delay τ in the echo for: (a) the I_2S spin system and (b) the I_3S spin system. The value of the delay τ is expressed in terms of the I–S coupling J.

The result of the calculation, Eq. 12.2 on the preceding page, is plotted in Fig. 12.10 (a). The graph shows how the amounts of $\hat{S}_y$, $2\hat{I}_{1z}\hat{S}_x$ and $4\hat{I}_{1z}\hat{I}_{2z}\hat{S}_y$ vary as a function of the delay τ, expressed in terms of the coupling constant J ($2\hat{I}_{2z}\hat{S}_x$ behaves in the same way as $2\hat{I}_{1z}\hat{S}_x$). The behaviour is rather different to that of the IS spin system.

For example, there is no value of the delay which results in the generation only the singly anti-phase terms $2\hat{I}_{1z}\hat{S}_x$ and $2\hat{I}_{2z}\hat{S}_x$, whereas there is complete conversion to the doubly anti-phase term $4\hat{I}_{1z}\hat{I}_{2z}\hat{S}_y$ when $\tau = 1/(4J)$. In contrast to the IS spin system, a delay of $1/(2J)$ gives the same result as a delay of zero. The maximum amount of the singly anti-phase terms $2\hat{I}_{1z}\hat{S}_x$ and $2\hat{I}_{2z}\hat{S}_x$ are generated when $\tau = 1/(8J)$, whereas for the IS spin system the maximum is at a delay of $1/(4J)$. This result may be interpreted by noting that the outer lines of the triplet are in some ways analogous to a doublet with *twice* the coupling constant (the lines are separated by $2J$), and so the anti-phase terms develops twice as quickly.

12.4.3 I_3S spin system

For the I_3S spin system we have one more coupling to evolve which gives rise to a total of eight terms. Starting from Eq. 12.2 on the preceding page and using the compact notation the evolution is

$$-c^2\,\hat{S}_y + cs\,2\hat{I}_{2z}\hat{S}_x + cs\,2\hat{I}_{1z}\hat{S}_x + s^2\,4\hat{I}_{1z}\hat{I}_{2z}\hat{S}_y$$

$$\xrightarrow{\;2\pi J(2\tau)\,\hat{I}_{3z}\hat{S}_z\;}$$

$$-c^3\,\hat{S}_y + c^2s\,2\hat{I}_{3z}\hat{S}_x + c^2s\,2\hat{I}_{2z}\hat{S}_x + cs^2\,4\hat{I}_{2z}\hat{I}_{3z}\hat{S}_y$$
$$+ c^2s\,2\hat{I}_{1z}\hat{S}_x + cs^2\,4\hat{I}_{1z}\hat{I}_{3z}\hat{S}_y + cs^2\,4\hat{I}_{1z}\hat{I}_{2z}\hat{S}_y - s^3\,8\hat{I}_{1z}\hat{I}_{2z}\hat{I}_{3z}\hat{S}_x$$

$$\xrightarrow{\;\pi(\hat{I}_{1x}+\hat{I}_{2x}+\hat{I}_{3x}+\hat{S}_x)\;}$$

$$c^3\,\hat{S}_y - c^2s\,2\hat{I}_{3z}\hat{S}_x - c^2s\,2\hat{I}_{2z}\hat{S}_x - cs^2\,4\hat{I}_{2z}\hat{I}_{3z}\hat{S}_y$$
$$- c^2s\,2\hat{I}_{1z}\hat{S}_x - cs^2\,4\hat{I}_{1z}\hat{I}_{3z}\hat{S}_y - cs^2\,4\hat{I}_{1z}\hat{I}_{2z}\hat{S}_y + s^3\,8\hat{I}_{1z}\hat{I}_{2z}\hat{I}_{3z}\hat{S}_x.$$

Collecting terms together gives the final result

$$c^3\,\hat{S}_y - c^2 s\left(2\hat{I}_{1z}\hat{S}_x + 2\hat{I}_{2z}\hat{S}_x + 2\hat{I}_{3z}\hat{S}_x\right) \tag{12.3}$$
$$-cs^2\left(4\hat{I}_{1z}\hat{I}_{2z}\hat{S}_y + 4\hat{I}_{2z}\hat{I}_{3z}\hat{S}_y + 4\hat{I}_{1z}\hat{I}_{3z}\hat{S}_y\right) + s^3\,8\hat{I}_{1z}\hat{I}_{2z}\hat{I}_{3z}\hat{S}_x.$$

This result mirrors the pattern already seen for the I_2S spin system: a cosine term is replaced by a sine for each $\hat{I}_{iz}$ operator present in the product. Figure 12.10 (b) on the preceding page shows plots of the amount of each of the different types of operator as a function of the delay τ.

A delay of $1/(4J)$ generates just the triply anti-phase term $8\hat{I}_{1z}\hat{I}_{2z}\hat{I}_{3z}\hat{S}_x$. Setting the delay to $1/J$ gives the same result as for a delay of zero, and $\tau = 1/(2J)$ gives just the in-phase term, but with opposite sign to that for $\tau = 0$. All other values of τ give contributions from each of the different operators.

The delay which gives the greatest amount of the singly anti-phase term can be found by locating the turning points in the function $\cos^2(2\pi J\tau)\sin(2\pi J\tau)$, and for the doubly anti-phase term we have to consider the function $\cos(2\pi J\tau)\sin^2(2\pi J\tau)$. Finding these turning points is one of the Exercises at the end of the chapter; the lowest values of τ which give the maximum amounts of these two terms turn out to be at $\tau = 0.0980/J$ and $\tau = 0.152/J$. In each case the term is at its most negative, as can be seen from Fig. 12.10 (b).

12.4.4 Attached proton test (APT)

The APT experiment gives a very simple way of determining whether a line in a proton-decoupled ^{13}C spectrum is due to a C, CH, CH_2 or CH_3 group. The experiment itself is just a simple spin echo, like the one we have been discussing, but with the addition of broadband I spin (proton) decoupling during acquisition. The pulse sequence is shown in Fig. 12.11 (a) on the following page.

APT: Attached Proton Test

As was discussed in section 7.10.3 on page 169, broadband I spin decoupling effectively sets the I–S coupling to zero and as a result the lines in any anti-phase multiplets simply cancel one another out making such terms unobservable. The only observable operators are therefore $\hat{S}_x$ and $\hat{S}_y$, each of which gives, under decoupled conditions, a single line at the offset of the S spin. Referring back to our calculations in the previous section, we can therefore see that the intensity of the observed signal is simply proportional to the trigonometric term multiplying the operator $\hat{S}_y$. This term is different for each spin system:

spin system	intensity
S	$+1$
IS	$\cos(2\pi J\tau)$
I_2S	$\cos^2(2\pi J\tau)$
I_3S	$\cos^3(2\pi J\tau)$

In the table we have added an entry for just a single S spin which is not coupled to any I spins. It is trivial to see that in such a case the spin echo

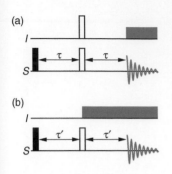

Fig. 12.11 Two alternative APT pulse sequences. Sequence (a) is a *J*-modulated spin echo with broadband *I* spin decoupling (indicated by the blue rectangle) during acquisition. As a result only in-phase terms lead to observable signals. Sequence (b) achieves the same result, but rather than a 180° pulse being applied to the *I* spins, broadband decoupling is switched on half-way through the echo, effectively setting the coupling to zero. The coupling therefore only evolves for the first period τ' and so is not refocused. To have the same overall modulation it is necessary to set $\tau' = 2 \times \tau$.

just gives the term $\hat{S}_y$. Inspection of the table on the previous page gives us a very nice result: in this experiment the intensity of the signal from an I_nS spin system goes as $\cos^n(2\pi J\tau)$.

In the APT experiment we choose $\tau = 1/(2J)$ which means that $\cos(2\pi J\tau) = -1$. The peaks from S and I_2S spin systems will therefore be positive, while those from IS and I_3S spin systems will be negative. In other words, peaks from I_nS spin systems with n *even* will be *positive*, and those with n odd will be *negative*.

Thus, simply by inspecting the sign of the peaks in the APT experiment we can obtain an indication as to whether a peak in a ^{13}C spectrum is from a C or CH_2 group on the one hand, or from a CH or CH_3 group on the other. Although the experiment does not distinguish between C and CH_2, nor between CH or CH_3, the information that is provided is often sufficient to be a considerable help in assigning the spectrum. So simple and reliable is the APT experiment that it is often recorded as a matter of routine alongside a conventional ^{13}C spectrum.

In practice the pulse sequence of Fig. 12.11 (b) is generally used, rather than the simple echo sequence (a). Sequence (b) is a spin echo on the S spin, but there is no 180° pulse on the I spin. Rather, broadband I-spin decoupling starts half way through the echo and then carries on into the acquisition period. The I–S coupling evolves during the first period τ', but during the second period the coupling is effectively set to zero and so does not evolve. A spin echo can only refocus a coupling if it evolves equally in the two delays, which is certainly not the case here, and as a result the coupling is *not* refocused. Effectively, sequence (b) is *J*-modulated spin echo.

In sequence (a) the coupling evolves for both the periods τ, giving a total evolution time of 2τ. However, in (b) the coupling only evolves for τ', so we need to set $\tau' = 2 \times \tau$ to obtain a comparable result i.e. τ' has to be set to $1/J$ in order to achieve the result described above. Sequence (b) is generally preferred as it is somewhat simpler, having only one 180° pulse which needs to be calibrated carefully.

Of course, the time τ (or τ') has to be set according to the value of the I–S coupling. In the case of CH_n groups this is the one-bond C–H coupling which does not vary greatly and so it is relatively easy to select a single value for the delay which will give acceptable intensity for all the different carbons in the molecule. Figure 12.12 on the facing page shows the APT spectrum of quinine.

12.5   INEPT in I_nS spin systems

In section 7.10 on page 167 we discussed the INEPT experiment which is used to transfer magnetization from the I spins to the S spins by first generating an anti-phase state on the I spin and then applying two 90° pulses to transfer this to the S spin. The pulse sequence, in the form which utilizes I-spin decoupling during acquisition, is shown in Fig. 12.13 on the facing page.

If we start out with equilibrium magnetization on the first I spin, $\hat{I}_{1z}$, the 90° pulse generates $-\hat{I}_{1y}$. Period **A** is a spin echo during which the

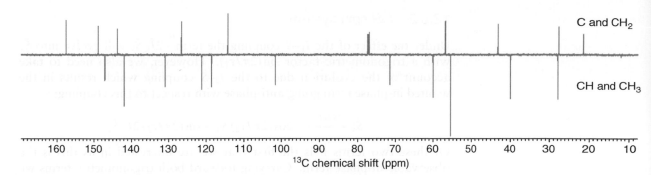

Fig. 12.12 APT spectrum of quinine recorded at 500 MHz for protons. The spectrum has been phased such that signals from C and CH_2 groups are positive whereas those from CH and CH_3 groups are negative. The multiplet from $CDCl_3$ appears as a positive peak as the carbon is not attached to any protons: it therefore behaves like a quaternary carbon.

I–S coupling evolves for a total time $2\tau_1$ and the I-spin offset is refocused. We do not need to worry about any couplings amongst the magnetically equivalent I spins, so our calculation is valid for any I_nS spin system.

We have worked out the operators present at the end of period **A** before, and so can simply quote these as

$$\cos(2\pi J\tau_1)\,\hat{I}_{1y} - \sin(2\pi J\tau_1)\,2\hat{I}_{1x}\hat{S}_z.$$

Equilibrium magnetization on the second or third I spin (if present) gives exactly analogous terms – all we need to do is change the index of the I-spin operator from 1 to 2, or to 3.

Since we know that only the anti-phase term will be transferred to the S spin, we can right away see that the optimum value for τ_1 is $1/(4J)$ and from now on we will assume that this is the case. After the two transfer pulses, period **B**, we therefore have

$$-2\hat{I}_{1z}\hat{S}_y.$$

Period **C** is a spin echo, the purpose of which is to allow this anti-phase term to evolve into an in-phase term which will be observable under conditions of broadband I-spin decoupling. What we are about to discover is that the details of this process depend on the number of equivalent I spins in the I_nS spin system.

12.5.1   *IS* spin system

It is not really necessary to do a full calculation of the effect of the final spin echo since it is only the in-phase term which is observable. All we need to do is concentrate on how this term arises. During the echo the term $-2\hat{I}_{1z}\hat{S}_y$ will evolve into $\hat{S}_x$, multiplied by the trigonometric factor $\sin(2\pi J\tau_2)$. We must not forget to take account of the two 180° pulses, but in this case they have no effect on this term since a pulse to the I spin has no effect on $\hat{S}_x$, and an x-pulse to the S spin also has no effect on $\hat{S}_x$. The final result is that the observable signal goes as $\sin(2\pi J\tau_2)$.

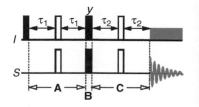

Fig. 12.13 Pulse sequence for the INEPT experiment with decoupled acquisition.

12.5.2 I_2S spin system

Under the effect of the I_1–S coupling the term $-2\hat{I}_{1z}\hat{S}_y$ will evolve into $\hat{S}_x$ with a trigonometric factor $\sin(2\pi J\tau_2)$. However, we also need to take account of the evolution due to the I_2–S coupling which results in the wanted in-phase term going anti-phase with respect to this coupling:

$$\hat{S}_x \xrightarrow{\;2\pi Jt\,\hat{I}_{2z}\hat{S}_z\;} \cos(2\pi J\tau_2)\,\hat{S}_x + \sin(2\pi J\tau_2)\,2\hat{I}_{2z}\hat{S}_y.$$

Of these two terms it is the first that we are interested in as this is the observable in-phase term. Carrying forward both trigonometric terms we see that the observable term is

$$\sin(2\pi J\tau_2)\cos(2\pi J\tau_2)\,\hat{S}_x.$$

Again, this is unaffected by the 180° pulses.

Finally, we need to recall that there are two I spins, the equilibrium magnetization from each of which will contribute a term identical to that just given. As a result, the final intensity goes as $2\sin(2\pi J\tau_2)\cos(2\pi J\tau_2)$ which, using the usual trigonometric identities, can be rewritten $\sin(4\pi J\tau_2)$. The immediate conclusion from this calculation is that whereas the optimum value for τ_2 for the IS spin system is $1/(4J)$, for the I_2S spin system the optimum value is $1/(8J)$.

$\sin(2\theta) \equiv 2\sin(\theta)\cos(\theta)$.

12.5.3 I_3S spin system

If we add a further I spin then the evolution of the I_3–S coupling has to be considered. As before, it is the in-phase part that we are interested in, so all that happens is that we acquire an additional cosine factor in complete analogy to that for the second I spin. The intensity of the in-phase term therefore goes as $3\sin(2\pi J\tau_2)\cos^2(2\pi J\tau_2)$, where the three arises from the fact that there are three spins.

It is necessary to use calculus to find the maximum in this function, which turns out to be at $\tau_2 = 0.0980/J$ (with a value of 1.15). Written in decimals for ease of comparison, the optimum value of τ_2 for an IS spin system is $0.25/J$, and for an I_2S spin system the value is $0.125/J$. All three values are significantly different.

12.5.4 Comparison

Figure 12.14 on the next page shows in graphical form the results of our calculation as to how the intensity of the in-phase signal varies with the delay τ_2. It is only in the IS spin system that all of the magnetization from the I spin can be transferred to the S spin. In the I_2S system if all of the magnetization were to be transferred we would expect an intensity of 2 since there are two I spins coupled to the S spin. In fact, the maximum transfer is just 1. The reason for this reduced intensity is that although the transfer into the anti-phase term $2\hat{I}_{1z}\hat{S}_y$ is complete, under the influence of the I_1–S and I_2–S couplings this term cannot evolve completely into the observable in-phase operator.

The story is the same for the I_3S spin system. Complete transfer would result in an intensity of 3 due to the presence of the three I spins. However,

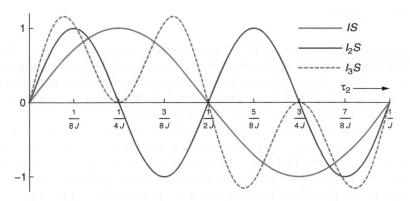

once again the presence of three I–S couplings prevents the complete conversion of the anti-phase term into the in-phase term.

In summary, for the I_nS spin system we expect the intensity of the in-phase signal to have a factor of $\sin(2\pi J\tau_2)$ representing the interconversion of the anti-phase to in-phase terms, and then $(n-1)$ factors of $\cos(2\pi J\tau_2)$ arising from the evolution of the other couplings. Finally, there is a factor of n representing the transfer from the n I spins. Overall, the intensity goes as

$$ n \times \sin(2\pi J\tau_2) \times [\cos(2\pi J\tau_2)]^{n-1} . $$

12.5.5 Consequences

One immediate consequence of this analysis is that if we wish to use the INEPT sequence to enhance the intensity of the S-spin signal then the choice of τ_2 will depend on the spin system present. If, as would be the case for ^{13}C NMR, there are IS, I_2S and I_3S spin systems present, then no single value of τ_2 is ideal for all three spin systems, and so a compromise value has to be sought.

Another consequence of the different behaviour of I_nS spin systems is the possibility of differentiating between them, much in the same way as the APT experiment. For example, if we ran an INEPT experiment with $\tau_2 = 1/(4J)$ then, as is obvious from Fig. 12.14, there will be no signal at all from I_2S and I_3S spin systems. In the case of ^{13}C, the resulting spectrum would therefore show *only* those peaks from CH groups, which might well be useful information to have when working on an assignment.

From the graph we can see that setting the delay to some value around $3/(8J)$ would result in CH and CH$_3$ groups giving positive signals, and CH$_2$ groups giving negative signals. Thus, by recording two separate INEPT experiments with these two different values for τ_2 it ought to be possible to identify the lines from CH, CH$_2$ and CH$_3$ groups. The main difficulty with this approach is that the value of τ_2 depends on the value of the coupling. As has already been commented on, in the case of C–H one-bond couplings there is not that much variation in these values. However, any variation that there is will mean that setting $\tau_2 = 1/(4J)$ for an average value of J will result in some signals from CH$_2$ and CH$_3$ groups appearing for those carbons in which the coupling is significantly different from the mean.

This idea of separating out the lines from different CH$_n$ groups in a ^{13}C spectrum is developed further in the DEPT experiment.

12.6   DEPT

DEPT: Distortionless Enhancement
by Polarization Transfer

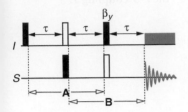

Fig. 12.15 The DEPT pulse
sequence. Note that the third
I-spin pulse has flip angle β and
is applied about the y-axis. The
optimum value of τ is $1/(2J)$.

The pulse sequence for the DEPT experiment is shown in Fig. 12.15. Like
INEPT, this sequence results in an overall transfer of I-spin magnetization
to the S spins. However, quite how it achieves this is, at first sight, not at
all obvious. The best thing to do is to analyse the sequence using product
operators: then it will become clear how it works.

Before starting on our analysis it is helpful to spot that over period **A** the
I-spin offset is refocused by the centrally placed 180° pulse to those spins.
Likewise, the other 180° pulse refocuses the S-spin offset over period **B**.
Therefore, during **A** we can ignore the I-spin offset, and during **B** we can
ignore the S-spin offset. It is true that a 90° pulse to the S spins intervenes
during period **A**, but as was the case in the HMQC experiment (section 8.8
on page 212) this does not prevent the I-spin offset being refocused.

12.6.1   *IS* spin system

Starting with equilibrium magnetization on the I spin, the first pulse
generates $-\hat{I}_{1y}$. This evolves under the coupling during the first delay τ,
and then the 180° pulse to I, in the following way

$$-\hat{I}_{1y} \xrightarrow{2\pi J\tau \hat{I}_{1z}\hat{S}_z} -\cos{(\pi J\tau)}\,\hat{I}_{1y} + \sin{(\pi J\tau)}\,2\hat{I}_{1x}\hat{S}_z$$

$$\xrightarrow{\pi\hat{I}_{1x}} \cos{(\pi J\tau)}\,\hat{I}_{1y} + \sin{(\pi J\tau)}\,2\hat{I}_{1x}\hat{S}_z.$$

Next comes the 90° pulse to the S spin:

$$\cos{(\pi J\tau)}\,\hat{I}_{1y} + \sin{(\pi J\tau)}\,2\hat{I}_{1x}\hat{S}_z \xrightarrow{(\pi/2)\hat{S}_x} \cos{(\pi J\tau)}\,\hat{I}_{1y} - \sin{(\pi J\tau)}\,2\hat{I}_{1x}\hat{S}_y.$$

We see that the anti-phase term has been transferred into (heteronuclear)
multiple-quantum coherence, $2\hat{I}_{1x}\hat{S}_y$. Since there are no further S-spin
pulses, other than a 180° pulse which cannot cause transfer, it is not
possible for the term $\hat{I}_{1y}$ to become observable on the S spin, so we will
drop it from the calculation.

During the second period τ nothing happens to the multiple-quantum
term since we have already decided that the offsets can be ignored and,
in addition, it is a property of I–S multiple-quantum coherence that it
does not evolve due to the coupling between the two nuclei involved (see
section 7.12.3 on page 176). We therefore can move on to consider the
effect of the 180° pulse to S and the β pulse (applied about the y-axis) to I

$$-\sin{(\pi J\tau)}\,2\hat{I}_{1x}\hat{S}_y \xrightarrow{\pi\hat{S}_x} +\sin{(\pi J\tau)}\,2\hat{I}_{1x}\hat{S}_y$$

$$\xrightarrow{\beta\hat{I}_{1y}} +\cos{(\beta)}\sin{(\pi J\tau)}\,2\hat{I}_{1x}\hat{S}_y - \sin{(\beta)}\sin{(\pi J\tau)}\,2\hat{I}_{1z}\hat{S}_y.$$

The effect of the β pulse is to transfer part of the multiple-quantum
coherence into an anti-phase term on the I spin. Since there are no further
pulses in the sequence, only this latter term can go on to give an observable
signal; we will therefore carry on the calculation with just this term.

In the final delay τ the anti-phase term evolves under the I–S coupling

$$- \sin(\beta) \sin(\pi J \tau) \, 2\hat{I}_{1z}\hat{S}_y \xrightarrow{\ 2\pi J \tau \, \hat{I}_{1z}\hat{S}_z\ }$$

$$- \cos(\pi J \tau) \sin(\beta) \sin(\pi J \tau) \, 2\hat{I}_{1z}\hat{S}_y + \sin(\pi J \tau) \sin(\beta) \sin(\pi J \tau) \, \hat{S}_x.$$

Since we are using I-spin decoupling during acquisition, only the in-phase term $\hat{S}_x$ is observable, so the final signal intensity goes as

$$\sin^2(\pi J \tau) \sin(\beta).$$

It immediately follows that the optimum value for τ is $1/(2J)$ and that the optimum value for β is $\pi/2$ (90°). With these values there is complete transfer from I to S.

Having completed the analysis it is now clearer how the sequence 'works'.

(a) During the first delay τ anti-phase magnetization is generated on the I spin.

(b) The 90° pulse to S transfers this to heteronuclear multiple-quantum coherence.

(c) In the case of the IS spin system, this multiple-quantum coherence does not evolve during the second delay τ (we shall see shortly that this is not the case for I_2S and I_3S spin systems).

(d) The β pulse to the I spins transfers some of the multiple-quantum coherence to anti-phase magnetization on the S spin; for this to happen, the pulse must be about the y-axis.

(e) During the third delay τ the anti-phase magnetization evolves into in-phase magnetization which is the only term observable under conditions of broadband I-spin decoupling.

12.6.2 I_2S spin system

For the I_2S spin system the result of the first delay τ, the 180° pulse and the 90° pulse to the S spin is the same as for the IS spin system. Therefore, at the start of the second delay τ we have

$$\cos(\pi J \tau) \, \hat{I}_{1y} - \sin(\pi J \tau) \, 2\hat{I}_{1x}\hat{S}_y.$$

Since there are no further pulses to the S spin (other than a 180° pulse), the $\hat{I}_{1y}$ term will not be transferred to the S spin and so we can ignore it.

The multiple-quantum term $2\hat{I}_{1x}\hat{S}_y$ does not evolve under the influence of the I_1–S coupling, but it does evolve under the I_2–S coupling. We can work out the result by noting that $\hat{I}_{1x}$ will be unaffected by this coupling, so we can write $2\hat{I}_{1x}$ as a constant A. The evolution then follows the usual rule:

$$- \sin(\pi J \tau) A\hat{S}_y \xrightarrow{\ 2\pi J \tau \, \hat{I}_{2z}\hat{S}_z\ }$$

$$- \cos(\pi J \tau) \sin(\pi J \tau) A\hat{S}_y + \sin(\pi J \tau) \sin(\pi J \tau) A2\hat{I}_{2z}\hat{S}_x.$$

Replacing A by $2\hat{I}_{1x}$ gives the final result as

$$-\cos(\pi J\tau)\sin(\pi J\tau)\,2\hat{I}_{1x}\hat{S}_y + \sin(\pi J\tau)\sin(\pi J\tau)\,4\hat{I}_{1x}\hat{I}_{2z}\hat{S}_x,$$

which can be interpreted as the multiple-quantum term $2\hat{I}_{1x}\hat{S}_y$ evolving into a multiple-quantum term $4\hat{I}_{1x}\hat{I}_{2z}\hat{S}_x$ in which the presence of the $\hat{I}_{2z}$ term indicates that the multiple-quantum coherence is anti-phase with respect to the I_2–S coupling.

The $180°$ pulse (about x) to the S spin simply changes the sign of the first term to give

$$\cos(\pi J\tau)\sin(\pi J\tau)\,2\hat{I}_{1x}\hat{S}_y + \sin(\pi J\tau)\sin(\pi J\tau)\,4\hat{I}_{1x}\hat{I}_{2z}\hat{S}_x.$$

Next comes the β_y pulse to the I spins. First, consider the effect of the pulse to spin I_1:

$$\cos(\pi J\tau)\sin(\pi J\tau)\,2\hat{I}_{1x}\hat{S}_y + \sin(\pi J\tau)\sin(\pi J\tau)\,4\hat{I}_{1x}\hat{I}_{2z}\hat{S}_x \xrightarrow{\beta\hat{I}_{1y}}$$

$$\cos(\beta)\cos(\pi J\tau)\sin(\pi J\tau)\,2\hat{I}_{1x}\hat{S}_y - \sin(\beta)\cos(\pi J\tau)\sin(\pi J\tau)\,\boxed{2\hat{I}_{1z}\hat{S}_y}$$

$$+\cos(\beta)\sin(\pi J\tau)\sin(\pi J\tau)\,4\hat{I}_{1x}\hat{I}_{2z}\hat{S}_x - \sin(\beta)\sin(\pi J\tau)\sin(\pi J\tau)\,4\hat{I}_{1z}\hat{I}_{2z}\hat{S}_x.$$

The pulse to I_2 has no effect on the first two terms, but does on the second two

$$+\cos(\beta)\sin(\pi J\tau)\sin(\pi J\tau)\,4\hat{I}_{1x}\hat{I}_{2z}\hat{S}_x - \sin(\beta)\sin(\pi J\tau)\sin(\pi J\tau)\,4\hat{I}_{1z}\hat{I}_{2z}\hat{S}_x \xrightarrow{\beta\hat{I}_{2y}}$$

$$+\cos(\beta)\cos(\beta)\sin(\pi J\tau)\sin(\pi J\tau)\,4\hat{I}_{1x}\hat{I}_{2z}\hat{S}_x + \sin(\beta)\cos(\beta)\sin(\pi J\tau)\sin(\pi J\tau)\,4\hat{I}_{1x}\hat{I}_{2x}\hat{S}_x$$

$$-\cos(\beta)\sin(\beta)\sin(\pi J\tau)\sin(\pi J\tau)\,\boxed{4\hat{I}_{1z}\hat{I}_{2z}\hat{S}_x} - \sin(\beta)\sin(\beta)\sin(\pi J\tau)\sin(\pi J\tau)\,4\hat{I}_{1z}\hat{I}_{2x}\hat{S}_x.$$

Of these terms, only those in $\boxed{\text{boxes}}$ are observable on the S spins. However these terms need to evolve into in-phase terms under the I_1–S and I_2–S couplings during the third τ delay for them to be observable under conditions of broad band I-spin decoupling.

For $2\hat{I}_{1z}\hat{S}_y$ to become in phase it must evolve with respect to the I_1–S coupling, thus acquiring a factor of $\sin(\pi J\tau)$, but must remain unaffected by the I_2–S coupling, thus acquiring a factor $\cos(\pi J\tau)$. The resulting in-phase term is therefore

$$\cos(\pi J\tau)\sin(\pi J\tau)\sin(\beta)\cos(\pi J\tau)\sin(\pi J\tau)\,\hat{S}_x.$$

For $4\hat{I}_{1z}\hat{I}_{2z}\hat{S}_x$ to become in phase it must evolve with respect to the I_1–S coupling, thus acquiring a factor of $\sin(\pi J\tau)$, and similarly evolve with respect to the I_2–S coupling, thus acquiring a further factor of $\sin(\pi J\tau)$. The resulting in-phase term is therefore

$$\sin(\pi J\tau)\sin(\pi J\tau)\cos(\beta)\sin(\beta)\sin(\pi J\tau)\sin(\pi J\tau)\,\hat{S}_x.$$

Assuming that $\tau = 1/(2J)$ as before, the former term goes to zero, whereas the latter becomes

$$\cos(\beta)\sin(\beta)\,\hat{S}_x.$$

A similar calculation can be completed starting out with equilibrium magnetization on the I_2 spin, and this gives exactly the same result. The overall intensity is therefore $2\cos(\beta)\sin(\beta)$, which can be written as $\sin(2\beta)$ and is therefore a maximum when $\beta = \pi/4$ ($45°$).

12.6.3 I_3S spin system

To simplify the calculation we will assume from the start that $\tau = 1/(2J)$. Therefore at the start of the second period τ we have just the term $-2\hat{I}_{1x}\hat{S}_y$. During this period the term is first converted to anti-phase with respect to the I_2–S coupling

$$-2\hat{I}_{1x}\hat{S}_y \xrightarrow{\ 2\pi J\tau\, \hat{I}_{2z}\hat{S}_z,\ \tau=1/(2J)\ } 4\hat{I}_{1x}\hat{I}_{2z}\hat{S}_x,$$

and then to anti-phase with respect to the I_3–S coupling

$$4\hat{I}_{1x}\hat{I}_{2z}\hat{S}_x \xrightarrow{\ 2\pi J\tau\, \hat{I}_{3z}\hat{S}_z,\ \tau=1/(2J)\ } 8\hat{I}_{1x}\hat{I}_{2z}\hat{I}_{3z}\hat{S}_y.$$

The 180° pulse to the S spin changes the sign of this term.

For the β_y pulse to the I spins to make this term observable $\hat{I}_{1x}$ must be rotated to $\hat{I}_{1z}$, but the other two I-spin operators must remain as $\hat{I}_{2z}$ and $\hat{I}_{3z}$. The rotation therefore has a $\sin(\beta)$ factor associated with the first process, and two factors of $\cos(\beta)$ associated with the second:

$$-8\hat{I}_{1x}\hat{I}_{2z}\hat{I}_{3z}\hat{S}_y \xrightarrow{\ \beta(\hat{I}_{1y}+\hat{I}_{2y}+\hat{I}_{3y})\ } \sin(\beta)\cos(\beta)\cos(\beta)\,8\hat{I}_{1z}\hat{I}_{2z}\hat{I}_{3z}\hat{S}_y.$$

Finally, during the third delay τ the triply anti-phase term becomes in phase as a result of evolution of the three couplings. Assuming that $\tau = 1/(2J)$ we have

$$8\hat{I}_{1z}\hat{I}_{2z}\hat{I}_{3z}\hat{S}_y \xrightarrow{\ 2\pi J\tau\, \hat{I}_{1z}\hat{S}_z,\ \tau=1/(2J)\ } -4\hat{I}_{2z}\hat{I}_{3z}\hat{S}_x$$

$$\xrightarrow{\ 2\pi J\tau\, \hat{I}_{2z}\hat{S}_z,\ \tau=1/(2J)\ } -2\hat{I}_{3z}\hat{S}_y$$

$$\xrightarrow{\ 2\pi J\tau\, \hat{I}_{3z}\hat{S}_z,\ \tau=1/(2J)\ } \hat{S}_x.$$

Taking into account that each of the three I spins can contribute equally we find that the intensity of the transferred signal therefore goes as

$$3\sin(\beta)\cos^2(\beta).$$

Summarising these results, we see that each spin system has a different dependence on the flip angle β, as is shown in the following table:

spin system	intensity
IS	$\sin(\beta)$
I_2S	$2\cos(\beta)\sin(\beta)$
I_3S	$3\cos^2(\beta)\sin(\beta)$

This behaviour results from the fact that at the end of the second delay τ the heteronuclear multiple quantum which is present has become anti-phase with respect to one I spin in the case of the I_2S spin system, and anti-phase with respect to two I spins in the case of the I_3S spin system. To convert this multiple quantum in observable coherence requires rotation of the spin I_1 but the other I spins (which are passive) must be left along z. The rotation of spin I_1 results in a factor $\sin(\beta)$, whereas factors of $\cos(\beta)$ are found for each of the spins which must be left along z.

Fig. 12.16 Plot showing how
the intensity of the
transferred signal in a DEPT
experiment varies with the
flip angle β of the final pulse
applied to the I spins. It is
assumed that $\tau = 1/(2J)$.

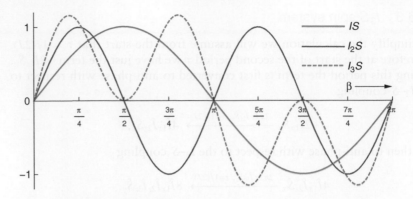

Fig. 12.16 Plot showing how the intensity of the transferred signal in a DEPT experiment varies with the flip angle β of the final pulse applied to the I spins. It is assumed that $\tau = 1/(2J)$.

12.6.4   Editing with DEPT

Figure 12.16 shows how the intensity of the signal in a DEPT experiment varies with the flip angle β for I_nS spin systems i.e. plots of the functions given in the table on the previous page. This plot is of exactly the same form as that shown in Fig. 12.14 on page 461, which shows how the intensity in an INEPT experiment varies with the value of the delay τ_2. Indeed, looking at the results of our calculations we see that if we identify $(2\pi J\tau_2)$ with the angle β the two pulse sequences give precisely the same intensities.

There is, however, an important difference between INEPT and DEPT. In the former the intensity depends on the delay τ_2 and the coupling constant, whereas in the latter it varies just with the flip angle β. The fact that in DEPT the dependence is on the flip angle, and not on the coupling (to the first approximation), means that a cleaner separation of the different spin systems is possible. For example, selecting $\beta = \pi/2$ will give a DEPT spectrum containing just resonances from IS spin systems *regardless* of the value of the coupling, whereas in INEPT it is not possible to choose $\tau_2 = 1/(4J)$ for all values of J. As a result, in the INEPT spectrum lines from I_2S and I_3S spin systems may be seen at low intensity if their couplings deviate from the assumed average value used to set τ_2.

The DEPT experiment is not perfect, however, in that it also contains delays which have to be set according to the value of the couplings. However, the fact that a compromise value has to be used for these delays has only a small effect on the separation of the different kinds of spin system.

It is possible to generate 'subspectra' from IS, I_2S and I_3S spin systems by combining a number of DEPT spectra recorded with different flip angles. There are a number of different ways of achieving this, one of which is set out in the table below. This gives the intensities, for each spin system, for three experiments recorded with $\beta = \pi/4$, $\pi/2$ and $3\pi/4$ (i.e. 45°, 90° and 135°).

		intensity		
experiment	flip angle β / rad	IS	I_2S	I_3S
(a)	$\pi/4$	0.707	1.00	1.06
(b)	$\pi/2$	1.00	0.00	0.0
(c)	$3\pi/4$	0.707	−1.00	1.06

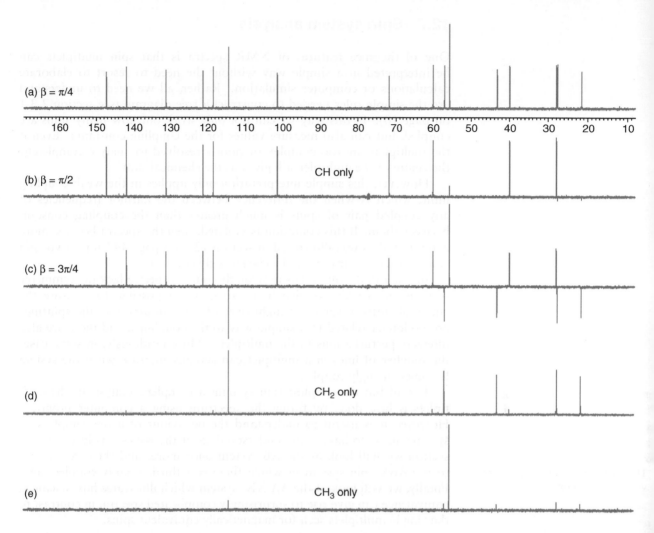

Fig. 12.17 Experimental DEPT spectra of quinine recorded at 500 MHz for proton. Spectra (a), (b) and (c) were recorded with the flip angle of the final I-spin pulse (β) set to $\pi/4$, $\pi/2$ and $3\pi/4$, respectively. In principle, only lines from CH groups should appear in spectrum (b). Spectra (d) and (e) are formed by combining (a)–(c) in the way decribed in the text. Only lines from CH_2 groups should appear in spectrum (d), and only lines from CH_3 groups should appear in spectrum (e). The separation is not quite perfect, but nevertheless the experiment is a very useful way of identifying the type of CH_n group responsible for each peak, Spectra (a)–(c) are plotted on the same scale, but the scale has been altered for (d) and (e).

The three subspectra are generated by making the following combinations:

IS: spectrum (b)

I_2S: spectrum (a) − spectrum (c)

I_3S: spectrum (a) − 1.414 × spectrum (b) + spectrum (c)

Figure 12.17 shows the result of this process for the experimental spectrum of quinine.

12.7   Spin system analysis

One of the nice features of NMR spectra is that spin multiplets can be interpreted in a simple way without the need to resort to elaborate calculations or computer simulation. Rather, all we need to understand are the simple rules needed to construct a tree diagram (see section 2.3.1 on page 10). In this way we can not only work out how many spins are coupled, but can also measure values for the coupling constants. Even if the multiplets are too complex or poorly resolved to analyse completely, the centre of the multiplet still gives us the chemical shift.

However, this simple interpretation only applies in the weak coupling limit, which is when the difference between the Larmor frequencies of any coupled pair of spins is much greater than the coupling constant between them. If this condition is violated, then the spectra become more complicated as was illustrated in section 12.1 on page 442 for a two-spin system: such spectra are said to be strongly coupled.

In general, the multiplets in strongly coupled spectra lack the symmetry seen in the weakly coupled limit, so it is not possible to measure the chemical shifts in such a straightforward way. In addition, the splittings are no longer related in a simple way to the couplings, and there are also intensity perturbations in the multiplets. More insidiously, in some cases the number of lines in a multiplet can actually increase when the system becomes strongly coupled.

For all but the simplest spin systems a complete analysis of the multiplets is best done by fitting the spectrum using a computer program. However, it is useful to understand the behaviour of a few simple spin systems so as to have some understanding of the issues involved. In this section we will look at the AB system once more, and then extend this to the ABX spin system in which there is a third, weakly coupled spin. Finally, we will look at the AA′XX′ system which illustrates how chemical equivalence can give rise to surprisingly complicated spectra, in contrast to the simple multiplets seen for magnetically equivalent spins.

This 'ABX' notation for spin systems was introduced on page 449.

12.7.1   AB spin system

In section 12.1 on page 442 we have already looked in some detail at the spectra arising from two strongly coupled spins. Figure 12.1 on page 445 illustrates how, as the difference in Larmor frequencies becomes smaller, one of the lines in each doublet diminishes in intensity while the other line becomes stronger. In addition, the midpoint between the two lines in the 'doublet' is no longer at the Larmor frequency.

The frequencies and intensities of the lines in the spectrum were given in Table 12.2 on page 444, but for convenience these are repeated here in Table 12.3 on the facing page in a slightly different form. For this spin system the two Larmor frequencies are $v_{0,A}$ and $v_{0,B}$, the coupling constant is J_{AB}, and the following definitions are also used

$$\Sigma = v_{0,A} + v_{0,B} \quad D = \sqrt{(v_{0,A} - v_{0,B})^2 + J_{AB}^2} \quad \tan \xi = \frac{J_{AB}}{v_{0,A} - v_{0,B}}. \quad (12.4)$$

Table 12.3 Frequencies and intensities of the lines in a strongly coupled AB spin system.

transition	eigenfunctions	frequency	intensity
1	$\alpha\alpha \to (\alpha\beta, \beta\alpha)$	$+\frac{1}{2}D - \frac{1}{2}\Sigma - \frac{1}{2}J_{AB}$	$\frac{1}{2}(1 + \sin\xi)$
2	$(\beta\alpha, \alpha\beta) \to \beta\beta$	$+\frac{1}{2}D - \frac{1}{2}\Sigma + \frac{1}{2}J_{AB}$	$\frac{1}{2}(1 - \sin\xi)$
3	$\alpha\alpha \to (\beta\alpha, \alpha\beta)$	$-\frac{1}{2}D - \frac{1}{2}\Sigma - \frac{1}{2}J_{AB}$	$\frac{1}{2}(1 - \sin\xi)$
4	$(\alpha\beta, \beta\alpha) \to \beta\beta$	$-\frac{1}{2}D - \frac{1}{2}\Sigma + \frac{1}{2}J_{AB}$	$\frac{1}{2}(1 + \sin\xi)$

The eigenfunctions (i.e. energy levels) involved in each of the four transitions are also given in the table. As we saw in Table 12.1 on page 443 some of these eigenfunctions are mixtures of the eigenfunctions for the weakly coupled system e.g. eigenfunction b, which is $\cos\left(\frac{1}{2}\xi\right)\psi_{\alpha,1}\psi_{\beta,2} + \sin\left(\frac{1}{2}\xi\right)\psi_{\beta,1}\psi_{\alpha,2}$. In Table 12.3 such an eigenfunction is denoted by the shorthand notation $(\alpha\beta, \beta\alpha)$, where the first entry in the bracket is the eigenfunction in the case of weak coupling (i.e. $\xi = 0$) and the second entry is the weakly coupled eigenfunction which is mixed in.

If we have a spectrum in which we identify the four lines of an AB pattern, then it is easy to use the information in the table to work out the parameters of the spin system i.e. the Larmor frequencies (shifts) and the coupling constant. The process is illustrated in Fig. 12.18.

The first thing to note is that the frequency difference between lines 1 and 2, and between lines 3 and 4, is simply J_{AB}: it is therefore trivial to measure the value of the coupling constant. From the table we can also see that the frequency separation of lines 2 and 4, and of lines 1 and 3, is D (recall that D is a positive quantity). Finally, the average frequency of lines 1 and 4, and of lines 2 and 3, is $-\frac{1}{2}\Sigma$.

Thus, from the spectrum it is possible to measure J_{AB}, D and Σ. Armed with these data, and the definitions given in Eq. 12.4 on the preceding page, it is possible to find the two Larmor frequencies. Note that the spectrum is invariant to the sign of J_{AB}, so this cannot be determined.

This is a good opportunity to point out that some care is needed in calculating the angle ξ. In Eq. 12.4 on the facing page we see how to compute $\tan\xi$, so to find the angle itself we need to compute an inverse tangent (or arc tangent). However, it is important that the resulting angle is placed in the correct quadrant of the circle, which can be achieved using the geometric construction shown in Fig. 12.19 on the next page.

The basis of this is that, since $\tan\xi = J_{AB}/(v_{0,A} - v_{0,B})$, the angle ξ can be constructed by placing a point at x-coordinate $(v_{0,A} - v_{0,B})$ and

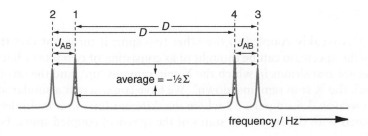

Fig. 12.18 Illustration of how the parameters of an AB system can be extracted from the spectrum. J_{AB}, D and $-\frac{1}{2}\Sigma$ can all be measured as shown, and manipulation of these quantities will lead to the shifts of A and B.

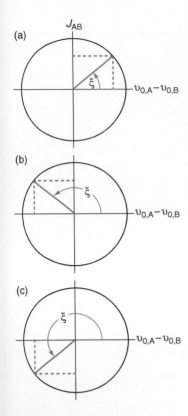

Fig. 12.19 The correct value of
the angle ξ is found by placing a
point at x-coordinate $(v_{0,A} - v_{0,B})$
and y-coordinate J_{AB}. The angle
ξ is measured anti-clockwise
from the x-axis to the line
joining the point to the origin.

y-coordinate J_{AB}. Recall that in a right-angle triangle the tangent of an angle is the length of the side of the triangle *opposite* to the angle divided by the length of the side *adjacent* to the angle. Therefore if a line is taken from this point to the origin, then ξ is the angle measured anti-clockwise from the x-axis.

If J_{AB} and $(v_{0,A} - v_{0,B})$ are both positive, ξ must be in the range 0–90°, as shown in (a). For example, if $J_{AB} = +10$ Hz and $(v_{0,A} - v_{0,B}) = +12$ Hz, which is the situation depicted in (a), $\tan \xi$ is $+0.833$ and pressing the 'inverse tan' key on a calculator will give the value 39.8°, which is correct.

Now suppose that $J_{AB} = +10$ Hz but that $(v_{0,A} - v_{0,B}) = -12$ Hz. This is the situation depicted in (b) and we therefore conclude that ξ must be in the range 90–180°. For the parameters given $\tan \xi = -0.833$ and the calculator tells us that the angle is $-39.8°$. Mathematically this is correct, but the angle is not in the correct quadrant. The best way to sort this out is first to compute the angle assuming that both J_{AB} and $(v_{0,A} - v_{0,B})$ are positive; this gives the value 39.8°. We then refer to (b) and deduce that the correct value for ξ is therefore $(180 - 39.8) = 140.2°$.

Finally, consider the case where $J_{AB} = -10$ Hz and $(v_{0,A} - v_{0,B}) = -12$ Hz, as depicted in (c). It follows that the angle must be between 180° and 270°. For these parameters $\tan \xi = +0.833$ and so the calculator gives us the value of 39.8° for the angle. However, it is clear from (c) that the correct value is $(180 + 39.8) = 219.8°$.

This may all seem incredibly fussy, but if you do not follow this recipe and then go on to use the value of ξ to compute line intensities you will often obtain incorrect results. Mathematically oriented computer programs usually have a version of the inverse tangent function which takes two arguments (i.e. the x- and y-coordinates of the point such that $\tan \theta = y/x$) and thus places the angle in the correct quadrant without further work on our part. In EXCEL the function is `ATAN2(x,y)`, in *Mathematica* the function is `ArcTan[x,y]`.

12.7.2 ABX spin system

The next spin system we will look at contains three coupled spins. Two of the spins, A and B, have shifts which are close enough to make them strongly coupled. In contrast, the shift of the third spin X is sufficiently separated from both A and B that weak coupling between A and X, and between B and X, can be assumed. The spectrum of this spin system separates into two parts: the AB part, containing lines associated with transitions of the A and B spins, and the X part, containing lines associated with transitions of the X spin. We will discuss these two parts separately.

The AB part

Since X is weakly coupled to the other two spins it turns out that the AB part of the spectrum can be thought of as consisting of two sets of lines: the first set are transitions in which the X spin remains 'up', and the second set in which the X spin remains 'down'. We have come across a similar idea to this in section 3.6 on page 38 where the different lines in a multiplet were associated with different spin states of the (passive) coupled spins. For the

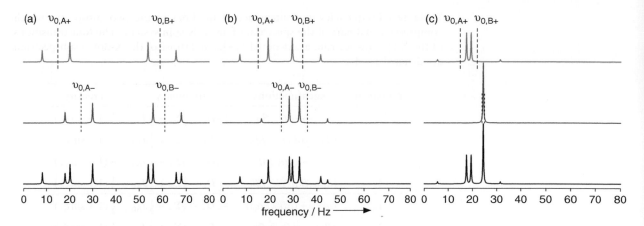

Fig. 12.20 Illustration of how the AB part of the spectrum of an ABX spin system can be decomposed into two subspectra, one in which the X spin is up (shown in blue), and one in which the X spin is down (shown in dark grey). The complete AB spectrum is shown in black beneath the two subspectra. The parameters used for these simulations were $\upsilon_{0,A}$ = −20 Hz, J_{AX} = 10 Hz, J_{BX} = 2 Hz and J_{AB} = 12 Hz. The Larmor frequency of spin B was set to −60 Hz, −35 Hz and −23 Hz in spectra (a), (b) and (c), respectively. In each case the effective Larmor frequencies $\upsilon_{0,A\pm}$ and $\upsilon_{0,B\pm}$ are indicated by the dashed lines. In going from (a) to (c) the degree of strong coupling increases, but in each case the X-down subspectrum is more strongly coupled than the X-up subspectrum.

ABX spin system, we are not associating a particular line with a spin state of X but a whole set of lines, called a *subspectrum*.

The AB part of the spectrum therefore consists of two subspectra, each of which is a four-line pattern just like that expected for a simple AB spin system (e.g. of the type shown in Fig. 12.1 on page 445). The two subspectra are different because in each the Larmor frequencies of A and B can be thought of as being modified by the coupling to the X spin. In the first subspectrum, associated with the X spin being up, the Larmor frequency of A is modified from $\upsilon_{0,A}$ to $\upsilon_{0,A} + \frac{1}{2}J_{AX}$, where J_{AX} is the A–X coupling; likewise the Larmor frequency of B is modified to $\upsilon_{0,B} + \frac{1}{2}J_{BX}$, where J_{BX} is the B–X coupling. The second subspectrum is associated with the X spin being down, and in this the A and B Larmor frequencies become $\upsilon_{0,A} - \frac{1}{2}J_{AX}$ and $\upsilon_{0,B} - \frac{1}{2}J_{BX}$; note the minus signs. These *effective Larmor frequencies* are written $\upsilon_{0,A\pm}$ and $\upsilon_{0,B\pm}$, with the + sign for the X-up subspectrum and the − sign for the X-down subspectrum:

$$\upsilon_{0,A\pm} = \upsilon_{0,A} \pm \tfrac{1}{2}J_{AX} \qquad \upsilon_{0,B\pm} = \upsilon_{0,B} \pm \tfrac{1}{2}J_{BX}.$$

We can predict the frequencies and intensities of the lines in the X-up subspectrum simply by replacing $\upsilon_{0,A}$ and $\upsilon_{0,B}$ in Table 12.3 on page 469 and in Eq. 12.4 by $\upsilon_{0,A+}$ and $\upsilon_{0,B+}$. Similarly, the form of the X-down subspectrum can be found by replacing $\upsilon_{0,A}$ and $\upsilon_{0,B}$ with $\upsilon_{0,A-}$ and $\upsilon_{0,B-}$. Some typical examples of the resulting subspectra and the complete AB part of the spectrum are shown in Fig. 12.20.

In the figure the X-up subspectra are shown in blue and the X-down subspectra are shown in dark grey. Note that each subspectrum consists of four lines whose intensities show the roofing effect characteristic of an

Table 12.4 Frequencies and intensities of the lines in the two subspectra which comprise the AB part of the spectrum of an ABX spin system. The four transitions of the X-up subspectrum are denoted 1_+–4_+, and those of the X-down subspectrum are denoted 1_-–4_-.

transition	eigenfunctions	frequency	intensity
1_+	$\alpha\alpha\alpha \rightarrow (\alpha\beta\alpha, \beta\alpha\alpha)$	$+\frac{1}{2}D_+ - \frac{1}{2}\Sigma_+ - \frac{1}{2}J_{AB}$	$\frac{1}{2}(1 + \sin\xi_+)$
2_+	$(\beta\alpha\alpha, \alpha\beta\alpha) \rightarrow \beta\beta\alpha$	$+\frac{1}{2}D_+ - \frac{1}{2}\Sigma_+ + \frac{1}{2}J_{AB}$	$\frac{1}{2}(1 - \sin\xi_+)$
3_+	$\alpha\alpha\alpha \rightarrow (\beta\alpha\alpha, \alpha\beta\alpha)$	$-\frac{1}{2}D_+ - \frac{1}{2}\Sigma_+ - \frac{1}{2}J_{AB}$	$\frac{1}{2}(1 - \sin\xi_+)$
4_+	$(\alpha\beta\alpha, \beta\alpha\alpha) \rightarrow \beta\beta\alpha$	$-\frac{1}{2}D_+ - \frac{1}{2}\Sigma_+ + \frac{1}{2}J_{AB}$	$\frac{1}{2}(1 + \sin\xi_+)$
1_-	$\alpha\alpha\beta \rightarrow (\alpha\beta\beta, \beta\alpha\beta)$	$+\frac{1}{2}D_- - \frac{1}{2}\Sigma_- - \frac{1}{2}J_{AB}$	$\frac{1}{2}(1 + \sin\xi_-)$
2_-	$(\beta\alpha\beta, \alpha\beta\beta) \rightarrow \beta\beta\beta$	$+\frac{1}{2}D_- - \frac{1}{2}\Sigma_- + \frac{1}{2}J_{AB}$	$\frac{1}{2}(1 - \sin\xi_-)$
3_-	$\alpha\alpha\beta \rightarrow (\beta\alpha\beta, \alpha\beta\beta)$	$-\frac{1}{2}D_- - \frac{1}{2}\Sigma_- - \frac{1}{2}J_{AB}$	$\frac{1}{2}(1 - \sin\xi_-)$
4_-	$(\alpha\beta\beta, \beta\alpha\beta) \rightarrow \beta\beta\beta$	$-\frac{1}{2}D_- - \frac{1}{2}\Sigma_- + \frac{1}{2}J_{AB}$	$\frac{1}{2}(1 + \sin\xi_-)$

AB spectrum, but that the pattern of intensities is different in each case. Beneath, in black, is shown the complete AB part of the spectrum which is simply the sum of the two subspectra. In these spectra the values of the coupling constants and the Larmor frequency of spin A are held constant, but as we go from (a) to (c) the Larmor frequency of spin B is brought progressively closer to that of A. As a result, in all of the subspectra the degree of strong coupling increases and the patterns become more roofed.

In the spectra J_{AX} has been chosen to be larger than J_{BX} (with both positive). Therefore, the separation between $\upsilon_{0,A-}$ and $\upsilon_{0,B-}$ is smaller than that between $\upsilon_{0,A+}$ and $\upsilon_{0,B+}$. A consequence of this is that the X-down subspectrum (grey) is more strongly coupled than the X-up subspectrum (blue), as is clearly visible in Fig. 12.20 on the preceding page. Indeed, in case (c) the X-down subspectrum has become so strongly coupled that the two inner lines of the AB pattern have just about merged, whereas in the X-up subspectrum these lines are still clearly separate. In this case the complete AB part of the spectrum (black) looks rather odd until we realize the way it can be disentangled into two AB subspectra.

Analysing these spectra is simply a matter of identifying the lines which belong to each of the two AB subspectra, something which is readily achieved since the pattern of four lines is very characteristic. Each subspectrum is then analysed using the process described in the previous section for a simple AB spectrum. This analysis gives $\upsilon_{0,A+}$ and $\upsilon_{0,B+}$ from one subspectrum, and $\upsilon_{0,A-}$ and $\upsilon_{0,B-}$ from the other. From these it is then possible to determine $\upsilon_{0,A}$, $\upsilon_{0,B}$ and the two coupling constants J_{AX} and J_{BX}. Applying this procedure is the subject of one of the Exercises at the end of this chapter.

It is interesting to note that although the sign of the A–B coupling constant has no effect on the spectrum, the relative signs of the A–X and B–X couplings does. For example, consider the parameters used to generate Fig. 12.20 (a). These give the following effective Larmor frequencies in the two subspectra:

$$\nu_{0,A+} = \nu_{0,A} + \tfrac{1}{2}J_{AX} = -20 + \tfrac{1}{2} \times 10 = -15 \text{ Hz}$$
$$\nu_{0,B+} = \nu_{0,B} + \tfrac{1}{2}J_{BX} = -60 + \tfrac{1}{2} \times 2 = -59 \text{ Hz}$$
$$\nu_{0,A-} = \nu_{0,A} - \tfrac{1}{2}J_{AX} = -20 - \tfrac{1}{2} \times 10 = -25 \text{ Hz}$$
$$\nu_{0,B-} = \nu_{0,B} - \tfrac{1}{2}J_{BX} = -60 - \tfrac{1}{2} \times 2 = -61 \text{ Hz}.$$

The separation of the effective Larmor frequencies in the X-up subspectrum is therefore $-15 - (-59) = 44$ Hz, whereas in the X-down subspectrum it is 36 Hz. If the sign of the A–X coupling is reversed, such that $J_{AX} = -10$ Hz, then repeating the above calculations gives the separation of the effective Larmor frequencies as 34 Hz and 46 Hz. The AB patterns will look different in each case, so the spectrum is therefore sensitive to the sign of the A–X coupling. However, it is only the *relative* signs of the A–X and B–X couplings which can be determined i.e. having both coupling constants positive gives an identical result to having both couplings negative. This point is explored further in the Exercises at the end of this chapter.

For completeness, Table 12.4 on the preceding page gives the frequencies and intensities of the eight lines in the AB part of the spectrum. In analogy to the simple AB spectrum we define the quantities $\Sigma_\pm$, $D_\pm$ and $\xi_\pm$ for each subspectrum using the corresponding effective shifts:

$$\Sigma_\pm = \nu_{0,A\pm} + \nu_{0,B\pm} \qquad D_\pm = \sqrt{(\nu_{0,A\pm} - \nu_{0,B\pm})^2 + J_{AB}^2}$$
$$\tan\xi_\pm = \frac{J_{AB}}{\nu_{0,A\pm} - \nu_{0,B\pm}}. \tag{12.5}$$

In the table the eigenfunctions involved in each transition are denoted in the same way that was used for the AB spectrum, with the spin states being given in the order ABX; for contrast the state of the X spin is shown in blue. Note that in the four transitions which comprise the X-up subspectrum (transitions 1_+–4_+) the X spin is always in the α state: this is what it means for these lines to be described as the X-up subspectrum. Similarly, in the other four transitions the X spin is always in the β (down) state.

The X part

Since the shift of the X spin is well-separated from that of the A and B spins, it might, quite reasonably, be thought that the X part of the spectrum will be unaffected by the strong coupling between A and B. However, this is not the case since the eigenfunctions between which the transitions take place are affected by strong coupling.

Figure 12.21 shows the X part of the ABX spectrum for the same set of parameters used to generate the spectra shown in Fig. 12.20 on page 471. In (a) we see what appears at first sight to be a doublet of doublets, which is what we would expect for a weakly coupled spin system. However, closer inspection shows that the two inner lines in the multiplet are weaker than the outer lines, an effect which is slightly more pronounced in spectrum (b). This is a result of strong coupling between A and B.

The multiplet shown in (c) also shows two additional weak lines flanking the multiplet i.e. a total of *six* lines, as opposed to the four

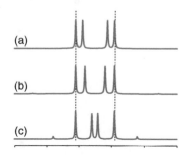

(a)

(b)

(c)

Fig. 12.21 The X part of the ABX spectrum simulated for the same set of parameters used in Fig. 12.20 on page 471; the tick marks are spaced by 10 Hz. Note that the middle lines of the multiplet are weaker than the outer two lines, and that when the coupling becomes very strong in one of the subspectra, as in (c), two weaker lines appear at the edges of the multiplet. The separation between the two strong lines, indicated by the dashed lines, is always $(J_{AX} + J_{BX})$.

Table 12.5 Frequencies and intensities of the lines in the X part of the spectrum of an ABX spin system.

transition	eigenfunctions	frequency	intensity
X_1	$\beta\beta\alpha \rightarrow \beta\beta\beta$	$-\upsilon_{0,X} - \frac{1}{2}J_{AX} - \frac{1}{2}J_{BX}$	$\frac{1}{4}$
X_2	$(\alpha\beta\alpha, \beta\alpha\alpha) \rightarrow (\alpha\beta\beta, \beta\alpha\beta)$	$-\upsilon_{0,X} + \frac{1}{2}(D_+ - D_-)$	$\frac{1}{4}\cos^2\left(\frac{1}{2}\xi_+ - \frac{1}{2}\xi_-\right)$
X_3	$(\beta\alpha\alpha, \alpha\beta\alpha) \rightarrow (\beta\alpha\beta, \alpha\beta\beta)$	$-\upsilon_{0,X} + \frac{1}{2}(-D_+ + D_-)$	$\frac{1}{4}\cos^2\left(\frac{1}{2}\xi_+ - \frac{1}{2}\xi_-\right)$
X_4	$\alpha\alpha\alpha \rightarrow \alpha\alpha\beta$	$-\upsilon_{0,X} + \frac{1}{2}J_{AX} + \frac{1}{2}J_{BX}$	$\frac{1}{4}$
X_5	$(\beta\alpha\alpha, \alpha\beta\alpha) \rightarrow (\alpha\beta\beta, \beta\alpha\beta)$	$-\upsilon_{0,X} + \frac{1}{2}(-D_+ - D_-)$	$\frac{1}{4}\sin^2\left(\frac{1}{2}\xi_+ - \frac{1}{2}\xi_-\right)$
X_6	$(\alpha\beta\alpha, \beta\alpha\alpha) \rightarrow (\beta\alpha\beta, \alpha\beta\beta)$	$-\upsilon_{0,X} + \frac{1}{2}(D_+ + D_-)$	$\frac{1}{4}\sin^2\left(\frac{1}{2}\xi_+ - \frac{1}{2}\xi_-\right)$

expected in the case of weak coupling; indeed, close inspection reveals that corresponding lines are also visible in (b). These lines only have significant intensity if one (or both) of the X subspectra shows very strong coupling, as is the case for the X-down subspectrum in Fig. 12.20 (c). Note that although the size of J_{AX} and J_{BX} are the same for (a)–(c), the inner lines in the multiplet are changing in frequency.

Table 12.5 gives the frequencies and intensities of the six lines which comprise the X part of the spectrum, along with the eigenfunctions involved in each transition. Both the intensities and frequencies of transitions X_1 and X_4 are independent of the strength of coupling: these are indicated by the dashed lines in Fig. 12.21 on the previous page. Transitions X_2 and X_3 correspond to the inner two lines of the multiplet – their frequencies and intensities are affected by the amount of strong coupling present.

Transitions X_5 and X_6 are rather different to the others. In these transitions all three spins flip e.g. $(\beta\alpha\alpha, \alpha\beta\alpha) \rightarrow (\alpha\beta\beta, \beta\alpha\beta)$, in contrast to all of the other transitions in which only one spin flips. Despite all three spins being flipped, the quantum number M changes by ± 1 as it does in all of the other transitions.

Transitions which have $\Delta M = \pm 1$ but in which three spins flip lead to what are called *combination lines* in the spectrum. We came across these in section 3.7.4 on page 43 where it was noted that in weakly coupled spectra such transitions are not allowed. However, in the case of strong coupling they can acquire some intensity, which is exactly what happens in the ABX spectrum. These combination lines appear as the weak lines flanking the multiplet in spectra (b) and (c).

The X part of the spectrum is invariant to the sign of the A–B coupling constant but, like the AB part of the spectrum, it is sensitive to the relative signs of the A–X and B–X couplings. Again, this behaviour is explored in the Exercises at the end of the chapter. Under some circumstances the X part of the spectrum can be very sensitive to the precise separation between the Larmor frequencies of the A and B spins. This point is illustrated in Fig. 12.22 on the next page which shows how the X-spin multiplet changes as $\upsilon_{0,B}$ is moved in steps of 2 Hz.

In (a) we see four strong lines flanked by the two rather weak combination lines. In (b) $\upsilon_{0,B}$ is set to −24 Hz which makes the effective Larmor frequencies of A and B in the X-down subspectrum equal i.e. infinite strong

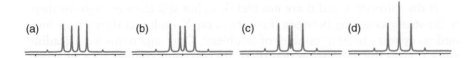

Fig. 12.22 Plots of the X part of an ABX spectrum for the case where the Larmor frequency of the B spin is (a) −26 Hz, (b) −24 Hz, (c) −22 Hz and (d) −20 Hz. The Larmor frequency of the A spin is −20 Hz and all the other parameters are as in Fig. 12.20 on page 471. The tick marks are spaced by 10 Hz. Note how even these small changes in Larmor frequency affect the appearance of the multiplet.

coupling in this subspectrum. However, this only makes a relatively small difference to the spectrum. Further increasing $\nu_{0,B}$ to −22 Hz and then to −20 Hz causes a more dramatic change in the multiplet, and in the latter case the two central lines merge into one. In fact, the special thing about $\nu_{0,B}$ = −20 Hz is that it is midway between the value of −24 Hz, in which the X-down subspectrum is infinitely strongly coupled, and the value of −16 Hz which makes the X-up subspectrum infinitely strongly coupled.

If the weak peaks flanking the X-spin multiplet are ignored or obscured by noise, then it would be easy to misinterpret each of the multiplets in Fig. 12.22 as a doublet of doublets and then go on to measure the values of the two couplings present in the usual way. The values obtained in this way would be entirely wrong since the splittings are not related to the couplings in a simple way. Remember that J_{AX} and J_{BX} are in fact the *same* for all four multiplets (a)–(d), despite the fact that the multiplets appear to be rather different.

Virtual coupling

As we have seen, although the X spin is well-separated from A and B, strong coupling between the latter two spins still has an influence on the X-spin multiplet. Unless we are aware of this, it is possible to be misled by the appearance of this multiplet. In this section we are going to describe a particular case, which is encountered quite often, in which the X-spin multiplet is very misleading.

Imagine that the A–X coupling is zero, but the B–X and A–B couplings are still present. If the Larmor frequencies of A and B are identical (i.e. an AA′ system) then it turns out that the X-spin multiplet appears to be a 'triplet', as illustrated in Fig. 12.23. It is thus easy to be misled into thinking that the X spin must have equal couplings to A and A′, whereas in fact there is only a coupling to *one* of these spins. This phenomenon is called *virtual coupling* since it appears that a coupling is present, whereas in fact there is none.

Strong coupling effects, such as those seen in AB or ABX spin systems, generally decrease as the static magnetic field is increased since this increases the separation between the Larmor frequencies of A and B. However, in the case of an AA′X spin system A and A′ have the *same* Larmor frequency – a situation which does not change when the static field is increased. Therefore virtual coupling effects *do not* disappear when we move to a higher magnetic field.

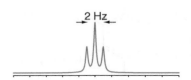

Fig. 12.23 X-spin multiplet for the case where the Larmor frequencies of A and B are the same (i.e. an AA′ system) and in which J_{AX} = 0 Hz and J_{BX} = 2 Hz; as before J_{AB} = 12 Hz. Despite the fact that X is only coupled to one spin, the X-spin multiplet appears as a triplet, with splitting J_{BX} between the outer lines. This is the phenomenon of *virtual coupling*. The tick marks on the axis are spaced by 2 Hz.

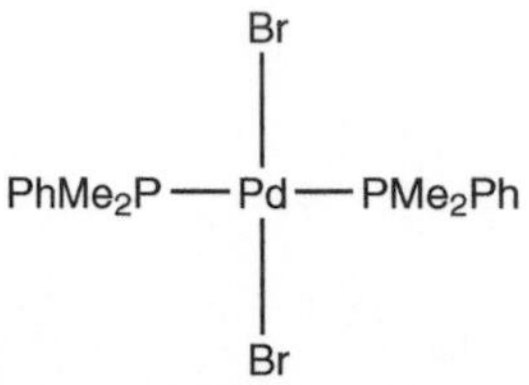

Fig. 12.24 A typical palladium phosphine complex which exhibits virtual coupling in its proton spectrum.

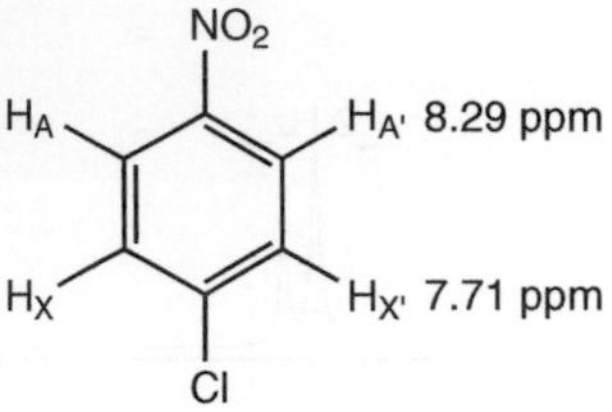

$J_{AA'}$ = 2.8 Hz
$J_{XX'}$ = 2.5 Hz
$J_{A'X'} = J_{AX}$ = 8.7 Hz
$J_{A'X} = J_{AX'}$ = 0.3 Hz

Fig. 12.25 The molecule 1,4-chloronitrobenzene is an example of an AA′XX′ system. The parameters for the spin system are shown.

If the shifts of A and B are not identical, but still close enough for there to be strong coupling between them, then the X-multiplet shows four lines and so appears to be a 'doublet of doublets'. Once again this is misleading since it gives the impression that there are two spins coupled to X, whereas there is in fact only one. As the degree of strong coupling between A and B decreases further the four lines merge into two, giving the simple doublet expected in the case of weak coupling.

The palladium phosphine complex shown in Fig. 12.24 is a good example of how virtual coupling can lead to what are at first sight somewhat unexpected spectra. In the proton NMR of this compound the resonance from the methyl groups appears as a 'triplet' with a separation of around 7 Hz between the outer lines. This splitting must be due to coupling to ^{31}P, since the protons on the benzene ring are too far away to be coupled to a significant extent. The fact that the multiplet is a triplet might be seen to imply that the two-bond P–H and six-bond P–H couplings have the same magnitude, which seems rather unlikely.

The correct explanation is that the two ^{31}P nuclei are chemically (but not magnetically) equivalent and that there is a significant coupling between them. Assuming that all six methyl protons in the –PMe$_2$Ph ligand are equivalent, these protons and the two phosphorus nuclei thus form an AA′X$_6$ system. The X spins are only coupled to the ^{31}P which is two-bonds away, so what we have here is analogous to the AA′X system we have just been discussing. The X part of the spectrum thus appears to be a triplet, despite the fact that the protons are only coupled to one ^{31}P.

12.7.3   The AA′XX′ spin system

The final spin system we will look at is the AA′XX′ system consisting of two pairs of chemically equivalent spins. It is well-exemplified by a 1,4-substituted benzene ring, such as 1,4-chloronitrobenzene, as shown in Fig. 12.25. Spins A and A′ have the same shift, but A has a different coupling to X than it does to X′, which is why A and A′ are not magnetically equivalent.

What we have here is in part a strongly coupled spin system since A and A′ have the same shift and are coupled together, as is also the case for X and X′. From our experience with the ABX spin system we should not be surprised to find that the multiplet for the X spins is affected by the strong coupling between A and A′, and vice versa.

Figure 12.26 on the next page shows a simulation of the multiplet from the A spins for this molecule; the multiplet for the X spins is identical. The multiplet is rather complex and there is no immediately recognizable pattern to the lines. The two strongest peaks are separated by J_{AX}, and the multiplet is symmetrically placed about the shift of the A spins. This multiplet is not simplified by moving to higher magnetic field strengths since, as explained above, A and A′ have the chemical shift and hence the same Larmor frequency, regardless of the applied field.

12.7.4 More complex spin systems

It is pretty tedious work to use the expressions in Table 12.4 on page 472 and Table 12.5 on page 474 to compute out the form of an ABX spectrum for a particular combination of shifts and couplings. For more complex spin systems the task becomes even more tedious, and it may be that expressions for the expected line frequencies are simply not available.

In such cases it is best to resort to computer-based numerical calculations. There are a range of programs available which will simulate a spectrum for a more-or-less arbitrary spin system. Often these programs also make it possible to compare a simulated spectrum with an experimental one and then to adjust, automatically, the values of the shifts and coupling constants so as to obtain the best match.

Such programs are often part of the suite provided along with a spectrometer. There are also some freely available versions, some of which are listed under Further reading.

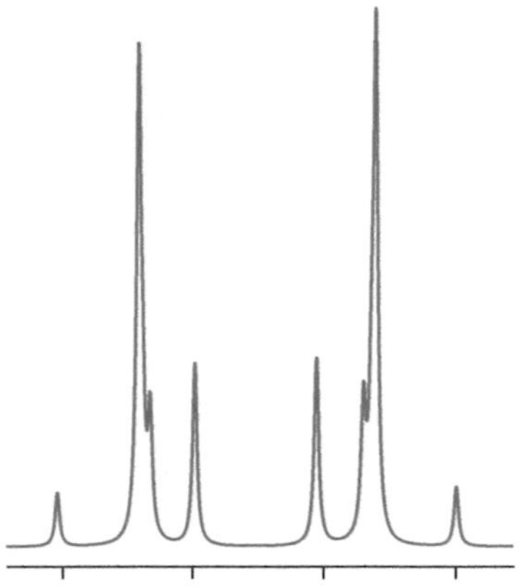

12.8 Further reading

Chemical and magnetic equivalence; spin system analysis; strong coupling:

Chapter 2 from R. K. Harris, *Nuclear Magnetic Resonance Spectroscopy, A Physiochemical View* (Longman, 1983).

Chapter 17 from M. H. Levitt, *Spin Dynamics* (2nd edition, John Wiley & Sons, Ltd, 2008).

Chapter 5 from H. Günther, *NMR Spectroscopy – An Introduction* (2nd edition, John Wiley & Sons, Ltd, 1992).

Products operators for equivalent spins:

O. W. Sørensen, G. W. Eich, M. H. Levitt, G. Bodenhausen and R. R. Ernst, *Progress in Nuclear Magnetic Resonance Spectroscopy*, **16**, 163–192 (1983).

Spin system simulation programs:

SpinWorks (Dr K. Marat, University of Manitoba NMR Laboratory)
`http://www.umanitoba.ca/chemistry/nmr/spinworks/index.html`

WINDNMR (Prof. Hans J. Reich, University of Wisconsin)
`http://www.chem.wisc.edu/areas/reich/plt/windnmr.htm`

Fig. 12.26 Simulated multiplet for the A spins in an AA′XX′ system using the parameters appropriate for 1,4-chloronitrobenzene; the multiplet for the X spins is identical. The tick marks are spaced by 5 Hz, and the linewidth is 0.2 Hz.

12.9 Exercises

You may find it convenient to make the necessary calculations for strongly coupled spin systems using a spreadsheet or some other mathematical program.

12.1 For each of the following molecules determine which groups of protons or ^{19}F nuclei (if any) are chemically or magnetically equivalent, giving your reasons. Hence describe each spin system using the conventional notation (you may assume that all spins, other than those whose shifts are identical by symmetry, are weakly coupled). In (e) and (f) ignore couplings over more than four bonds, and in (g) and (h) only consider those protons attached to carbon.

12.2 This question is concerned with an I_2S spin system.

(a) Show that, under the influence of the I–S couplings and the offset of the S spin, an initial operator $\hat{S}_x$ gives rise to the following term in $\hat{S}_x$ (there are, of course, several other product operators generated, but these are not of interest here):

$$\cos\left(\Omega_S t\right)\cos^2\left(\pi J t\right)\hat{S}_x.$$

(b) By first applying the identity $\cos^2\left(A\right) \equiv \frac{1}{2}[1 + \cos\left(2A\right)]$, and then the identity $\cos\left(A\right)\cos\left(B\right) \equiv \frac{1}{2}[\cos\left(A + B\right) + \cos\left(A - B\right)]$, show that the product of trigonometric terms in front of the operator $\hat{S}_x$ can be written as the following sum of trigonometric terms; you may also need to use $\cos\left(-A\right) \equiv \cos\left(A\right)$

$$\frac{1}{4}\left[2\cos\left(\Omega_S t\right) + \cos\left(\Omega_S t + 2\pi J t\right) + \cos\left(\Omega_S t - 2\pi J t\right)\right].$$

(c) Assuming that only the x-magnetization is detected and the resulting signal is subject to a cosine Fourier transform, explain why your result predicts that the spectrum will consist of a 1:2:1 triplet, centred at Ω_S. What is the spacing of the outer two lines?

(d) Repeat the above calculation for the case where the initial state is $2\hat{I}_{1z}\hat{S}_x$ and show that the resulting term in $\hat{S}_x$ is

$$-\sin(\Omega_S t)\cos(\pi J t)\sin(\pi J t)\,\hat{S}_x.$$

By using the identities $\sin(A)\cos(A) \equiv \frac{1}{2}\sin(2A)$ and then $\sin(A)\sin(B) \equiv \frac{1}{2}[\cos(A-B) - \cos(A+B)]$, show that the product of trigonometric terms can be written as

$$\tfrac{1}{4}\left[\cos(\Omega_S t + 2\pi J t) - \cos(\Omega_S t - 2\pi J t)\right].$$

Hence explain that the resulting spectrum is a $-1{:}0{:}+1$ triplet.

(e) Repeat the above calculation for the case where the initial state is $4\hat{I}_{1z}\hat{I}_{2z}\hat{S}_x$ and show that the resulting spectrum is a $+1{:}-2{:}+1$ triplet. You may need to use the identity $\sin^2(A) \equiv \frac{1}{2}[1 - \cos(2A)]$.

12.3 Using the same approach as in the previous exercise, predict the form of the multiplets arising from the terms $\hat{S}_x$, $2\hat{I}_{1z}\hat{S}_x$, $4\hat{I}_{1z}\hat{I}_{2z}\hat{S}_x$ and $8\hat{I}_{1z}\hat{I}_{2z}\hat{I}_{3z}\hat{S}_x$ in an I_3S spin system; compare your results with Fig. 12.6 on page 452. You will need to apply repeatedly several of the following trigonometric identities

$$
\begin{aligned}
\cos(A)\cos(B) &\equiv \tfrac{1}{2}[\cos(A+B) + \cos(A-B)] \\
\sin(A)\sin(B) &\equiv \tfrac{1}{2}[\cos(A-B) - \cos(A+B)] \\
\sin(A)\cos(B) &\equiv \tfrac{1}{2}[\sin(A+B) + \sin(A-B)]
\end{aligned}
$$

$$\sin(2A) \equiv 2\sin(A)\cos(A) \qquad \cos(2A) \equiv \cos^2(A) - \sin^2(A)$$
$$\cos^2(A) \equiv \tfrac{1}{2}[1 + \cos(2A)] \qquad \sin^2(A) \equiv \tfrac{1}{2}[1 - \cos(2A)]$$
$$\cos(-A) \equiv \cos(A) \qquad \sin(-A) \equiv -\sin(A).$$

12.4 (a) Compute the evolution of the operator $\hat{S}_y$ under the influence of the $I{-}S$ couplings in (i) an IS, (ii) an I_2S and (iii) an I_3S spin system.

(b) Compute the evolution of the operator $8\hat{I}_{1z}\hat{I}_{2z}\hat{I}_{3z}\hat{S}_y$ under the influence of the $I{-}S$ couplings in an I_3S spin system. After what time will this triply anti-phase operator become entirely in phase?

12.5 On page 457 it was shown that for a spin echo applied to an I_3S spin system the amount of the singly anti-phase terms goes as $\cos^2(2\pi J\tau)\sin(2\pi J\tau)$. Maximizing this function is equivalent to finding the maximum in $\cos^2(\theta)\sin(\theta)$ where $\theta = 2\pi J\tau$. By differentiating the function $\cos^2(\theta)\sin(\theta)$ with respect to θ show that one of the extrema is at $\sin\theta = \sqrt{1/3}$, and hence show that the greatest amount of the singly anti-phase term is found when

$\tau = 0.0980/J$. [The calculation is made easier if you use the identity $\sin^2(\theta) + \cos^2(\theta) \equiv 1$ so that the expressions can be written entirely in terms of $\cos(\theta)$ or $\sin(\theta)$, as is convenient.]

Similarly show that the maximum amount of the doubly anti-phase term, which goes as $\cos(2\pi J\tau)\sin^2(2\pi J\tau)$, occurs when $\tau = 0.152/J$.

12.6 In an APT experiment the delay τ has to be set to a compromise value based on the average value of a one-bond C–H coupling. Suppose that we take this value to be 130 Hz, what will be the value for the delay τ? The one-bond C–H coupling in an ethyne (acetylene) group –C≡C–H can be as large as 250 Hz. In an APT spectrum with τ set to the value you have computed, how would the line from such a carbon be affected? Does the line appear with the expected sign?

12.7 Predict the form of the cross-peak multiplet that would be seen in a COSY spectrum of an I_2S and of an I_3S spin system, assuming that the I and S spins are affected by both 90° pulses. [The calculation is simplest if you start with $\hat{I}_{1z}$ as the equilibrium magnetization, but you should also attempt it starting from $\hat{S}_z$. When predicting the form of the multiplets use the graphical approach illustrated in Fig. 8.10 on page 194.]

12.8 In an AB spectrum the two weaker lines are found to have frequencies of 40.8 Hz and 9.18 Hz, and the two stronger lines are found to have frequencies of 31.1 Hz and 18.7 Hz, measured relative to some arbitrary origin.

(a) Using the method described in section 12.7.1 on page 468, determine the Larmor frequencies of the two spins and the coupling constant between them.

(b) Use these data to compute the intensities of the lines and hence sketch the spectrum.

(c) Suppose that the static magnetic field is doubled. Compute the frequencies and intensities of the lines you would then expect to see.

12.9 (a) Using EXCEL (or some other suitable program) set up a spreadsheet which computes the line frequencies and intensities of both the AB and X parts of an ABX system for an arbitrary set of parameters.

(b) Use your spreadsheet to compute the spectrum for the following parameters: $\nu_{0,A} = -20$ Hz, $\nu_{0,B} = -27$ Hz, $\nu_{0,X} = -1000$ Hz, $J_{AB} = 5$ Hz, $J_{AX} = 12$ Hz and $J_{BX} = 3$ Hz. Make a sketch of the spectrum (or do so using a computer) and note any features of interest.

(c) Investigate the effect on the spectrum of changing the signs of the various coupling constants.

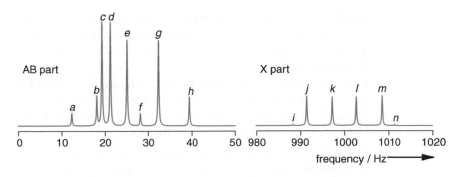

12.10 For the ABX spectrum shown in Fig. 12.27 the frequencies of the lines, relative to some arbitrary origin, are given in the following table.

line	frequency / Hz	line	frequency / Hz
a	12.3	i	988.4
b	18.1	j	991.5
c	19.3	k	997.3
d	21.2	l	1002.7
e	25.1	m	1008.5
f	28.2	n	1011.6
g	32.4		
h	39.4		

(a) From the AB part, pick out the lines which correspond to the two subspectra. In doing this it is useful to remember that in an AB pattern: (i) the separation between the two low-frequency lines, and the separation between the two high-frequency lines, is J_{AB}; (ii) the two outer lines have the same intensity as one another, as do the two inner lines. In addition the same splitting by J_{AB} occurs in both subspectra.

(b) Analyse each subspectrum using the method described in section 12.7.1 on page 468 and hence obtain the effective Larmor frequencies of A and B in each of the subspectra (i.e. $\nu_{0,A\pm}$ and $\nu_{0,B\pm}$), as well as the value of the coupling constant J_{AB}. Since it is not possible to determine the sign of this coupling constant, simply assume that it is positive. Check your answers by using the values of the parameters you have determined to predict the frequencies and intensities of the lines in the spectrum and then comparing these with the data. Be careful to follow the guidelines on page 469 when computing the angles $\xi_{\pm}$.

(c) You should now have four effective Larmor frequencies, two for each subspectrum: let us call these ν_1 and ν_2 for the first subspectrum, and ν_3 and ν_4 for the second subspectrum. The problem is that at this stage there is no way of knowing which pair of these frequencies correspond to $\nu_{0,A\pm}$ and which pair correspond to $\nu_{0,B\pm}$. There are basically two distinct possibilities: (1) ν_1 and ν_3 are $\nu_{0,A\pm}$, and ν_2 and ν_4 are $\nu_{0,B\pm}$; (2) ν_1 and ν_4 are $\nu_{0,A\pm}$, and ν_2 and ν_3 are $\nu_{0,B\pm}$. In the first case it does not matter whether ν_1 or ν_3 is $\nu_{0,A+}$, and so on for the other combinations.

Make a choice corresponding to (1) and hence find values for $\nu_{0,A}$, $\nu_{0,B}$, J_{AX} and J_{BX} (the signs of the coupling constants should be retained). Then make a choice corresponding to (2) and hence find a second set of parameters.

(d)   The only way of distinguishing between the two choices is to compute the form of the X part of the spectrum. Assuming that ν_X is -1000 Hz do this for both possibilities and, by comparing your result with Fig. 12.27 on the previous page, determine which possibility is the correct one.

12.11   Consider a spin system consisting of a ^{13}C, spin X, and two protons, A and B. The carbon has a one-bond coupling of 150 Hz to proton A, and a long-range coupling of 15 Hz to proton B. Protons A and B are also coupled.

(a)   Describe the circumstances in which degree of strong coupling in the two AB subspectra will be substantially different.

(b)   Assuming that the proton Larmor frequency is 500 MHz, compute the difference in chemical shift (in ppm) between protons A and B which will result in infinite strong coupling in first one subspectrum and then in the other. [Hint: compute the frequency separation in Hz and then convert this to ppm.]

(c)   Explain why your answers will be the same if the sign of the A–B coupling is reversed, but will change if the sign of the B–X coupling is reversed.

(d)   Explain why it is that in a homonuclear AB spin system the degree of strong coupling always decreases as the static field is increased, whereas in an ABX system of the type described at the start of the question increasing the static field can, under some circumstances, increase the degree of strong coupling.

13

How the spectrometer works

NMR spectrometers have now become very complex instruments capable of performing an almost limitless number of sophisticated experiments. We certainly do not need to understand the details of how the spectrometer works in order to be able to use it effectively, but it is helpful to have a broad understanding of the basic components which make up the spectrometer, and the way in which they work together.

Broken down to its simplest form, the spectrometer consists of the following:

- An intense, homogeneous and stable magnetic field.

- A 'probe' in which the coils used to excite and detect the NMR signal are held close to the sample.

- An RF transmitter capable of delivering short high-power pulses, and longer low-power pulses for selective excitation.

- A sensitive RF receiver to capture and amplify the NMR signals.

- A digitizer to convert the NMR signals into a form which can be stored in computer memory.

- A 'pulse programmer' to produce precisely timed pulses and delays.

- A computer to control everything and to process the data.

We will consider each of these in turn.

13.1 The magnet

Modern NMR spectrometers use persistent superconducting magnets to generate the B_0 field. Basically such a magnet consists of a coil of wire through which a current passes, thereby generating a magnetic field. The wire is held at a sufficiently low temperature (typically < 6 K) for it to become *superconducting*, meaning that its resistance goes to zero. Thus, once the current is set running in the coil it will persist for ever, thereby

Understanding NMR Spectroscopy, Second Edition James Keeler
© 2010 John Wiley & Sons, Ltd

generating a magnetic field without consuming any electrical power. Super-conducting magnets tend to be very stable and so are ideal for NMR.

One particular difficulty in constructing superconducting magnets is that there is a physical effect which causes the wire to cease to be super-conducting once the magnetic field exceeds a certain critical value. Simple copper wires do not remain superconducting at the kinds of field strengths we need for NMR. However, such high fields can be achieved by using special wires in which filaments of one metal or alloy are embedded in a matrix of another; typical combinations include copper, niobium and tin.

To maintain the wire in its superconducting state the coil is immersed in a bath of liquid helium. Surrounding this is usually a 'heat shield' kept at 77 K by contact with a bath of liquid nitrogen. This reduces the amount of (expensive) liquid helium which boils off due to heat flowing in from the surroundings. The whole assembly is constructed in a vacuum flask so as to further reduce the heat flow. The cost of maintaining the magnetic field is basically the cost of the liquid helium and liquid nitrogen needed to keep the magnet cool.

Of course, we do not want the sample to be at liquid helium tempera-tures, therefore a room temperature region – accessible to the outside world – has to be engineered as part of the design of the magnet. Usually this room temperature zone takes the form of a vertical tube passing through the magnet (called the *bore tube* of the magnet).

13.1.1   Shims

The lines in NMR spectra are very narrow – linewidths of 1 Hz or less are not uncommon – and so the magnetic field has to be very homogeneous. Just how homogeneous the field has to be is best illustrated by an example.

Consider a proton spectrum recorded at 500 MHz, which corresponds to a magnetic field of 11.75 T. The Larmor frequency is given by

$$v_0 = -\frac{1}{2\pi}\gamma B_0 \tag{13.1}$$

where γ is the gyromagnetic ratio (2.67×10^8 rad s^{-1} for protons). We need to limit the variation in the magnetic field across the sample so that the corresponding variation in the Larmor frequency is much less than the width of the line, say by a factor of ten.

Suppose that the maximum acceptable change in Larmor frequency across the sample is 0.1 Hz. Using Eq. 13.1 we can compute the change in the magnetic field as $(0.1 \times 2\pi)/\gamma = 2.4 \times 10^{-9}$ T. Expressed as a fraction of the main magnetic field this variation is about 2×10^{-10}. We can see that we need to have an extremely homogeneous magnetic field for work at high resolution.

On its own, no superconducting magnet can produce such a homoge-neous field. What we have to do is to surround the sample with a set of *shim coils*, each of which produces a tiny magnetic field with a particular spatial profile which can be used to cancel out the inhomogeneities in the main magnetic field. The current through each of these coils is adjusted until the magnetic field has the required homogeneity, something we can easily assess by recording the spectrum of a sample which has a sharp line.

Modern spectrometers might have up to 40 different shim coils, so adjusting them is an involved task. However, once set it is usually only necessary on a day to day basis to alter a few of the shims which generate the simplest field profiles.

The shims are labelled according to the field profiles they generate. So, for example, there are usually shims labelled x, y and z, which generate magnetic fields varying in the corresponding directions. The shim z^2 generates a field that varies quadratically along the z direction, which is the direction of B_0. There are more shims whose labels you might recognize as corresponding to the names of the hydrogen atomic orbitals. This is no coincidence since the magnetic field profiles that the shims coils create are in fact the spherical harmonic functions, which are the angular parts of the atomic orbitals.

13.1.2 The lock

Although the field produced by a superconducting magnet is very stable, there will nevertheless be some drift in the field which is certainly significant for the very narrow lines we see in NMR. This slow drift is compensated for by the *field–frequency lock*, which is a feedback system designed to keep the field at a steady value.

The lock uses the ^{2}H NMR signal from a deuterated solvent used to prepare the sample (most commonly $CDCl_3$, or D_2O in the case of biological samples). The magnetic field is adjusted by small amounts in such a way as to keep the deuterium resonance at a *fixed* frequency thus ensuring that the field is held at a constant value. These adjustments to the field are made by varying the current through a coil rather like the shim coils, but this time designed to produce a homogeneous field profile.

The deuterium NMR signal is monitored using a continuous wave (CW) NMR experiment, rather than the usual pulse–acquire experiment. The reason for using a CW experiment is that it is by far the simplest way of monitoring the frequency of a single line, which is all we want to do in this case.

The lock is a feedback system: if the field changes, the deuterium line shifts, resulting in an error signal which in turn alters the field in such a way as to bring the line back to its original position. As we are not expecting the field to change quickly, this feedback loop is given a long *time constant* which means that it integrates the error signal over a long time. The advantage of this approach is that any noise in the system is also integrated over this long time, thus diminishing its effect.

13.2   The probe

The probe is a cylindrical metal tube which is inserted into the bore of the magnet. The small coil used both to excite and detect the NMR signal (see section 4.2 on page 50 and section 4.4 on page 52) is held in the top of this assembly in such a way that the sample can come down from the top of the magnet and drop into the coil. Various other pieces of electronics are contained in the probe, along with some arrangements for heating or cooling the sample.

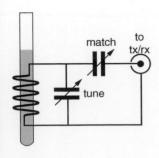

Fig. 13.1 Schematic of the key parts of the probe. The coil is shown on the left (with the sample tube in blue) which forms a tuned circuit with the capacitor marked 'tune'. The power transfer to the transmitter and receiver is optimized by adjusting the capacitor marked 'match'. Note that the coil geometry as shown is not suitable for a superconducting magnet in which the main field is parallel to the sample axis.

The coil is connected in parallel with a capacitor to form a *tuned circuit*, which has a particular resonant frequency, depending on the inductance of the coil and capacitance of the capacitor. The signal which a given amount of magnetization gives rise to is greatly increased when the resonance frequency of the tuned circuit matches the Larmor frequency. Therefore to optimize the sensitivity, it is vital to make sure that the tuned circuit is resonant at the Larmor frequency: this is what we do when we 'tune the probe'.

Tuning the probe means adjusting the capacitor until the tuned circuit is resonant at the Larmor frequency. Usually we also need to 'match the probe' which involves further adjustments designed to maximize the power transfer between the probe and the transmitter and receiver. Figure 13.1 shows a typical arrangement in which variable capacitors are used for tuning and matching. The two adjustments tend to interact rather, so tuning the probe can be a tricky business. To aid us, the instrument manufacturers provide various indicators and displays so that the tuning and matching can be optimized. We expect the tuning of the probe to be particularly sensitive to changing solvent or to changing the concentration of ions in the solvent, as such things affect the inductance of the coil.

In NMR, the main source of noise in a well-designed spectrometer is actually from the coil itself. This is thermal noise, which arises from the thermal motion of the electrons in the metal. Therefore, cooling the coil will reduce the noise – the lower we go in temperature, the less noise there will be. Technologically, cooling the coil down to temperatures of a few kelvin, while keeping the sample at room temperature, is a very challenging problem, but it is one that has been solved. Such *cryo probes*, as they are know, are now readily available and the increase in signal-to-noise ratio which these have when compared with conventional probes is very significant.

13.3   The transmitter

The RF transmitter is the part of the spectrometer which generates the pulses. We start with an RF source, such as a *frequency synthesizer*, which produces a stable frequency which can be set precisely via a computer interface. We also need to be able to shift the phase of the RF source in order to generate phase-shifted RF pulses.

As we only need the RF to be applied for a short time, the output of the synthesizer has to be 'gated' so as to create a pulse of RF energy. Such a gate will be under computer control so that the length and timing of the pulse can be controlled.

The RF source will be at a low level (a few mW) and so needs to be boosted by a high-power amplifier to provide the 100 W or so needed to create hard (non-selective) pulses. However, we may not always want the pulses to be at full power; for example, we might want to generate selective pulses, which require much lower power. To allow for this option, an attenuator, under computer control, is placed between the RF source and the amplifier. The amplifier has a fixed gain, but by using the attenuator to alter the power going into the amplifier we can alter its power output. The complete arrangement is illustrated in Fig. 13.2 on the facing page.

The more power that is applied to the probe the more intense the B_1 field will become and so the shorter the 90° pulse length. However, there is a limit to the amount of power which can be applied because of the high voltages which are generated in the probe, especially across the coil and tuning capacitor. Eventually, the voltage will reach a point where it is sufficient to ionize the air, thus generating a discharge or arc. Not only does this *probe arcing* have the potential to destroy the coil and capacitor, but it also results in unpredictable and erratic B_1 fields.

13.3.1 Power levels and 'dB'

As we have seen, the attenuator between the RF source and the amplifier gives us a way of altering the output power of the transmitter. The attenuation is normally expressed in *decibels* (abbreviated dB and pronounced 'dee bee'). If the power of the signal going into the attenuator is P_{in} and power at the output is P_{out}, then the attenuation in dB is

$$10 \times \log_{10} \frac{P_{out}}{P_{in}}.$$

Note that the logarithm is to the base 10, not the natural logarithm; the factor of 10 is the 'deci' part in the dB.

For example, if the output power is half the input power, i.e. $P_{in} = 2 \times P_{out}$, the power ratio in dB is

$$10 \times \log_{10} \frac{P_{out}}{P_{in}} - 10 \times \log_{10} \frac{1}{2} = -3.0$$

So, halving the power corresponds to a change of -3.0 dB, the minus indicating that there is a power reduction i.e. an attenuation. An attenuator which achieves this effect would be called 'a 3 dB attenuator'.

Likewise, a power reduction by a factor of four corresponds to -6.0 dB. In fact, because of the logarithmic relationship we can see that each 3 dB of attenuation will halve the power. So, a 12 dB attenuator will reduce the power by a factor of sixteen.

The B_1 field strength is proportional to the *square root* of the power applied. The reason for this is that it is the current in the coil which is responsible for generating the B_1 field, and current and power are related by power = resistance × current^2. Thus the current is proportional to the square root of the power.

In order to double the B_1 field we need to double the current, which means multiplying the power by a factor of four: this corresponds to a power ratio of 6 dB. Therefore decreasing the attenuation by 6 dB will cause the B_1 field to double, and decreasing by a further 6 dB will cause a further doubling of the field, and so on.

Usually the attenuator is under computer control and its value can be set in dB. This is very helpful to us as we can determine the attenuation needed for different B_1 field strengths. Suppose that with a certain setting of the attenuator we have determined the B_1 field strength to be ω_1^{init} (in angular frequency units). However, in another experiment we want the field strength to be ω_1^{new}. The ratio of the powers needed to achieve these

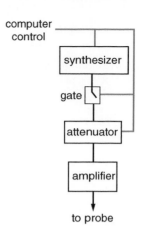

computer control

Fig. 13.2 Typical arrangement of the RF transmitter. The synthesizer, which is the source of the RF, produces a low-level output. This is fed, via a gate and an attenuator, to a high-power amplifier. The power output is controlled by using the attenuator to vary the input to the amplifier. The gate is used to switch on the RF power when a pulse is required. All of the components are under computer control.

two field strengths is equal to the square of the ratio of the field strengths:

$$\text{power ratio} = \left(\frac{\omega_1^{\text{new}}}{\omega_1^{\text{init}}}\right)^2.$$

Expressed in dB this is

$$
\begin{aligned}
\text{power ratio in dB} &= 10\log_{10}\left(\frac{\omega_1^{\text{new}}}{\omega_1^{\text{init}}}\right)^2 \\
&= 20\log_{10}\left(\frac{\omega_1^{\text{new}}}{\omega_1^{\text{init}}}\right). \quad (13.2)
\end{aligned}
$$

This expression can be used to find the correct setting for the attenuator.

As the duration of a pulse of a given flip angle is *inversely* proportional to ω_1 the relationship can be expressed in terms of the initial and new pulse widths, $t_{\text{p, init}}$ and $t_{\text{p, new}}$:

$$\text{power ratio in dB} = 20\log_{10}\left(\frac{t_{\text{p, init}}}{t_{\text{p, new}}}\right).$$

For example, suppose we have calibrated the pulse width for a 90° pulse to be 15 μs, but now we want a 90° pulse of 25 μs. The required attenuation would be:

$$\text{power ratio in dB} = 20\log_{10}\left(\frac{15}{25}\right) = -4.4 \text{ dB.}$$

We would therefore need to increase the attenuator setting by 4.4 dB.

13.4   The receiver

The NMR signal emanating from the probe is very small (of the order of μV), but there is no problem in amplifying this signal to a level where it can be digitized. The amplifiers need to be designed so that they introduce a minimum of extra noise i.e. they should be *low-noise* amplifiers.

The first of these amplifiers, called the *pre-amplifier* or *pre-amp* is usually placed as close to the probe as possible (you will often see it resting by the foot of the magnet). This is so that the weak signal is boosted before being sent down a cable to the spectrometer console.

One additional problem which needs to be solved comes about because the coil in the probe is used for both exciting the spins and detecting the signal. This means that at one moment 100 W of RF power are being applied, and the next we are trying to detect a signal at the μV level. We need to ensure that the high-power pulse does not end up in the sensitive receiver, thereby destroying it!

This separation of the receiver and transmitter is achieved by a device known as a *diplexer*. There are various different ways of constructing such a device, but at the simplest level it is just a fast acting switch set up so that when the pulse is on the high power RF is routed to the probe, and the receiver is protected by disconnecting its input. When the pulse is off the receiver is connected to the probe and the transmitter is disconnected.

Some diplexers are passive, in the sense that they require no external power to achieve the required switching. Other designs use fast electronic switches (rather like the gate in the transmitter) which are under the control of the pulse programmer so that the receiver or transmitter is connected to the probe at the right times.

13.5 Digitizing the signal

13.5.1 The analogue to digital converter

A device known as an *analogue to digital converter* or ADC is used to convert the NMR signal from a voltage to a binary number which can be stored in computer memory. The ADC samples the signal at regular intervals, resulting in a representation of the FID as *data points*.

The output from the ADC is just a number, and the range of different numbers that the ADC can output is set by the number of binary 'bits' that the ADC uses. For example, the output of a three-bit ADC can take just eight values: the binary numbers 000, 001, 010, 011, 100, 101, 110 and 111. The total number of possibilities is 2 raised to the power of the number of bits.

The waveform which the ADC is digitizing is varying continuously, but output of the three-bit ADC is restricted to one of eight levels, so what the device has to do is simply pick which of these levels is closest to the input, as is illustrated in Fig. 13.3. The output of the ADC is therefore an approximation to the actual waveform.

The accuracy of the digital representation of the signal can be improved by increasing the number of bits, as this gives more levels. At present, ADCs with between 16 and 32 bits are commonly in use in NMR spectrometers; a further increase in the numbers of bits is limited by technical considerations.

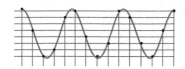

Fig. 13.3 Digitization of a waveform using an ADC with eight levels (three bits). The output of the ADC can only be one of the eight levels, so the smoothly varying waveform has to be represented by data points at one of the eight levels. The data points, indicated by black dots, are therefore an approximation to the true waveform. Note that the waveform is sampled at regular intervals, as indicated by the blue vertical lines.

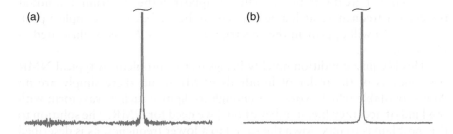

(a)

(b)

Fig. 13.4 Illustration of the effect of increasing the resolution of the ADC on the size of digitization sidebands. Spectrum (a) is from a FID which has been digitized using a six-bit ADC (i.e. 64 levels); the vertical scale has been expanded ten-fold so that the digitization sidebands are clearly visible. Spectrum (b) is from a FID which has been digitized using an eight-bit ADC (256 levels); the improvement over (a) is evident.

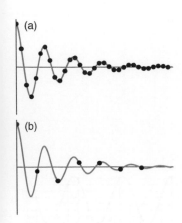

Fig. 13.5 Illustration of the effect of sampling rate on the representation of the FID. In (a) the data points (shown by black dots) are quite a good representation of the signal (shown by the blue line). In (b) the data points are too widely separated and so are a very poor representation of the signal.

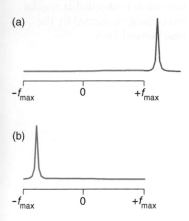

Fig. 13.6 Illustration of the concept of folding. In spectrum (a) the peak (shown in dark grey) is at a higher frequency than the maximum set by the Nyquist condition. In practice, such a peak would appear in the position shown in (b).

The main consequence of the approximation inherent in the ADC is the generation of a forest of small sidebands – called *digitization sidebands* – around the base of the peaks in the spectrum. Usually these are not a problem as they are likely to be swamped by thermal noise. However, if the spectrum contains a very strong peak the sidebands from it can swamp a nearby weak peak. Increasing the number of bits used by the ADC results in a better approximation of the signal, and hence reduced digitization sidebands. This point is illustrated in Fig. 13.4 on the preceding page.

13.5.2   Sampling rates

Given that the ADC is only going to sample the signal at regular intervals, the question arises as to how frequently it is necessary to sample the FID i.e. what the time interval between the data points should be. Clearly, if the time interval is too long we will miss important features of the waveform, and so the digitized points will be a poor representation of the signal. This is illustrated in Fig. 13.5.

It turns out that, if the interval between the points is Δ, the highest frequency which can be represented correctly, f_{max}, is given by

$$f_{max} = \frac{1}{2\Delta}.$$

f_{max} is called the *Nyquist frequency*, and a signal at this frequency will have two data points per cycle. Usually we think of this relationship the other way round i.e. if we wish to represent correctly frequencies up to f_{max} the sampling interval is given by:

$$\Delta = \frac{1}{2 f_{max}}.$$

The sampling interval Δ is often called the *dwell time*.

We will see in section 13.6 on the facing page that we are able to distinguish positive and negative frequencies, so if the dwell time is Δ, it means that the range of frequencies from $-f_{max}$ to $+f_{max}$ are represented correctly.

A signal at greater than f_{max} will still appear in the spectrum, but not at the correct frequency; such a peak is said to be *folded*. For example a peak at $(f_{max} + F)$ will appear in the spectrum at $(-f_{max} + F)$, as is illustrated in Fig. 13.6.

This Nyquist condition quickly brings us to a problem. A typical NMR frequency is of the order of hundreds of MHz, but there simply are no ADCs available which work fast enough to digitize such a waveform with the kind of accuracy (i.e. number of bits) we need for NMR. The solution to this problem is to mix down the signal to a lower frequency, as is described in the next section.

13.5.3   Mixing down to a lower frequency

The range of frequencies that a typical NMR spectrum covers is rather small, usually no more than a few tens of kHz. So, if we choose a frequency in the middle of this range, and then *subtract* this from the frequencies of the NMR signals, we will end up with signals whose frequencies are no more than a few tens of kHz, rather than hundreds of MHz. Digitizing such low frequencies is easily within the capabilities of typical ADCs.

This frequency we subtract from the NMR signals is called the *receiver reference frequency* or sometimes just the *receiver frequency*. Shifting the frequencies in this way is the equivalent of detection in a rotating frame, as was described in section 4.6 on page 60.

The subtraction process is carried out by a *mixer*. Such a device takes two signal inputs at frequencies f_1 and f_2, and produces an output which contains signals at the *sum* of the two input frequencies, $(f_1 + f_2)$, and at the *difference* $(f_1 - f_2)$. The process is visualized in Fig. 13.7.

One of the inputs to the mixer will be the locally generated receiver reference frequency, and the other will be the NMR signal from the probe. Since we choose the reference frequency to be close to the NMR frequency, the difference of these two will be at a low frequency, whereas the sum will be at around twice the Larmor frequency. This high-frequency signal is easily separated from the required low-frequency signal by passing the output of the mixer though a *low-pass filter*. The filtered signal is then passed to the ADC.

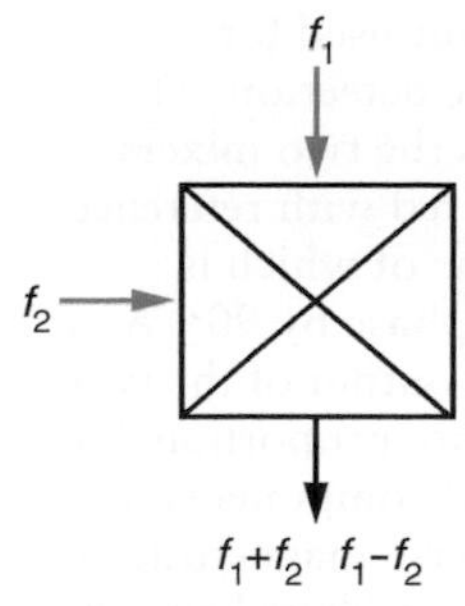

Fig. 13.7 A radiofrequency mixer takes inputs at two different frequencies, and produces an output which contains signals at the sum and difference of the frequencies of the two inputs.

13.6 Quadrature detection

In discussing the vector model, we noted that it was possible to detect both the x- and y-components of the precessing magnetization (section 4.6 on page 60). These two signals are then used to construct a complex time-domain signal:

$$S(t) = S_x + i\,S_y.$$

It is $S(t)$ which we subject to a Fourier transform in order to generate the spectrum (section 5.2 on page 82). Typically, $S(t)$ is a damped oscillation

$$S(t) = S_0 \exp(i\Omega t) \exp(-Rt).$$

This time-domain signal is sensitive to the *sign* of Ω: $\exp(+i\Omega t)$ and $\exp(-i\Omega t)$ are different functions, and upon Fourier transformation will give a peak at $+\Omega$ and $-\Omega$, respectively. The spectrum is said to have frequency discrimination. It is very important that our spectrum is discriminated in this way since, as we place the receiver reference frequency in the middle of the spectrum, there will be peaks at both positive and negative offsets.

The question is, how is it possible to detect the x- and y-components of the magnetization? One possibility is to have two coils in the probe, one aligned along x and one along y; these would detect the x- and y-components of the magnetization. In practice, it turns out to be very hard to achieve such an arrangement, partly because of the confined space in the probe and partly because of the difficulties in making the two coils electrically isolated from one another.

The same effect as having two coils can be achieved by feeding the output from *one* coil into *two* mixers, which have different phases for the

Fig. 13.8 The schematic arrangement used for quadrature detection. The key part is the two mixers which are fed with reference signals, one of which is shifted in phase by 90°. As a result, the output of the two detectors are proportional to orthogonal components of the transverse magnetization. The low-pass filters between the mixers and the ADCs are there to ensure that only the low-frequency difference signal is digitized.

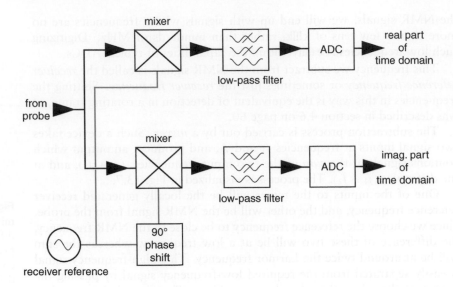

receiver reference frequency. The way this works can be understood as follows.

A mixer works by literally 'multiplying together' the two inputs. So, if one input is the NMR signal of the form $S \cos(\omega_0 t)$, and the other is the receiver reference, represented by $\cos(\omega_{rx} t)$, multiplying these two together gives

$$\underbrace{S\cos(\omega_0 t)}_{\text{NMR}} \times \underbrace{\cos(\omega_{rx} t)}_{\text{reference}} = \tfrac{1}{2}S\left[\cos(\omega_0 + \omega_{rx})t + \cos(\omega_0 - \omega_{rx})t\right],$$

where we have used the identity $\cos A \cos B \equiv \tfrac{1}{2}[\cos(A+B) + \cos(A-B)]$. The low frequency signal, $\cos(\omega_0 - \omega_{rx})t$, is the one passed to the ADC. The difference $\omega_0 - \omega_{rx}$ is the offset, Ω, so the ADC digitizes the signal $\cos(\Omega t)$.

Now suppose we shift the phase of the receiver reference frequency by 90° ($\pi/2$ radians). The signal applied to the mixer is now $\cos(\omega_{rx} t + \pi/2)$ which is the same as $-\sin(\omega_{rx} t)$. Multiplying this by the NMR signal gives

$$\underbrace{S\cos(\omega_0 t)}_{\text{NMR}} \times \underbrace{-\sin(\omega_{rx} t)}_{\text{reference}} = \tfrac{1}{2}S\left[-\sin(\omega_0 + \omega_{rx})t + \sin(\omega_0 - \omega_{rx})t\right],$$

where we have used the identity $\cos A \sin B \equiv \tfrac{1}{2}[\sin(A+B) - \sin(A-B)]$. The low-frequency signal is now $\sin(\omega_0 - \omega_{rx})t$, which is $\sin(\Omega t)$.

Therefore, by changing the phase of the receiver reference we can alter the output of the detector from $\tfrac{1}{2}S\cos(\Omega t)$ to $\tfrac{1}{2}S\sin(\Omega t)$, which we recognize as the two orthogonal components of the precessing transverse magnetization. We do not need two coils, therefore, but just two mixers fed with reference frequencies which differ in phase by 90°.

This method of generating the two orthogonal components is called *quadrature detection*. Figure 13.8 shows a typical practical implementation of this scheme. The NMR signal from the probe is split into two and fed to two separate mixers, and the receiver reference signal fed to one of the mixers is shifted by 90° relative to that fed to the other. As a result, the outputs of the two mixers are proportional to the two orthogonal components of the magnetization. These two outputs are digitized separately and become the real and imaginary parts of a complex time-domain signal.

13.7 The pulse programmer

The pulse programmer has become an immensely sophisticated piece of computer hardware, controlling as it does all of the functions of the spectrometer. As the pulse programmer needs to produce very precisely timed events, often in rapid succession, it is usual for it to run independently of the main computer. Typically, the pulse program is specified in the main computer and then, when the experiment is started, the instructions are loaded into the pulse programmer and then executed there.

The acquisition of data is usually also handled by the pulse programmer, again separately from the main computer. Only when the experiment is finished are the data passed back to the main computer.

13.8 Further reading

NMR data processing, including issues relating to digitization:

> J. C. Lindon and A. G. Ferrige, *Progress in Nuclear Magnetic Resonance Spectroscopy*, **14**, 27–66 (1980).

A more detailed description of the RF hardware used in spectrometers:

> Chapter 5 from E. Fukushima and S. B. W. Roeder, *Experimental Pulse NMR: a Nuts and Bolts Approach*, Addison–Wesley (1981).

> Chapter 4 from M. H. Levitt, *Spin Dynamics* (2nd edition, John Wiley & Sons, Ltd, 2008).

13.9   Exercises

13.1   You have been offered a superconducting magnet which claims to have a homogeneity of '1 part in 10^8'. Your intention is to use it to record ^{31}P spectra at Larmor frequency of 180 MHz, and you know that your typical linewidths are likely to be of the order of 25 Hz. Is the magnet sufficiently homogeneous to be of use? For ^{31}P $\gamma = 1.08 \times 10^8$ rad s^{-1} T^{-1}.

13.2   A careful pulse calibration experiment has determined that the 180° pulse length is 24.8 μs. How much attenuation, in dB, would have to be introduced into the transmitter in order to give an RF field strength, $(\omega_1/2\pi)$, of 2 kHz?

13.3   A spectrometer is equipped with a transmitter capable of generating a maximum of 100 W of RF power at the frequency of ^{13}C. Using this transmitter at full power, the 90° pulse width is found to be 20 μs. What power would be needed to reduce the 90° pulse width to 7.5 μs? Would you have any reservations about using this amount of power?

13.4   Explain what is meant by 'a two-bit ADC' and draw a diagram to illustrate the outcome of such a ADC being used to digitize a sine wave.

Why is it generally desirable to use an ADC with the largest number of bits available?

13.5   A spectrometer operates at 800 MHz for proton, and it is desired to record a spectrum covering a shift range of 15 ppm. Assuming that the receiver reference frequency is placed in the middle of this range, what range of frequencies (in Hz) is covered by the spectrum, and what would the sampling interval (dwell time) have to be?

Some mathematical topics

This appendix contains brief outlines of some mathematical concepts which are used frequently in the text. These outlines are more by way of a reminder of the key ideas, rather than a full exposition of the topic.

A.1   The exponential function and logarithms

The exponential function arises in the mathematical description of all sorts of physical processes; in NMR it is encountered particularly in the theory of relaxation. This function can be written in one of two ways:

$$e^{-Ax} \quad \text{or} \quad \exp(-Ax)$$

where A is a constant and x is the variable; we tend to use the latter version in this book. Figure A.1 on the next page shows a plot of this function for three different positive values of A.

When $x = 0$ the function takes the value 1 for all values of A, and then as x increases the function decays away towards zero; the larger the constant A, the faster the decay rate. For negative values of x, $\exp(-Ax)$ is greater than one and increases steadily the more negative x becomes. This kind of behaviour is not usually encountered in physical systems.

The *natural logarithm*, denoted ln, is closely related to the exponential:

$$\text{if} \quad C = \exp(D) \quad \text{then} \quad \ln(C) = D.$$

It follows from this definition of the logarithm that

$$\exp(\ln[C]) = C.$$

Any number raised to the power of zero is 1, thus $\exp(0) = 1$ and so it follows that $\ln(1) = 0$.

Exponentials and natural logarithms have a number of properties which

Understanding NMR Spectroscopy, Second Edition James Keeler
© 2010 John Wiley & Sons, Ltd

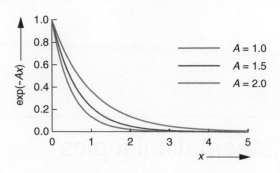

Fig. A.1 Plots of the exponential function for three different values of the constant A; note that, for all values of A, the function is equal to one when $x = 0$, but that as A increases, the rate of decay of the function increases.

we use frequently in manipulations:

$$\text{multiplication:} \quad \exp(L) \times \exp(M) \equiv \exp(L + M)$$

$$\text{division:} \quad \frac{\exp(L)}{\exp(M)} \equiv \exp(L - M)$$

$$\text{reciprocal:} \quad \exp(-M) \equiv \frac{1}{\exp(M)}$$

$$\text{addition:} \quad \ln(L) + \ln(M) \equiv \ln(L \times M)$$

$$\text{subtraction:} \quad \ln(L) - \ln(M) \equiv \ln\left(\frac{L}{M}\right)$$

From the final relationship we have the following special case when $L = 1$:

$$\ln(1) - \ln(M) \quad \equiv \quad \ln\left(\frac{1}{M}\right)$$

$$\text{hence} \quad \ln\left(\frac{1}{M}\right) \quad \equiv \quad -\ln(M),$$

where we have used $\ln(1) = 0$.

The relevant differentials and integrals are:

$$\frac{d}{dx} \exp(-Ax) \quad = \quad -A \exp(-Ax)$$

$$\int \exp(-Ax)\, dx \quad = \quad \frac{1}{-A} \exp(-Ax) + \text{const.}$$

$$\frac{d}{dx} \ln(x) \quad = \quad \frac{1}{x}$$

$$\int \frac{1}{x}\, dx \quad = \quad \ln(x) + \text{const.}$$

Note that as $\ln(Ax) \equiv \ln(A) + \ln(x)$, the derivative of $\ln(Ax)$ is $1/x$ as the $\ln(A)$ term is a constant.

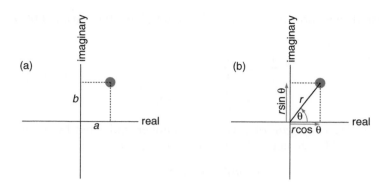

Fig. A.2 Visualization of the complex plane, in which the two orthogonal axes are labelled 'real' and 'imaginary'. In (a) we see how a complex number, the blue dot, can be thought of as a point in the complex plane, with components a and b along the real and imaginary axes. The position of the point can also be specified by the distance r from the origin, and the angle θ, as shown in (b). Note that r is real and positive.

A.2 Complex numbers

Complex numbers, particularly when combined with the exponential function, occur in the mathematical description of all sorts of physical phenomena associated with oscillations and other kinds of periodic motion. In quantum mechanics, wavefunctions and the matrix representations of operators frequently involve complex numbers.

One way to think about ordinary numbers is to consider them as falling on a line, which extends from minus infinity, through zero, to plus infinity. Complex numbers are an extension of this idea in which the numbers lie in a *plane*, the horizontal axis of which gives the *real* part of the number, and the orthogonal vertical axis gives the *imaginary* part of the number.

Figure A.2 (a) illustrates this complex plane. The blue dot represents the number, and its coordinate along the two axes are a and b. Therefore a is the real part of the number and b is the imaginary part. Such a complex number is written

$$a + \mathrm{i}b,$$

where i is the 'complex i'. This quantity has a number of properties which are very important when it comes to manipulating complex numbers:

$$\mathrm{i}^2 \equiv -1$$
$$\mathrm{i}^3 \equiv \mathrm{i}^2 \times \mathrm{i} \equiv -\mathrm{i}$$
$$\mathrm{i}^4 \equiv \mathrm{i}^2 \times \mathrm{i}^2 \equiv -1 \times -1 \equiv +1$$
$$\frac{1}{\mathrm{i}} \equiv -\mathrm{i}.$$

The last of these properties is proved by multiplying the top and bottom of $1/i$ by i:

$$\frac{1}{i} \equiv \frac{1}{i} \times \frac{i}{i}$$

$$\equiv \frac{i}{i^2} \equiv \frac{i}{-1} \equiv -i.$$

The *complex conjugate* of a complex number, indicated by a super-script $\star$, is found by changing the sign of the imaginary part:

$$(a + ib)^\star \equiv (a - ib)$$

If a complex number is multiplied by its complex conjugate, the result is a real positive number:

$$\begin{aligned}(a + ib) \times (a + ib)^\star &= (a + ib) \times (a - ib) \\ &= a^2 - iab + iba - i^2 b^2 \\ &= a^2 + b^2,\end{aligned}$$

where, to go to the last line we have used $i^2 = -1$. For a general complex number z, $zz^\star$ is known as the square modulus of z, $|z|^2$:

$$\text{square modulus:} \quad |z|^2 = zz^\star.$$

Another way of thinking about a complex number is shown in Fig. A.2 (b) on the previous page. Here we specify a position in the complex plane in terms of the distance r of the point from the origin, and an angle θ, measured as shown in the diagram. The distance r is, by definition, real and positive. It follows from simple trigonometry that the real and imaginary parts are given by:

$$\text{real: } a = r \cos \theta \qquad \text{imaginary: } b = r \sin \theta. \tag{A.1}$$

Using this representation, the square modulus is computed as follows

$$\begin{aligned}(a + ib)(a + ib)^\star &= (r \cos \theta + i\, r \sin \theta) \times (r \cos \theta - i\, r \sin \theta) \\ &= (r \cos \theta)^2 - i(r \cos \theta)(r \sin \theta) + i(r \sin \theta)(r \cos \theta) - i^2 (r \sin \theta)^2 \\ &= r^2 (\cos^2 \theta + \sin^2 \theta) \\ &= r^2.\end{aligned}$$

To go to the last line, we have used the identity $\cos^2 \theta + \sin^2 \theta \equiv 1$. What we have shown here is that the square modulus is just r^2, the square of the distance from the origin.

A.2.1   The complex exponential

It can be shown that the exponential of an imaginary number (ix) obeys

$$\exp(ix) \equiv \cos x + i \sin x. \tag{A.2}$$

If we express the real and imaginary parts of a complex number $a + ib$ in terms of r and θ, as in Eq. A.1, we find:

$$a + ib = r \cos \theta + i\, r \sin \theta. \tag{A.3}$$

Comparison of Eq. A.3 with Eq. A.2 shows that we can write the complex number as

$$a + \mathrm{i}\,b = r\exp(\mathrm{i}\theta).$$

This representation of a complex number is very convenient for many purposes.

The complex conjugate of $r\exp(\mathrm{i}\theta)$ is found in the usual way by changing the sign of the imaginary part (remember that r is real):

$$[r\exp(\mathrm{i}\theta)]^{\star} = r\exp(-\mathrm{i}\theta).$$

Using this, the square modulus is

$$
\begin{aligned}
(a+\mathrm{i}b)\times(a+\mathrm{i}b)^{\star} &= r\exp(\mathrm{i}\theta)\times[r\exp(\mathrm{i}\theta)]^{\star} \\
&= r\exp(\mathrm{i}\theta)\,r\exp(-\mathrm{i}\theta) \\
&= r^{2}\exp(\mathrm{i}\theta - \mathrm{i}\theta) \\
&= r^{2},
\end{aligned}
$$

where to go to the last line we have used $\exp(0) = 1$. As we found before, the square modulus is simply r^{2}:

$$|r\exp(\mathrm{i}\theta)|^{2} = r^{2}.$$

Since we have defined r to be positive, it follows that the modulus of a complex number written in the form $r\exp(\mathrm{i}\theta)$ is r.

The complex exponential obeys all the rules which apply to the regular exponential function, so manipulation of complex numbers represented in this r/θ format is straightforward.

From Eq. A.2 on the preceding page it is clear that the complex exponential is closely related to trigonometric functions. This leads to a number of useful identities, which can be developed in the following way:

$$
\begin{aligned}
\exp(\mathrm{i}\theta) + \exp(-\mathrm{i}\theta) &\equiv (\cos\theta + \mathrm{i}\,\sin\theta) + (\cos\theta - \mathrm{i}\,\sin\theta) \\
&\equiv 2\cos\theta.
\end{aligned}
$$

Similarly, $\exp(\mathrm{i}\theta) - \exp(-\mathrm{i}\theta) \equiv 2\mathrm{i}\sin\theta$. It is therefore possible to express sine and cosine in terms of complex exponentials. These relationships are usually written as

$$\cos\theta \equiv \tfrac{1}{2}\left[\exp(\mathrm{i}\theta) + \exp(-\mathrm{i}\theta)\right], \tag{A.4}$$

and

$$\sin\theta \equiv \tfrac{1}{2\mathrm{i}}\left[\exp(\mathrm{i}\theta) - \exp(-\mathrm{i}\theta)\right]. \tag{A.5}$$

A.3   Trigonometric identities

The sine and cosine functions obey the following:

$$\sin(-A) \equiv -\sin(A) \qquad \cos(-A) \equiv \cos(A).$$

Products of sine and cosine functions arise every time we make a calculation using product operators, so we frequently need to manipulate such products. For convenience, the set of identities needed for such manipulations are repeated here:

$$\cos A \cos B \equiv \tfrac{1}{2} \left[\cos (A + B) + \cos (A - B) \right]$$
$$\sin A \sin B \equiv \tfrac{1}{2} \left[\cos (A - B) - \cos (A + B) \right]$$
$$\sin A \cos B \equiv \tfrac{1}{2} \left[\sin (A + B) + \sin (A - B) \right]$$
$$\cos A \sin B \equiv \tfrac{1}{2} \left[\sin (A + B) - \sin (A - B) \right].$$

The fourth of these is the same as the third with A and B swapped, but it is included for convenience. These identities are easily proved by expressing the product on the left using complex exponentials i.e. using Eqs A.4 and A.5 from the preceding page. For example:

$$\cos A \cos B \equiv \tfrac{1}{2} \left[\exp (iA) + \exp (-iA) \right] \tfrac{1}{2} \left[\exp (iB) + \exp (-iB) \right]$$
$$\equiv \tfrac{1}{4} \Big\{ \underbrace{\exp (i\,[A + B]) + \exp (-i\,[A + B])}_{= 2 \cos (A+B)}$$
$$+ \underbrace{\exp (i\,[A - B]) + \exp (-i\,[A - B])}_{= 2 \cos (A-B)} \Big\}$$
$$\equiv \tfrac{1}{2} \left[\cos (A + B) + \cos (A - B) \right].$$

The sine and cosine of the sum and difference of angles have the following identities:

$$\sin (A \pm B) \equiv \sin A \cos B \pm \cos A \sin B$$
$$\cos (A \pm B) \equiv \cos A \cos B \mp \sin A \sin B.$$

Again, these are easily proved using complex exponentials. Of special interest is the case where $A = B$, which gives

$$\sin (2A) \equiv 2 \sin A \cos B$$
$$\cos (2A) \equiv \cos^2 A - \sin^2 A.$$

From the definition of sine and cosine it follows that

$$\cos^2 A + \sin^2 A = 1,$$

from which it follows that

$$\cos^2 A = 1 - \sin^2 A \quad \text{and} \quad \sin^2 A = 1 - \cos^2 A.$$

These can be used to rewrite $\cos (2A)$ in two different ways

$$\cos (2A) \equiv 2 \cos^2 A - 1 \quad \text{or} \quad \cos (2A) \equiv 1 - 2 \sin^2 A.$$

These identities can be used to express $\sin^2 A$ and $\cos^2 A$ in terms of $\cos (2A)$:

$$\cos^2 A \equiv \tfrac{1}{2} \left[1 + \cos (2A) \right] \quad \text{and} \quad \sin^2 A \equiv \tfrac{1}{2} \left[1 - \cos (2A) \right].$$

A.4   Further reading

D. S. Sivia and S. G. Rawlings, *Foundations of Science Mathematics* (Oxford University Press, 1999).

Index